Lecture Notes in Computer Science 11661

More information about this series at http://www.springer.com/series/7407

Matthew England · Wolfram Koepf ·
Timur M. Sadykov · Werner M. Seiler ·
Evgenii V. Vorozhtsov (Eds.)

Computer Algebra in Scientific Computing

21st International Workshop, CASC 2019
Moscow, Russia, August 26–30, 2019
Proceedings

Editors
Matthew England (ID)
Coventry University
Coventry, UK

Wolfram Koepf
University of Kassel
Kassel, Germany

Timur M. Sadykov (ID)
Plekhanov Russian University of Economics
Moscow, Russia

Werner M. Seiler (ID)
University of Kassel
Kassel, Germany

Evgenii V. Vorozhtsov
Russian Academy of Sciences
Novosibirsk, Russia

ISSN 0302-9743   ISSN 1611-3349 (electronic)
Lecture Notes in Computer Science
ISBN 978-3-030-26830-5   ISBN 978-3-030-26831-2 (eBook)
https://doi.org/10.1007/978-3-030-26831-2

LNCS Sublibrary: SL1 – Theoretical Computer Science and General Issues

This Springer imprint is published by the registered company Springer Nature Switzerland AG
The registered company address is: Gewerbestrasse 11, 6330 Cham, Switzerland

Preface

The International Workshop on Computer Algebra in Scientific Computing (CASC) is an annual conference offering a unique opportunity for researchers and developers from academia and industry to share ideas in the areas of computer algebra and in various application areas that apply pioneering methods of computer algebra in sciences such as physics, chemistry, celestial mechanics, life sciences, and engineering.

This year the 21st CASC conference was held in Moscow (Russia). The first (not only in Russia, then in the USSR, but in the world!) results in computer algebra were obtained in the mid-1950s by the St. Petersburg (then Leningrad) mathematician and economist L. V. Kantorovich, who in 1975 was awarded the Nobel Prize in Economics for his mathematical theory of optimal allocation of resources. A few years before that, he had moved to Moscow to become one of the founders of the Department of Mathematical Methods in Economics at Plekhanov Russian University of Economics. In his report, prepared in collaboration with his student L. E. Petrova, and presented at the III All-Union Congress of Mathematicians held in Moscow (June–July 1956), and then published in the same year in the proceedings of this congress, entitled "A Mathematical Symbolism Convenient When Computing on Machines," the issues of data representation, allowing one to produce on a computer analytical operations on mathematical expressions, including differentiation, were discussed.

The rapid development of computer algebra in Moscow and the Moscow region began in the 1970s due to the spread of the REDUCE system, performed by the Joint Institute for Nuclear Research (JINR), the permission for which was given by the system creator A. Hearn. Although by that time in Russia (USSR) a number of specialized systems of computer algebra were created, REDUCE, as a general purpose system, was very popular in scientific and educational organizations of Russia. In total, about 130 copies of the Reduce system were transferred to other organizations.

With regard to the development and application of mathematical methods, algorithms, and software packages of computer algebra, JINR played a leading role in the development of this scientific direction, which organized a series of international conferences held in Dubna in 1980, 1983, 1985, and 1990, as well as at Moscow State University, in which, in the 1980s, a regular, monthly seminar on computer algebra was held, which was first arranged at the Physics Department, and then at the Faculty of Computational Mathematics and Cybernetics and is held to date under the leadership of S. A. Abramov. Every May for the past decade, an extended, two-day meeting of the seminar is held in Dubna.

This background has influenced the choice of the Plekhanov Russian University of Economics in Moscow as the venue for the CASC 2019 workshop.

This volume contains 28 full papers submitted to the workshop by the participants and accepted by the Program Committee after a thorough reviewing process with usually two independent referee reports. The volume also includes two invited talks.

In addition to talks about the papers in these proceedings, CASC 2019 also hosted talks of two other submission types. There were a group of talks from authors on work that was submitted as extended an abstract, only for distribution locally at the conference. Such work was either already published or not yet ready for publication, but of interest to the CASC audience and selected following review by the CASC Program Committee. Another group of talks was offered from authors who have extended pieces of work accepted or in revision for a special CASC issue of the Springer journal *Mathematics in Computer Science*.

In his invited talk, Stanislav Poslavsky describes the open-source library RINGS written in the JAVA and SCALA programming languages, which implements basic concepts and algorithms from computational commutative algebra. The goal of the RINGS library is to provide a high-performance implementation packed into a lightweight *library* (not a full-featured CAS) with a clean application programming interface (API), which meets modern standards of software development. Polynomial arithmetic, GCDs, factorization, and Gröbner bases are implemented with the use of fast modern algorithms. RINGS provides a simple API with a fully typed hierarchy of algebraic structures and algorithms for commutative algebra. The use of the SCALA language brings a rather novel powerful, strongly typed functional programming model allowing one to write short, expressive, and fast code for applications.

The other invited talk by Chee Yap addresses the issue of correct and practical numerical algorithms for geometric problems. Exactness in geometric problems comes from an approach called exact geometric computation (EGC) that requires an algorithm to compute the exact combinatorial or topological structure underlying the geometric problem. This requires the solution of various zero problems inherent in the problem. Such zero problems may have high complexity and are possibly undecidable. The author aims to introduce notions of "soft-ε correctness" in order to avoid these zero problems. The talk offers a bird's eye view of the author's recent work with collaborators in two principle areas: computing zero sets and robot path planning. They share a common subdivision framework, resulting in algorithms with adaptive complexity, and which are practical and effective. Here, "effective algorithm" means it is easily and correctly implementable from standardized algorithmic components. The talk outlines these components and suggests new components to be developed. A systematic pathway to derive effective algorithms in the subdivision framework is discussed.

Polynomial algebra, which is at the core of computer algebra, is represented by contributions devoted to: some new root finders for univariate polynomials, the root-finding with implicit deflation, the application of parametric standard Gröbner bases in catastrophe and singularity theories and automated geometric theorem discovery, the use of the computer algebra system (CAS) PARI/GP for solving a quartic diophantine equation satisfying Runge's condition, the reduction of the black-box multivariate polynomial interpolation to the black-box univariate polynomial interpolation over any ring, the robust Schur stability of a polynomial matrix family, the solution of polynomial systems with the aid of a reduced lexicographic Gröbner basis and characteristic and quasi-characteristic decompositions, the theoretical investigation of the relation between the Berlekamp–Massey algorithm from coding theory and its generalized version—the Berlekamp–Massey–Sakata algorithm.

Several papers deal with applications of symbolic and the symbolic-numerical computations for: the solution of problems in nuclear physics, quantum mechanics, symbolic-numerical solution of the incompressible Navier–Stokes equations of fluid mechanics, the construction of a new implicit difference scheme for the 2D Boussinesq paradigm equation, investigation of optical and electromagnetic waveguides with the aid of the CAS MAPLE, and the computation of invariant projectors in the representation of wreath products for quantum mechanics problems.

Four papers deal with the application of symbolic computations for investigating and solving ordinary differential equations (ODEs): the derivation of new exponential integrators, explicit difference schemes for autonomous systems of ODEs on manifolds, the integrability of an autonomous planar polynomial system of ODEs with a degenerate singular point at the origin depending on five parameters, and a search for symmetries of ODEs with a small parameter.

Two papers deal with applications of CASs at the investigation and solution of celestial mechanics problems: obtaining and analysis of the necessary conditions of stability of an orbital gyrostat with the aid of the CAS MATHEMATICA, and study of the satellite-stabilizer dynamics under the influence of gravitational torque with the aid of Gröbner bases and the CAS MAPLE.

Applications of CASs in mechanics, physics, and robotics are represented by the following themes: the study of the problem of rotation of a rigid body with a fixed point in a magnetic field with the aid of the CAS MATHEMATICA and the implementation of the HuPf algorithm for the inverse kinematics of general 6R/P manipulators.

The remaining topics include the solution of Hadamard's maximal determinant problem with the aid of a periodic autocorrelation function reconstruction, the development of a high-performance algorithm for the work with databases involving huge amounts of data rows, the investigation of the analytic complexity of hypergeometric functions satisfying systems with holonomic rank two, the solution of the Heilbronn triangle problem by solving a group of non-linear optimization problems via symbolic computation, and the use of the generalized arithmetic-geometric mean for the calculation of complete elliptic integrals.

The CASC 2019 workshop was supported financially by the Plekhanov Russian University of Economics in Moscow. We gratefully acknowledge the research atmosphere and excellent conference facilities provided by this institution.

Our particular thanks are due to the members of the CASC 2019 local Organizing Committee at the Plekhanov Russian University of Economics, i.e., Vitaly Minashkin, Timur Sadykov, Vitaly Krasikov, Olga Kitova, and Timur Bosenko, who ably handled all the local arrangements in Moscow. In addition, Prof. V. P. Gerdt provided us with the information on the computer algebra activities in Moscow and in the Moscow region.

Furthermore, we want to thank all the members of the Program Committee for their thorough work. We also thank the external referees who provided reviews.

We are grateful to the members of the group headed by T. Sadykov for their technical help in the preparation of the camera-ready manuscript for this volume. We are grateful to the CASC publicity chair, Andreas Weber (Rheinische Friedrich-Wilhelms-Universität Bonn), and his assistant Hassan Errami for the management of the conference website: http://www.casc-conference.org. Finally, we are grateful to Dr. Dominik Michels (King Abdullah University, Jeddah, Saudi Arabia) for the design of the conference poster.

June 2019

Matthew England
Wolfram Koepf
Timur M. Sadykov
Werner M. Seiler
Evgenii V. Vorozhtsov

Organization

CASC 2019 was organized jointly by the Institute of Mathematics at Kassel University and the Plekhanov Russian University of Economics, Moscow, Russia.

Workshop General Chairs

François Boulier, Lille
Vladimir P. Gerdt, Dubna
Werner M. Seiler, Kassel

Program Committee Chairs

Matthew England, Coventry
Wolfram Koepf, Kassel
Evgenii V. Vorozhtsov, Novosibirsk

Program Committee

Moulay Barkatou, Limoges
François Boulier, Lille
Changbo Chen, Chongqing
Jin-San Cheng, Beijing
Victor F. Edneral, Moscow
Jaime Gutierrez, Santander
Sergey A. Gutnik, Moscow
Thomas Hahn, Munich
Jeremy Johnson, Philadelphia
Dominik L. Michels, Thuwal
Michael Monagan, Burnaby, Canada
Marc Moreno Maza, London, Canada

Veronika Pillwein, Linz
Alexander Prokopenya, Warsaw
Georg Regensburger, Linz
Eugenio Roanes-Lozano, Madrid
Valery Romanovski, Maribor
Timur M. Sadykov, Moscow
Doru Stefanescu, Bucharest
Thomas Sturm, Nancy
Akira Terui, Tsukuba
Elias Tsigaridas, Paris
Jan Verschelde, Chicago
Stephen M. Watt, Waterloo, Canada

Local Organization

Vitaly G. Minashkin, Moscow
Vitaly A. Krasikov, Moscow
Timur M. Bosenko, Moscow
Timur M. Sadykov, Moscow
Olga V. Kitova, Moscow
Dina E. Bortsova, Moscow

Publicity Chairs

Dominik L. Michels, Thuwal
Andreas Weber, Bonn

Website

http://www.casc.cs.uni-bonn.de/2019
(Webmaster: Hassan Errami)

Contents

Rings: An Efficient JVM Library
for Commutative Algebra (*Invited Talk*)

Stanislav Poslavsky[✉][iD]

NRC "Kurchatov Institute" - IHEP, Nauki square 1, Protvino 142281, Russia
stvlpos@mail.ru

Abstract. RINGS is an open-source library, written in Java and Scala programming languages, which implements basic concepts and algorithms from computational commutative algebra. The goal of the RINGS library is to provide a high-performance implementation packed into a lightweight *library* (not a full-featured CAS) with a clean application programming interface (API), which meets modern standards of software development. Polynomial arithmetic, GCDs, factorization, and Gröbner bases are implemented with the use of modern fast algorithms. RINGS provides a simple API with a fully typed hierarchy of algebraic structures and algorithms for commutative algebra. The use of the Scala language brings a quite novel powerful, strongly typed functional programming model allowing to write short, expressive, and fast code for applications.

Keywords: Computer algebra software · Commutative algebra

1 Overview

Efficient implementation of polynomial rings and related concepts is crucial for modern computational algebra. RINGS [1] is an open-source library which provides a high-performance implementation of basic concepts and algorithms from computational commutative algebra.

Java is *perhaps* the most widely used language in industry today and combines several programming paradigms including object-oriented, generic, and functional programming. Scala, which is fully interoperable with Java, additionally implements several advanced concepts like pattern matching, an advanced type system, and type enrichment. Use of these concepts in RINGS made it possible to implement mathematics in a quite natural and expressive way directly inside the programming environment offered by Java and Scala.

In a nutshell, RINGS allows to construct different rings and perform arithmetic in them, including both very basic math operations and advanced methods like polynomial factorization, linear systems solving, and Gröbner bases. The built-in rings provided by the library include: integers $\mathbb{Z}$, modular integers $\mathbb{Z}_p$, finite fields $\mathbb{GF}(p^k)$ (with arbitrary large p and $k < 2^{31}$), algebraic field extensions $F(\alpha_1, \ldots, \alpha_s)$, fractions $Frac(R)$, univariate $R[x]$ and multivariate $R[X]$ polynomial rings, where R is an arbitrary ground ring (which may be either one or any combination of the listed rings).

© Springer Nature Switzerland AG 2019
M. England et al. (Eds.): CASC 2019, LNCS 11661, pp. 1–11, 2019.
https://doi.org/10.1007/978-3-030-26831-2_1

In further sections, we will illustrate the key features of RINGS by a few examples given in Scala language. They can be evaluated directly in RINGS REPL. The source code of the library is hosted at GitHub github.com/PoslavskySV/rings. Installation instructions and comprehensive online documentation (with both Java and Scala examples) can be found at rings.readthedocs.io.

2 Basic Concepts

The high-level architecture of the RINGS library is designed based on two key concepts: the concept of mathematical ring and the concept of generic programming. The use of generic programming allows one to systematically translate abstract mathematical constructions into machine data structures and algorithms. At the same time, the library remains completely type-safe due to the deep use of strong type model of the Scala language.

In RINGS, an abstract mathematical ring has parameterized type `Ring[E]`—a ring of elements of abstract type E. Trait `Ring[E]` is a supertype for all particular rings, for example:

$$\mathtt{Frac[C]} <: \mathtt{Ring[Rational[C]]},$$
$$\mathtt{UnivariateRing[C]} <: \mathtt{Ring[UnivariatePolynomial[C]]},$$
$$\mathtt{MultivariateRing[C]} <: \mathtt{Ring[MultivariatePolynomial[C]]},$$

where <: denotes the subtyping relation.

Generic programming, powered by the advanced type system of Scala language, provides a great level of abstraction when working with different rings. For example, consider the following generic implementation of Euclidean algorithm:

```
def gcd[E](a: E, b: E)(implicit ring: Ring[E]): E =
    if (b == ring(0)) a else gcd(b, a % b)
```

It works with elements of any (Euclidean) ring. E.g. apply it to elements of $\mathbb{Z}$:

```
implicit val zRing = Z // ring of (arbitrary precision) integers
val i1 = zRing(16) // convert machine number to element of ring
val i2 = zRing(18)
val iGcd = gcd(i1, i2)
assert ( iGcd == zRing(2) )
```

E.g. apply it to elements of $\mathbb{Q}[x]$:

```
implicit val pRing = UnivariateRing(Q, "x") // polynomials Q[x]
val p1 = pRing("1 - x^8") // parse poly from string
val p2 = pRing("1 + x^5")
val pGcd = gcd(p1, p2)
assert ( pGcd == pRing("1+x") )
```

Importantly, each object from the above example has complete compile-time type, which is just omitted for shortness, but inferred automatically by the compiler. So in fact, the above lines are effectively expanded to:

```
val pRing : UnivariateRing[Rational[IntZ]] = ...
val p1    : UnivariatePolynomial[Rational[IntZ]] = ...
val pGcd  : UnivariatePolynomial[Rational[IntZ]] = ...
```
⋮

Another key point is the use of `implicit` variables in connection with the Scala concept of "type enrichment". In RINGS , it is used to add operator overloading for elements of arbitrary rings in an elegant way: all math operators (like modulo operator `%` used in the above `gcd` definition) work for arbitrary type `E`, provided that there is an `implicit` instance of `Ring[E]` in the scope:

```
implicit val ring : Ring[E] = ...    // implicit ring instance
val t1 : E = ... ; val t2 : E = ... // some ring elements
t1 % t2 // compiles to ring.remainder(t1, t2)
t1 / t2 // compiles to ring.divide(t1, t2)
t1 + t2 // compiles to ring.add(t1, t2)
t1 * t2 // compiles to ring.multiply(t1, t2)
```
⋮

The following example shows how the presence of an `implicit` ring changes the behaviour of math operators:

```
// some arbitrary precision integers
val t1 : IntZ = 12 ; val t2 : IntZ = 13
assert (t1 * t2 == Z(156)) // multiply integers
{
   implicit val ring = Zp(2)
   assert (t1 * t2 == ring(0)) // multiply modulo 2
}
{
   implicit val ring = Zp(17)
   assert (t1 * t2 == ring(3)) // multiply modulo 17
}
```

3 Polynomials, GCDs, and Factorization

3.1 Polynomial Types

Polynomials are the central objects in computational commutative algebra. RINGS provides separate implementations for univariate (dense) and multivariate (sparse) polynomials: univariate are represented by dense arrays, while multivariate polynomials are represented by binary trees (currently Java's built-in red-black `TreeMap` is used). This way, there are univariate rings and multivariate rings:

```
// univariate ring Z[t]
val uRing = UnivariateRing(Z, "t")
// multivariate ring Z[x, y, z]
val mRing = MultivariateRing(Z, Array("x", "y", "z"))
```

Additionally, there are special implementations for polynomials over $\mathbb{Z}_p$ with $p < 2^{64}$ (fits in machine word). These implementations are highly optimized to achieve the best possible performance using machine intrinsics and special CPU instructions. For example, for univariate polynomials:

```
// univariate ring Z/p[t] with arbitrary large p
val mPrime = Z("2^521 - 1") // Mersenne prime
val uRingZp = UnivariateRing(Zp(mPrime), "t")
// optimized univariate ring Z/p[t] with machine p
val uRingZp64 = UnivariateRingZp64(17, "t")
```

Elements of these two rings have correspondingly different types:

```
val p1 : UnivariatePolynomial[IntZ] = uRingZp("(1 + t)^100")
val p2 : UnivariatePolynomialZp64 = uRingZp64("(1 + t)^100")
```

3.2   Example: Working with Polynomials

Let us proceed with an example of some particular polynomial ring. As a coefficient ring we will take the Galois field $\mathbb{GF}(17^3)$:

```
1   //Galois field GF(17^3) ("t" is the generator)
2   implicit val gf = GF(17, 3, "t")
3   val t  = gf("t")
4   val t1 = 3 + t - t.pow(22)/(1 + t + t.pow(9))
5   //compute e.g. minimal polynomial of t1
6   val mpoly = gf.minimalPolynomial(t1)
7   // assert that t1 is a root of mpoly
8   assert( gf(mpoly.composition(t1)).isZero )
```

Elements of this Galois field are internally represented as univariate polynomials over $\mathbb{Z}_{17}$.

Define multivariate polynomial ring over the ground ring $\mathbb{GF}(17^3)$:

```
 9   // multivariate ring GF(17^3)[x, y, z]
10   implicit val ring = MultivariateRing(gf, Array("x","y","z"), LEX)
11   // construct some multivariate polynomials
12   val p1 = ring("(1 + t + x + y + z)^3 - (1 - x/t - y/t)^3")
13   val p2 = ring("(1 - x^2 - y^2 - z^2)^4 + (1 + z/t)^4 - 1")
14   val p3 = (p1 + p2).pow(2) - 1
```

Again the **ring** instance is defined **implicit**, so all math operations with multivariate polynomials, which have the type

    MultivariatePolynomial[UnivariatePolynomialZp64]

will be delegated to that instance.

In line 10, we explicitly specified to use **LEX** monomial order for multivariate polynomials. This choice affects some algorithms like multivariate division and Gröbner bases. The explicit order may be omitted (**GREVLEX** will be used by default).

Polynomial greatest common divisors and polynomial factorization work for polynomials over all available built-in rings. Now we continue our example:

```
15   // GCD of polynomials from GF(17³)[x, y, z]
16   val gcd1 = ring.gcd(p1 * p3, p2 * p3)
17   assert ( gcd1 % p3 == ring(0) )
18   val gcd2 = ring.gcd(p1 * p3, p2 * p3 + 1)
19   assert ( gcd2.isConstant )

20   // large polynomial from GF(17, 3)[x, y, z]
21   // with more than 10⁴ terms and total degree of 123
22   val bigPoly = p1.pow(3) * p2.pow(2) * p3
23   // factorize it
24   val factors = ring.factor(bigPoly)
```

One of the key features of RINGS library is that it does polynomial GCD and factorization of really huge polynomials over different ground rings robustly and fast (see Sect. 3.4).

3.3   Example: Writing Algorithms

With the use of Scala function programming, RINGS allows to write short and expressive code. Consider the following example, which implements a solver of Diophantine equations—a straightforward generalization of the extended GCD on more than two arguments (algorithm from Sec. 4.5 in [2]):

```
25   /**
26    * Solves equation ∑ fᵢsᵢ = gcd(f₁, ..., f_N) for given fᵢ
27    * @return a tuple (gcd, solution)
28    */
29   def solveDiophantine[E](fi: Seq[E])(implicit ring: Ring[E]) =
30     fi.foldLeft((ring(0), Seq.empty[E])) { case ((gcd, seq), f) =>
31       val xgcd = ring.extendedGCD(gcd, f)
32       (xgcd(0), seq.map(_ * xgcd(1)) :+ xgcd(2))
33     }
```

With this function, it is quite easy to implement, for example, an efficient algorithm for partial fraction decomposition with just a few lines of code. The resulting function will work with elements of arbitrary fields of fractions:

```
34   /** Computes partial fraction decomposition of given rational */
35   def apart[E](frac: Rational[E]) = {
36     implicit val ring: Ring[E] = frac.ring
37     // compute factors
38     val facs = ring.factor(frac.denominator).map {case (f,e) => f^e}
39     // compute co-factors
40     val (gcd, nums) = solveDiophantine(facs.map(frac.denominator/_))
41     val (ints, rats) = (nums zip facs)
42       .map { case (num, den) =>
43         Rational(frac.numerator * num, den * gcd)
44       }
45       .flatMap(_.normal)          // extract fractions integral parts
```

```
46        .partition(_.isIntegral) // separate integrals and fractions
47     rats :+ ints.foldLeft(Rational(ring(0)))(_ + _)
48  }
```

This generic function may be applied to rationals over any Euclidean rings, e.g. to rational number:

```
49  // partial fraction decomposition for rational numbers
50  // gives List(184/479, (-10)/13, 1/8, (-10)/47, 1)
51  val qFracs = apart( Q("1234213 / 2341352") )
```

or to rational function over $\mathbb{Z}_p$:

```
52  // partial fraction decomposition for rational functions
53  val ufRing = Frac(UnivariateRingZp64(17, "x"))
54  // gives List(4/(16+x), 1/(10+x), 15/(1+x), (14*x)/(15+7*x+x^2))
55  val pFracs = apart( ufRing("1 / (3 - 3*x^2 - x^3 + x^5)") )
```

The input may be complicated up to any extent. Returning to our initial example we have constructed ring $\mathbb{GF}(17,3)[x,y,z]$, let us define a field of fractions over this domain, adjoin a new variable, say W, and construct the partial fraction decomposition in this complicated ring $Frac(\mathbb{GF}(17^3)[x,y,z])[W]$:

```
56  // partial fraction decomposition of rational functions
57  // in the ring Frac(GF(17^3)[x,y,z])[W]
58  implicit val uRing = UnivariateRing(Frac(ring), "W")
59  val num = uRing("W + 1")
60  val den = uRing("(x/y + W^2) * (z/x + W^3)")
61  val fracs = apart(Rational(num, den))
```

The function call on the last line involves nearly all main components of the RINGS library: from very basic algebra to multivariate factorization over sophisticated rings.

3.4 Benchmarks

Much attention in the library is paid to the performance of core algorithms. One of the main goals of RINGS is to provide really fast implementations of modern algorithms.

To compare the speed of GCD with other tools, the following benchmark was used. Polynomials a, b, and g were generated at random and time needed to compute $gcd(ag, bg)$ was measured. Each polynomial had 40 terms (so the products ag and bg had at most 1600 terms each), and monomial exponents were generated using two strategies. In the first one (uniform), exponent of each variable in monomial was taken uniformly in $0 \leq \exp \leq 30$. In the second strategy (sharp) the total degree of each monomial was fixed and equal to 50 (so input polynomials were homogeneous). Benchmarking was performed for different numbers of variables. The performance of RINGS (v2.3.2) was compared to MATHEMATICA (v11.1.1.0), SINGULAR (v4.1.0), FORM (v4.2.0) [4] and FERMAT (v6.19) [3].

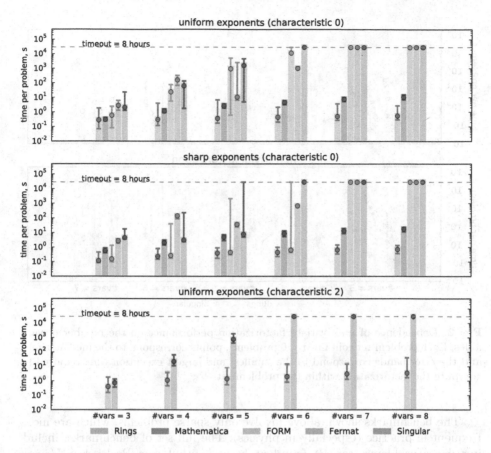

Fig. 1. Dependence of multivariate GCD performance on the number of variables. Each problem set contains 110 problems, points correspond to the median times and the error bands correspond to the smallest and largest execution time required to compute the GCD within the problem set. If computation of a single GCD took more than 8 h (timeout) it was aborted and the timeout value was adjoined to the statistics.

Figure 1 shows how the performance of different libraries behaves with the increase of the number of variables. In all considered problems performance of RINGS was unmatched. Notably, its performance almost doesn't depend on the number of variables in such sparse problems.

Performance of polynomial factorization was tested using the following benchmark. Polynomials a, b, and c were generated at random and time needed to compute $factor(abc + 1)$ (trivial) and $factor(abc)$ (non trivial) was measured. Each polynomial had 20 terms (so the products abc had at most 8000 terms each). The exponent of each variable in monomials was chosen uniformly in $0 \leq exp \leq 30$.

Figure 2 shows how the performance of multivariate factorization depends on the number of variables. It follows that the median time required to compute factorization changes quite slowly, while some outstanding points (typically ten times slower than median values) appear, when the number of variables becomes large.

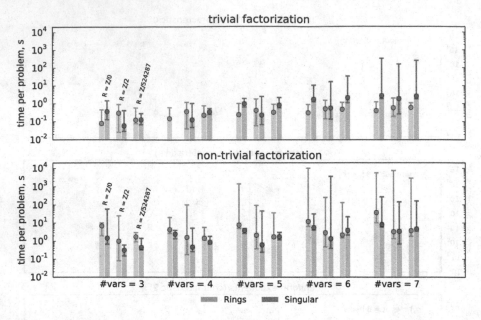

Fig. 2. Dependence of multivariate factorization performance on the number of variables. Each problem set contained 110 problems, points correspond to the median times and the error bands correspond to the smallest and largest execution time required to compute the factorization within the problem set.

The benchmarks shown above involve only sparse problems, which are more frequent in practice (especially in physics). The full set of benchmarks, including dense problems, can be found at https://github.com/PoslavskySV/rings.benchmarks.

3.5   Implementation Notes

To achieve the high performance of polynomial GCD and factorization, RINGS uses different algorithms depending on the type of input.

Univariate GCD uses the Half-GCD algorithm for polynomials over finite fields and modular algorithms in other cases (i.e., $\mathbb{Q}[x]$ and $\mathbb{Q}(\alpha)[x]$). Univariate polynomial factorization is implemented with the use of the Cantor–Zassenhaus method with optional use of Shoup's baby-step giant-step algorithm [5] (for large polynomials or for finite fields with large characteristic).

Multivariate GCD switches between Zippel's sparse algorithms and Enhanced Extended Zassenhaus algorithm (EEZ-GCD). The latter is used only on very dense problems. Zippel's algorithms require that the ground ring contains a sufficient number of elements (so they will always fail in e.g. $\mathbb{Z}_2[\boldsymbol{X}]$). When the cardinality of a ground ring is not sufficiently large, RINGS switches to a Kaltofen–Monagan generic modular algorithm [6]. For polynomials over algebraic number fields, the modular approach with either sparse (Zippel) or dense (EEZ-GCD) interpolation is used with further rational number reconstruction.

Multivariate factorization uses Kaltofen's algorithm [7], with major modifications due to Lee [8]. For factoring bivariate polynomials, the very efficient Bernardin's algorithm [9] is used. Additionally, RINGS performs some fast early checks based on Newton polygons to ensure that there is a nontrivial factorization pattern. Multivariate Hensel lifting is done via Zippel-like sparse method: the problem of lifting is reduced to a system of (in general non-linear) equations which may be solved efficiently in many cases. For factorization of polynomials over algebraic number fields $\mathbb{Q}(\alpha)$, RINGS uses Trager's algorithm [10].

4 Ideals and Gröbner Bases

The concept of mathematical ideal is implemented by the `Ideal` class, which computes corresponding Gröbner basis automatically at instantiation. The following code snippet continues our example from the previous section with polynomial ring $\mathbb{GF}(17,3)[x,y,z]$ and illustrates the main methods provided by the `Ideal` class:

```
62   val (x, y, z) = ring("x", "y", "z")
63   // define a set of polynomial generators
64   val (i1,i2,i3) = (x.pow(16) +y + z, x-y-z, y.pow(8) - z.pow(8))
65   // construct Ideal from a set of generators
66   // (Groebner basis with GREVLEX order will be computed)
67   val ideal = Ideal(Seq(i1, i2, i3), GREVLEX)
68   // print Groebner basis
69   println( ideal.groebnerBasis )
70   // print dimension of ideal
71   println( ideal.dimension )
72   // print degree of ideal
73   println( ideal.degree )
74   // print Hilbert series of ideal
75   println( ideal.hilbertSeries )
76   // reduce poly modulo ideal
77   val p4 = p2 %% ideal
```

RINGS also provides built-in algorithms for manipulating ideals:

```
78   val othIdeal = Ideal(Seq(p1, p2), GREVLEX)
79   // union of ideals
80   val union = ideal + othIdeal
81   // product of ideals
82   val prod = ideal * othIdeal
83   // intersection of ideals
84   val in = ideal intersection othIdeal
85   // quotient of ideals
86   val quot = othIdeal :/ ideal
```

4.1 Implementation and Benchmarks

RINGS implements Faugere's F4 and Buchberger's algorithms for computing Gröbner bases. These implementations show sufficient performance on small and medium problems. Table 1 shows time needed to compute a Gröbner bases of classical Katsura and cyclic systems for RINGS , MATHEMATICA and SINGULAR. Timings are in general comparable between RINGS and SINGULAR for polynomial ideals over $\mathbb{Z}_p$ while for $\mathbb{Q}$ RINGS behaves worse. It should be noted that for very hard problems, much more efficient dedicated tools like FGB [11] (proprietary) or OPENF4 [12] (open source) exist.

Table 1. Time required to compute Gröbner basis in graded reverse lexicographic order. In case of $\mathbb{Z}_p$ coefficient ring, value of $p = 1000003$ was used.

Problem	Ring	RINGS	MATHEMATICA	SINGULAR
c-7	$\mathbb{Z}_p$	3 s	26 s	N/A
c-8	$\mathbb{Z}_p$	51 s	897 s	39 s
c-9	$\mathbb{Z}_p$	14603 s	∞	8523 s
k-7	$\mathbb{Z}_p$	0.5 s	2.4 s	0.1 s
k-8	$\mathbb{Z}_p$	2 s	24 s	1 s
k-9	$\mathbb{Z}_p$	2 s	22 s	1 s
k-10	$\mathbb{Z}_p$	9 s	216 s	9 s
k-11	$\mathbb{Z}_p$	54 s	2295 s	65 s
k-12	$\mathbb{Z}_p$	363 s	28234 s	677 s
k-7	$\mathbb{Q}$	5 s	4 s	1.2 s
k-8	$\mathbb{Q}$	39 s	27 s	10 s
k-9	$\mathbb{Q}$	40 s	29 s	10 s
k-10	$\mathbb{Q}$	1045 s	251 s	124 s

5   Conclusion and Future Work

RINGS is a high-performance and lightweight library for commutative algebra that provides both basic methods for manipulating with polynomials and high-level methods including polynomial GCD, factorization, and Gröbner bases over sophisticated ground rings. Special attention in the library is paid to high performance and a well-designed API. High performance is crucial for today's computational problems that arise in many research areas including high-energy physics, commutative algebra, cryptography, etc. RINGS' performance is comparable to many advanced open-source and commercial software packages. The API provided by the library allows to write short and expressive code on top of the library, using both object-oriented and functional programming paradigms in a completely type-safe manner.

Some of the planned future work for RINGS includes improvement of Gröbner bases algorithms (better implementation of "change of ordering algorithm" and some special improvements for polynomials over $\mathbb{Q}$), optimization of univariate polynomials with more advanced methods for fast multiplication, specific optimized implementation of $\mathbb{GF}(2, k)$ fields which are frequently arise in cryptography, and better built-in support for polynomials over arbitrary-precision real numbers ($\mathbb{R}[X]$) and over 64-bit machine floating-point numbers ($\mathbb{R}64[X]$).

RINGS is an open-source library licensed under Apache 2.0. The source code and comprehensive online manual can be found at http://ringsalgebra.io.

Acknowledgements. The author would like to thank the organizers of CASC 2019. The work was supported by the Russian Science Foundation grant #18–72–00070.

References

1. Poslavsky, S.: Rings: an efficient Java/Scala library for polynomial rings. Comput. Phys. Commun. **235**, 400–413 (2019)
2. von zur Gathen, J., Gerhard, J.: Modern Computer Algebra, 3rd edn. Cambridge University Press, New York (2013)
3. Lewis, R.: Computer Algebra System Fermat (2018). http://home.bway.net/lewis
4. Ruijl, B., Ueda, T., Vermaseren, J. : FORM version 4.2 (2017). arXiv:1707.06453
5. Shoup, V.: A new polynomial factorization algorithm and its implementation. J. Symb. Comput. **4**(20), 363–397 (1995)
6. Kaltofen, E., Monagan, M.B.: On the genericity of the modular polynomial GCD algorithm. In: Proceedings of ISSAC 1999, pp. 59–66. ACM Press, New York (1999)
7. Kaltofen, E.: Sparse Hensel lifting. In: Caviness, B.F. (ed.) EUROCAL 1985. LNCS, vol. 204, pp. 4–17. Springer, Heidelberg (1985). https://doi.org/10.1007/3-540-15984-3_230
8. Lee, M.M.-D.: Factorization of multivariate polynomials. Ph.D. thesis, University of Kaiserslautern (2013)
9. Bernardin, L., Monagan, M.B.: Efficient multivariate factorization over finite fields. In: Mora, T., Mattson, H. (eds.) AAECC 1997. LNCS, vol. 1255, pp. 15–28. Springer, Heidelberg (1997). https://doi.org/10.1007/3-540-63163-1_2
10. Trager, B.M.: Algebraic factoring and rational function integration. In: Proceedings of Third ACM Symposium on Symbolic and Algebraic Computation, SYMSAC 1976, pp. 219–226. ACM, New York (1976)
11. Faugère, J.-C.: FGb: a library for computing Gröbner bases. In: Fukuda, K., Hoeven, J., Joswig, M., Takayama, N. (eds.) ICMS 2010. LNCS, vol. 6327, pp. 84–87. Springer, Heidelberg (2010). https://doi.org/10.1007/978-3-642-15582-6_17
12. Coladon, T.: OpenF4 implementation (2018). https://github.com/nauotit/openf4

Towards Soft Exact Computation
(*Invited Talk*)

Chee Yap[✉]

Department of Computer Science, Courant Institute, NYU,
New York, NY 10012, USA
yap@cs.nyu.edu

Abstract. Exact geometric computation (EGC) is a general approach
for achieving robust numerical algorithms that satisfy geometric constraints. At the heart of EGC are various Zero Problems, some of which
are not-known to be decidable and others have high computational complexity. Our current goal is to introduce notions of "soft-ε correctness"
in order to avoid Zero Problems. We give a bird's eye view of our recent
work with collaborators in two principle areas: computing zero sets and
robot path planning. They share a common Subdivision Framework.
Such algorithms (a) have adaptive complexity, (b) are practical, and
(c) are effective. Here, "effective algorithm" means it is easily and correctly implementable from standardized algorithmic components. Our
goals are to outline these components and to suggest new components to
be developed. We discuss a systematic pathway to go from the abstract
algorithmic description to an effective algorithm in the subdivision framework.

1 Introduction

We are interested in computations involving the continuum and the reals. Most
algorithms in scientific computation and engineering are of this nature (e.g.,
[40]). In practice, they fall under the domain of numerical computing [27,48].
Numerical algorithms are expected to make errors and the question of their
correctness takes on a much more subtle meaning than the typical discrete or
algebraic algorithms. One way to avoid these errors is to reformulate these problems algebraically with exact algorithms. This is often possible but not always
desirable or practical [56]. So we aim at solutions that are fundamentally numerical.

The most widely used procedure for constructing numerical algorithms is
to first construct an algorithm, say A, based on a real RAM computational
model ([50, Section 9.7] or [39]) and then implement A as an algorithm $\widetilde{A}$ of the
Standard Model of Numerical Analysis ([21, Section 2.2] or [47]). All operations
in A are exact, but in $\widetilde{A}$, each numerical operation $x \circ y$ is replaced by an

This work is supported by NSF Grants CCF-1423228 and CCF-1564132.

M. England et al. (Eds.): CASC 2019, LNCS 11661, pp. 12–36, 2019.
https://doi.org/10.1007/978-3-030-26831-2_2

approximation $x \tilde{\circ} y$ whose relative error is at most $u > 0$ (the unit round-off error). In principle, such an $\tilde{A}$ (unlike A) can be implemented on a Turing machine. For simplicity, we assume no overflow in the Standard Model. In the simplest case, $\tilde{A}$ is just a copy of A except for the $\circ \mapsto \tilde{\circ}$ transformation. In the analysis of algorithms, our first task is to prove the correctness of a given algorithm. In the present setting, we are faced with a pair $(A, \tilde{A})$ of algorithms. The modus operandi is to (i) show correctness of A and (ii) do error analysis of $\tilde{A}$. There are issues with this procedure. It is the correctness of $\tilde{A}$ that we need. Correctness of A is a necessary, but not sufficient condition. The translation $A \mapsto \tilde{A}$ hits a snag if there is branch-at-zero step in A: we must decide if a pivot value is exactly zero. This situation arises, for instance, in Gaussian elimination with partial pivoting (GEPP). In this paper, the problem of deciding if a numerical value x is equal to zero will be called "the Zero Problem". In reality, there are various Zero Problems – see [45] for a formal definition of these problems. When x is algebraic, the Zero Problem is decidable, but otherwise, it is generally not-known to be decidable. Partial solutions include replacing the standard model by arbitrary precision arithmetic ("BigNums"), or using the modified statement: "$\tilde{A}$ is correct if u is small enough." What is the status of $\tilde{A}$ if u is not small enough? There is also no guarantee that such a u exists. Basically, there are still Zero Problems lurking beneath such reformulations.

In this paper, we wish to avoid the Zero Problems by modifying the notion of correctness of the given computational problem P: instead of seeking algorithms that are (unconditionally) correct, we seek algorithms that are ε-**correct** where $\varepsilon > 0$ is an extra input, called the **resolution parameter**. Let P_ε denote this modified problem. As $\varepsilon \to 0$, then P_ε converges to the original problem. Unlike the "correct when u is small enough" criteria, we want our ε-correctness criterion to be met for each $\varepsilon > 0$. What we seek is an algorithm for P_ε that is **uniform** in ε. In discrete optimization algorithms, this is called an "approximation scheme" [43]. The polynomial-time versions of such schemes are called PTAS ("polynomial-time approximation schemes"). It is known that unless $P = NP$, the "hardest" problems in the complexity class APX do not have PTAS's. Although our continuum problems do not fall under such discrete complexity classes, our Zero Problems represent fundamental intractability analogous to the $P = NP$ barrier. The difference is that discrete intractability leads to exhaustive or exponential search, but continuum intractability leads to a halting problem.

Besides the viewpoint of combinatorial optimization, we briefly note other ways of using ε in the literature. In numerical computation, ε is commonly interpreted as an *a priori* guaranteed upper bound on the forward and/or backwards error of the algorithm's output. Depending on whether the error is taken in the absolute or relative sense of error, this gives at least 6 distinct notions of "ε-correct". As in the above Standard Model (with unit roundoff error u) such interpretations do not automatically escape the Zero Problem. Such interpretations of ε may be extended to geometry. For example, instead of bound on numerical errors, we interpret ε as bound on deviations from ideal geometric objects such as points, curves or surfaces. Suppose the output of the algorithm is a finite set $S = \{p_1, \ldots, p_n\}$ of points (e.g., S are the extreme points of a convex hull), one might define S to be ε-correct if each $p_i \in B_\varepsilon(p_i^*)$ where $S^* = \{p_1^*, \ldots, p_n^*\}$ is the exact solution. Here, $B_\varepsilon(p)$ denotes the ball centered at p of radius ε. Unfortunately, such a view still encodes a (deferred) Zero Problem because the condition $p_i \in B_\varepsilon(p_i^*)$ is a "hard predicate". Intuitively, the ball $B_\varepsilon(p_i)$ has a hard boundary (this is a "hard-ε"). Our goal is to soften such boundaries, using suitable **soft-ε** criteria. This will be illustrated through some non-trivial problems.

In this paper, we give a bird's eye view of a collection of papers over the last decade with our collaborators, from computing zero sets to path planning in robotics. We will attempt to put them all under a single subdivision rubric. We are less interested in the specific results or algorithms than in the conceptual framework they suggest.

1.1   From Zero Problems to Predicates

The "soft <u>exact</u> computation" in our title is an apparent oxymoron since softness suggests numerical approximation in opposition to exact computation. The notion of "exactness" here comes from computational geometry [8] where it is assumed that algorithms must compute geometric objects with the exact combinatorial or topological structure. The most successful way to achieve such algorithms is called "Exact Geometric Computation" (EGC) [45]. In EGC, we explicitly reduce our computation to various Zero Problems. An example of a Zero Problem is to decide if a determinant $D(\boldsymbol{x})$ is zero where $\boldsymbol{x}$ are the entries of a $n \times n$ matrix. This may arise as the so-called orientation predicate in which $\boldsymbol{x}$ represent n vectors of the form $\boldsymbol{a}_i - \boldsymbol{a}_0$ $(i = 1, \ldots, n)$ with $\boldsymbol{a}_j \in \mathbb{R}^n$ $(j = 0, \ldots, n)$. For real geometry, we usually need a bit (sic) more than just deciding zero or not-zero: we need $\texttt{sign}(D(\boldsymbol{x})) \in \{-1, 0, +1\}$. If this sign computation is error-free, then the combinatorial structure is guaranteed to be exact. Practitioners avoid the Zero Problem by defining an **approximate sign function**, $\widetilde{\texttt{sign}}(D(\boldsymbol{a})) \in \{-1, \widetilde{0}, 1\}$ where the "approximate zero sign" $\widetilde{0}$ is determined by the condition $|D(\boldsymbol{a})| < \varepsilon$. This is called "$\varepsilon$-tweaking" (using different multiples of ε's in different parts of the code) to reduce the possibility of failure. This tweaking is rarely justified (presumably it introduces some ε-correctness criteria, but what is it?).

What is the correct way to use this ε? We need a different perspective on Zero Problems: each Zero Problem arises from the evaluation of a predicate. We distinguish two kinds of predicates: **logical predicates** are 2-valued (`true` or `false`) but **geometric predicates** are typically 3-valued $(-1, 0, +1)$. Thus we view $\texttt{sign}(D(\boldsymbol{x}))$ above as a geometric predicate. Calling the sign function a "predicate" imbues it with geometric meaning: thus when we call $\texttt{sign}(D(\boldsymbol{x}))$ an orientation predicate, we know that we are dealing with a geometrically meaningful property of the n vectors arising from $n + 1$ points. These 3 sign values are not fully interchangeable: we call 0 the **indefinite value** and the other two values are **definite values**.

To continue this discussion, let us fix a geometric predicate C on $\mathbb{R}^m$,

$$C : \mathbb{R}^m \to \{-1, 0, +1\}. \tag{1}$$

We assume C has[1] the **Bolzano property** in the sense that if S is a connected set and there exists $a, b \in S$ such that $C(a) = -1$ and $C(b) = +1$ then there exists $c \in S$ such that $C(c) = 0$. For example, if $C(a) := \mathtt{sign}(D(a))$ where $D(x)$ is the above determinant (with $m = n^2$), then C has the Bolzano property. More generally, if $D(x)$ is any continuous function, its sign predicate is Bolzano.

Next we take a critical step by extending the predicate C on points $a \in \mathbb{R}^m$ to sets of points $S \subseteq \mathbb{R}^m$. The significance is that we have moved from algebra to analysis: we could treat $D(a)$ algebraically since it amounts to polynomial evaluation, but the analytic properties come to the forefront when we consider the set $D(S)$. And soft-ε concepts are fundamentally analytic.

We define the **set extension** of C as follows: for $S \subseteq \mathbb{R}^m$, define $C(S) = 0$ if there exists $a \in S$ such that $C(a) = 0$; otherwise, $C(S)$ may be defined to be $C(a)$ for any $a \in S$. The Bolzano property implies that $C(a)$ is well-defined. Thus the set extension of C is the predicate $C : 2^{\mathbb{R}^m} \to \{-1, 0, 1\}$ where 2^X denotes the power set of any set X. If we are serious about computation, we know that S must be suitably restricted to "nice" subsets of $\mathbb{R}^m$. Following the lead of interval analysis [33], we interpret "nice" to mean axes-aligned full-dimensional boxes in $\mathbb{R}^m$. Let $\square\mathbb{R}^m$ be the collection of such boxes. When the domain of the set extension of C is restricted to such boxes, we have this **box predicate**

$$C : \square\mathbb{R}^m \to \{-1, 0, +1\} \tag{2}$$

where the symbol 'C' from the point predicate in (1) is reused. This reuse is justified if we regard $\mathbb{R}^m$ as a subset of $\square\mathbb{R}^m$. In other words, each element of $\square\mathbb{R}^m$ is either a full-dimensional box or a point.

Our next goal is to approximate the box predicate C. Consider another box predicate

$$\widetilde{C} : \square\mathbb{R}^m \to \left\{-1, \widetilde{0}, +1\right\}. \tag{3}$$

Call $\widetilde{C}$ a **soft version** of (2) if it is conservative and convergent: **conservative** means $\widetilde{C}(B) \neq \widetilde{0}$ implies $\widetilde{C}(B) = C(B)$; **convergent** means if $\{B_i : i \geq 0\}$ is an infinite monotone[2] sequence of boxes that converges to a point a, then $\widetilde{C}(B_i)$ converges to $C(a)$, i.e., $\widetilde{C}(B_i) = C(a)$ for i large enough. We say $\{B_i : i \geq 0\}$ is **firmly convergent** if there is some $\sigma > 0$ such that $B_{i+1} \subseteq B_i/\sigma$ for all $i \geq 0$. We say $\widetilde{C}$ is **firm** relative to C if $C(B) \neq 0$ implies $\widetilde{C}(B/2) \neq \widetilde{0}$. The "2" in this definition may be replaced by any **firmness factor** $\sigma > 1$, if desired. As $\sigma \to 1$, the computational cost of C would increases. For resolution-exact path planning, we only need[3] half of the properties of firmness, namely, $C(B) = 1$ implies $\widetilde{C}(B/\sigma) = 1$ [49]. But even path planning may exploit the other half of

---

[1] After Bernard Bolzano (1817). Bolzano's Theorem states that if $a < b$ and $\mathtt{sign}(f(a)f(b)) < 0$ then there is some $c \in (a, b)$ such that $f(c) = 0$. See also [3, 42] for this principle in real root isolation.

[2] Monotone means $B_{i+1} \subseteq B_i$ for all i.

[3] The factor $\sigma > 1$ was call the "effectivity factor" in [49]. In the present paper, we avoid this terminology since it conflicts with our notion of "effectivity" of this paper.

firmness (i.e., $C(B) = 0$ implies $\widetilde{C}(B/\sigma) = 0$) because it could lead to faster determination of NO-PATH.

From any geometric predicate C, we derive three logical predicates $C_+, C_-,$ C_0 in a natural way: $C_+(a) := [[C(a) > 0]]$. The notation "$[[S]]$" denotes the truth value of any sentence S: for instance $[[1 > 2]]$ is equal to **false** but $[[1+2 = 3]]$ is equal to **true**. We call "$[[S]]$" a **test**. In general, $S = S(x)$ depends on variables x, and our test $[[S(x)]]$ represents a logical function. Similarly, $C_-(a) := [[C(a) < 0]]$ and $C_0(a) := [[C(a) \neq 0]]$. This last predicate is called the **exclusion predicate** and is very important for us: it is used in all of our algorithms. Again we extend these predicates naturally to sets or boxes as above. In particular, we have $C_0(B) = [[0 \notin C(B)]]$.

2   Two Illustrative Classes of Problems

We introduce two classes of geometric problems to serve as running examples:

(A) Computing a Zero Set Zero(f).
     Here, $f = (f_1, \ldots, f_m)$, $f_i : \mathbb{R}^n \to \mathbb{R}$, and Zero($f$) := $\{a \in \mathbb{R}^n : f_i(a) = 0,$ $i = 1, \ldots, m\}$. We can also define this problem for complex zeros, i.e., Zero(f) $\subseteq \mathbb{C}^n$. In the case f_i are integer polynomials, Zero(f) is an algebraic variety where an exact algebraic solution is often interpreted to mean computing some nice representative (e.g., a Gröbner basis) of the ideal generated by f. But we are literally interested in the continuum: we seek some "explicit" representation of Zero(f) as a subset of $\mathbb{R}^n$. Invariably, "explicit" has to be numerical, not symbolic. For example, an explicit solution to Zero(f_1) where $f_1(x) = x^2 - 2$ may be 1.4 but not the expression "$\sqrt{2}$". For our discussion, let us interpret an explicit representation to mean a simplicial complex K [9, Chap. 7] of the same dimension as Zero(f), and whose support $\underline{K} \subseteq \mathbb{R}^n$ is ε-isotopic to Zero(f). This definition implies that their Hausdorff distance satisfies $d_H(\underline{K}, \text{Zero}(f)) < \varepsilon$, but more is needed: an ε-isotopy maps points in Zero(f) to points in $\underline{K}$ within a distance ε. It is important that K has the same dimension as Zero(f): for instance, if Zero(f) is a curve in $\mathbb{R}^3$, we really want the output K to represent a polygonal curve $\underline{K}$. In contrast, a common output criteria asks for an ε-tubular path containing Zero(f). Unfortunately, this allows the curve to have unexpected behavior within the tube (e.g., doubling back arbitrarily far on itself within the tube). Most of the current research are aimed at cases where Zero(f) is zero-dimensional (finite set) or co-dimension one (hypersurface). We mostly focus on the zero-dimensional case in this survey.
(B) Robot path planning.
     Suppose a robot R_0 is fixed. Then the problem is: given a polyhedral obstacle set $\Omega \subseteq \mathbb{R}^k$ ($k = 2, 3$) and start α and goal β configurations, find an Ω-avoiding path π of R_0 from α to β; or declare NO-PATH if such π's do not exist. Let $\mathcal{C}space = \mathcal{C}space(R_0)$ denote the configuration space, and $\mathcal{C}free = \mathcal{C}free(R_0, \Omega)$ denote the Ω-free configurations in $\mathcal{C}space$. For

instance, if R_0 is a rigid spatial robot, then configurations are elements of $SE(3) = \mathbb{R}^3 \times SO(3)$ where $SO(3)$ are the orthogonal 3×3 matrices representing rotations. One challenge in this area is to produce implemented algorithms that are rigorous yet competitive with the practical approaches based on sampling.

Both these problems have large literatures. Problem (A) is a highly classical problem in mathematics with applications in geometric, numeric and symbolic computation. Problem (B) is central to robotics. In both cases, there are many available algorithms, and thus there are high standards for any proposed new algorithm, both theoretically and practically. In particular, they need to be implemented and compared to existing ones: subdivision algorithms appear to be able to meet both criteria. In robotics, to be "practical" includes an informal requirement of being "real time" for standard size input instances. In contrast, exact algorithms (especially for Problem (B)) are rarely implemented.

There are Zero Problems in both (A) and (B) as formulated above. For (A), even for the case $n = m = 1$ (univariate roots) where the input $f = (f_1)$ is polynomial, we face Zero Problems. These Zero Problems are not an issue when f_1 has rational coefficients; but we are interested in coefficients that are algebraic numbers or number oracles [4,28]. For real roots, there are many complete real RAM algorithms based on Sturm sequences, on Descartes rule of sign or on Newton-bisection. In each case, the algorithms call for testing if $f_1(a) = 0$ for various points $a \in \mathbb{R}$ (we may assume a is a dyadic number, but this does not make the test any easier when the coefficients of f_1 are irrational). The very formulation of root isolation requires the output interval to have exactly one root, possibly a multiple root. Distinguishing between two simple roots that are close together from a single double root is again a Zero Problem. There are difficult Zero Problems in higher dimensional problems (even for hypersurfaces) that remain open: most current correctness criteria is conditioned on non-singularity of $\text{Zero}(f)$. For Problem (B), there is also a Zero Problem corresponding to the sharp transition from path to NO-PATH. We now introduce soft-ε criteria to circumvent these Zero Problems:

(a) For root isolation [52], we introduce the ε-**clustering problem**: given a region-of-interest $B_0 \in \square\mathbb{R}^n$ and $\varepsilon > 0$, output a set $\Delta_1, \ldots, \Delta_k \subseteq 2B_0$ of disjoint balls with radii $< \varepsilon$, and output multiplicities $\mu_1, \ldots, \mu_k$ satisfying

$$\mu_i := \#_f(\Delta_i) = \#_f(3\Delta_i) \geq 1$$

where $\#_f(S)$ is the total multiplicities of the roots in S. The union of these balls must cover all the roots of f in B_0 but they may not include any root outside of $2B_0$. Each Δ_i represents a cluster of roots and the requirement $\#_f(\Delta_i) = \#_f(3\Delta_i)$ (which is our definition of "natural" clusters) can be viewed as a robustness property.

(b) For path planning [49], we say that the planner is **resolution-exact** if it satisfies two conditions:

(Path) If there is a path of clearance $K\varepsilon$, the algorithm must return some path π;

(Nopath) If there is no path of clearance ε/K. the algorithm must output NO-PATH.

Here $K > 1$ (called the **accuracy constant**) depends only on the algorithm and is independent of the input instance. The key is that (Path) and (Nopath) are not exhaustive because they do not cover input instances where the largest clearance of paths is strictly between ε/K and $K\varepsilon$. The planner may output either a path or NO-PATH in such instances. Since we require halting algorithms, the planner would produce an **indeterminate** answer in these cases. As we will see, indeterminacy (as opposed to determinacy) is a characteristic feature of soft-ε algorithms.

Based on criteria (a), we achieved the most general setting for root clustering algorithms – when the polynomials have number oracles as coefficients. Based on criteria (b), we developed and implemented path planners for various planar robots, culminating in our planners for rods and rings in 3D [22]. This is the first practical, non-heuristic algorithm for spatial robots with 5 degrees-of-freedom (DOFs). We remark that although we assumed that the robot R_0 is fixed, all our subdivision path planners can uniformly treat robots from a parametric family $R_0(p_0, \ldots, p_k)$. For instance, if R_0 is a 2-link robot, we may define $R_0(p_0, p_1, p_2)$ as the 2-link robot whose first two links have lengths p_1 and p_2, and these links have thickness p_0 (see [54]). Links are line segments, and they are thickened by forming a Minkowski sum with a ball of radius p_0. The thickness parameter is extremely useful in practice. Treating parametric families of robots is a feat that few exact algorithms are able to do; the only exception we know of is when R_0 is a ball, and here, the exact path planners based on Voronoi diagram can allow the radius of the ball as a parameter [34].

In general, besides the extra ε input, our subdivision algorithms also accept an input box B_0 called the region-of-interest (ROI), meaning that we wish to restrict the solutions to B_0. Specifying B_0 is not generally a burden, and is often a useful feature. In the case of root clustering, this meaning is clear – we must account for all the roots in B_0. There are Zero Problems associated with the boundary of B_0. To avoid this issue in root clustering, we allow the output clusters to include roots outside of B_0, but still within an expanded box $2B_0$ (or $(1 + \varepsilon)B_0$ if so desired).

3   Effectivity of the Subdivision Framework

We have noted that the usual pathway to a numerical algorithm $\widetilde{A}$ is through an intermediate real RAM algorithm A. This $\widetilde{A}$ amounts to specifying a suitable precision for each arithmetic operation in A. The difficulty of this pathway is illustrated by the **benchmark problem** in root isolation: this is the problem of isolating all the roots of a univariate polynomial $p(x)$ with integer coefficients [5]. It has been known for about 30 years that there is an explicit real RAM algorithm A with transformation $A \mapsto \widetilde{A}$ such that $\widetilde{A}$ is a near-optimal algorithm for the benchmark problem. The algorithm A is from Schönhage-Pan (1981-92) [20]. Here "near-optimal" means[4] a bit complexity of $\widetilde{O}(n^2 L)$ where $p(x)$ has degree n and L-bit coefficients. Although the construction of such an $\widetilde{A}$ from a real RAM A remains open, there are now several implementations of near-optimal algorithms based on subdivision, all shortly after the appearance of the subdivision algorithms [23]. See [51] for an account of this development (there are two parallel accounts, for complex roots and for real roots). We may ask why? Intuitively, it is because subdivision computation is reduced to operations on individual boxes (i.e., locally) and we can adjust the precision to increase as the box size decreases. In contrast, controlling the precision of arithmetic operations in a real RAM algorithm for some target resolution in the output appears to be hopelessly complicated at present.

There is no formal "subdivision model of computation". We intend our algorithms to be ultimately Turing-computable. So we only speak of the "subdivision approach or paradigm". Nevertheless, it is useful to introduce a **Subdivision Framework** which can be instantiated to produce many different algorithms.

In the simplest terms, we may describe it as follows: first assume that we are computing in $\mathbb{R}^m$, where $\square\mathbb{R}^m$ is the set of full-dimensional axes-aligned boxes. Let S be a subset of $\square\mathbb{R}^m$. Its **support** is the set $\underline{S} := \bigcup_{B \in S} B$. We call S a **subdivision** (of its support $\underline{S}$) if the interiors of any two boxes in S are disjoint. The subdivision process is typically controlled by two box predicates $C_0, C_1 : \square\mathbb{R}^m \to \{\text{true}, \text{false}\}$. Here C_0 is the standard **exclusion predicate**, and C_1 the **inclusion predicate** (which varies with the application).

Our central problem is this: given a box B_0, to recursively split B_0 into subboxes until each subbox B satisfies $C_0(B) \vee C_1(B)$. The recursive splitting forms a tree $\mathcal{T}(B_0)$ of boxes with B_0 at the root, with each internal node B failing $C_0(B) \vee C_1(B)$. We assume some scheme for splitting a box B into a subset $B_1, \ldots, B_k$ where $\{B_1, \ldots, B_k\}$ is a subdivision of B. A simple scheme is to let $k = 2^m$ and the B_i's are congruent to each other. There are also various binary schemes where $k = 2$. In the binary schemes, it is necessary to ensure that the aspect ratios of the subboxes remain bounded. Assuming that k is a constant in the splits, each internal node in $\mathcal{T}(B_0)$ has degree k. If $\mathcal{T}(B_0)$ is finite, then the leaves of $\mathcal{T}(B_0)$ form a subdivision of B_0. We are mainly interested in the subdivision $\mathcal{S}(B_0)$, comprising those leaves that fail $C_0(B)$ (thus satisfying $C_1(B)$).

Let Q_0, Q_1 be queues of boxes, with the usual queue operations (push and pop) for adding and removing boxes. Consider the following subroutine to compute $\mathcal{S}(Q_1)$. We may, for instance, initialize Q_1 to $\{B_0\}$.

---

[4] The $\widetilde{O}$-notation is like the O-notation except that logarithm factors in n and in L are ignored. In the subdivision setting, "near-optimality" may be taken to be $\widetilde{O}(n^2(n + L))$.

```
Subdivide Routine
    INPUT: Q_1
    OUTPUT: Q_2
    Q_2 ← ∅
    While Q_1 is non-empty
        B ← Q_1.pop()
        If C_0(B) fails
            If C_1(B) holds,
                Q_2.push(B) //i.e., output B
            Else
                Q_1.push(split(B))
```

The main correctness question about the Subdivide routine is termination: does Q_1 eventually become empty? This is equivalent to every box B eventually satisfying $C_0(B) \vee C_1(B)$ (if a box is split, this consideration is transferred to its children). For instance, in real root isolation, B_0 is an interval and we have termination iff there are no multiple roots in B_0. For path planning, we modify $C_0(B)$ to $C_0(B) \vee C_\varepsilon(B)$ where $C_\varepsilon(B)$ holds if the width of B is less than ε. Therefore, to ensure termination, we must either restrict the input (e.g., there are no multiple roots), or introduce suitable ε-correctness concepts (such as resolution-exactness in path planning in Sect. 2(b)).

We view Subdivide Routine as the centerpiece of our algorithm. Its output is the queue Q_2 containing the subdivision $\mathcal{S}(B_0)$. For instance, the EVAL and CEVAL algorithms in [41] are basically this subroutine. But in general, we expect to do some post processing of Q_2 to obtain the final result. For example, we may have to construct the simplicial complex K representing the zero set Zero(f) [30, 31, 38]. Likewise, we may need to do some initialization to prepare for subdivision. For example, in path planning, we need to first ensure that the start α and goal β configurations are free [49]. This suggests that we need an initialization phase before the Subdivide Routine, and we need a construction phase after. Following [30], we may assume that input and output for each phase are appropriate queues. We are ready to present a simple form of this framework:

```
Simple Subdivision Framework
    INPUT: B_0, ε, ...
    OUTPUT: Q_3
        I. Initialization Phase
                Q_1 ← Preprocessing(B_0)
       II. Subdivision Phase
                Q_2 ← Subdivide(Q_1)
      III. Construction Phase
                Q_3 ← Construct(Q_2)
```

We can derive algorithms for the illustrative problems (A) and (B) using this framework. This amounts to instantiating the routines in the three phases. A key idea in our design of these routines is to make the subdivision phase do most of the work, i.e., its complexity ought to dominate that of the other two phases. This is not true for all subdivision algorithms: an example is Snyder's approach to isotopic curves and surfaces [46] (see [9, Chap. 5.2.3]). The plausibility of our key idea comes from the fact that when subdivision is fine enough, everything would be "as simple as possible", modulo singularities. Singularities, even isolated ones, can be arbitrarily complex. For example, the neighborhood of a degenerate Voronoi vertex can have arbitrarily high degree. We may simply exclude singularities by fiat (as in isotopic curves [30,38] or in arrangement of curves [29]). But our ultimate goal is not to avoid singularities but to introduce soft-ε notions (as in root isolation [4] or in Voronoi diagrams [6]). We design the $C_0(B)$ and $C_1(B)$ to capture the non-degenerate situations outside of such singularities. We say that output Q_2 of Subdivide(Q_1) is "fine enough" if *the cost of constructing the final output is $O(1)$ per box in Q_2*. In the problems of isotopic curves and surfaces [30,31,38], the output is a planar embedded graph (for curves) or a triangulation (for surfaces). When the subdivision is "fine enough", we only need to construct simple, almost-trivial, graphs or triangulations $G(B)$ in $O(1)$ time for each $B \in \mathcal{S}(B_0)$. The output is the union of these $G(B)$'s. Thus, the global complexity of these algorithms is indeed dominated by the subdivision process. This key idea ensures that the resulting algorithm is easy implementable or practical. A caveat is that the complexity may become a bottle neck in higher dimensions. Nevertheless, it ensures that we could solve such problems, at least in small regions-of-interest.

How good is the proposed framework? For real root isolation of integer polynomials, the size of the subdivision tree $\mathcal{T}(B_0)$ is near-optimal [12,44]; the analysis can be greatly generalized [14], including accounting for bit complexity. The complexity of the PV algorithm in higher dimensions has also been analyzed [15,16]. For top performance in univariate complex root isolation [4,5] it is necessary to introduce Newton iteration and to maintain more complicated data structures ("components") in order to achieve near-optimal bounds. See [51, §1.1] for a subdivision framework that incorporates Newton iteration. Newton iteration will produce non-aligned boxes, i.e., boxes that do not come from repeated splits of B_0. This is not an issue for root isolation but in geometric problems such as arrangement of curves [29] and Voronoi diagrams [6], non-aligned boxes (called root boxes) arise where it was necessary to provide "plumbing" so that the non-aligned boxes "conforms" with the rest of the aligned boxes.

We generally need to maintain adjacency relations among boxes in $\mathcal{S}(B_0)$, especially for the construction phase. Two boxes B, B' are **adjacent** if $B \cap B'$ has codimension 1. There is a general technique to efficiently maintain such information, namely to ensure that the subdivision $\mathcal{S}(B_0)$ is **smooth** [7]. Smoothness means that if $B, B' \in \mathcal{S}(B_0)$ are two adjacent boxes, then their depths in the tree $\mathcal{T}(B_0)$ differ by at most 1. This can be done systematically by (1) maintaining "principal neighbor" pointers for each box and (2) perform smoothSplit(B)

instead of split(B) in Subdivide(Q_1). In smoothSplit(B), we split B and recursively split any adjacent boxes necessary to maintain smoothness. Although a single smoothSplit(B) can be linear in the size of Q_1, we show in [7] that this operation has amortized $O(1)$ complexity, and hence does not change the overall complexity.

So far, we have assumed subdivision in $\mathbb{R}^n$. What about subdivision in non-Euclidean spaces? Burr [14] has provided an account of subdivision in abstract measure space, aimed at amortized complexity analysis. We take a different approach, with an eye towards implementation rather than analysis: in path planning, we need to perform subdivision in configuration spaces $Cspace$. Such spaces are typically non-Euclidean: $Cspace = \mathbb{R}^2 \times \mathbb{T}$ where $\mathbb{T}$ is the torus [54], $Cspace = \mathbb{R}^3 \times S^2$ [22], and $Cspace = SE(3) = \mathbb{R}^3 \times SO(3)$ [53]. Using the analogy of charts and atlases in manifold theory, we define charts and atlases for subdivision. Furthermore, we generalize boxes to general shapes called "test cells" that include simplices or convex polytopes which have bounded aspect ratios. Resolution-exact planners (Sect. 2(b)) can be achieved in such settings and with an accuracy constant given by $K = C_0 D_0 L_0 (1 + \sigma)$ where C_0, D_0, L_0, σ are constants associated with (respectively) the atlas, subdivision scheme, a Lipshitz constant and effectivity factor [53]. It is also clear that we could extend subdivision atlases to projective spaces ($\mathbb{RP}^n$ and $\mathbb{CP}^n$).

In our abstract, we said that algorithms in the subdivision framework are "effective" in the sense of *easily and correctly implementable from standard algorithmic components*. The preceding outline exposes some of these algorithmic components: queues, subdivision structures, boxes with adjacency links, union find data structure, etc. But the critical issue of numerical approximation is deferred to the next section.

4   Numerical Precision in Subdivision Framework

The main problem of subdivision is when to stop, and this is controlled by predicates. In our Subdivide Subroutine, we used two logical box predicates $C_0(B)$ and $C_1(B)$. Both are typically reduced to some form of sign computation: in the PV algorithm [38], $C_0(B)$ is defined as $[[0 \notin f(B)]]$ for some continuous function $f : \mathbb{R}^n \to \mathbb{R}$. As for $C_1(B)$, we follow a nice device of [15] for describing this predicate: first define

$$\nabla^{(2)} f : \mathbb{R}^n \times \mathbb{R}^n \to \mathbb{R}$$

where $\nabla^{(2)} f(\boldsymbol{x}, \boldsymbol{y}) = \langle \nabla f(\boldsymbol{x}), \nabla f(\boldsymbol{y}) \rangle$ and $\langle \cdot, \cdot \rangle$ denotes the dot product. For instance, for $n = 2$, $\nabla^{(2)} f(\boldsymbol{x}, \boldsymbol{y}) = \partial_1 f(\boldsymbol{x}) \cdot \partial_1 f(\boldsymbol{y}) + \partial_2 f(\boldsymbol{x}) \cdot \partial_2 f(\boldsymbol{y})$ where ∂_i denotes partial derivative with respect to x_i. Then $C_1(B)$ is $[[0 \notin \nabla^{(2)} f(B, B)]]$. Both $C_0(B)$ and $C_1(B)$ are mathematically exact formulations, but far from effective.

We now sketch a 3 stage development to systematically derive an implementable form, following [51]. The outline may be illustrated by using the $C_0(B)$ predicate: we first define an interval version of $C_0(B)$ denoted $\square C_0(B)$. Then we modify the interval version to an "effective" version denoted $\widetilde{\Box}C_0(B)$. These version are connected through a chain of logical implications:

$$\widetilde{\Box}C_0(B) \Rightarrow \Box C_0(B) \Rightarrow C_0(B).$$

Each of these predicates are, in turn, based on underlying functions on boxes: $\Box C_0(B)$ is $[[0 \notin \Box f(B)]]$ and $\widetilde{\Box}C_0(B)$ is $[[0 \notin \widetilde{\Box}f(B)]]$. We must now define the functions $\Box f$ and $\widetilde{\Box}f$ for any f.

Our numerical algorithms are intended to be certified in the sense of interval arithmetic [25, 33]. But we wish to carry our subdivision algorithms in a slightly more general setting, say in normed vector spaces X, Y. Here, we can do differentiation (as in $\nabla^{(2)}f$) and do dilation of boxes or balls (as in $B \mapsto 2B$). Suppose we have a function $f : X \to Y$. Define the **natural set extension** of f to be

$$f : 2^X \to 2^Y \tag{4}$$

where $f(S) := \{f(x) : x \in S\}$ for $S \in 2^X$. We are[5] "overloading" the symbol f in (4). But if we identify the elements of X with the singletons in 2^X, we see that this extension is natural, and justifies reuse of the symbol f. Again, 2^X is too big and we restrict f to the nice subsets of X. Let $\Box X$ and $\Box Y$ be the collection of nice subsets of X and Y. Note that even if $B \subseteq X$ is a nice set, $f(B)$ need not be nice (except when $Y = \mathbb{R}$). In other words, the function (4) does not naturally induce a function of the form

$$F : \Box X \to \Box Y. \tag{5}$$

Thus we are obliged to explicitly define the function F in (5). What is the relation between f and F? We call F a **box form** of f provided it is **conservative relative to** f (i.e., $f(B) \subseteq F(B)$) and **convergent to** f (i.e., if $\{B_i : i \geq 0\}$ converges to a point $p \in X$, then $\lim_{i \to 0} F(B_i) = f(p)$). We may write "$F \to f$" if F is convergent to a point function f. This parallels our definition of soft predicates. We write "$\Box f$" for a generic box form of f. If it is necessary to distinguish different box forms, we use subscripts such as $\Box_2 f$. The function (5) is called a **box function** when it is the box form of some f. The interval literature defines many box forms for $f : \mathbb{R}^n \to \mathbb{R}$. For example, the **mean value form** of f given by

$$\Box_{\mathtt{M}} f(B) := f(m(B)) + \Box \nabla f(B)^T \cdot (B - m(B)) \tag{6}$$

where $m(B)$ is the midpoint of B and $\nabla f = (f_1, \ldots, f_n)^T$ is the gradient of f, with $f_i = \partial_i f$. Our definition of mean value form invokes another box form $\Box \nabla f(B) = (\Box f_1, \ldots, \Box f_n)^T$. Since this second box form is generic, $\Box_{\mathtt{M}}$ is still not fully unspecified.

Suppose F is a box form of f. By regarding X as a subset of $\Box X$, we can view f as the restriction of F to X, i.e., $f = F|_X$. Let $F_i : \Box X \to \Box Y$ ($i = 1, 2$)

---

[5] Some authors introduce a new symbol, say F, to signal this change.

be two functions (not necessarily box forms). Write $F_1 \subseteq F_2$ if for all $B \in \square X$, $F_1(B) \subseteq F_2(B)$. Then we have

> Let $F_1 \subseteq F_2$. If F_2 is a box form, then F_1 is a box form.

Of course, we also have $F_1|_X = F_2|_X$. What we need in our application, however, is the "converse": if $F_1 \subseteq F_2$ and F_1 is a box form, then F_2 is a box form. To motivate this application, consider the mean value form $\square_{\mathsf{M}}$: its definition (6) calls for an exact evaluation $f(m(B))$, which we must approximate. In general, for any interval form $\square f$, we need to approximate it by some function of the type $\widetilde{\square} f : \square \mathbb{R}^m \to \square \mathbb{R}$. But how are $\widetilde{\square} f$ and $\square f$ related? We will say $\widetilde{\square} f$ an **effective** form of $\square f$ provided these properties hold:

(i) (Inclusion) $\square f \subseteq \widetilde{\square} f$ and
(ii) (Precision) $q(\square f(B), \widetilde{\square} f(B)) \leq w(B)$ where $w(B)$ is the width of B and $q(I, J)$ is the Hausdorff metric on closed intervals.
(iii) (Exactness) $\square f$ is dyadically exact.

We will discuss the third property (iii) below. But first, we note that properties (i) and (ii) ensure our desired converse:

Lemma 1. *If $\widetilde{\square} f$ satisfies (i) and (ii), then $\widetilde{\square} f$ is a box form of f.*

To compute $\widetilde{\square} f(B)$, this lemma says that, provided our numerical approximation is rounded correctly to satisfy Property (i), then we only have to ensure that the error is bounded by the width of B as in Property (ii). Although the boxes B are distributed over space and time, the global correctness is guaranteed by the nature of our predicates.

We now turn to Property (iii). This requirement is connected to general ideas about efficiency and effectivity of numerical computation. For this, we assume that $X = \mathbb{R}^n$ and $Y = \mathbb{R}$. In practice, real numbers are most efficiently approximated by dyadic numbers, $\mathbb{Z}[\frac{1}{2}]$ or BigFloats (see [55]). Our definition of $\widetilde{\square} f$ serves the fiction that it could accept every box in $\square \mathbb{R}^n$. This is useful fiction because it cleanly fits into mathematical analysis. But in implementations, these box functions only need to accept **dyadic boxes**, i.e., boxes whose corners have dyadic coordinates. We say a box function $F : \square \mathbb{R}^n \to \square \mathbb{R}$ is **dyadically exact** if its restriction to dyadic boxes outputs dyadic intervals. This explains our Property (iii). Evidently, it is not hard satisfy all 3 properties of effectivity.

Literate[6] Algorithmic Development. In [51], we developed a subdivision algorithm for isolating the simple real roots of a real system

$$\boldsymbol{f} = (f_1, \ldots, f_n) : \mathbb{R}^n \to \mathbb{R}^n.$$

As a subdivision algorithm, it has several predicates: the centerpiece is the Miranda Test $\mathtt{MK}(B)$ for existence of real roots in B. We have our ubiquitous exclusion test, but defined as $C_0(B) := [[(\exists i = 1, \ldots, n)(0 \notin f_i(B))]]$. We also

---

[6] In the spirit of Knuth's "Literate Programming".

need a Jacobian Test $\mathrm{JC}(B)$ to confirm at most one root. Each predicate C is first defined mathematically, then as box predicates $\square C$, and finally as effective predicates $\widetilde{\square} C$. Thus, there are three levels of description:

(A) **Abstract** C, f
(I) **Interval** $\square C, \square f$
(E) **Effective** $\widetilde{\square} C, \widetilde{\square} f$

As expressed by Burr et al. [15, §2.3], the goal is to delay the introduction of $\square C$ (and hence $\widetilde{\square} C$). The motivation comes from the fact that the theory is cleanest at Level (A), and less so at later levels.

In effect, we have three algorithms:

$$A, \quad \square A, \quad \widetilde{\square} A \tag{7}$$

each being an instantiation of a common algorithmic scheme by predicates and functions of the appropriate level. This is analogous to standard construction of numerical algorithms from $A \mapsto \widetilde{A}$ (see the Introduction); the difference is that our starting point A is in the Subdivision Framework. We then prove the algorithms correct at each level. At each level, we bring in new details but are able to rely on the properties already proved in the previous level. For instance, an important phenomenon when we transition from A to $\square A$ is the appearance of Lipshitz constants inherent in interval methods. This approach ("AIE methodology") displays a continuity of ideas and exposes the issues unique to each level. The clarity and confidence in the correctness of $\widetilde{\square} A$ are surely much better than if we had attempted[7] an *ab initio* correctness proof of $\widetilde{\square} A$. Quoting Knuth:

> *"Beware of bugs in the above code; I have only proved it correct, not tried it."*

5   On Oracle Objects

In our **Simple Subdivision Framework**, we pass queues from one phase to the next. Such queues serve to represent intermediate states of our ultimate output (the simplex K for Problem A or the path π for Problem B). In this section, we explore the idea of representing computational objects that encode states and other information. The term "object" suggests connection to Object Oriented Programming Languages (OOPL) since, in order to make our algorithms effective, it must be ultimately implemented in a programming language. See the recent paper of Brauße et al. [10] that also brings programming semantics into the theory of real computation.

---

[7] It is possible that such proofs contribute to the poor reputation of error analysis as a topic.

5.1 Soft Tests

We begin by discussing the "ur-predicate", the comparison of two real numbers x, y. We may write the comparison as a logical predicate $[[x < y]]$, using the test notation of Sect. 1.1. If true, we branch to point A, and otherwise we branch to point B. In exact computation, the two points A and B in the program encode the respective assertions $[x < y]$ and $[x \geq y]$, where the notation "$[S]$" is now an **assertion** that S is true. In numerical computation, we might need point C in the program to encode the assertion $[x \, ? \, y]$ (don't know). To simplify our primitives, let us reduce this 3-valued test to a 2-valued version, denoted $[(x < y)]$ where point A represents the assertion $[x < y]$ (as before) but point B asserts $[x \geq ? y]$. But outcome "$[x \geq ? y]$" suggests the assertion $[x \geq y] \vee [x \, ? \, y]$ (we will explore this more carefully below). In reality, we reached the point B because the test $[(x < y)]$ was done with limited precision p. Thus we may explicitly indicate this precision[8] by writing $[(x < y)]_p$. We call $[(x < y)]_p$ a **one-way test** because the failure to assert $[x < y]$ does not imply the negated assertion $[x \geq y]$.

The 2-valued exact tests $[[x < y]]$ and $[[x \geq y]]$ are equivalent in the sense that one is obtained from the other by switching truth values. But $[(x < y)]_p$ and $[(x \geq y)]_p$ have no such symmetry. This suggests that we could define another form of $[(x < y)]_p$ in which the point B encodes the assertion $[x \geq y]$, but point A encodes $[x < ? y]$. These two versions of the one-way test have their respective uses – the first version is aimed at confirming the assertion "$[x < y]$", and the second version is aimed at falsifying it. To distinguish them, let us write $Con[x < y]_p$ for confirmation test and $Fal[x < y]_p$ for other. Unless otherwise stated, we continue to view the test $[(x < y)]$ in the confirmation mode. It might appear that we are splitting hairs by reducing a 3-way test to two 2-way tests. But since these tests may involve heavy computations (such as testing if a robot configuration is free), this split may be useful. Alternatively, the numbers x, y may represent complicated expressions (see below).

To implement such one-way tests, we need to assume that x, y are **number oracles** (see [28]). That means for each $p \in \mathbb{Z}$, we can ask for a p-bit approximation of x, denoted $(x)_p$. This[9] means $(x)_p = x \pm 2^{-p}$. We may represent $(x)_p$ by a dyadic number with at most p bits after the binary point. For instance, we can implement the one-way $[(x < y)]_p$ as follows:

```
if (x)_p + 2^-p < (y)_p - 2^-p
    return [x < y]
else
    return [x ≥?y]
```

---

[8] We use "precision" for the *a priori* user-specified bound. The algorithm delivers a value whose *a posteriori* error is at most this precision.

[9] We write $a = b \pm c$ to mean there exists a constant $\theta \in [-1, 1]$ such that $a = b + \theta \cdot c$. Alternatively, $|a - b| \leq |c|$.

Observe that this algorithm is indeterminate (see Sect. 2) because $(x)_p$ does not identify a specific value, but depends on the oracle for x. The exact test $[[x < y]]$ can be reduced to two one-way tests as follows:

```
Subroutine [[x < y]]:
   For p = 0, 1, . . .
      if Con[x < y]_p
         return [x < y]
      else if Fal[x < y]_p
         return [x ≥ y]
```

In the case $x = y$, this subroutine is non-halting. Unfortunately, this is the best we can do without more information about x or y. It turns out that we can modify the loop above to produce a halting subroutine. That is the **Soft Zero Test** in [52] which has three outcomes: $[x < y]$, $[x > y]$ and $[x \simeq y]$. The last outcome is new, and is defined[10] to mean

$$\left[\frac{1}{2}x < y < 2x\right] \quad \vee \quad \left[\frac{1}{2}x > y > 2x\right]. \tag{8}$$

The first disjunct implies that x, y are both positive, and the second implies both are negative. We denote this test by $[\{x : y\}]$. What makes this test decidable (halting) is the introduction of the new outcome. But we also need a "mild" assumption: *either x or y is non-zero*. It is assumed that both x and y are non-negative in [52]. That is justified by the intended application where both x and y are sums of absolute values (from the Pellet Test). Essentially, this Soft Zero Test is at the heart of our soft-ε criteria for roots. In exact computation, comparing two numbers $x : y$ is equivalent to the computing the sign of the single number $x - y$. The Soft Zero Test shows that you can do a bit more by keeping x and y separate.

What is the logical status of the intuitive formula "$[x >?y] = [x > y] \vee [x?y]$"? The truth-values $[x < y]_p$ are parametrized by x, y and also p. It is enough to consider the non-parametric setting where, in addition to true, false, we add a third logical value, true (false-or-true). Then we have these truth tables:

∧	true	frue	false
true	true	frue	false
frue	frue	frue	false
false	false	false	false

	true	frue	false
¬	false	frue	true

∨	true	frue	false
true	true	true	true
frue	true	frue	frue
false	true	frue	false

So far, we looked at point-based comparisons. We now consider interval-based comparisons. The ur-predicate here is the **Membership Test** $[[x \in I]]$ where I is an interval. Here, we view I as a dyadic interval and x is the usual oracle. Let $[\{x \in I\}]$ denote the **Soft Membership Test** with two outcomes, $[x \widetilde{\in} I]$ and

---

[10] Despite the appearance of asymmetry, x and y are treated symmetrically by this definition.

$[x \tilde{\notin} I]$. We define them as $[x \in 2I]$ and $[x \notin I/2]$, respectively. If desired, we may replace '2' by $1 + 2^{-p}$ and denote it by $[\{x \in I\}]_p$. It is indeterminate because, in case $x \in 2I \setminus I/2$, both outcomes are acceptable. This can be implemented as the exact test

$$[[(x)_p \in I]] \tag{9}$$

since $p = -\lceil 1 + \log_2(w(I)) \rceil$ is easily computable. Exactness of (9) follows from the fact that I and $(x)_p$ are dyadic. If $[(x)_p \in I]$, we output $[x \tilde{\in} I]$, and otherwise, we output $[x \tilde{\notin} I]$. The Soft Membership Test is unconditional. These ideas can be generalized to produced soft membership in balls or boxes: the predicates in [5,41] are examples of such tests.

5.2   Whence Number Oracles?

Number oracles are ubiquitous in the theory of real computation [28,50]. Their availability is largely assumed. Perhaps it is generally assumed that they come from well-known mathematical series, and all we need to do is to evaluate such series to enough precision. But even this problem deserves careful investigation from the viewpoint of complexity. For instance, the family of hypergeometric functions provide us with a rich class of series that include the elementary functions and much more. But if we are given a function in terms of its hypergeometric parameters, there are issues of transforming them to speed up the convergence. From the work of Richard Brent in the 1970s, it is well-known that to evaluate such functions at a fixed point is extremely fast (basically the speed of integer multiplication, perhaps with extra log factors). But this tells us little about global or uniform complexity of these approximation algorithms. We refer the interested reader to [18,19]. In algebraic computation, we do not have such series. But it is easy to provide an oracle for any algebraic number α if we have a defining integer polynomial, $p(\alpha) = 0$ with $p'(\alpha) \neq 0$. If α is real, we can find an isolating interval $[a, b]$ for α. Thereafter, we can use bisection to produce a convergent sequence of intervals. Following Dekker and Brent [11], we can use a Newton-bisection iteration to speed up this process. A recent variant of Newton-bisection from Abbot, Sagraloff and Kerber [1,26] led to the complexity analysis of such speedups. There are analogous procedures to produce oracles for complex α. It turns out that in geometric computations, we seldom begin with algebraic numbers: instead we typically start with rational numbers and α is built up as an expression using different algebraic operators to produce arbitrary algebraic numbers. Let us briefly describe this class of oracles.

Let Ω be a set of real algebraic operators. Typically, Ω contains at least $\{\pm, \times, \div, \sqrt{\cdot}\} \cup \mathbb{Z}$. Assume that each operator $\omega \in \Omega$ has an approximation algorithm [55]. Let $E(\boldsymbol{x}) = E(x_1, \ldots, x_n)$ be an algebraic expression over $\Omega \cup \{x_1, \ldots, x_n\}$. E.g., $E(\boldsymbol{x}) = \sqrt{x^2 - 2y + 1} - \sqrt[3]{xy - y^2}$. If $\boldsymbol{a} = (a_1, \ldots, a_n)$ is a sequence of number oracles, then there are general mechanisms to construct an oracle for the number $E(\boldsymbol{a})$ (see [32,57]). Note that this description is more general than the usual setting for EGC where the expression E is a constant; but the extension to expressions with arguments is relatively direct straightforward.

Zero Problems arise from the fact that some operators such as $\div, \sqrt{\cdot}$ or $\log$ are partial functions, and so $E(\boldsymbol{a})$ may be undefined. We need to detect this situation and halt. This is a hopeless case, even for $E(\boldsymbol{a}) = a_1 - a_2$, unless we have prior information such as a_1, a_2 are algebraic with degree and height bounds (or some height substitute).

5.3   Cluster Oracles

Oracles arising from algebraic expressions can be generalized to geometric expressions in the sense of Constructive Solid Geometry (CSG) [2]. Here, the expressions are built from primitive geometric shapes such as numbers, points, balls and half-spaces, using boolean operators such as intersection and union. Such expressions can be new sources for number oracles. But in this subsection, we will focus on a recent extension of number oracles to "cluster oracles". It arose in our root clustering algorithms [4,5], and its extension to solving triangular systems in $\mathbb{C}^n$ [24]. Intuitively, Cauchy sequences must be generalized to "Cauchy trees" because clusters may split upon request for more precision.

Multisets arise naturally when we consider the zero sets of functions: let $D \subseteq \mathbb{C}^n$ and $\boldsymbol{f} : \mathbb{C}^n \to \mathbb{C}^n$. Assume that $\boldsymbol{f}^{-1}(\boldsymbol{0}) \cap D$ is a finite set, and for each $\boldsymbol{a} \in \boldsymbol{f}^{-1}(\boldsymbol{0})$, we can assign an integer $\mu(\boldsymbol{a}) \geq 1$ called its multiplicity. We introduce two useful notations: let $\mathtt{Zero}_f(D) := D \cap \boldsymbol{f}^{-1}(\boldsymbol{0})$ and $\#_f(D)$ be the total multiplicities of the roots in $\mathtt{Zero}_f(D)$. The pair $(\mathtt{Zero}_F(D), \#_F(D))$ is an example of a multiset.

In general, a **multiset** S is a pair $(\underline{S}, \mu)$ where $\underline{S}$ is[11] an ordinary set (called the **underlying set** of S) and $\mu = \mu_S$ assigns a **multiplicity**, a positive integer $\mu(x)$, to each $x \in \underline{S}$. Also let $\mu(S) := \sum_{x \in \underline{S}} \mu(x)$ be the (total) **multiplicity** of S, assumed to be finite. Let $|S|$ denote the cardinality of $\underline{S}$; so $|S| \leq \mu(S)$. If $|S| = 1$ (resp., $|S| = 0$) we say S is a **singleton** (resp., **empty**). We denote the empty multiset as well (ordinary) empty set by the same symbol $\emptyset$. If T is another multiset, we write $S \subseteq T$ and call S a **subset** of T if $\underline{S} \subseteq \underline{T}$ and $\mu_S(x) \leq \mu_T(x)$ for all $x \in \underline{S}$. In this paper, we assume[12] equality, i.e., $\mu_S(x) = \mu_T(x)$, in subset relations. The intersection $S \cap T$ is the largest multiset R such that $R \subseteq S$ and $R \subseteq T$.

Our multisets interact with the world of ordinary sets: let X be an ordinary set. Then '$S \subseteq X$' means that $\underline{S} \subseteq X$. Likewise '$S \cap X$' denotes the multiset $T \subseteq S$ where $x \in \underline{T}$ iff $x \in X$.

We are interested in the concept of a "cluster" C. Informally, a cluster C is a multiset in a larger multiset U which is nicely separated from $U \setminus C$. Let us formulate this concept in the context of a normed linear space X with norm $\|\cdot\|$. Let $\Delta = \Delta(m, r) \subseteq X$ denote the **ball** centered at m of radius $r \geq 0$. For real $\alpha > 0$, let $\alpha\Delta(m, r)$ denote the ball $\Delta(m, \alpha r)$. Let us fix a multiset $U \subseteq X$. A

---

[11]   There should no confusion with the notion support of a simplicial complex $\underline{K}$.

[12]   Strict inequality may arise in subsets of zero sets: if $F = (F_1, \ldots, F_n)$ where F_i are polynomials, and $G = (G_1, \ldots, G_n)$ where each G_i divides F_i, then $\mathtt{Zero}_G(D) \subseteq \mathtt{Zero}_F(D)$ might exhibit this phenomenon.

cluster (of U) is a non-empty set $C \subseteq U$ of the form $C = U \cap \Delta$ for some ball $\Delta \subseteq X$. Call such a Δ an **isolator** of cluster C; this isolator is **natural** if, in addition, $C = U \cap 3\Delta$. If C has a natural isolator, we call it a **natural cluster**. The fundamental property of natural clusters is this:

Lemma 2. *Let X be a normed linear space and $U \subseteq X$ be a finite multiset. Then any two natural clusters C, C' of U are either disjoint or have a subset relation, i.e., either $C \cap C' = \emptyset$ or $C \subseteq C'$ or $C' \subseteq C$.*

Basically the proof works because of the triangle inequality in X. As a corollary, there are at most $2|U| - 1$ natural clusters of U, and they can be organized into a **cluster tree**: each node in the tree is a natural cluster of U and the child-parent relation is just $C \subset C'$. The original cluster concept in [4,52] assumes $X = \mathbb{C}$. In [24] it was extended to $X = \mathbb{C}^n$, for complex roots of triangular systems. Computing natural clusters may be regarded as the soft-ε criterion for root isolation; as we shall see, it is effective and can completely remove the Zero Problems associated with multiple roots.

But how do we compute such clusters? We need predicates to check if a given Δ is an isolator and to determine its total multiplicity. For $X = \mathbb{C}^n$, we could use some multidimensional form of Pellet's test, and for $X = \mathbb{R}^n$, there are similar tools such as multidimensional Sturm theory based on quadratic forms [35–37]. Unfortunately, at present, these tools do not appear to be practical. In lieu of this, we take another route in [24]: we first reduce the input system into triangular systems using known algebraic techniques. In the triangular form, we can compute the multivariate clusters and their multiplicities using the efficient univariate multiplicity tests of [4,23].

The main tool in [4,23] is a test from Pellet (1881). Fix a complex polynomial $f(z) \in \mathbb{C}[z]$. First consider the test

$$T_k(f) := \left[\!\left[\, |a_k| > \sum_{i \neq k} |a_i| \,\right]\!\right] \tag{10}$$

where $f(z) = \sum_{i=0}^{n} a_i z^i$. This test is defined for any $k = 0, \dots, n$. Pellet's theorem says that

$$\text{if } T_k(f) \text{ succeeds then } \#_f(\Delta(0,1)) = k.$$

So this test, which is a simple application of Rouché's Theorem, can confirm that the total multiplicity of the complex roots of f in the unit disc $\Delta(0,1)$ centered at the origin is exactly k. The case $k = 0$ is interesting – it is an exclusion test! It is more expensive than the standard C_0 test, but we shall see that its failure provides some partial converse information. We can confirm that the disc $\Delta(0,r)$ of radius $r > 0$ has k roots by applying the T_k-test to the polynomial $f(rz) = \sum_i b_i z^i$ where $b_i = a_i r^i$. Similarly, we can confirm that $\Delta(m,1)$ (the unit disc centered at $m \in \mathbb{C}$) has k roots by applying the T_k-test to the Taylor-shifted polynomial $f(z + m)$. Combining these two operations, we obtain a test for an arbitrary disk $\Delta(m,r)$. Let $T_k(f, m, r)$, or simply T_k, denote such a test.

The next question is crucial: *when is the success of T_k test assured?* It is shown that there are positive numbers $c < 1 < C$ such that if

$$\#_f(c \cdot \Delta(m,r)) = \#_f(C \cdot \Delta(m,r))$$

then $T_k(f, m, r)$ will succeed. In other words, this gives a partial converse to Pellet's test. Unfortunately, these numbers depend on the degree: $c = \Omega(1/n)$ and $C = O(n^3)$. By applying Graeffe iteration $5 + \log \log n$ times to f, we reduce these numbers to $c = 2\sqrt{2}/3 \simeq 0.94$ and $C = 4/3$. Suppose B is a subdivision box and $\Delta(B)$ is its circumscribing disc. Let $T_k^G(f; B)$ denote the application of the T_k test to the Graeffe-transformed Δ-shifted polynomial f. Choose $k = 0$:

$$\text{If } T_0^G(f; B) \, succeeds, \ \#_f(B) = 0.$$
$$\text{If } T_0^G(f; B) \, fails, \ \#_f(2B) > 0.$$

We classify boxes as **excluded** if this test succeeds, and **included** otherwise. We now have a very powerful test that is analogous to the Soft Membership Test earlier. The remaining issues treated in the paper are:

- We need approximate versions of these tests: thus we use $\widetilde{T}_k^G$ instead of T_k^G. The Soft Zero Test is used to make numerical comparison of (10).
- We use $\widetilde{T}_0^G(B)$ as exclusion test. We maintain the connected components of those B's that are included (i.e., fail the exclusion test). These components are potentially cluster. We refine a component by splitting each of its constituent boxes, and applying the $\widetilde{T}_0^G$ tests again.
- To obtain near-optimal bounds, we use the Abbot form of Newton-Bisection [1,26] on a connected component C. If C is sufficiently separated from the other components, then we could use the $\widetilde{T}_k^G$ test to determine that $\#(C) = k$, and even apply the order k Newton iteration successfully.
- For complexity analysis (in the bench mark case), we need charging schemes that charge these tests to roots of f in $2B_0$. It turns out that for the non-integer polynomials, we can provide some complexity estimates based on the root geometry.

We hope this overview may make the original papers more accessible. In [24], we package the above structures into **cluster oracles** in order to compute multi-dimensional clusters inductively. Such cluster oracles, viewed as objects in the sense of OOPL, can provide an efficient mechanism for other similar applications.

6   Conclusion and Open Problems

The foundations of subdivision computation is a wide-open area of research, with promises of new and effective algorithms that have mild (or no) conditions on the input. Our illustrative examples suggest that such algorithms can be practical and compare favorably with less-rigorous solutions or symbolic or exact solutions.

Our soft-ε criteria for two key problems seems to have achieved a satisfactory level of completeness: (a) complex root clustering for polynomials with oracle coefficients [4] and (b) resolution-exact path planning [49]. Of course, success creates its own (new) set of problems: for (a), we would like to treat more general functions such as analytic or harmonic functions. For (b), the challenge is to design a practical nonheuristic planner for spatial robots with 6 degrees-of-freedom. This natural but elusive quest appears to be reachable within our framework. Finally, we pose some open problems:

1. Algorithms with soft-ε correctness is the continuous analogue of "approximation schemes" in discrete optimization algorithms. Just as the barrier to polynomial-time schemes (PTAS) are located in NP-hardness or similar complexity classes, the barrier in the continuous case are various Zero Problems. We would like a complexity theory of such Zero Problems.
   See also the recent paper [17].
2. It is generally challenging to remove *all* Zero Problems. A prime example is the PV-type algorithms [38]. Such algorithms are based on the Marching Cube paradigm, and require the exact sign evaluation of a function at the corners of subdivision boxes. How do we soften this?
3. Interval methods are central to all our algorithms. We would like to develop interval methods in more abstract spaces than Euclidean ones. Are normed vector spaces or metric spaces the natural home for such extensions? As usual, we need good problems on which to cut our teeth.
4. Path planning in very high dimensions is an open problem. An example of a currently out-of-reach path planning problem: a planar snake with 10 joints. The configuration space is $\mathbb{R}^2 \times (S^1)^{10}$. This requires new paradigms, but we believe they can be built upon a subdivision framework.
5. Path planning is only the simplest of motion planning problems. What do soft-ε algorithms mean in non-holonomic planning, or kino-dynamic planning? A good problem is to try subdivision in state space: imagine a point robot in the plane amidst obstacles. Its state or coordinates are $(x, y, \dot{x}, \dot{y})$ representing position and velocity. We want to plan a minimum time trajectory from some start to goal states, subject to acceleration bounds.
6. The notion of natural root clusters suggests other applications and extensions. How do we cluster matrix eigenvalues? It seems that other considerations come into play: the invariant subspaces associated with eigenvalues should play a role in defining "natural clusters of eigenvalues".
7. Complexity analysis of subdivision is largely open. The case of univariate zeros is reasonably well-understood, but there are many open problems even for zero-dimensional problems in higher dimensions. A key tool is continuous amortization [13], but recently Cucker et al. [17] initiated a Smale-type average case analysis for subdivision algorithms.

Acknowledgements. The author is deeply grateful for the feedback and bug reports from Michael Burr, Matthew England, Rémi Imbach, Juan Xu and Bo Huang.

References

1. Abbott, J.: Quadratic interval refinement for real roots. ACM Commun. Comput. Algebra **48**(1/2), 3–12 (2014). https://doi.org/10.1145/2644288.2644291. http://doi.acm.org/10.1145/2644288.2644291
2. Agrawal, A., Requicha, A.: A paradigm for the robust design of algorithms for geometric modeling. In: Computer Graphics Forum, vol. 13, no. 3, pp. 33–44 (1994). 15th Annual Conference and Exhibition. EUROGRAPHICS 1994
3. Becker, R.: The Bolzano method to isolate the real roots of a bitstream polynomial. Bachelor thesis, University of Saarland, Saarbruecken, Germany, May 2012
4. Becker, R., Sagraloff, M., Sharma, V., Xu, J., Yap, C.: Complexity analysis of root clustering for a complex polynomial. In: 41st International Symposium on Symbolic and Algebraic Computation, iSSAC 2016, Wilfrid Laurier University, Waterloo, Canada, 20–22 July, pp. 71–78 (2016)
5. Becker, R., Sagraloff, M., Sharma, V., Yap, C.: A near-optimal subdivision algorithm for complex root isolation based on Pellet test and Newton iteration. J. Symbolic Comput. **86**, 51–96 (2018)
6. Bennett, H., Papadopoulou, E., Yap, C.: Planar minimization diagrams via subdivision with applications to anisotropic Voronoi diagrams. In: Eurographics Symposium on Geometric Processing, SGP 2016, Berlin, Germany, 20–24 June 2016, vol. 35, no. 5 (2016)
7. Bennett, H., Yap, C.: Amortized analysis of smooth quadtrees in all dimensions. Comput. Geom. Theory Appl. **63**, 20–39 (2017). Also, in Proceedings SWAT 2014
8. de Berg, M., Cheong, O., van Kreveld, M., Overmars, M.: Computational Geometry: Algorithms and Applications, 3rd edn. Springer, Berlin (2008). https://doi.org/10.1007/978-3-540-77974-2
9. Boissonnat, J.D., Teillaud, M. (eds.): Effective Computational Geometry for Curves and Surfaces. Springer, Heidelberg (2007). https://doi.org/10.1007/978-3-540-33259-6
10. Brauße, F., et al.: Semantics, logic, and verification of "exact real computation" (2019)
11. Brent, R.P.: Algorithms for Minimization Without Derivatives. Prentice Hall, Englewood Cliffs (1973)
12. Burr, M., Krahmer, F.: SqFreeEVAL: an (almost) optimal real-root isolation algorithm. J. Symbolic Comput. **47**(2), 153–166 (2012)
13. Burr, M., Krahmer, F., Yap, C.: Continuous amortization: a non-probabilistic adaptive analysis technique. Electronic Colloquium on Computational Complexity (ECCC) TR09(136), December 2009. http://eccc.hpi-web.de/report/2009/136/
14. Burr, M.A.: Continuous amortization and extensions: with applications to bisection-based root isolation. J. Symb. Comput. **77**, 78–126 (2016). https://doi.org/10.1016/j.jsc.2016.01.007
15. Burr, M.A., Gao, S., Tsigaridas, E.: The complexity of an adaptive subdivision method for approximating real curves. In: 42nd International Symposium on Symbolic and Algebraic Computation (ISSAC), ISSAC 2017, pp. 61–68. ACM, New York (2017). https://doi.org/10.1145/3087604.3087654
16. Burr, M.A., Gao, S., Tsigaridas, E.: The complexity of subdivision for diameter-distance tests. J. Symbolic Computation (2019, to appear)
17. Cucker, F., Ergür, A.A., Tonelli-Cueto, J.: Plantinga-vegter algorithm takes average polynomial time. arXiv:1901.09234 [cs.CG] (2019)

18. Du, Z., Eleftheriou, M., Moreira, J., Yap, C.: Hypergeometric functions in exact geometric computation. In: Brattka, V., Schoeder, M., Weihrauch, K. (eds.) Proceedings of 5th Workshop on Computability and Complexity in Analysis, Malaga, Spain, 12–13 July 2002, pp. 55–66 (2002). In Electronic Notes in Theoretical Computer Science 66:1 (2002). http://www.elsevier.nl/locate/entcs/volume66.html

19. Du, Z., Yap, C.: Uniform complexity of approximating hypergeometric functions with absolute error. In: Pae, S., Park, H. (eds.) Proceedings of 7th Asian Symposium on Computer Mathematics, ASCM 2005, pp. 246–249 (2006)

20. Emiris, I.Z., Pan, V.Y., Tsigaridas, E.P.: Algebraic algorithms. In: Gonzalez, T., Diaz-Herrera, J., Tucker, A. (eds.) Computing Handbook: Computer Science and Software Engineering, 3rd edn., pp. 10: 1–30. Chapman and Hall/CRC, Boca Raton (2014)

21. Higham, N.J.: Accuracy and Stability of Numerical Algorithms, 2nd edn. Society for Industrial and Applied Mathematics, Philadelphia (2002)

22. Hsu, C.H., Chiang, Y.J., Yap, C.: Rods and rings: soft subdivision planner for R^3 x S^2. In: Proceedings of 35th International Symposium on Computational Geometry, SoCG 2019, 18–21 June 2019. CG Week 2019, Portland Oregon. Also in arXiv:1903.09416

23. Imbach, R., Pan, V.Y., Yap, C.: Implementation of a near-optimal complex root clustering algorithm. In: Davenport, J.H., Kauers, M., Labahn, G., Urban, J. (eds.) ICMS 2018. LNCS, vol. 10931, pp. 235–244. Springer, Cham (2018). https://doi.org/10.1007/978-3-319-96418-8_28

24. Imbach, R., Pouget, M., Yap, C.: Effective subdivision algorithm for isolating zeros of real systems of equations, with complexity analysis, 21st CASC, Moscow (2019, to appear)

25. Kearfott, R.B.: Rigorous Global Search: Continuous Problems, vol. 13. Springer, Dordrecht (2013). https://doi.org/10.1007/978-1-4757-2495-0

26. Kerber, M., Sagraloff, M.: Efficient real root approximation. In: Schost, É., Emiris, I.Z. (eds.) ISSAC, pp. 209–216. ACM (2011)

27. Kincaid, D., Cheney, W.: Numerical Analysis: Mathematics of Scientific Computing, 3rd edn. Brooks/Cole, Boston (2002)

28. Ko, K.I.: Complexity Theory of Real Functions. Progress in Theoretical Computer Science. Birkhäuser, Boston (1991)

29. Lien, J.-M., Sharma, V., Vegter, G., Yap, C.: Isotopic arrangement of simple curves: an exact numerical approach based on subdivision. In: Hong, H., Yap, C. (eds.) ICMS 2014. LNCS, vol. 8592, pp. 277–282. Springer, Heidelberg (2014). https://doi.org/10.1007/978-3-662-44199-2_43

30. Lin, L., Yap, C.: Adaptive isotopic approximation of nonsingular curves: the parameterizability and nonlocal isotopy approach. Discrete Comp. Geom. **45**(4), 760–795 (2011)

31. Lin, L., Yap, C., Yu, J.: Non-local isotopic approximation of nonsingular surfaces. Comput. Aided Des. **45**(2), 451–462 (2012). Symposium on Solid and Physical Modeling (SPM). U. of Burgundy, Dijon, France, 29–31 October 2012

32. Mehlhorn, K., Schirra, S.: Exact computation with `leda_real` - theory and geometric applications. In: Alefeld, G., Rohn, J., Rump, S., Yamamoto, T. (eds.) Symbolic Algebraic Methods and Verification Methods, pp. 163–172. Springer, Vienna (2001). https://doi.org/10.1007/978-3-7091-6280-4_16

33. Moore, R.E.: Interval Analysis. Prentice Hall, Englewood Cliffs (1966)

34. Ó'Dúnlaing, C., Yap, C.K.: A "retraction" method for planning the motion of a disc. J. Algorithms **6**, 104–111 (1985). Also, Chapter 6 in Planning, Geometry, and Complexity, eds. Schwartz, Sharir and Hopcroft, Ablex Pub. Corp., Norwood, NJ 1987

35. Pedersen, P.: Counting real zeros. In: Proceedings of Conference on Algebraic Algorithms and Error Correcting Codes. LNCS, vol. 539, pp. 318–332. Springer (1991)

36. Pedersen, P.: Counting real zeros. Ph.D. thesis, New York University (1991). Also, Courant Institute Computer Science Technical Report 545 (Robotics Report R243)

37. Pedersen, P., Roy, M.-F., Szpirglas, A.: Counting real zeros in the multivariate case. In: Eyssette, F., Galligo, A. (eds.) Computational Algebraic Geometry. PM, vol. 109, pp. 203–224. Birkhäuser, Boston (1993). https://doi.org/10.1007/978-1-4612-2752-6_15

38. Plantinga, S., Vegter, G.: Isotopic approximation of implicit curves and surfaces. In: Proceedings of Eurographics Symposium on Geometry Processing, pp. 245–254. ACM Press, New York (2004)

39. Preparata, F.P., Shamos, M.I.: Computational Geometry. Springer, New York (1985). https://doi.org/10.1007/978-1-4612-1098-6

40. Riley, K., Hopson, M., Bence, S.: Mathematical Methods for Physics and Engineering, 3rd edn. Cambridge University Press, New York (2006)

41. Sagraloff, M., Yap, C.K.: A simple but exact and efficient algorithm for complex root isolation. In: Emiris, I.Z. (ed.) 36th International Symposium on Symbolic and Algebraic Computing, San Jose, California, 8–11 June, pp. 353–360 (2011)

42. Sagraloff, M., Yap, C.K.: An efficient exact subdivision algorithm for isolating complex roots of a polynomial and its complexity analysis, July 2009, submitted. Full paper from http://cs.nyu.edu/exact/ or http://www.mpi-inf.mpg.de/~msagralo/

43. Schuurman, P., Woeginger, G.: Approximation schemes: a tutorial. In: Möhring, R., Potts, C., Schulz, A., Woeginger, G., Wolsey, L. (eds.) Lectures in Scheduling (2007, to appear)

44. Sharma, V., Yap, C.: Near optimal tree size bounds on a simple real root isolation algorithm. In: 37th International Symposium on Symbolic and Algebraic Computing, ISSAC 2012, Grenoble, France, 22–25 July 2012, pp. 319–326 (2012)

45. Sharma, V., Yap, C.K.: Robust geometric computation. In: Goodman, J.E., O'Rourke, J., Tóth, C. (eds.) Handbook of Discrete and Computational Geometry, 3rd edn., chap. 45, pp. 1189–1224. Chapman & Hall/CRC, Boca Raton (2017)

46. Snyder, J.: Generative Modeling for Computer Graphics and CAD. Symbolic Shape Design Using Interval Analysis. Academic Press Professional Inc., San Diego (1992)

47. Trefethen, L.N., Bau, D.: Numerical Linear Algebra. Society for Industrial and Applied Mathematics, Philadelphia (1997)

48. Ueberhuber, C.W.: Numerical Computation 2: Methods, Software, and Analysis. Springer, Berlin (1997)

49. Wang, C., Chiang, Y.J., Yap, C.: On soft predicates in subdivision motion planning. Comput. Geom. Theory Appl. **48**(8), 589–605 (2015). (Special Issue for SoCG 2013)

50. Weihrauch, K.: Computable Analysis. Springer, Berlin (2000). https://doi.org/10.1007/978-3-642-56999-9

51. Xu, J., Yap, C.: Effective subdivision algorithm for isolating zeros of real systems of equations, with complexity analysis. In: 44th International Symposium Symbolic and Algebraic Computing, Beihang University, Beijing, 15–18 July (2019)

52. Yap, C., Sagraloff, M., Sharma, V.: Analytic root clustering: a complete algorithm using soft zero tests. In: Bonizzoni, P., Brattka, V., Löwe, B. (eds.) CiE 2013. LNCS, vol. 7921, pp. 434–444. Springer, Heidelberg (2013). https://doi.org/10.1007/978-3-642-39053-1_51

53. Yap, C.K.: Soft subdivision search in motion planning, II: axiomatics. In: Wang, J., Yap, C. (eds.) FAW 2015. LNCS, vol. 9130, pp. 7–22. Springer, Cham (2015). https://doi.org/10.1007/978-3-319-19647-3_2

54. Yap, C., Luo, Z., Hsu, C.H.: Resolution-exact planner for thick non-crossing 2-link robots. In: Proceedings of 12th International Workshop on Algorithmic Foundations of Robotics, WAFR 2016, San Francisco, 13–16 December 2016 (2016). The appendix in the full paper (and arXiv from http://cs.nyu.edu/exact/ (and arXiv:1704.05123 [cs.CG]) contains proofs and additional experimental data

55. Yap, C.K.: On guaranteed accuracy computation. In: Chen, F., Wang, D. (eds.) Geometric Computation, chap. 12, pp. 322–373. World Scientific Publishing Co., Singapore (2004)

56. Yap, C.K.: In praise of numerical computation. In: Albers, S., Alt, H., Näher, S. (eds.) Efficient Algorithms. LNCS, vol. 5760, pp. 380–407. Springer, Heidelberg (2009). https://doi.org/10.1007/978-3-642-03456-5_26

57. Yu, J., Yap, C., Du, Z., Pion, S., Brönnimann, H.: The design of core 2: a library for exact numeric computation in geometry and algebra. In: Fukuda, K., Hoeven, J., Joswig, M., Takayama, N. (eds.) ICMS 2010. LNCS, vol. 6327, pp. 121–141. Springer, Heidelberg (2010). https://doi.org/10.1007/978-3-642-15582-6_24

An Arithmetic-Geometric Mean
of a Third Kind!

Semjon Adlaj[(⊠)]

Federal Research Center "Informatics and Control"
of the Russian Academy of Sciences, Vavilov St. 44, Moscow 119333, Russia
SemjonAdlaj@gmail.com

Abstract. The concept of the generalized arithmetic-geometric mean (GAGM) embraces both the arithmetic-geometric mean (AGM) and the modified arithmetic-geometric mean (MAGM) as two special concepts. The GAGM is applied for attaining a unifying formula for calculating complete elliptic integrals (CEI), including those of the third kind, thereby providing a conceptual basis for their exploration and exact evaluation, bypassing typical troubles of common software in calculating CEI. Detailed clarifying examples are provided.

Keywords: Generalized arithmetic-geometric mean ·
Linear fractional transformation · Quadratic convergence ·
Complete elliptic integral

1 Introduction

The arithmetic-geometric mean (AGM) is the key for attaining a "perfect" formula for calculating complete elliptic integrals (which we shall abbreviate as CEI whether singular or plural). The first perfect formula for calculating CEI of the first kind was obtained by Gauss. Aside from conciseness and exactness, it gave rise to an iterative sequence of intervals, swiftly converging to their common point. A termination at any step requires no additional calculations of error estimates, as other (imperfect) formulas usually require, since the exact value is guaranteed to lie inside its corresponding interval. The same process, based on Landen transformations, turned out being generalizable to calculating CEI, of any kind, via a quadratically convergent procedure. Surprisingly, however, the second perfect formula (possessing all the virtues of the first) for calculating CEI of the second kind had skipped the attention of all for over two centuries after discovering the first.[1] But only a few additional years were required to attain the third (general) perfect formula for calculating CEI of the third (or any) kind. As was the case with the two formulas, preceding it, the general formula gives rise to an iterative sequence of intervals, quadratically collapsing onto their common point. And, as before, aside from basic arithmetic operations, only a single square-root operation is required at each iteration!

---

[1] Leading some to allege (in desperation) that no simple exact formula for calculating the perimeter of an ellipse existed. Nevertheless, one ought not overestimate the significance of the second formula which must remain secondary to the first, without which it could not have been conceived. The two formulas "resonate" one with other, and the second, borrowing a word from [22], "echoes" the first.

© Springer Nature Switzerland AG 2019
M. England et al. (Eds.): CASC 2019, LNCS 11661, pp. 37–56, 2019.
https://doi.org/10.1007/978-3-030-26831-2_3

2   An Historical Overview of Elliptic Integrals

A dramatic struggle for efficiently calculating (complete and incomplete) elliptic integrals emerged with their inoculation by Fagnano.[2] Fagnano's contribution [15] to the division of elliptic arcs constitutes a most remarkable and never fading jewel of mathematics of all time![3] But it even brighter highlighted the necessity for efficiently calculating CEI, since it clarified how calculating incomplete elliptic integrals incessantly depended upon calculating CEI. A breakthrough was carried out by Gauss, who recorded the discovery of his unsurpassable arithmetic-geometric mean (AGM) method for calculating CEI of the first kind, in his diary on May 30, 1799 [21],[4] thereby laying the foundation for a distinctly novel and superb approach for calculating CEI (of any kind). Nevertheless, a formula as simple and powerful for calculating CEI of the second kind had to await December 16, 2011 to be discovered! The modified arithmetic-geometric mean (MAGM), being the necessary concept for attaining the second formula, turned (moreover) being the basic concept, underlying the generalized arithmetic-geometric mean (GAGM), which enabled on September 2, 2015 attaining the third (general) formula for calculating CEI of third (and any) kind. The generalization of MAGM to GAGM was preceded by constructing the (so-called) elliptic and coelliptic polynomials for carrying out highly efficient arithmetic on elliptic curves, including division. Earlier, on May 30, 2011, a canonical fast inverse of the modular invariant was obtained [4], further unraveling a tight relationship between the modular invariant and CEI. Fourteen new special values of the modular invariant were calculated in 2014, and an infinite family of identities, called *modular polynomial symmetries*, were first presented on April 16, 2014 at the 7th PCA annual conference in St. Petersburg, Russia, and subsequently represented at a seminar at Moscow State University [6]. A crucial connection between calculating the roots of the modular equation of level p and calculating the p-torsion points, on a corresponding elliptic curve, must (surprisingly) be entirely attributed to Galois. Relevant details on Galois' amazing (yet far from fully appreciated) contribution to elliptic functions (and integrals) are given in [2,4]. Certainly, the idea, involving the action of the projective linear group in the main construction of this paper was guided by Galois,[5]

---

[2] The (highly successful) term "elliptic integral" in and of itself was apparently invented by Fagnano.

[3] According to Fricke [16, Vorwort], the day December 23, 1751 when Euler acknowledged the receipt of Fagnano's two-volume work was regarded by Jacobi as "the birthday of the theory of elliptic functions". On January 27, 1752 Euler, crediting Fagnano, made his first presentation (to the Berlin Academy of Sciences) on the addition theorem for elliptic integrals.

[4] Strangely, Gauss' method remained either unknown or unappreciated, until recently, as pointed out in [23, Appendix O: The Simple Plane Pendulum: Exact Solution] and further explained in [5].

[5] Those overly concerned that Galois' contribution has ever been overestimated must rest assured that it was not! Up to these days, Galois' last letter [17], which he wrote on the eve of his murder May 30, 1832, remains tragically untangled in spite of all efforts of those who never underestimated it!

whose abilities, as rightfully admitted in [18, 2.21. "L'unique" – ou le don de solitude], far exceeded ours.

3 The Generalized Arithmetic-Geometric Mean

We shall reserve the letter n to denote a natural number, including zero.

3.1 Construction and Definition

The modified arithmetic-geometric sequence was presented in [3,7–9,24]. It is the recursively defined triple sequence

$$x_{n+1} := \frac{x_n + y_n}{2}, \ y_{n+1} := z_n + \sqrt{(x_n - z_n)(y_n - z_n)},$$

$$z_{n+1} := z_n - \sqrt{(x_n - z_n)(y_n - z_n)}.$$

Given such a sequence $\{x_n, y_n, z_n\}_{n=0}^{\infty}$, we introduce (another) recursively defined sequence of (single-valued) parametric functions:[6]

$$u_{n+1} = u_{n+1}(t) = u_{n+1}(t, c, x_0, y_0, z_0) := \frac{c_n u_n - y_{n+1} z_{n+1}}{c_n + u_n - 2 z_n}, \ c_n := u_n(c),$$

where c is a fixed real parameter and the function u_0 is (naturally) presumed to coincide with the identity function: $u_0(t) = t$. We proceed to defining the functions

$$v_n = v_n(t) = v_n(t, a, c, x_0, y_0, z_0) := \frac{t - a_n}{t - c_n}, \ a_n := u_n(a),$$

$$w_n = w_n(t) = w_n(t, b, a, c, x_0, y_0, z_0) := \frac{v_n(t)}{v_n(b_n)}, \ b_n := u_n(b),$$

where a and b are (also) fixed real parameters distinct from c and each other.

We shall refer to the sextuple sequence

$$\{x_n, y_n, z_n, a_n, b_n, c_n\}_{n=0}^{\infty}$$

as the generalized arithmetic-geometric sequence (abbreviated as GAGS whether singular or plural).[7] The sequence $\{w_n\}_{n=0}^{\infty}$ is thereby seen as a sequence of linear fractional (Möbius) transformations, generated by GAGS, successively mapping the sequence of ordered triples $(a_n, b_n, c_n)_{n=0}^{\infty}$ to the (fixed) ordered triple $(0, 1, \infty)$.

Define the generalized arithmetic-geometric mean (GAGM) of two (strictly) positive numbers x and y, for a given pairwise distinct real parameters a, b and

---

[6] The adjective "parametric" is meant to indicate that each such (single-valued) function (of the argument t) does "depend" upon the (fixed) values of its parameters.

[7] Thus, the GAGS is an extended modified arithmetic-geometric sequence, with twice as many terms.

c, as the (common) limit of the sequence $\{\xi_n := w_n(x_n)\}_{n=0}^{\infty}$ and the sequence $\{\eta_n := w_n(y_n)\}_{n=0}^{\infty}$ with $x_0 = x$, $y_0 = y$ and $z_0 = 0$.

Later on, we extend the domain of the parameters a, b and c to include the point at (complex) infinity, so that a, b and c might be regarded as elements of the extended real line $\mathbb{R} \cup \infty$. However, we shall always require the parameter c to lie (strictly) outside the closed interval $[x, y]$.[8]

3.2 Basic Properties

Given a linear function $l(t) = \lambda\,(t - \mu)$, $\{\lambda \neq 0,\, \mu\} \subset \mathbb{R}$, we define an action of the function l upon the GAGS as

$$l \cdot \{x_n, y_n, z_n, a_n, b_n, c_n\}_{n=0}^{\infty} := \{l(x_n),\, l(y_n),\, l(z_n),\, l(a_n),\, l(b_n),\, l(c_n)\}_{n=0}^{\infty}, \quad (1)$$

thereby inducing an action upon the sequence $\{w_n\}_{n=0}^{\infty}$, which we shall denote by $l \cdot \{w_n\}_{n=0}^{\infty} := \{l \cdot w_n\}_{n=0}^{\infty}$, where $l \cdot w_n$ is the transformation mapping the ordered triple $(l(a_n),\, l(b_n),\, l(c_n))$ to the ordered triple $(0,\, 1,\, \infty)$. One might then verify that the sequence we have defined, in (1), is indeed a GAGS![9] Furthermore, neither the sequence $\{\xi_n\}_{n=0}^{\infty}$ nor $\{\eta_n\}_{n=0}^{\infty}$ is altered by this action, that is,

$$\xi_n = l \cdot w_n\big(l(x_n)\big) = w_n(x_n), \quad \eta_n = l \cdot w_n\big(l(y_n)\big) = w_n(y_n),$$

so that the GAGM is invariant under the action of linear functions upon the GAGS, permitting us to speak of *equivalence classes* of GAGS. So we shall say that a GAGS is equivalent to another if the GAGM is unaltered. In particular, The homogeneity degree of GAGM is zero (unlike the AGM and MAGM which are homogeneous of degree one), and we might exploit this property to extend the domain of GAGM, for fixed parameters,[10] to include (strictly) negative values of the arguments x and y. At each iteration, we might ensure the positivity of the product $(x_n - z_n)(y_n - z_n)$, before taking its square root, via acting upon the GAGS (at the required step whenever necessary) by the (constant) function -1.

We shall denote with the same letter N three functions, which we shall nevertheless distinguish by the (total) number of their arguments. The invariance of the GAGM under the action of linear functions upon the GAGS implies that four initial arguments suffice to determine the GAGM, so we designate $N(x, a, b, c)$ to denote the GAGM of 1 and x for parameters a, b and c.[11] Moreover, the expression

$$\left(\frac{(b - a)\, N(x, a, b, c)}{b - c} - 1\right)\Big/(c - a),$$

---

[8] This requirement is necessary for the GAGM to be well defined, as we shall soon find out.

[9] Being initiated by the sextuple $\{l(x_0),\, l(y_0),\, l(z_0),\, l(a_0),\, l(b_0),\, l(c_0)\}$, so (for all indices n) we have $l(x_0)_n = l(x_n)$, $l(y_0)_n = l(y_n)$, $l(z_0)_n = l(z_n)$, $l(a_0)_n = l(a_n)$, $l(b_0)_n = l(b_n)$, $l(c_0)_n = l(c_n)$.

[10] Generally speaking, the parameters might also be regarded as (special) arguments.

[11] An equivalence class of any GAGS might be represented by a sequence, where the initial values y_0 and z_0 are fixed at 1 and 0, respectively.

while seemingly dependent upon four arguments x, a, b and c, has x and c as its only "true" arguments. It actually depends neither upon a nor upon b. Consequently, we might define a bivariate function

$$N(x, c) := N(x, \infty, c+1, c),$$

and employ it in order to alternatively express the preceding quadrivariate function as

$$N(x, a, b, c) = \frac{b-c}{b-a}\Big((c-a)\,N(x, c) + 1\Big).$$

The latter formula extends not only to the case $c = 0$ but, as well, to the case $c = \infty$. In these two (dual) cases the GAGM "degenerates" to a (shifted) MAGM:

$$N(x, a, b, 0) = \frac{b}{a-b}\left(a\,N\left(\frac{1}{x}\right) - 1\right), \quad N(x, a, b, \infty) = \frac{N(x) - a}{b - a}, \qquad (2)$$

where the (univariate) function $N(x)$ is the modified arithmetic-geometric mean of 1 and x.

The equivalence of the latter two equations reflects a special (limiting) case of the relation

$$N(x, a, b, c) = N\left(\frac{1}{x}, \frac{1}{a}, \frac{1}{b}, \frac{1}{c}\right).^{12} \qquad (3)$$

3.3 Quadratic Convergence

The *difference sequence*

$$d_n := \xi_n - \eta_n = w_n(x_n) - w_n(y_n) = \frac{v_n(x_n) - v_n(y_n)}{v_n(b_n)} = s_n(x_n - y_n),$$

$$s_n := \frac{(c_n - b_n)(c_n - a_n)}{(b_n - a_n)(c_n - x_n)(c_n - y_n)},$$

depends upon all the (three) parameters a, b and c, while the ratio

$$\frac{s_{n+1}}{s_n} = \frac{c_n - z_{n+1}}{c_n - y_{n+1}} = \frac{c_n - c_{n+1}}{c_{n+1} - y_{n+1}} = \frac{c_{n+1} - z_{n+1}}{c_n - c_{n+1}}$$

---

[12] This relation suggests that the defining equality of the function $N(x, c)$ might be substituted with the equality

$$N(x, c) = N\left(\frac{1}{x}, 0, \frac{1}{c+1}, \frac{1}{c}\right),$$

which is suitable for explicit calculation, and is extendable to the case $c = 0$ as

$$N(x, 0) = N\left(\frac{1}{x}\right),$$

but, unlike the quadrivariate function, the bivariate function remains undefined for $c = \infty$.

depends upon c but neither upon a nor upon b.[13] In order to show that the GAGM is well-defined we must show that the sequence $\{d_n\}_{n=0}^{\infty}$ converges to zero. We already know that it does if $c = \infty$,[14] no matter what a and b are, since the GAGM for x and y would then coincide with the MAGM of $x/(b-a)$ and $y/(b-a)$, up to an additive constant $a/(b-a)$. The case when $c = 0$ might, as well, be reduced to the case when $c = \infty$, via identity (3) or by the first of formulas (2). The case $c = z_1 = -\sqrt{x\,y}$ would imply (whatever a and b are) that $a_1 = b_1 = c_1 = z_1$, forcing a termination of the GAGS with $d_1 = s_1 = 0$. The GAGM of x and y would then coincide with the value

$$\xi_1 = \eta_1 = \frac{1}{2}\left(1 + \frac{a\,b - x\,y}{(a-b)\sqrt{x\,y}}\right).[15]$$

The case $c < 0$ implies that $c_n < 0$ and $2\,c_{n+1} < c_n$ (for any index n), so

$$\left|\frac{s_{n+1}}{s_n}\right| < \left|\frac{c_n}{c_{n+1}} - 1\right| < 1 \Rightarrow \left|\frac{d_{n+1}}{d_n}\right| < \frac{x_{n+1} - y_{n+1}}{x_n - y_n},$$

and the GAGM would converge never (at any iteration) slower than the MAGM does, although unlike either the descending sequence $\{x_n\}_{n=1}^{\infty}$ or the ascending sequence $\{y_n\}_{n=1}^{\infty}$ neither the sequence $\{\xi_n\}_{n=0}^{\infty}$ nor the sequence $\{\eta_n\}_{n=0}^{\infty}$ is monotone.

The last case, for convergence to be considered, is the case $c > 0$. The sequence $\{c_n\}_{n=1}^{\infty}$ is then descending and, for all $n \geq 1$, $c_n > x_n$,[16] and

$$\frac{d_{n+1}}{d_n} = \frac{(c_{n+1} - z_{n+1})(x_{n+1} - x_{n+2})}{(c_n - c_{n+1})(x_n - x_{n+1})} \approx \frac{x_n - x_{n+1}}{2\,(c_n - c_{n+1})} \approx \left(\frac{x_{n-1} - x_n}{2\,(c_{n-1} - c_n)}\right)^2,$$

where the sign for approximate equality ($\approx$) must be interpreted here as an asymptotic (as n approaches infinity) equality. Consequently, the convergence is eventually (that is, asymptotically) quadratic.[17]

3.4   Alternative Calculations

The enlisted properties of GAGM enable endlessly many means of calculating it, but we shall indicate only two more. The first exploits the identity

$$N(x, a, b, c) = N\Big(\sigma(x, 1),\ \sigma(x, a, c),\ \sigma(x, b, c),\ \sigma(x, c)\Big),$$

---

[13] Elementary geometric constructions, involving mutually orthogonal circles as suggested in [10], might facilitate deriving the preceding triple-equation.

[14] We are alluding to the second formula of (2).

[15] The value on the rightmost side might be obtained by applying L'Hôpital's rule to either "undeterminate" $w_1(x_1)$ or $w_1(y_1)$.

[16] The condition that c lies (strictly) outside the closed interval, bounded by x and y must not be forgotten. We need not, however, require c to lie to the left of that interval, so c_0 need not exceed x_0. In other words, the inequality $c_n > x_n$ need not apply when $n = 0$.

[17] One might note, as well, that the monotonicity of the sequences $\{\xi_n\}_{n=1}^{\infty}$ and $\{\eta_n\}_{n=1}^{\infty}$ is restored, in this ($c > 0$) case.

$$\sigma(x,y) := \sigma(x,y,y), \ \ \sigma(x,y,z) := \frac{(\sqrt{x}+y)(\sqrt{x}+z)}{2(y+z)\sqrt{x}},$$

which allows introducing an *abbreviated* GAGS for which $y_n = 1$ and $z_n = 0$, for all n, and

$$\{x_{n+1} = \sigma(x_n, 1), a_{n+1} = \sigma(x_n, a_n, c_n), b_{n+1} = \sigma(x_n, b_n, c_n), c_{n+1} = \sigma(x_n, c_n)\}. \quad (4)$$

The second introduces a *truncated* GAGS

$$\left\{x_{n+1} = \sigma(x_n, 1) = \frac{(\sqrt{x_n}+1)^2}{4\sqrt{x_n}}, \ c_{n+1} = \sigma(x_n, c_n) = \frac{(\sqrt{x_n}+c_n)^2}{4 c_n \sqrt{x_n}}\right\},$$

for which we skip calculating a_n and b_n, but (instead) calculate the GAGM, recursively, on the basis of the identity

$$N(x, c) = \tau\Big(x, \ c, \ N(\sigma(x,1), \sigma(x,c))\Big), \quad (5)$$

$$\tau(x,y,z) := \frac{1}{2y}\left(\left(\frac{y}{\sqrt{x}} - \frac{\sqrt{x}}{y}\right)\frac{z}{4} - 1\right).$$

The truncated GAGS is not suitable for calculating the GAGM in the special case $c = 0$ or $c = \infty$ when the GAGM degenerates to MAGM, as given by formulas (2), but the abbreviated GAGS serves without exceptions. In particular, we readily infer from the limit formula, with $c = \infty$,

$$N(x^2, a, b, \infty) = N\left(\frac{(x+1)^2}{4x}, \frac{x+a}{2x}, \frac{x+b}{2x}, \infty\right)$$

a recursive formula for calculating MAGM:

$$N(x^2) = x\left(2 N(f(x)^2) - 1\right) = 2 f_n(x) N\big(f^{n+1}(x)^2\big) - \sum_{k=0}^{n} f_k(x) \approx$$

$$\approx f_n(x) - \sum_{k=0}^{n-1} f_k(x), \ \text{where}$$

$$f_n(x) := 2^n \prod_{k=0}^{n} f^k(x), \ f^{n+1}(x) := f(f^n(x)), \ f(x) := \frac{x+1}{2\sqrt{x}}, \ f_0(x) = f^0(x) = x.$$

Of course, we could have defined the GAGM via the abbreviated GAGS, as given by (4), at the cost of obscuring the origin of GAGM in MAGM.

4 Calculating Three Kinds and Three Types of CEI

Assume, unless indicated otherwise, that β and γ are two positive numbers which squares sum to one: $\beta^2 + \gamma^2 = 1$.

Before we apply GAGM, to calculating CEI, we shall further extend the domain of its parameters to include complex values, and we lift any remaining doubt that the GAGM is actually a generalized AGM by observing the identity

$$N\left(\beta^2, 1 - \gamma, 1 - \frac{\gamma^2}{2+\gamma}, 1+\gamma\right) = N(\beta^2, \beta^2 + i\beta\gamma, \beta, \beta^2 - i\beta\gamma) = M(\beta),\,^{18} \quad (6)$$

where $i := \sqrt{-1}$ and $M(x)$ is the AGM of 1 and x. The identity still holds if the sign of γ, which we shall refer to as *the elliptic modulus*, is flipped.[19]

4.1   Three Formulas for Calculating Three Kinds of CEI

A CEI of the first kind I_1 is defined and calculated as

$$I_1 = I_1(\gamma) := \int_0^1 \frac{dt}{\sqrt{(1-t^2)(1-\gamma^2 t^2)}} = \frac{\pi}{2\,M(\beta)}. \quad (1799.05.30)$$

A CEI of the second kind I_2 is defined and calculated as

$$I_2 = I_2(\gamma) := \int_0^1 \sqrt{\frac{1-\gamma^2 t^2}{1-t^2}} = \frac{\pi\,N(\beta^2)}{2\,M(\beta)}. \quad (2011.12.16)$$

Both formulas (1799.05.30) and (2011.12.16) apply at $\gamma = 0$, with $I_2(0) = I_1(0) = \pi/2$. The second applies, as well, at $\gamma = 1$, with $I_2(1) = 1$, as clarified in [7,8].

A CEI of the third kind I_3 is defined and calculated as

$$I_3 = I_3(\gamma, \delta) := \int_0^1 \frac{dt}{(t^2 - \delta)\sqrt{(1-t^2)(1-\gamma^2 t^2)}} = -\frac{\pi\,\gamma^2\,N(\beta^2, 1-\delta\gamma^2)}{2\,M(\beta)} =$$

$$= \frac{\pi\,N(\beta^2, \infty, \beta^2 - \delta\gamma^2, 1-\delta\gamma^2)}{2\,M(\beta)},\, \delta \in \mathbb{C}\backslash[0,1]. \quad (2015.09.02)$$

Put $\beta_{n+1} := \sqrt{\sigma(\beta_n^2, 1)}$ with $\beta_0 = \beta$. Thereby, the recursively defined sequence $\{\beta_n\}_{n=1}^\infty$ converges descendingly to one, whereas the sequence $\{\gamma_n^2 := 1 - \beta_n^2\}_{n=1}^\infty$ is a negative sequence, converging ascendingly to zero.[20] Define, recursively, the sequence

$$\delta_{n+1} := \frac{(1 - \delta_n(1+\beta_n))^2}{1 - \delta_n\gamma_n^2},\, \delta_0 = \delta.$$

---

[18] An equivalent GAGS has the initial values $x_0 = \beta$, $y_0 = 1/\beta$, $z_0 = 0$, $a_0 = \beta + i\gamma$, $b_0 = 1$, $c_0 = \beta - i\gamma$. Note here that if $x_0 = \beta$ and y_0 were the values at two (out of three) half-period of an essential elliptic function, as shown in [10, figures], then a_0, b_0 and c_0 are its values at three (out of six) quarter-periods.

[19] Whereas, flipping the sign of β leads to flipping the sign of $M(|\beta|)$.

[20] Alternatively, we might define the sequence of squares $\{\gamma_n^2\}_{n=0}^\infty$ recursively by putting $\gamma_{n+1}^2 := \sigma(\beta_n^2, -1)$ with $\gamma_0 = \gamma$.

The recursive relation (5) implies a recursive relation for I_3:

$$I_3(\gamma_n, \delta_n) = \lambda_n \left(I_1(\gamma_n) + \frac{\mu_n}{\sqrt{\beta_n}} I_3(\gamma_{n+1}, \delta_{n+1}) \right),^{21}$$

$$\lambda_n = \lambda(\gamma_n, \delta_n), \ \mu_n = \mu(\gamma_n, \delta_n),$$

$$\lambda(\gamma, \delta) := \frac{\gamma^2}{2(1 - \delta\gamma^2)}, \ \mu(\gamma, \delta) := \frac{\gamma^2(\gamma^2\delta^2 - 2\delta + 1)}{(1 - \beta)^2(\delta\gamma^2 - 1)},$$

which, along with the relation $\sqrt{\beta_n} \, I_1(\gamma_n) = I_1(\gamma_{n+1})$, implies the identity

$$I_3(\gamma, \delta) = \sum_{k=0}^{n} \eta_k \, I_1(\gamma) + \eta_n \, \mu_n \, I_3(\gamma_{n+1}, \delta_{n+1}) \Big/ \sqrt{\prod_{k=0}^{n} \beta_k}, \qquad (7)$$

$$\eta_n := \lambda_0 \prod_{k=1}^{n} \mu_{k-1} \lambda_k,$$

exhibiting that for infinitely many values of δ, satisfying (for any n) the relation $\delta_n = 1/(1 - \beta_n)$, the integral I_3 would degenerate to a multiple of I_1, by the coefficient $\sum_{k=0}^{n} \eta_k$, as μ_n vanishes. Identity (7) does not apply at $\delta = 1/\gamma^2$, where I_3 would degenerate to a multiple of CEI of the second kind, by a coefficient given in the latter of formulas (10). Observe here that the equality $\delta_n = 1/\gamma_n^2$ implies that $\lambda_n = \mu_n = \infty$. Moreover, the equivalence

$$\delta_n = \frac{1}{1 - \beta_n} \Leftrightarrow \delta_{n+1} = \frac{1}{\gamma_{n+1}^2}$$

holds.

The relations

$$I_3\left(\gamma, \frac{\pm 1}{\gamma}\right) = \frac{\gamma}{2}\left(\frac{\pi}{2(\gamma \mp 1)} \mp I_1(\gamma)\right),^{22}$$

$$I_3\left(\gamma, \frac{\gamma \pm i\beta}{\gamma}\right) = -\frac{\gamma}{2\beta}\left(\frac{(\gamma \mp i\beta)\pi}{2} \mp i I_1(\gamma)\right),$$

stemming from (6), would imply that for infinitely many values of δ, satisfying the relation $\delta = (\gamma \pm i\beta)/\gamma$ or (for any n) the relation $\delta_n^2 = 1/\gamma_n^2$, the integral I_3

---

[21] An analogous recursive relation for an elliptic integral of the second kind

$$I_2(\gamma) = 2\sqrt{\beta} \, I_2\left(\sqrt{\sigma(\beta^2, -1)}\right) - \beta \, I_1(\gamma)$$

is equivalent to formula (2011.12.16).

[22] Either the upper or the lower sign must be consistently taken throughout this or other equations in this paper.

would degenerate to a "linear combination" of the (ubiquitous) constant π and I_1.[23] Two equivalences are in order:

$$\delta_n^2 = \frac{1}{\gamma_n^2} \Leftrightarrow \delta_{n+1} = \frac{\gamma_{n+1} + i\,\beta_{n+1}}{\gamma_{n+1}}, \quad \delta_n = \frac{\gamma_n \pm i\,\beta_n}{\gamma_n} \Leftrightarrow \delta_{n+1} = \frac{\gamma_{n+1} - i\,\beta_{n+1}}{\gamma_{n+1}}.$$

Put

$$\delta_\pm(x) := \frac{1}{\left(1 - \sqrt{x}\right)\left(1 + x \pm \sqrt{x\,(1+x)}\right)}.$$

The preceding (primary) identities for I_3 might be applied to deriving two (secondary) identities, corresponding to $\delta_1 = 1/(1 - \beta_1)$ and $\delta_1 = 1/\gamma_1$,[24] respectively:

$$I_3(\gamma, \delta_\pm(\beta)) = \frac{\beta - 1}{4\sqrt{\beta^3}}\left((1 + \beta)^2 \pm \sqrt{1 + \beta}\left(1 + \sqrt{\beta^3}\right)\right) I_1(\gamma),$$

$$I_3(\gamma, \delta_\pm(-\beta)) = \pm\sqrt{\frac{1-\beta}{\beta}}\left(\frac{\sqrt{-\beta} \pm \sqrt{1-\beta}}{2\left(1 \mp \sqrt{1-\beta}\right)}\right)^2$$

$$\left((1 + \beta)\left(\beta + 3\sqrt{-\beta^3} - \left(3(1-\beta) + 4\sqrt{-\beta}\right)\left(1 \mp \sqrt{1-\beta}\right)\right) i\,I_1(\gamma)\right.$$

$$\left. + \left(1 + \sqrt{-\beta^3} \mp \sqrt{1-\beta}\,(1+\beta)\right)\pi\right).$$

4.2   An Unifying Formula for Calculating Three Types of CEI

For a given linear fractional transformation w, determined by three parameters a, b and c:

$$w(t) = w(t, a, b, c) := \frac{(b - c)(t - a)}{(b - a)(t - c)}, \quad \{a, b, c\} \subset \mathbb{C} \cup \infty,\text{[25]}$$

we might, as well, define a *proper* CEI I as the integral

$$I = I(\gamma, a, b, c) := \int_0^1 \frac{w(t^2)\,dt}{\sqrt{(1 - t^2)(1 - \gamma^2 t^2)}}, \tag{8}$$

in which we shall distinguish three types. The *first type* would correspond to the case when the transformation w has degenerated to a constant map, the

---

[23] We shall avoid specifying the algebraic properties of such "linear combination", leaving this (significant) issue to other papers and, perhaps, other authors.

[24] Note that the former value (of δ_1) is negative (real) and the latter is negative imaginary.

[25] The transformation w need not necessarily be Möbius transformation, since degenerate transformations are not (yet) excluded. In other words, the transformation w need not be a conformal automorphism of either the extended or unextended complex plane, and its determinant $(a - b)(b - c)(c - a)$ is allowed to vanish, be finite or infinite.

second type would correspond to the case when w is a linear function,[26] whereas the *third type* would correspond to the case when w is a linear fractional transformation which does not fix the point at (complex) infinity. Note, however, that the restriction upon c to be distinct from ∞ does not preclude a CEI from degenerating to a CEI of the first or the second type as are the instances

$$I\left(\gamma, a, b, \frac{1}{1-\beta}\right) = \frac{(1 - b(1-\beta))(1 - a(1+\beta)) I_1(\gamma)}{2(b-a)\beta}, \tag{9}$$

$$I\left(\gamma, a, b, \frac{1}{\gamma^2}\right) = \frac{1 - b\gamma^2}{(b-a)\gamma^2}\left(\frac{(1 - a\gamma^2) I_2(\gamma)}{\beta^2} - I_1(\gamma)\right).$$

In particular, the two special values

$$I\left(\gamma, \infty, \frac{2-\beta}{1-\beta}, \frac{1}{1-\beta}\right) = -\frac{\gamma^2 I_1(\gamma)}{2\beta},$$

$$I\left(\gamma, \infty, \frac{\gamma^2 + 1}{\gamma^2}, \frac{1}{\gamma^2}\right) = -\left(\frac{\gamma}{\beta}\right)^2 I_2(\gamma) \tag{10}$$

coincide with the values of I_3 if evaluated at $\delta = 1/(1-\beta)$ and $\delta = 1/\gamma^2$, respectively. The former of formulas (10) is, in fact, a special (first) case of identity (7).[27]

Whatever the type of I, as defined in (8), we might calculate it directly as

$$I(\gamma, a, b, c) = \frac{\pi N(\beta^2, 1 - a\gamma^2, 1 - b\gamma^2, 1 - c\gamma^2)}{2 M(\beta)}, ^{28} \quad c \in \mathbb{C}\setminus[0,1] \tag{11}$$

so that the case where $a = \infty$, $b = 1 + \delta$, $c = \delta$ is seen as the special case where I coincided with I_3. Identity (6) might now be translated to an identity for π:

$$I\left(\gamma, \pm\frac{1}{\gamma}, \frac{1}{2\pm\gamma}, \mp\frac{1}{\gamma}\right) = I\left(\gamma, \frac{\gamma \mp i\beta}{\gamma}, \frac{1}{1+\beta}, \frac{\gamma \pm i\beta}{\gamma}\right)$$

$$= -I\left(\gamma, \frac{\gamma \pm i\beta}{\gamma}, \frac{1}{1-\beta}, \frac{\gamma \mp i\beta}{\gamma}\right) \equiv \frac{\pi}{2}. \tag{12}$$

The identity is extendable (for all involved integrals) to the (limit) value of the elliptic modulus γ at 0, as well as, it is extendable for the first integral taken with upper signs to the (limit) value at $\gamma = 1$, that is,

$$\int_0^1 \frac{dt}{\sqrt{1-t^2}} = \int_0^1 \frac{2\,dt}{1+t^2} \equiv \int_0^1 \frac{(1+\gamma)(1-\gamma t^2)\,dt}{(1+\gamma t^2)\sqrt{(1-t^2)(1-\gamma^2 t^2)}}, \quad \gamma \in (0,1).$$

---

[26] Recall that the case $c = \infty$ was not excluded.

[27] Yet, even this (first) special case, where I_3 degenerates to (a multiple of) I_1 for $\delta = 1/(1-\beta)$, seems missing from standard sources on elliptic integrals.

[28] An equivalent GAGS leading to the GAGM that appears in the numerator has the initial values $x_0 = 0$, $y_0 = -1$, $z_0 = -1/\gamma^2$, $a_0 = -a$, $b_0 = -b$, $c_0 = -c$.

However, the second and the third integrals, in identity (12), are discontinuous at $\gamma = 1$. The upper signs correspond to the value $i\pi/2$, whereas the lower signs correspond to the value $-i\pi/2$.

We emphasize the methodological significance of a clear unifying formula (11) for calculating CEI (of any type). "The Handbook of Mathematical Functions" [1, ch. 17] fell short of accomplishing that task, as the section on "The Process of the Arithmetic-Geometric Mean" was not extended to calculating elliptic integrals of the third kind, which were left to appear in the next section of the chapter on "Elliptic Integrals" by Milne-Thomson. The current version of the latter chapter, written by Carlson [12], is amended with an expressions for calculating CEI of the third type, via AGM, in the section on "Quadratic Transformations", essentially providing yet another (as perfect) alternative for calculating the sequence ξ_n, converging to GAGM.[29] An enlightening succinct review of CEI is given in [20]. "Wolfram Mathematica" warns, in [25], that "more so than for other special functions, you need to be very careful about the arguments you give to elliptic integrals and elliptic functions" but exhibits insufficient care in evaluating the integral (8), where non-vanishing imaginary parts (occasionally!) appear for real parameters. A sample "notebook", exposing this and other typical troubles in calculating CEI, by "Mathematica 10.3",[30] is appended to this article.

4.3   The Formula for Calculating the Complementary CEI

The complementary CEI (denote by J) might as readily be calculated:

$$J = J(\gamma, a, b, c) := \int_{1}^{1/\gamma} \frac{w(t^2)\, dt}{\sqrt{(t^2 - 1)\,(1 - \gamma^2\, t^2)}} = \frac{\pi\, N(1/\gamma^2, a, b, c)}{2\, M(\gamma)}, \quad (13)$$

$$c \in \mathbb{C} \backslash [1,\, 1/\gamma].$$

The integral J, as was the case with I, would also degenerate to a CEI of the first or second type if w is, respectively, constant or linear. Furthermore,

$$J\left(\gamma, a, b, -\frac{1}{\gamma}\right) = \frac{1}{2}\left(1 + \frac{a\, b\, \gamma^2 - 1}{(a - b)\, \gamma}\right)\, I_1(\beta), \quad J(\gamma, a, b, 0) = \frac{b\, (a\, I_2(\beta) - I_1(\beta))}{a - b},$$

and, in particular,

---

[29] Each sequence element ξ_n is represented by a partial sum, as was the case with the (original) expression for calculating CEI of the second kind (given, as well, in the preceding chapter by Milne-Thomson). These expressions, involving infinite sums, do (most importantly) provide quadratically convergent procedures but, unlike the (first) formula for calculating CEI of the first kind, they do not produce a sequence of intervals, providing both (lower and upper) bounds.

[30] That version of "Mathematica" was released on October 15, 2015.

$$J\left(\gamma, \infty, \frac{1}{\gamma}, -\frac{1}{\gamma}\right) = I_1(\beta),\, ^{31} \quad J(\gamma, \infty, 1, 0) = I_2(\beta). \tag{14}$$

We rewrite the latter special case, with $c = 0$, explicitly as

$$\int_1^{1/\gamma} \frac{dt}{t^2 \sqrt{(t^2 - 1)(1 - \gamma^2 t^2)}} = \frac{\pi N(\gamma^2)}{2 M(\gamma)},$$

in order to emphasize that it was not excluded.[32]

5   Few Explicit Calculations of CEI via GAGM

Before we move on to numerical examples, we explicitly write down the iterative step for generating a (next) sextuple of the GAGS. It must be preceded by calculating the (temporary) values $r_2 = (x_n - z_n)(y_n - z_n)$, $r_1 = \sqrt{r_2}$, $t_2 = z_n^2 - r_2$, $t_1 = 2 z_n - c_n$. Then

$$(x_{n+1}, y_{n+1}, z_{n+1}, a_{n+1}, b_{n+1}, c_{n+1}) =$$

$$= \left(\frac{x_n + y_n}{2}, z_n + r_1, z_n - r_1, \frac{c_n a_n - t_2}{a_n - t_1}, \frac{c_n b_n - t_2}{b_n - t_1}, \frac{c_n^2 - t_2}{c_n - t_1}\right).$$

At the terminal step, one calculates

$$v_n(b_n) = \frac{b_n - a_n}{b_n - c_n}, \quad v_n(x_n) = \frac{x_n - a_n}{x_n - c_n}, \quad v_n(y_n) = \frac{y_n - a_n}{y_n - c_n},$$

$$(\xi_n, \eta_n) = \left(\frac{v_n(x_n)}{v_n(b_n)}, \frac{v_n(y_n)}{v_n(b_n)}\right).$$

Alternatively, one calculates the (same) values $(\xi_n, \eta_n) = (w_n(x_n), w_n(1))$ as they emerge from an equivalent abbreviated GAGS,[33] as given by (4), although (as we know) the transformation w_n in and of itself is not invariant under linear actions upon the GAGS.

Now, we shall presume that $\beta = \gamma = 1/\sqrt{2}$. Denote, for brevity, the values $M(\sqrt{2})$ and $N(2)$ by M and N, respectively, and put

$$L := \frac{\pi}{M} \approx 2.62205755429211981046.^{34}$$

---

[31] Thus,

$$J\left(\gamma, \frac{1}{\gamma}, \frac{1}{2 + \gamma}, -\frac{1}{\gamma}\right) = 0.$$

Note that the arguments of the (complementary) integral J coincide with the arguments of the first integral I from identity (12), taken with the upper signs.

[32] The inclusion of this case ($c = 0$) could not have been made possible had we chosen the conventional definition of the CEI of the third kind.

[33] Recall that for an abbreviated GAGS, $y_n = 1$ for all n.

[34] Assuming π is known with sufficient precision, the precision of the latter calculation is attained after four iterations towards the value of the constant M.

The constant L was referred to, in [7,8], as *the lemniscate constant*. It is the semi-length of the lemniscate of Bernoulli which focal distance is $\sqrt{2}$.[35]

Firstly, we calculate the first (exceptional) case $\delta = 1/(1 - \beta) = 2 + \sqrt{2}$ of formula (2015.09.02), via applying (9) or the first of equations (10),

$$\int_0^1 \frac{dt}{(t^2 - 2 - \sqrt{2})\sqrt{(1 - t^2)(1 - t^2/2)}} = I\left(\frac{1}{\sqrt{2}}, \infty, 3 + \sqrt{2}, 2 + \sqrt{2}\right)$$

$$= -I_1\left(\frac{1}{\sqrt{2}}\right)/\sqrt{8} = -\frac{L}{4}.$$

Secondly, we calculate two "mutually" complementary CEI, which share the same absolute value

$$J\left(\frac{1}{\sqrt{2}}, \infty, 1, 0\right) = -I\left(\frac{1}{\sqrt{2}}, \infty, 3, 2\right) = I_2\left(\frac{1}{\sqrt{2}}\right),[36]$$

where the first integral might be calculated via applying the second formula of (14), while the second integral might be calculated via applying the second formula of (10). The absolute values of both integrals turn out to coincide with the value of CEI of the second kind, which might be further evaluated as

$$I_2\left(\frac{1}{\sqrt{2}}\right) = \frac{\pi N}{2\sqrt{2}\,M} = \frac{L + M}{2\sqrt{2}} \approx 1.3506438810476755025.[34]$$

In other words, the absolute value of either of the aforesaid integrals coincides with the ratio of the semi-length of the perimeter of the self-complementary ellipse, as defined in [7,8], to the length of its diameter (that is, its major axis). The relationship of this ratio with the afore-defined constants M and L stems from the (central) case of Legendre relation, which was presented by Euler to the St. Petersburg Academy of Sciences on September 4, 1775 [14].[37] Here, we

---

[35] Such lemniscate is inscribed in a cocentered unit circle, as shown in [7,8, Fig. 2].

[36] The expression on the leftmost side is attained by applying formula (13) to the integral J and formula (11) to the integral I. Formula (2015.09.02) also applies at (the exception case) $\delta = 1/\gamma^2 = 2$.

[37] Another remarkable date when the first of two key ideas behind the "Gauss-Euler algorithm" was presented. Note that the combination of these two outstanding names is (nevertheless) as exceptionally rare as to require no further specification of the algorithm for calculating the constant π. Strangely, a few still argue that the term "Brent-Salamin algorithm" is preferable, being (as it seems to them) less ambiguous, "since" both names Brent and Salamin are much less frequently heard (than either Euler or Gauss). These few, including Brent [11], seem unaware that the frequency with which either the name Euler or Gauss is (separately) associated with so many methods does not imply that the two names (together) must be nearly as frequently associated with any other (or same) methods. In fact, Gauss-Euler algorithm is never confused with any other algorithm (whether or not related to calculating π), so there is no ambiguity here to be lessened.

might pause to express this beautiful relation with a marvelously simple and powerful formula

$$\pi = \frac{M^2}{N-1}, \quad 38$$

giving rise to a quadratically convergent algorithm for calculating π.[39] Such formula differs radically from power series representations of π. Combining iterations we might attain convergence to an arbitrarily high order, whereas no methods exist to *accelerate* a given linearly convergent algorithm to an algorithm which order of convergence (strictly) exceeds one.[40]

Thirdly, we calculate the CEI

$$I_3\left(\frac{1}{\sqrt{2}}, -1\right) = \frac{\pi\, N(2, 0, 1, 2/3)}{\sqrt{2}\, M} \approx 1.273127366749682458.$$

The precision of the last approximation is attained after the fifth iteration towards $N(2, 0, 1, 2/3)$ (assuming that π and M are known with sufficient precision). We list "chopping-off digits" approximations for the corresponding elements of GAGS:

$$x_1 = \frac{3}{2},\ y_1 = \sqrt{2},\ z_1 = -\sqrt{2},\ a_1 = 3,\ b_1 = \frac{8}{5},\ c_1 = \frac{11}{6},$$

$$x_2 \approx 1.4571067811865475244008443621048490392848359,$$
$$y_2 \approx 1.4567863831370551039780621988172076268033687,$$
$$z_2 \approx -4.2852135078832452015814396472366037839427124,$$
$$a_2 \approx 1.5326295766316171593518437666622521421080396,$$
$$b_2 \approx 1.4653984421606063564190843656326729981349874,$$
$$c_2 \approx 1.4786163382163143732381813974936920788385887,$$

$$x_3 \approx 1.4569465821618013141894532804610283330441023,$$
$$y_3 \approx 1.4569465799271259366148342272973271159626949,$$
$$z_3 \approx -10.027373595693616339777713521770534683848119,$$
$$a_3 \approx 1.4570881857430571212719577244909612749313210,$$
$$b_3 \approx 1.4569624860227001384221562624104465633839000,$$
$$c_3 \approx 1.4569873148583131939298920209737533559017021,$$

---

[38] This formula made its début in [8].

[39] Note that evaluating the square root at each iteration is best done via the quadratically convergent (so-called) Heron's method, which amounts to iteratively replacing a given approximation r of a square root of s by the arithmetic mean of r and s/r.

[40] For example, the Chudnovsky famously fast formula, for calculating π, converges (still) linearly [13].

$$x_4 \approx 1.4569465810444636254021437538791777245033986,\,^{41}$$
$$y_4 \approx 1.4569465810444636253477894912161889487201529,$$
$$z_4 \approx -21.5116937724316963049032165347572583164163 92,$$
$$a_4 \approx 1.4569465812955909691425417509597958735013842,$$
$$b_4 \approx 1.4569465810726702938839128147996788375690002,$$
$$c_4 \approx 1.4569465811167028907128642667070835586747654,$$

$$x_5 \approx 1.4569465810444636253749666225476833366117757,$$
$$y_5 \approx 1.4569465810444636253749666225476833366117596,$$
$$z_5 \approx -44.4803341259078562351813996920621999694 44545,$$
$$a_5 \approx 1.4569465810444636253753615361024413575571484,$$
$$b_5 \approx 1.4569465810444636253750109793093234100922159,$$
$$c_5 \approx 1.4569465810444636253750802233393765414321542,$$

as well as, approximations for the corresponding elements of the difference sequence:

$$d_1 \approx 0.119398062518129278742, \quad d_2 \approx 0.007245988895557086620,$$

$$d_3 \approx 0.000026834417169799896, \quad d_4 \approx 0.00000000036803770 6275,$$

$$d_5 \approx 0.000000000000000000069.$$

The GAGM is contained in the open interval (η_5, ξ_5), where

$$\xi_5 \approx 0.686664556900553064232, \quad \eta_5 \approx 0.686664556900553064163.$$

The same difference sequence and the same open interval, containing GAGM, arises had we calculated the abbreviated equivalent GAGS:

$$x_1 = \frac{4+3\sqrt{2}}{8}, \quad a_1 = \frac{2+3\sqrt{2}}{4}, \quad b_1 = \frac{5+4\sqrt{2}}{10}, \quad c_1 = \frac{12+11\sqrt{2}}{24},$$

$$x_2 \approx 1.0000557990344084608909536718021882886851740,$$
$$a_2 \approx 1.0132084978986451044553767711278338098480210,$$
$$b_2 \approx 1.0014998361523864412955349417792705697154935,$$
$$c_2 \approx 1.0038018034645730876221835149885411210608887,$$

$$x_3 \approx 1.0000000001945849073694805774440659743370935,$$
$$a_3 \approx 1.0000123303612025542191609858469557610928674,$$
$$b_3 \approx 1.0000013850271788806221085830365503548205435,$$
$$c_3 \approx 1.0000035470041381927545950431601339276865500,$$

---

[41] The values x_1 through x_4 were calculated earlier (with lesser precision) in [8] as successive approximations of N.

$$x_4 \approx 1.0000000000000000000023664553855388570440700,$$
$$a_4 \approx 1.0000000000109334875695742133189157444409144,$$
$$b_4 \approx 1.0000000000012280512952458860229688472649685,$$
$$c_4 \approx 1.0000000000003145125887007143411816504186\,4399,$$

$$x_5 \approx 1.0000000000000000000000000000000000000000003,$$
$$a_5 \approx 1.0000000000000000000000008596798693359601\,7383,$$
$$b_5 \approx 1.0000000000000000000000009655939785168947160,$$
$$c_5 \approx 1.0000000000000000000000002472954209412139\,8816.\ [42]$$

The truncated GAGS does not require calculating a_n and b_n. Instead, the transformation w_5 might be calculated, recursively, as

$$w_5(t) = \tau\left(x_0, c_0,\ \tau\left(x_1, c_1,\ \tau\left(x_2, c_2,\ \tau\left(x_3, c_3,\ \tau\left(x_4, c_4,\ \frac{1}{t - c_5}\right)\right)\right)\right)\right),$$

and so $\eta_5 = w_5(1)$ and $\xi_5 = w_5(x_5)$, where the value x_5 (and the transformation w_5) is the same whether the GAGS is abbreviated or truncated.

Whatever the case, one ought not confuse the definition of GAGM with the chosen method for calculating it. On the other hand, one must never forget that the GAGM might be calculated "independently" of the AGM.[43]

6 Conclusion

The concept of MAGM enables a "perfect" formula for calculating CEI of the second kind, as given by (2011.12.16), where a function of single variable (the elliptic modulus) appears in its numerator. A "perfect" formula for calculating CEI of the third kind requires constructing a bivariate function. Such function is constructed by "extending" the concept of MAGM to GAGM, and the formula for calculating CEI of the third kind is given by (2015.09.02). Moreover, the concept of GAGM permits constructing a quadrivariate function which is necessary for a general "perfect" formula for calculating any proper CEI, as given by (11).

Acknowledgment and Notification. The author supports an unrestricted access to knowledge, and grants his permission for using his algorithms and formulas to persons and non-profit-seeking organizations. Profit-seeking organizations, including commercial software companies and their representatives, must address the author for an explicit written permission, without which they are never permitted to use any formulas, algorithms or methods based on the concept of MAGM or GAGM.

---

[42] In our example, not only x_n, but a_n, b_n and c_n also converge to 1.

[43] So, of course, is the case with the MAGM which might be calculated, without the AGM, as for determnining the length of a thread in a linear parallel repelling force field [3,9].

A Worksheet on Typical Troubles with Calculating CEI

Printed from the Complete Wolfram Language Documentation, and addended to the article "An arithmetic-geometric mean of a third kind!" by S. Adlaj

(* **Sample problems in exact and numerical evaluations of elliptic integrals by "Mathematica 10.3"** *)

(* **"Mathematica 10.3" is unable to recognize that the following elliptic integral as identically zero** *)

$$\text{I1}[k\_] = \int_0^1 \frac{\frac{(1+k)\,t^2-1}{(1-k)\,t^2-1}\,dt}{\sqrt{(1-t^2)\,(1-(1-k^2)\,t^2)}};$$

(* **An approximation at $k = 1/\sqrt{2}$ is** *) $N\left[\text{I1}\left[1/\sqrt{2}\right]\right]$

NIntegrate::ncvb :

  NIntegrate failed to converge to prescribed accuracy after 9 recursive bisections in t near {t} = {0.99999999999999983355229740155489970544279013454540318735490767 7264}. NIntegrate
  obtained 2.366162821232365`*^-15 and 4.2827226680842307`*^-10 for the integral and error estimates. ≫

2.36616×10^{-15}

(* **Neither it is able to recognize the following elliptic integrals as equivalent to an
elliptic integral of the second kind** *) $\left\{\int_1^1 \frac{k^2\,t^2\,dt}{\sqrt{(t^2-1)\,(1-k^2\,t^2)}}, \int_1^1 \frac{dt}{t^2\,\sqrt{(t^2-1)\,(1-k^2\,t^2)}}\right\};$

$\text{NIntegrate}\left[\frac{t^2}{2\sqrt{(t^2-1)\,(1-t^2/2)}}, \{t, 1, \sqrt{2}\}, \text{WorkingPrecision} \to 36\right]$

NIntegrate::ncvb : NIntegrate failed to converge to prescribed accuracy after 9 recursive
  bisections in t near {t} = {1.4142135623730942003299277660615939758975378987916786520544997278778704162408871717061}.
  NIntegrate obtained 1.3506438810476755025201659822471160697182866641263270648893709152790751355061145585697`86. and
  2.571375717923436839774279633792891625430816941827454509177688212084834091567201156289`86.*^-16 for the integral and error estimates. ≫

$1.350643881047675502520165982247 11607$

$\text{NIntegrate}\left[\frac{1}{t^2\,\sqrt{(t^2-1)\,(1-t^2/2)}}, \{t, 1, \sqrt{2}\}, \text{WorkingPrecision} \to 69\right]$

NIntegrate::ncvb : NIntegrate failed to converge to prescribed accuracy after 9 recursive bisections in t near {t} =
  {1.0000223525117875735106570498516978245923452377501422920654175286660505491620748702709807729996958076317480295755259869434}. NIntegrate
  obtained 1.350643881047675502519712350739101597711072950974416176471104236556862519721188826827742197396961861628748882328470889`119.
  and 2.289670590278223045650222759267035567628696222386698051487038943896791043248462025439756237249411658728116159394450318061119.*^-14
  for the integral and error estimates. ≫

$1.350643881047675502519712350739101597711072950974416176471104236 55686$

(* **Both integrals might be accurately evaluated at $k = 1/\sqrt{2}$ as** *)

$N\left[\text{EllipticE}\left[\frac{1}{2}\right], 71\right]$

$1.350643881047675502520174735338725841349522366924354545 32325370885 78779$

(* **Things do not necessarily get any better if "Mathematica 10.3" comes up with an exact evaluation of a real-
valued elliptic integral, say, the integral** *)

$$\text{I3} = \int_0^1 \frac{t^2\,dt}{(t^2+1)\,\sqrt{(1-t^2)\,(1-t^2/4)}}$$

$$\frac{1}{60}\left(80\,i\,\sqrt{3} + 40\,\text{EllipticF}\left[\text{ArcCsc}\left[\sqrt{\frac{2}{3}}\right], \frac{8}{9}\right] + 60\,\text{EllipticK}\left[\frac{1}{4}\right] - \right.$$

$$30\,\sqrt{2}\,\text{EllipticK}\left[\frac{9}{8}\right] + (9-3\,i)\,\sqrt{2}\,\text{EllipticPi}\left[\frac{9}{10} - \frac{3\,i}{10}, \frac{9}{8}\right] + (9+3\,i)\,\sqrt{2}\,\text{EllipticPi}\left[\frac{9}{10} + \frac{3\,i}{10}, \frac{9}{8}\right] - $$

$$\left.(12-4\,i)\,\text{EllipticPi}\left[\frac{4}{5} - \frac{4\,i}{15}, \text{ArcCsc}\left[\sqrt{\frac{2}{3}}\right], \frac{8}{9}\right] - (12+4\,i)\,\text{EllipticPi}\left[\frac{4}{5} + \frac{4\,i}{15}, \text{ArcCsc}\left[\sqrt{\frac{2}{3}}\right], \frac{8}{9}\right]\right)$$

(* **where a non-vanising imaginary part appears!** *)

N[I3]

$2.49522 + 2.3094\,i$

References

1. Abramowitz, M., Stegun, I.A.: Handbook of Mathematical Functions with Formulas, Graphs and Mathematical Tables. Applied Mathematics Series, vol. 55, 10th Printing. National Bureau of Standards, Washington (1972)
2. Adlaj, S.: Galois elliptic function and its symmetries. In: Vassiliev, N.N. (ed.) 12th International Conference on Polynomial Computer Algebra, pp. 11–17. St. Petersburg department of Steklov Institute of Mathematics (2019)
3. Adlaj, S.: Thread Equilibrium in a Linear Parallel Force Field. LAP LAMBERT, Saarbrucken (2018). (in Russian)
4. Adlaj, S.: On the second memoir of Évariste Galois' last letter. Comput. Tools Sci. Educ. **4**, 5–20 (2018)
5. Adlaj, S.: An analytic unifying formula of oscillatory and rotary motion of a simple pendulum (dedicated to 70th birthday of Jan Jerzy Slawianowski). In: Mladenov, I., Ludu, A., Yoshioka, A. (eds.) Proceedings Interenational Conference on "Geometry, Integrability, Mechanics and Quantization" 2014, pp. 160–171. Avangard Prima, Sofia, Bulgaria (2015)
6. Adlaj, S.: Torsion points on elliptic curves and modular polynomial symmetries. Presented at the joint MSU-CCRAS Computer Algebra Seminar on September 24, 2014. http://www.ccas.ru/sabramov/seminar/lib/exe/fetch.php?media=adlaj140924.pdf
7. Adlaj, S.: A perfect formula for the perimeter of an ellipse. Math. Yilin **001**, 30–37 (2013). (in Chinese)
8. Adlaj, S.: An eloquent formula for the perimeter of an ellipse. Not. AMS **59**(8), 1094–1099 (2012)
9. Adlaj, S.: Mechanical interpretation of negative and imaginary tension of a tether in a linear parallel force field. In: Selected Works of the International Scientific Conference on Mechanics "Sixth Polyakhov Readings", pp. 13–18. Saint-Petersburg State University, Saint-Petersburg (2012)
10. Adlaj, S.: Eighth Lattice Points, arXiv:1110.1743 (2011)
11. Brent, R.: Old and new algorithms for pi. Letters to the Editor. Not. AMS **60**(1), 7 (2013)
12. Carlson, B.C.: Chapter 19: Elliptic Integrals. NIST Digital Library of Mathematical Functions (v. 1.0.10 released 2015–08–07). http://dlmf.nist.gov/19
13. Chudnovsky, D.V., Chudnovsky, G.V.: The computation of classical constants. Proc. Nat. Acad. Sci. U.S.A. **86**(21), 8178–8182 (1989)
14. Euler, L.: De miris proprietatibus curvae elasticae sub aequatione $y = \int \frac{xxdx}{\sqrt{1-x^4}}$ contentae. Presented to the St. Petersburg Academy of Sciences on September 4, 1775. Acta Academiae Scientarum Imperialis Petropolitinae, pp. 34–61 (1782)
15. Fagnano, G.C.: Produzioni Matematiche (1, 2). Stamperia Gavelliana, Pesaro (1750)
16. Fricke, R.: Die elliptischen Funktionen und ihre Anwendungen. Zweiter Band: Die algebraischen Ausführungen. Teubner, Leipzig (1922)
17. Galois, É.: Lettre de Galois á M. Auguste Chevalier. J. de Mathématiques Pures et Appliquées **XI**, 408–415 (1846)
18. Grothendieck, A.: Récoltes et Semailles: Réflexions et témoignage sur un passé de mathématicien. http://matematicas.unex.es/~navarro/res/res.pdf
19. Gauß, C.: Arithmetisch geometrisches Mittel. Werke **3**, 361–403 (1866). Königlichen Gesell. Wiss. Göttingen

20. He, K., Zhou, X., Lin, Q.: High accuracy complete elliptic integrals for solving the Hertzian elliptical contact problems. Comput. Math. Appl. **73**(1), 122–128 (2017)
21. Klein, F.: Gauß' wissenschaftliches Tagebuch 1796–1814. Math. Ann. **57**, 1–34 (1903)
22. Lord, N.: The moment of inertia of an elliptical wire. Math. Gaz. **98**(541), 121–125 (2014)
23. Simpson, D.G.: General Physics I: Classical Mechanics (updated on January 25, 2013 and August 26, 2014). http://www.pgccphy.net/1030/phy1030.pdf
24. Modified Arithmetic-Geometric Mean. Mathematics Stack Exchange. https://math.stackexchange.com/questions/391382/modified-arithmetic-geometric-mean. Accessed 3 June 2019
25. Wolfram Language Tutorial: Elliptic Integrals and Elliptic Functions. Wolfram Language & System Documentation Center. http://reference.wolfram.com/language/tutorial/EllipticIntegralsAndEllipticFunctions.html. Accessed 3 June 2019

Obtaining and Analysis of the Necessary Conditions of Stability of Orbital Gyrostat by Means of Computer Algebra

Andrei V. Banshchikov[✉]

Matrosov Institute for System Dynamics and Control Theory
of Siberian Branch of Russian Academy of Sciences,
PO Box 292, 134, Lermontov str., Irkutsk 664033, Russia
bav@icc.ru

Abstract. With the help of software developed on the basis of the "Mathematica" computer algebra system, the dynamics of a satellite-gyrostat moving in a Newtonian central field of forces along the circular Keplerian orbit was investigated. The linearized equations of perturbed motion in the vicinity of the relative equilibrium of the system are constructed in the symbolic form on PC and the necessary conditions for its stability are obtained. The parametric analysis of the inequalities considers one of the cases when the vector of the gyrostatic moment of the system is in one of the planes formed by the principal central axes of inertia. The obtained stability regions have an analytical form or a graphical representation in the form of 2D images.

1 Introduction

The study of stability and stabilization of nonlinear or linearized models of mechanical systems often leads to the problem of "parametric analysis" of the conditions (inequalities) obtained. In parametric analysis, it is important to have a possibility to estimate the range of parameter values at which the required state (property) of the system is provided. Naturally, it is hard to hope for obtaining any readable analytical results for the models which have high dimensions and contain many parameters. At this stage, one can efficiently use computer algebra system (CAS) as well as the corresponding software elaborated on the basis of CAS.

The rigid body with the fixed axis of a statically and dynamically balanced flywheel rotating about that axis with a constant relative angular velocity is a stationary gyrostat. The system moves along the circular Keplerian orbit in a central Newtonian field of forces around the gravitational center. It is accepted that the mutual influence of the motion of the gyrostat about its mass center and the displacement of the latter at a constant angular velocity ω along the above

---

The work has been partially supported by the Russian Foundation for Basic Research (grant No. 19-01-00301).

mentioned trajectory are neglected. This is a so-called restricted formulation of the problem of orbital motion. This problem has its prehistory (see the review [1]) and has so far attracted the attention of the researchers. The main attention of the authors was concentrated on (i) finding the relative equilibrium positions for various variants of positioning the flywheel's rotation axis in the gyrostat's shell and (ii) obtaining sufficient conditions of Lyapunov's stability from the analysis of the generalized energy integral. However, the analysis of necessary conditions of stability of the relative equilibria was carried out only for positioning of the system's gyrostatic moment vector along with any principal central axis of inertia of the system (see, for example, [2]).

2   Construction of a Symbolical Model and Stability Conditions

For the description of a motion of the system, two right rectangular Cartesian coordinate systems with the poles in the system's mass center O are introduced:

(1) $OZ_1 Z_2 Z_3$ is an orbital coordinate system (OCS), where OZ_3 axis is directed by the radius-vector drawn from the attracting center into the mass center of a gyrostat; OZ_2 axis is perpendicular to the plane of the orbit.

(2) The coordinate system $Oz_1 z_2 z_3$ rigidly connected to a body has the axes directed along the principal central axes of inertia of a gyrostat.
    A, B, and C are the moments of inertia concerning axes Oz_1, Oz_2, Oz_3, and h_j are the projections (onto the corresponding axis) of a vector of gyrostatic moment of system divided by ω. Here $\omega = |\boldsymbol{\omega}|$ is the module of orbital angular velocity. For definition of a relative positioning of the OZ_k and Oz_j axes, the directional cosines defined by aircraft angles α, β, γ are used (see, for example, [1]).

Consider the position of relative equilibrium ($\dot\alpha = 0, \dot\beta = 0, \dot\gamma = 0$) in general form:

$$\alpha = \alpha_0 = \text{const}, \quad \beta = \beta_0 = \text{const}, \quad \gamma = \gamma_0 = \text{const}. \tag{1}$$

With the help of developed software [3], which is described in [4], the following results in a symbolic form on PC are obtained: (a) kinetic and potential energy of a system; (b) nonlinear equations of motion of orbital gyrostat in the form of Lagrange of the 2nd kind; (c) existence conditions of equilibrium (1).

Let us write out the equations determining the equilibrium positions in regard to OCS of the gyrostat (conditions (c)):

$$\begin{cases} \sin 2\alpha_0 \left(2(2A - B - C)\cos^2\beta_0 - (B - C)(\cos 2\beta_0 - 3)\cos 2\gamma_0\right) \\ + 4(B - C)\cos 2\alpha_0 \sin\beta_0 \sin 2\gamma_0 = 0, \\ (B + C - 2A + (B - C)\cos 2\gamma_0)\sin 2\beta_0 (5 - 3\cos 2\alpha_0) \\ + 6(B - C)\cos\beta_0 \sin 2\alpha_0 \sin 2\gamma_0 \\ + 8(\sin\beta_0 (h_2 \cos\gamma_0 - h_3 \sin\gamma_0) - h_1 \cos\beta_0) = 0, \\ \frac{1}{2}(B - C)\left(\sin 2\gamma_0 (3\cos^2\alpha_0 + \cos^2\beta_0 - 3\sin^2\alpha_0 \sin^2\beta_0) \right. \\ \left. + 3\sin 2\alpha_0 \sin\beta_0 \cos 2\gamma_0\right) + \cos\beta_0 (h_2 \sin\gamma_0 + h_3 \cos\gamma_0) = 0. \end{cases} \tag{2}$$

Let us present the angles α, β, and γ in the perturbed motion in the form: $\alpha = \alpha_0 + \overline{\alpha}$, $\beta = \beta_0 + \overline{\beta}$, $\gamma = \gamma_0 + \overline{\gamma}$, where $\overline{\alpha}, \overline{\beta}$, and $\overline{\gamma}$ are small deviations from equilibrium (1).

The linearized equations of perturbed motion in vicinity of (1) look this way:

$$M\ddot{q} + G\dot{q} + Kq = 0, \tag{3}$$

where $q = (\overline{\alpha}, \overline{\beta}, \overline{\gamma})^T$ is a vector of deviations from unperturbed motion (1);

$M = \begin{pmatrix} M_{11} & M_{12} & M_{13} \\ M_{12} & M_{22} & 0 \\ M_{13} & 0 & M_{33} \end{pmatrix}$ is a positive definite symmetric matrix of kinetic energy;

$G = \begin{pmatrix} 0 & G_{12} & G_{13} \\ -G_{12} & 0 & G_{23} \\ -G_{13} & -G_{23} & 0 \end{pmatrix}$ is a skew-symmetric matrix of gyroscopic forces;

$K = \begin{pmatrix} K_{11} & K_{12} & K_{13} \\ K_{12} & K_{22} & K_{23} \\ K_{13} & K_{23} & K_{33} \end{pmatrix}$ is a symmetric matrix of potential forces.

Here

$M_{11} = A\sin^2\beta_0 + (B\cos^2\gamma_0 + C\sin^2\gamma_0)\cos^2\beta_0$,

$M_{22} = B\sin^2\gamma_0 + C\cos^2\gamma_0$, $\quad M_{12} = (B - C)\cos\beta_0 \sin\gamma_0 \cos\gamma_0$,

$M_{13} = A\sin\beta_0$, $\quad M_{33} = A$, $\quad \det M = ABC\cos^2\beta_0 > 0$;

$G_{12} = \sin 2\beta_0(A - B\cos^2\gamma_0 - C\sin^2\gamma_0) + h_1\cos\beta_0$
$\quad + \sin\beta_0(h_3\sin\gamma_0 - h_2\cos\gamma_0)$,

$G_{13} = -\cos\beta_0\left(\cos\beta_0(B - C)\sin 2\gamma_0 + h_2\sin\gamma_0 + h_3\cos\gamma_0\right)$,

$G_{23} = -\cos\beta_0\left(A + (C - B)\cos 2\gamma_0\right) - h_3\sin\gamma_0 + h_2\cos\gamma_0$;

$K_{11} = \dfrac{3}{4}\left(\cos 2\alpha_0\left((4A - 2(B + C))\cos^2\beta_0 - (B - C)(\cos 2\beta_0 - 3)\cos 2\gamma_0\right)\right)$
$\quad + 3(C - B)\sin 2\alpha_0 \sin\beta_0 \sin 2\gamma_0$,

$K_{12} = 3\cos\beta_0\left((B\cos^2\gamma_0 + C\sin^2\gamma_0 - A)\sin 2\alpha_0 \sin\beta_0\right.$
$\quad \left. + \dfrac{1}{2}(B - C)\cos 2\alpha_0 \sin 2\gamma_0\right)$,

$K_{13} = \dfrac{3}{4}(B - C)(4\cos 2\alpha_0 \sin\beta_0 \cos 2\gamma_0 + \sin 2\alpha_0(\cos 2\beta_0 - 3)\sin 2\gamma_0)$,

$K_{22} = \dfrac{1}{4}((2A + (C - B)\cos 2\gamma_0 - B - C)(3\cos 2\alpha_0 - 5)\cos 2\beta_0$
$\quad + 3(C - B)\sin 2\alpha_0 \sin\beta_0 \sin 2\gamma_0) + h_1\sin\beta_0 + (h_2\cos\gamma_0 - h_3\sin\gamma_0)\cos\beta_0$,

$K_{23} = \dfrac{1}{4}(B - C)(6\sin 2\alpha_0 \cos\beta_0 \cos 2\gamma_0 + (3\cos 2\alpha_0 - 5)\sin 2\beta_0 \sin 2\gamma_0)$
$\quad - (h_2\sin\gamma_0 + h_3\cos\gamma_0)\sin\beta_0$,

$K_{33} = \dfrac{1}{4}(B - C)\left(\cos 2\gamma_0(10\cos^2\beta_0 - 3\cos 2\alpha_0(\cos 2\beta_0 - 3))\right.$
$\quad \left. - 12\sin 2\alpha_0 \sin\beta_0 \sin 2\gamma_0\right) + (h_2\cos\gamma_0 - h_3\sin\gamma_0)\cos\beta_0$.

All derivatives in (3) are calculated by dimensionless time $\tau = \omega t$.

The characteristic equation of system (3): $\det\left(M\lambda^2 + G\lambda + K\right) = v_3\lambda^6 + v_2\lambda^4 + v_1\lambda^2 + v_0 = 0$ contains λ only in even degrees. The stability of a trivial solution of Eq. (3) takes place when all roots with respect to λ^2, being simple, will be real negative numbers. The algebraic conditions providing specified properties of roots (necessary conditions of stability), represent the system of inequalities [5]:

$$\begin{cases} v_3 \equiv \det M > 0, \quad v_2 > 0, \quad v_1 > 0, \quad v_0 \equiv \det K > 0, \\ Dis \equiv v_2^2 v_1^2 - 4v_1^3 v_3 - 4v_2^3 v_0 + 18v_3 v_2 v_1 v_0 - 27v_0^2 v_3^2 > 0. \end{cases} \tag{4}$$

It is worth noting that the first condition in (4) is always satisfied by virtue of the positive definiteness of the kinetic energy matrix.

Remark 1. The cases where the characteristic equation has zero roots and/or multiple purely imaginary roots are not analyzed here. If at least one of conditions (4) is replaced by a strict contrary inequality, the system will be unstable, according to the Lyapunov theorem on instability in the first approximation.

Emphasize that the construction of the symbolic linearized model (3) (i.e., obtaining in the analytical form the elements of the matrices M, G, K), the calculation of the coefficients v_i $(i = \overline{0,3})$ and the discriminant Dis from (4) was also performed using the software [3].

3   Relative Equilibria

The analytical or numerical determination of all the equilibrium positions of a satellite-gyrostat in regard to OCS has been described in detail in [6] for the general case $(h_1 \neq 0, h_2 \neq 0, h_3 \neq 0, A \neq B \neq C)$ or for the special cases in [7]. Depending on the effect of the gyrostatic moment, a bifurcation picture of equilibrium positions has been presented in the papers, and their number has been found.

Let us write out some particular solutions of system (2), for which the gyrostatic moment vector lies in one of the planes formed by the satellite's principal central axes of inertia.

In case 1 $(h_1 = 0, h_2 \neq 0, h_3 \neq 0)$, there are the following equilibrium positions:

$$\begin{cases} \alpha = \alpha_0 = 0, \quad \beta = \beta_0 = 0, \\ \gamma = \gamma_0 = \text{const}: \quad h_2 \sin\gamma_0 + \cos\gamma_0(h_3 + 4(B - C)\sin\gamma_0) = 0; \end{cases} \tag{5}$$

$$\begin{cases} \alpha = \alpha_0 = \pi/2, \quad \beta = \beta_0 = 0, \\ \gamma = \gamma_0 = \text{const}: \quad h_2 \sin\gamma_0 + \cos\gamma_0(h_3 + (B - C)\sin\gamma_0) = 0. \end{cases} \tag{6}$$

In case 2 $(h_3 = 0, h_1 \neq 0, h_2 \neq 0)$, there are the following equilibrium positions:

$$\begin{cases} \alpha = \alpha_0 = 0, \quad \gamma = \gamma_0 = 0, \\ \beta = \beta_0 = \text{const}: \quad h_2 \sin\beta_0 - \cos\beta_0(h_1 + (A - B)\sin\beta_0) = 0; \end{cases} \tag{7}$$

$$\begin{cases} \alpha = \alpha_0 = \pi/2, \quad \gamma = \gamma_0 = 0, \\ \beta = \beta_0 = \text{const}: \quad h_2 \sin\beta_0 - \cos\beta_0(h_1 + 4(A - B)\sin\beta_0) = 0. \end{cases} \quad (8)$$

In case 3 ($h_2 = 0$, $h_1 \neq 0$, $h_3 \neq 0$), there are the following equilibrium positions:

$$\begin{cases} \alpha = \alpha_0 = 0, \quad \gamma = \gamma_0 = \pi/2, \\ \beta = \beta_0 = \text{const}: \quad h_1 \cos\beta_0 + \sin\beta_0(h_3 + (A - C)\cos\beta_0) = 0; \end{cases} \quad (9)$$

$$\begin{cases} \alpha = \alpha_0 = \pi/2, \quad \gamma = \gamma_0 = \pi/2, \\ \beta = \beta_0 = \text{const}: \quad h_1 \cos\beta_0 + \sin\beta_0(h_3 + 4(A - C)\cos\beta_0) = 0. \end{cases} \quad (10)$$

The equilibrium positions (5) and (7) can be found in [1], the equilibrium positions (6) and (8) are the classes of equilibrium orientations in the cases, respectively, $h_1 = 0$, $a_{31}^2 = 1$ and $h_3 = 0$, $a_{33} = 0$ (see [7]), the equilibrium positions (9) and (10) belong, respectively, to the group of solutions I and III in [8].

It has been ascertained that the conditions of positive definiteness of the matrix of potential forces K for solutions (5) and (7) coincide with the sufficient conditions of stability given in [1], and in case of solutions (9) and (10) – with those obtained and studied in [8].

Obtaining and analysis of the necessary stability conditions of relative equilibria on the basis of equations of the first approximation were conducted by the author earlier for the oblate [9] and prolate [10] axisymmetric orbital gyrostats, in the case when the system's gyrostatic moment vector lies in one of the principal central inertia planes.

4 Parametric Analysis of Stability Conditions

For example, consider the parametric analysis of the necessary stability conditions (4) in case 3: $h_2 = 0$, $h_1 \neq 0$, $h_3 \neq 0$.

Without loss of generality, let $h_i > 0$ ($i = 1, 3$) and $B > A > C$.

Let us enter dimensionless parameters:

$$H_1 \equiv \frac{h_1}{B}; \quad H_3 \equiv \frac{h_3}{B}; \quad J_A \equiv \frac{A}{B}; \quad J_C \equiv \frac{C}{B}; \quad p_c \equiv \cos\beta_0; \quad p_s \equiv -\sin\beta_0. \quad (11)$$

The values of the parameters belong to the intervals:

$$\frac{1}{2} < J_A < 1, \quad 1 - J_A < J_C < J_A; \quad -1 < p_c < 1, \left(p_c \neq 0, \ p_s = \pm\sqrt{1 - p_c^2}\right). \quad (12)$$

The restrictions on the parameters J_A, J_C come from the conditions $B > A > C$, $B < A + C$. With the values of $p_c = 0$ (or $p_s = 0$), from equations in (9) and (10), it then follows that $h_3 = 0$ (or $h_1 = 0$), what contradicts the conditions $h_1 > 0$, $h_3 > 0$.

4.1   Necessary Stability Conditions of Equilibrium (10).

Let us start with the equilibrium position (10). Using (11), let us resolve the equation from (10) with respect to the parameter H_1:

$$H_1 = p_s \left(4 \left(J_A - J_C \right) + \frac{H_3}{p_c} \right). \tag{13}$$

Taking into account notations (11) and expression (13), equations (3) have matrices:

$$M = \begin{pmatrix} J_C p_c^2 + J_A p_s^2 & 0 & -J_A p_s \\ 0 & 1 & 0 \\ -J_A p_s & 0 & J_A \end{pmatrix};$$

$$G = \begin{pmatrix} 0 & -2 \left(J_C - J_A \right) p_c p_s & 0 \\ 2 \left(J_C - J_A \right) p_c p_s & 0 & \left(J_C - J_A - 1 \right) p_c - H_3 \\ 0 & H_3 + \left(J_A - J_C + 1 \right) p_c & 0 \end{pmatrix};$$

$$K = \begin{pmatrix} 3 \left(1 - J_A p_c^2 - J_C p_s^2 \right) & 0 & -3 \left(1 - J_C \right) p_s \\ 0 & 4 \left(J_C - J_A \right) p_c^2 - \frac{H_3}{p_c} & 0 \\ -3 \left(1 - J_C \right) p_s & 0 & K_{33} \end{pmatrix}, \tag{14}$$

where $K_{33} = \left(1 - J_C \right) \left(1 - 2 \left(p_c^2 - p_s^2 \right) \right) - H_3 p_c$.

The parameter p_s enters the coefficients of the system's characteristic equation only in even degrees. Let us eliminate it, considering $p_c^2 + p_s^2 = 1$. Let us write down these coefficients depending on four parameters J_A, J_C, p_c, H_3 in an explicit form:

$$v_3 \equiv \det M = J_A J_C p_c^2 \ ; \qquad v_2 = H_3^2 \left(p_c^2 \left(J_C - J_A \right) + J_A \right)$$
$$+ H_3 p_c \left(J_A \left(1 + 6 J_A - 7 J_C \right) - p_c^2 \left(1 + 6 J_A - 2 J_C \right) \left(J_A - J_C \right) \right)$$
$$+ p_c^2 \left(3 \left(3 J_A - 1 \right) J_C^2 + \left(3 - 2 J_A \left(1 + 9 J_A \right) \right) J_C + 3 \left(1 + 3 J_A \right) J_A^2 \right)$$
$$- p_c^4 \left(J_A - J_C \right) \left(9 J_A^2 - 6 J_A \left(J_C - 1 \right) + \left(J_C - 1 \right) \left(J_C + 3 \right) \right);$$
$$v_1 = H_3^2 \left(4 p_c^2 \left(J_C - J_A \right) + J_A - 3 J_C + 3 \right) + H_3 p_c \left(3 - 22 J_A J_C \right.$$
$$- 24 J_C + 3 J_A^2 + 19 J_A + 21 J_C^2 - p_c^2 \left(6 J_A - 26 J_C + 19 \right) \left(J_A - J_C \right))$$
$$+ p_c^4 \left(J_A - J_C \right) \left(6 J_A \left(5 J_C - 7 \right) + 9 J_A^2 - 31 J_C^2 + 34 J_C - 3 \right)$$
$$- 3 p_c^2 \left(J_C - 1 \right) \left(J_A \left(3 - 18 J_C \right) + 9 J_A^2 + J_C \left(9 J_C - 2 \right) \right);$$
$$v_0 \equiv \det K = 3 \left(H_3 + 4 p_c^3 \left(J_A - J_C \right) \right) \times \left(H_3 \left(p_c^2 \left(J_C - J_A \right) - J_C + 1 \right) \right.$$
$$+ p_c \left(J_C - 1 \right) \left(J_A \left(4 p_c^2 - 3 \right) - 4 \left(p_c^2 - 1 \right) J_C - 1 \right)). \tag{15}$$

The discriminant Dis (see (4)) of a cubic equation is an 8th degree polynomial in regard to H_3 with the coefficients depending in a complicated manner on the parameters J_A, J_C, and p_c. This polynomial is not presented in an explicit analytical form due to being immense.

According to Kelvin–Chetaev's theorems [11], studying the questions on stability of equilibria begins with the analysis of a matrix of potential forces. With the help of "Mathematica" function

$Reduce[\{ 1/2 < J_A < 1,\ 1 - J_A < J_C < J_A,\ -1 < p_c < 1,\ p_c \neq 0,\ H_3 > 0,$
$1 - J_A p_c^2 - J_C(1 - p_c^2) > 0,\ 4(J_C - J_A) p_c^2 - H_3/p_c > 0,\ \det K > 0 \},$
$\{ J_A,\ J_C,\ p_c,\ H_3 \},\ \mathrm{Reals}\ ]$

designed to find the symbolic (analytical) solution of the inequalities systems, the conditions of positive definiteness of a matrix K from (14) are obtained. Due to the solution bulkiness, its presentation is omitted here. An analysis of the solution obtained allows us to conclude the following conclusion.

Proposition 1. *The matrix of potential forces (14) for the equilibrium (10) with the values of parameters from (12) can be positively determined only in the interval* $-1 < p_c < 0$.

It is not possible to obtain an analytical solution for the entire system of inequalities (4) (with the coefficients v_i ($i = \overline{0,3}$) from (15)) because of the large number of parameters and the complexity of the expressions being analyzed. Therefore, to simplify the analysis, let us move on to symbolic-numerical analysis for fixed values of some parameters.

Let us find the graphic solution of the system of inequalities (4) for two variants of fixed parameter values:

(1) $J_A = 2/3$, $J_C = 1/2$, the conditions for the parameters p_c and H_3 are to be obtained;
(2) $p_c = -1/\sqrt{2}$, $H_3 = 1$, the conditions for the parameters J_A and J_C are to be obtained.

Let us construct the regions of necessary conditions of stability in the parameter plane p_c, H_3 (or J_A, J_C) using "Mathematica" function

$RegionPlot[-1 < p_c < 1 \wedge p_c \neq 0 \wedge H_3 > 0 \wedge v_0 > 0 \wedge v_1 > 0 \wedge v_2 > 0 \wedge Dis > 0,$
$\{ p_c, -1, 1 \}, \{ H_3, 0, 1.6 \}\ ]$

$RegionPlot[1/2 < J_A < 1 \wedge 1 - J_A < J_C < J_A \wedge v_0 > 0 \wedge v_1 > 0 \wedge v_2 > 0 \wedge Dis > 0,$
$\{ J_A, 1/2, 1 \}, \{ J_C, 0, 1 \}\ ]$

designed for a graphical representation of the solution of the system of inequalities. The results obtained are shown with shaded regions in Figs. 1 and 2.

The darker areas obtained in Fig. 1 (or Fig. 2) belong to the region where the matrix K is positively definite.

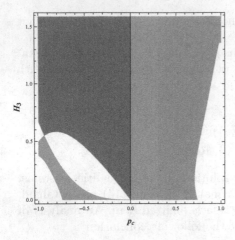

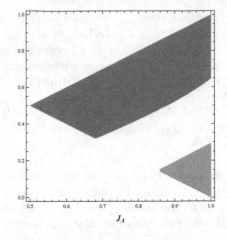

Fig. 1. Region of necessary conditions of stability for variant (1)

Fig. 2. Region of necessary conditions of stability for variant (2)

4.2   About the Gyroscopic Stabilization of the Equilibrium (9)

The equation from (9) in parameters (11) is solved for H_1 as follows:

$$H_1 = p_s \left((J_A - J_C) + \frac{H_3}{p_c} \right).$$

Considering last relation and notations (11), the matrix of potential forces takes the form:

$$K = \begin{pmatrix} 3\left(J_A p_c^2 + J_C p_s^2 - 1\right) & 0 & -3\left(J_C - 1\right)p_s \\ 0 & \left(J_C - J_A\right)p_c^2 - \frac{H_3}{p_c} & 0 \\ -3\left(J_C - 1\right)p_s & 0 & K_{33} \end{pmatrix}, \qquad (16)$$

where $K_{33} = 3\left(J_C - 1\right) - p_c\left(H_3 + (1 - J_C)p_c\right).$

The coefficients of the characteristic equation (after the elimination of parameter p_s) become:

$$v_2 = H_3^2 \left(p_c^2\left(J_C - J_A\right) + J_A\right) + H_3 p_c \left(p_c^2\left(2J_C - 1\right)\left(J_A - J_C\right) - J_A\left(J_C - 1\right)\right)$$
$$+ p_c^2 \left(J_A\left(p_c^2\left(1 - J_C\right)J_C + J_C - 3\right) + 3J_A^2 + J_C\left(J_C - 1\right)\left(p_c^2 J_C + 3\right)\right);$$

$$v_1 = H_3^2 \left(2p_c^2\left(J_A - J_C\right) + J_A + 3\left(J_C - 1\right)\right) + H_3 p_c \left(3J_A^2\left(2p_c^2 - 1\right)\right.$$
$$- J_A\left(4p_c^2\left(J_C + 1\right) + J_C - 4\right) - J_C\left(2p_c^2 + 3\right)\left(J_C - 2\right) - 3\right)$$
$$+ 3p_c^2\left(J_C - 1\right)\left(3J_A + J_C - 3\right)$$
$$- p_c^4\left(J_A - J_C\right)\left(3J_A\left(J_C - 2\right) + 4J_C^2 - 10J_C + 6\right);$$

$$v_3 \equiv \det M = J_A J_C p_c^2;$$

$$v_0 \equiv \det K = 3\left(p_c^3\left(J_C - J_A\right) - H_3\right) \times \left(H_3\left(1 - J_C + p_c^2\left(J_C - J_A\right)\right)\right.$$
$$- p_c\left(J_C - 1\right)\left(4\left(1 - J_A\right) + \left(J_C - J_A\right)\left(p_c^2 - 1\right)\right)\right). \qquad (17)$$

The main diagonal first-order minor of the matrix K from (16) on intervals (12) is negative. Therefore, matrix (16) is not positive definite and equilibrium (9) will be unstable.

It is known that if the equilibrium position is unstable at potential forces, Kelvin–Chetaev's theorem [11] of influence of gyroscopic forces tells us that gyroscopic stabilization is possible only for systems with an even degree of instability. The evenness (or oddness) of the degree of instability according to Poincaré is determined by positivity (or negativity) of the determinant of the matrix of potential forces.

Let us pose the question of the possibility of gyroscopic stabilization of unstable equilibrium (9) under condition $\det K > 0$. The set in the parameter space that meets the last-mentioned inequality determines the region with an even degree of instability:

$$\frac{1}{2} < J_A < 1 \wedge 1 - J_A < J_C < J_A \wedge -1 < p_c < 0 \wedge$$

$$\wedge \, p_c^3 (J_C - J_A) < H_3 < \frac{p_c (J_C - 1) \left(4 (1 - J_A) + (J_C - J_A) (p_c^2 - 1) \right)}{1 - J_C + p_c^2 (J_C - J_A)}.$$

For the detection of a property of gyroscopic stabilization, it is necessary to find in what part of the region with an even degree of instability the remaining inequalities from (4) are fulfilled (except for $v_3 \equiv \det M > 0$, $v_0 \equiv \det K > 0$).

Instead of solving system (4), let us first consider a simpler problem. The symbolic solution of the system of inequalities that determines the positivity of coefficients (17) obtained with the help of the function *Reduce* is FALSE. Obviously, the system of inequalities (4) will also appear to be inconsistent. As a result of the analysis, the following proposition can be formulated.

Proposition 2. *The unstable equilibrium (9) for parameters from the range (12) cannot be stabilized by gyroscopic forces.*

5   Conclusion

Based on the analogy with the parametric analysis presented above, the necessary stability conditions for the equilibrium positions (5)–(8) were also investigated. Note that under other conditions for the inertia moments of gyrostat, it is possible to introduce new parameters and perform the analysis again, applying the same functions (*Reduce, RegionPlot, RegionPlot3D*) of symbolic-numerical modeling.

Generally speaking, the necessary conditions do not guarantee the Lyapunov stability in general, but only for those parts of the stability region where the condition of the Lagrange theorem (see, for example, [11]) holds true. In the considered problem, these are the darker areas obtained in Figs. 1 and 2. But obtaining the necessary stability conditions allows, for example, to consider the possibility of gyroscopic stabilization for an unstable system under the action of potential forces.

Once again we emphasize that all computations on PC were carried out in the symbolic or symbolic-numerical form and, moreover, the differential equations of motion and coefficients of the characteristic equation in an automatic mode were obtained with the software created on the basis of the CAS "Mathematica".

References

1. Sarychev, V.A.: Problems of orientation of satellites. Itogi Nauki i Tekhniki. Series "Space Research" **11**, 5–224 (1978). VINITI Publication, Moscow (in Russian)
2. Anchev, A.A., Atanasov, V.A.: Analysis of the necessary and sufficient conditions for the stability of the equilibrium of a gyrostatic satellite. Kosm. Issled. **28**(6), 831–836 (1990). (in Russian)
3. Banshchikov, A.V., Irtegov, V.D., Titorenko, T.N.: Software package for modeling in symbolic form of mechanical systems and electrical circuits. In: Certificate of State Registration of Computer Software No. 2016618253. Federal Service for Intellectual Property. Issued 25 July 2016. (in Russian)
4. Banshchikov, A.V., Burlakova, L.A., Irtegov, V.D., Titorenko, T.N.: Symbolic computation in modelling and qualitative analysis of dynamic systems. Comput. Technol. **19**(6), 3–18 (2014). (in Russian)
5. Kozlov, V.V.: Stabilization of the unstable equilibria of charges by intense magnetic fields. J. Appl. Math. Mech. **61**(3), 377–384 (1997)
6. Gutnik, S.A., Santos, L., Sarychev, V.A., Silva, A.: Dynamics of a gyrostat satellite subjected to the action of gravity moment. Equilibrium attitudes and their stability. J. Comput. Syst. Sci. Int. **54**(3), 469–482 (2015)
7. Gutnik, S.A., Sarychev, V.A.: Application of computer algebra methods for investigation of stationary motions of a gyrostat satellite. Program. Comput. Softw. **43**(2), 90–97 (2017)
8. Sarychev, V.A., Mirer, S.A., Degtyarev, A.A.: Dynamics of a gyrostat satellite with the vector of gyrostatic moment in the principal plane of inertia. Cosm. Res. **46**(1), 60–73 (2008)
9. Banshchikov, A.V.: Research on the stability of relative equilibria of oblate axisymmetric gyrostat by means of symbolic-numerical modelling. In: Gerdt, V.P., Koepf, W., Seiler, W.M., Vorozhtsov, E.V. (eds.) CASC 2015. LNCS, vol. 9301, pp. 61–71. Springer, Cham (2015). https://doi.org/10.1007/978-3-319-24021-3_5
10. Banshchikov, A.V., Chaikin, S.V.: Analysis of the stability of relative equilibriums of a prolate axisymmetric gyrostat by symbolic-numerical modeling. Cosm. Res. **53**(5), 378–384 (2015)
11. Chetaev, N.G.: Stability of Motion. Works on Analytical Mechanics. AS USSR, Moscow (1962). (in Russian)

Efficient Exact Algorithm for Count Distinct Problem

Nikolay Golov[1], Alexander Filatov[2], and Sergey Bruskin[3(✉)]

[1] Avito, Higher School of Economics, Moscow, Russia
azathot.mail@gmail.com
[2] Avito, Moscow Engineering Physics Institute MEPhI, Moscow, Russia
philatalexander@gmail.com
[3] Higher School of Economics, Moscow, Russia
sergey.n.bruskin@gmail.com
http://www.avito.ru

Abstract. This paper describes and analyses optimization approaches, which make possible the exact calculation of millions of hierarchical count distinct measures over hundreds of billions data rows. Described approach evolved for several years, in parallel with the growth of tasks from a fast growing internet company, and was finally implemented as a PEAPM (Pipelined Exact Accumulation for Paralleled Measures) algorithm. Current version of an algorithm outputs exact values (not estimates), works in a single thread, in minutes using a general commodity hardware, and requires volume of RAM equal to the doubled size of required measures.

Keywords: Big Data · MPP · Database · Analytics ·
Cardinality estimation · Distinct elements problem ·
Clickstream analysis · Performance

1 Introduction

Big Data analytics is rapidly becoming a commonplace task for many companies, especially for electronic (internet) commerce ones. Such companies rely on its audience as a main asset. Electronic commerce companies with a large and loyal audience are attractive for investors, even if their current cash flow is negative. Audience of electronic commerce service is traditionally estimated by DAU (daily active users) and MAU metrics (monthly active users) [2]. These metrics are essentially a count distinct aggregates over all user actions performed inside the electronic commerce service. Avito is a Russian classified site, very big and extremely profitable [6]. Daily volume of daily unique visitors exceeds 10 mln [6], user actions in Avito exceeds few billions events a day, number of count distinct aggregates-per-day and -per-month exceeds few millions. Estimation and probabilistic approaches were rejected because of multiple reasons. Engineering team of Avito successfully implemented an efficient exact count distinct algorithm, which calculates millions of count distinct aggregates over (hundreds) billions of events in minutes (in total), using only one processor thread and

M. England et al. (Eds.): CASC 2019, LNCS 11661, pp. 67–77, 2019.
https://doi.org/10.1007/978-3-030-26831-2_5

minimum amount of RAM. Algorithm is described, real-world data experiments and optimization steps are listed. Link to the open-source implementation of an algorithm is also provided.

2   Note and Issues

Internet companies collect data about user activity by recording logs of detailed user actions, click rates, visits, and other property records of web users. Data about user activity is usually collected inside Data Warehouse or Data Lake. The first version of the Data Warehouse (DW) at Avito was built in 2013 using Anchor modeling methodology [4], it contained 10 TB of data, and ran on a Vertica cluster of 3 servers. It loaded data from two data sources: the back office system at Avito and clickstream web logs. Since then, the DW has grown, and the current size of the Avito data warehouse has been limited to 276 Tb for licensing reasons (176 TB at 2018, 276 Tb at 2019). It now contains years of consistent historical data from 40+ data sources (back office, Google DFP/AdSense, MDM system, CRM system, RTB systems, among others), and a rolling half year of detailed clickstream data. The cluster has been increased from 3 to 20 nodes in order to scale up performance [5]. The sequence of pages visited by within a particular website is known as the *clickstream* of the users. Clickstreams are analyzed to understand traffic, the number of unique visitors, sessions, and page views [3]. Clickstream is one of the biggest and most important data domains of Avito. It is a huge sequence of events (up to few billions a day), where each event is a vector of few dozens (hundreds) of attributes.

Clickstream data are loaded every 15 min. At the beginning of 2014, each such batch contained 5 million rows ($\approx$1.5 GB) and 15 million ($\approx$5 Gb) one year later. Avito has evolved their data model over the years. The clickstream records originally had less than 30 attributes, while now containing more than 200. Clickstream data has grown many times, both in terms of velocity (number of rows per minute), volume (size), and variety (number of attributes). The growth was successfully handled through scaling up the cluster by the addition of nodes.

At the beginning of 2019, Avito had billions of raw click stream events per day. Those raw events are being preliminarily filtered to remove non-human activity and combined into final, "wide" events. For example, single search request of a user produces few raw front-end events (pixels, banners shown) and few raw backend events (call of search engine, call of recommendations engine), but, for analytical purpose, those events have to be combined into a single wide search event, which contains all required fields. Each final wide event must contain cookie_id field (universal code of a user) and dimension attributes (example given at Sect. 2). Preliminary filtering and combining produce hundreds of millions of final events from billions of raw events. Onward we will use the term "event" for the final wide events, after filtering and combining.

Typical attributes are:

- cookieID - unique identifier of a user (even unlogged ones).
- eventDT - date and time of event.

- eventType - type of action performed by a user, search, page view, contact, banner click, starting a chat and so on.
- platform - type of device used by a user: IOS, Android, PC or tablet.
- locationID - city or region, where action was performed.
- categoryID - category, part of a site, where action is performed, such as Autos, RealEstate, CV and so on.
- isRegisteredUser - analytical metric, which is true for a user who performed log in.
- isNewVisitor - analytical metric, which is true for new arrivals (for freshly issued cookies).
- ....

Above-listed basic attributes. Really some attributes are a group of attributes. For example, locationID can be observed as a set of three attributes: City, Region, and Country. All the same, eventType is also augmented with secondary attributes: isContactEvent, isPageViewEvent, isPaymentEvent. Such augmentation helps to analyze hundreds of events, thousands of categories, and tens of thousands of locations. Typical clickstream analysis requires calculation of count of unique users (cookies) for a given set of attributes. This calculation can be described as a typical SQL query:

```
select count(distinct cookieID)
from ClickStream
where EventType=... and eventDT = ... and locationID=...
```

This SQL query can be processed by almost any database, with different performance. In any case, database has to filter clickstream table (of billions event-a-day) and perform count distinct aggregation over filtered values (maybe all values, if filters are omitted and total count distinct values are required). Modern column-based database can execute such query in up to few seconds (4s or more [7]). But real-world analytical scenarios require not a single aggregation, but dozens and hundreds of aggregations: for example, to illustrate the dynamics of daily count distinct values through a year, or to compare cities to find the most rapidly growing ones. Hundreds of aggregation (required for a single analysis) will take many minutes and load database hardware a lot. If there are dozens of analysts performing such sorts of analysis in parallel, database will suffer. Also, if a company want to retain an ability to perform count distinct analysis for any combination of filters, then it has to store all clickstream events for the required period of time. For Avito it means up to 1 bln. events a day, for 5 years, about 1.5 trln. events in total. So, if the number of parallel count distinct based analysis is high, some pre-aggregation step sounds reasonable, it can speed up analysis and reduce storage space requirements.

The main issue of count distinct metric is that it is non-additive one. So it is impossible to calculate elementary values (for smallest possible combination of attributes), and simply combine them all together to get a total value. If a single cookie performed actions for Auto, Real Estate and General categories (value of count distinct cookies is 1 for each category), total count distinct value is also 1, which can be obtained by combination by per-category values. Therefore, to calculate all possible pre-aggregates, one has to calculate all possible combinations of attributes using both existing values of the attributes and a special value ANY. For example, for attributes is RegisteredUser=[True, False] and is NewVisitor=[True, False], there are 9 possible combinations rather than 4: {Any, Any},{Any, True},{Any, False}, {True, Any},{True, True},{True, False}, {False, Any},{False, True},{False, False}. Combination {True, Any} means that count distinct have to be calculated with condition isRegisteredUser = True, and without condition of isNewVisitor. Therefore, Cartesian product of just the values of attributes listed above will give us up to 100 mln. of possible combinations, which will lead to 3 years of calculations, if each calculation is only lasting 1 s, but has to be performed sequentially.

2.1 Methods

Direct approach to count distinct calculation described in the section above. This approach assumes two possible scenarios:

- **Scenario 1:** "On demand" calculation - store all raw events, calculate metrics in real-time, on demand, when analysis is required. As was mentioned above, single click-stream event contains ~100 attributes, and, therefore, single event requires ~1 kb. of disc space. Therefore, 1 bln. events (average single day traffic) require 1 Tb of disc space. 1.5 trln. events (combined 5 year traffic) require 1.5 Pb of disc space. 1.5 Pb of disc space - it is hundreds of servers to store. Exact numbers depend on replication factor (2 or 3), compression, storage format, size of single server (6, 9 or 10 Tb per. server), but, either way, the total number of servers is no less than a hundred and a half. One or two hundred of servers is an affordable amount for a profitable internet company, but it is extremely expensive. Contemporary Big Data databases, such as Vertica, GreenPlum, ClickHouse or Hadoop (HDFS+Hive) can operate such volumes of data and process given on-demand calculations in reasonable time: few minutes [7]. But if the performance of an installation becomes insufficient (because of quantity of parallel count distinct requests), company has to double or triple its set of hundreds of servers. **So, the scenario 1 is realistic, it is rather simple (because of modern Big Data databases), but it is very-very expensive.**
- Scenario 2: preliminary calculations. This scenario requires storage of raw events for ~2 months (previous one plus current one, up to 60 Tb if single day traffic is ~1 Tb), and a preliminary calculation of all daily metrics (on a daily basis) plus all monthly metrics once a month. As it was described in the previous section, Cartesian product over all values of general attributes of web traffic produce up to 100 mln. of possible metrics. Real cases of Avito demonstrate that 90 percentiles of metrics are empty almost every time and can be omitted. Next 9 percentiles are also not very high (value less than 100 for daily metrics), and can be deliberately ignored. Last percentile, 1 mln. of the biggest metrics, cannot be further reduced. Therefore, this scenario,

in case of maximal possible reduction of metrics, for Avito figures, means 1 mln. of daily metric calculations over 1 bln. events, plus 1 mln. of monthly metric calculations over 30 bln. of events. **Second scenario is possible, if it is possible to calculate monthly metrics in 2.5 seconds per metric (month is 2.5 mln s, for 1 mln. metrics), and daily metrics in 0.08 second per metric.**

Let us illustrate the above-mentioned estimate of $\sim$100 mln. metrics as a result of Cartesian product of all possible values of critical attributes:

```
(2+"ANY"=3 for isRegisteredUser)*(4+"ANY"=5 for Platform)
*(2+"ANY"=3 for isNewVisitor)*(97+"ANY"=98 for Regions)
*(300+"ANY"=301 for EventType)*(75+"ANY"=76 for CategoryID)
=100 548 000.
```

All figures in given formula are real for Avito, they can differ for another internet company. They are listed in this paper only to illustrate how Cartesian product works. How rather small numbers of unique values (less than 300) for just 6 attributes can lead to 100 mln. of combinations. Let us illustrate how figures change, if "ANY" special value is removed from all attributes:

```
(2 for isRegisteredUser)*(4 for Platform)
*(2 for isNewVisitor)*(97 for Regions)
*(300 for EventType)*(75 for CategoryID)
=34 803 600.
```

Therefore, if it is somehow possible to make metrics additive, only lowest level aggregates can be precalculated and stored, reducing the number of required metric calculations several times ($\sim$3 times for example above, better reduction is also possible for more attributes with lower number of unique values).

Count distinct metrics can be estimated by a HyperLogLog [1] algorithm, which is somehow additive (with some limitations). HyperLogLog makes possible a Scenario 3.

– **Scenario 3:** perform aggregations only for lowest level metrics using Hyper-LogLog algorithms, which require several times less aggregations than Scenario 2. Store raw aggregation buffers (auxiliary memory filled in the first part of HyperLogLog algorithms, [1]). In case of count distinct analysis, one has to load elementary raw aggregation buffers (auxiliary memory blocks) and combine them to get required aggregates, using additivity introduced by HyperLogLog algorithm. *Example: to calculate count distinct measure for ANY isRegisteredUser;* with additivity, you can take auxiliary memory buffers for isRegisteredUser==True;* and isRegisteredUser=False;*, combine them using HyperLogLog and estimate total count distinct value using HyperLogLog.* Drawbacks of scenario 3 are as follows: existing papers recommend to use 1.5 kb or 2 kb auxiliary memory buffers. Therefore, although Scenario 3 requires to calculate several times less metrics (only lower level ones), each metric has to be stored not as a single value (big integer value, 8 bytes), but as an auxiliary memory buffer of 2 kb. Therefore, taking 1 mln. metrics from Scenario 2 (deliberate ignoring of empty-ones), and reducing them, for example, 3 times, one will have to store 0.6 Gb per day, or ~1.5 Tb per 5 years, which is cheap enough. Second drawback is caused by the probabilistic nature of the HyperLogLog algorithm. HyperLogLog with a 1.5 kb auxiliary memory buffer gives a 2% of errors for metrics with cardinality less than 1 bln. of unique values [1]. Tests of Avito showed that combining of auxiliary memory buffers leads to multiplication of errors, if the total number of unique values is less than 1 bln., and event worthier errors, if the total number of unique values for the metric exceeds 1 bln. **So, finally, Scenario 3 is few times better in terms of the number of required preliminary aggregations, requires more storage space (non-critical growth), but it gives not an exact number, but an estimate with an unpredictable level of error, which worsens with the growth of estimated values.**

Three scenarios were described and estimated. First one is most convenient for analysts, but extremely expensive in support and scaling. Third one is cheaper, more computationally expensive, but provides approximate rather than exact values. Second scenario is the cheapest one in terms of storage, provides exact results, but requires a huge volume of computations. Further sections describe an approach, which gave Avito the ability to perform all computations required for Scenario 2, with a single thread, for 15 minutes per day.

2.2 Solution to the Problem of Fast and Efficient Count Distinct Calculation

Let the basic terms be defined.

Definition 1. Attribute is a field of events with a fixed set of values. Examples: isNewVisitor [True, False], platform [IOS, Android, Desktop].

Definition 2. Event is a cookie_id plus Date plus list of Attributes: Event::= [cookie_id, Date, A_1, A_2,...,A_N].

Definition 3. Dimension is a copy of attribute with one additional technical value "ANY". Attribute isNewVisitor [True, False] corresponds to Dimension isNewVisitor [True,False, ANY], Attribute platform [IOS, Android, Desktop] corresponds to Dimension platform [IOS, Android, Desktop, ANY].

Definition 4. Accumulator is a list of Dimensions and correspondent values. Accum::=[D_1_val, D_2_val,...,D_N_val]. For example, if we take attributes isNewVisitor [True, False] and platform [IOS, Android, Desktop], there are 2*3=6 possible value combinations. The correspondent dimensions isNewVisitor [True, False, ANY] and platform [IOS, Android, Desktop, ANY] produce 3*4=12 possible accumulators.

Therefore, if there are set of events E, E=[cookie_id, Date, A_1, A_2,...,A_N] and one accumulator [D_1_val, D_2_val,...,D_N_val], then count distinct measure value for the accumulator can be defined as follows:

```
SELECT count(distinct cookie_id) from {E} e
WHERE
    ((e.A_1 = D_1_val) OR (D_1_val='ANY')) AND
    ...
    ((e.A_N = D_N_val) OR (D_N_val='ANY'))
```

If events are stored on disk ordered by cookie_id column, than ordering can be utilized to calculate distinct count for a given single accumulator in a pipelined manner:

1. result = 0, prev_cookie_id = *NULL*
2. *read next* event E *from* disc according to filters (*WHERE* clause *in* the sample above)
3. if *no* more event − return result
4. if E.cookie_id <> prev_cookie_id *then*
   prev_cookie_id = E.cookie_id
   result = result+1
5. goto 2

Listing 1.1. Basic count distinct in a pipelined manner, pseudo-code.

This algorithm is well known, it is extremely efficient in terms of memory and speed, it reads data from a disc only once, but it calculates only one accumulator at a time.

Section 2.1 describes business case of Avito, which requires the calculation of hundreds of millions of accumulators. Even efficient algorithm above, repeated hundred of millions times, becomes inefficient. Hereinafter we describe two versions of efficient algorithm, which does absolutely the same, as a hundred millions times repetition of the simple algorithm above, but in a much more efficient and fast way.

1. Sort *all* required events *by* a cookie_id field (*or initially* store them sorted).
2. *Read all unique* combination of attribute *values from* events (A_1, A_2,...,A_N). Produce *unique* list of possible accumulators according to available attribute *values*. For each accumulator *in* a list (Accum, Accum=[D_1_val, D_2_val,...,D_N_val]), calculate a 64 *bit* hash code, accum_hash = hash64(D_1_val, D_2_val,...,D_N_val]). Therefore, each accumulator can be encoded as a single BigInt rather than a *full* list of dimension's values. Therefore, (accum_hash) of hundred millions of accumulators will take only 800 Mb of space. Same for feature hashes.
3. Create matrix of correspondence between hash(A_1, A_2,...,A_N), or a feature_hash, and a list of accum_hash. This matrix shall be like this:
   CorrMatrix=(feature_hash,(accum_hash)).
   This matrix can be rather big, up to hundred of Gb of space.
4. Create a Result data structure, which stores for each accumulator the metric counter and the last cookie: Result=[accum_hash, counter=0, prev_cookie_id=NULL]. For a hundred millions of accumulators, this structure will take approximately 2.5 Gb of space.
5. Read next event E=[cookie_id, Date, A_1, A_2,...,A_N] from a disc according to filters (filters can also affect features of events not used inside dimensions of accumulators, such as date of event).
6. Calculate feature_hash = hash(A_1, A_2,...,A_N).
7. Get a list of correspondent accumulator hashes from a CorrMatrix according to feature_hash.
   (accum_hash) = CorrMatrix[feature_hash]
8. For each accum_hash in a set, compare E.cookie_id with Result[accum_hash].prev_cookie_id. If equals −− skip, if not:
   Result[accum_hash].prev_cookie_id = cookie_id
   Result[accum_hash].counter += 1
9. Return to step 5, if there are any other events, or output the Result data structure.

Listing 1.2. Efficient algorithm with an excessive correspondent matrix, or a proto-PEAPM (Pipelined Exact Accumulation for Paralleled Measures).

If all events are initially stored ordered by cookie_id (they are stored this way in a Data Warehouse of Avito, no problems caused), then the given algorithm just reads all events twice (steps 2 and 5) and calculates all distinct counters in a single run. There are no other operations beside reading data twice and looking up two hash tables. This approach innovates in combining hash join and pipelined group-by together. Main issue of an algorithm 1 is caused by the necessity to store rather big correspondent matrix, which can be hundreds of Gb. Second issue is the necessity to read events twice (maybe it is possible to read them once?).

Given issues led to the creation of optimized version of an algorithm, which does not require the correspondent matrix (CorrMatrix) and requires only a single read of events. This optimized version was called PEAPM (Pipelined Exact Accumulation for Paralleled Measures).

Until now, we generated hash out of event attributes using non-invertible hash function. In order to get all combinations of attributes for given hash we needed a specific correspondent matrix. This matrix contained for every event feature_hash=hash(A_1, A_2,...,A_N) correlated accumulator hashes we want cookie_id to be counted in. So basically the correlation looks like this:

```
hash(A_1, A_2,...,A_N) -> [ hash(0  ,A_2,...,A_N),
                            hash(A_1,  0,...,A_N),
                            hash(0  ,  0,...,A_N),
                            ... ,
                            hash(0  ,  0,...,  0) ]
```

We just put zero for subset of attributes and calculate hash.

Let us define ihash function as follows:

```
ihash = concat([encode(A_1),encode(A_2),...,encode(A_N)])
```

Now we can interpret hash value itself and there is no need for CorrMatrix anymore as we expose parts of hash corresponding to specific event attribute. With given hash function for each feature_hash = ihash(A_1, A_2,...,A_N) we can generate correspondent accumulator hashes on the fly simply by multiplying hash with mask of 00 and FF for each byte in a hash, so

```
ihash(0, A_2,...,A_N) = ihash(A_1, A_2,...,A_N) * [00,FF,...,FF]
```

With given hash function, the algorithm can be described as follows:

1. Sort events by a cookie_id (or initially store them sorted).
2. **Read next** event E=[cookie_id, *Date*, A_1, A_2,...,A_N] *from* a disc according to filters (filters can also affect features of events, *not* used inside dimensions of accumulators, such *as date* of event.
3. Calculate feature_hash = ihash(A_1, A_2,...,A_N).
4. For each mask_i *in* a constant *set* of masks calculate accumulator product:
   accum_hash = feature_hash * mask_i
5. For each accum_hash obtained *at* the previous step, compare E.cookie_id with
   Result[accum_hash].prev_cookie_id. If equals − skip, if *not*:
   Result[accum_hash].prev_cookie_id = cookie_id
   Result[accum_hash].counter += 1
6. Return to step 2, if there *are any* other events, *or output* the Result data structure.

Listing 1.3. PEAPM (Pipelined Exact Accumulation for Paralleled Measures), efficient algorithm without correspondent matrix, which reads events only once.

Real life depersonalized sample given below.

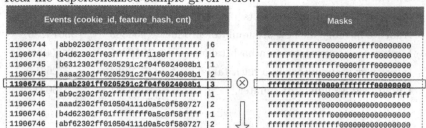

User 11906745 made tree similar actions featuring combination of dimensions of 0xaaab2301ff0205291c2f04f6024008b1. After exposing it to mask 0xffffffff ffff0000ffffffff00000000 we have accumulator of 0xaaab2301ff0200001c2f04f6000 00000:

– aaab for metric contacts,
– 23 for real visitors first time identified by cookie,
– 01 for unauthorized,
– ff for undefined user segment,
– 02 for mobile site,
– 0000 for any business category,
– 1c2f for search category of toys,
– 04f6 for Moscow region,
– 0000 for any city in region,
– 0000 for any specific type of toy.

We check that the last seen id for given accumulator is not equal to 11906745 and advance counter of unique visitors at that specific combination.

2.3 Key Results

C++ implementation of PEAPM (Pipelined Exact Accumulation for Paralleled Measures) is regularly used in Avito to calculate daily unique visitors from a fact table of 700 millions rows. Those rows are combined into 10 million count distinct measures. Such calculation works 15 min on a Vertica database, installed on 20 physical servers. Result of profiling the algorithm run for few seconds is provided below.

Operation	mics systime	ops cnt
Generate accumulator products	372028	33074280
Allocate memory for new accumulator product	454383	33074280
Lookup for accumulator in a hash table	1350395	33074280
Insert new accumulator in a hash table	776984	2726355
Update old accumulator counter	105541	9360217

The most heavy operation there is to perform lookup in a hash table of accumulators. So algorithm performance is based on performance of hash table with short string keys (16–32 bytes long). You are lucky if your hash size is shorter than 16 bytes so you can apply short string optimization and store just two numbers instead of strings. If not you end up with wider hash string you can improve performance by storing hash print generated by hash function alongside with raw hash and reimplementing equality operation.

Comparing with naive materializing of all combinations of dimensions and running built-in pipelined group by algorithm, we have 10+ times optimization on disk IO (100% to around zero value) and CPU usage (20% to 2%) by throwing away unnecessary disk operations.

Alter implementing and testing it in real life on production servers for monthly active users as most heavy metric we have 2.5 h versus 3.5 h on naive algorithm. But we utilize just 1 thread on each node instead of 128 parallel executions we used before, which generated 100% disk utilization, 4 GB RAM versus 40 GB RAM per node. Now it is just a regular select query generating a regular load and runs faster. If we decide to run it even faster we can scale it horizontally by splitting data set on chunks by cookie_id and running calculations in parallel independently or by adding new nodes in a cluster. Code is publicly available at github [8].

3   Conclusions and Future Work

Described business case of Avito, as well as observed count distinct calculation algorithm PEAPM, both demonstrate the following conclusions. The first one, although modern analytic databases theoretically solve Big Data tasks automatically, there are still some issues, especially when the problem cannot be obviously horizontally scaled (because of non-additive nature of count distinct metrics, for example). In such cases, thoughtless usage of modern databases leads to tremendous expenses, delays or inappropriate level of errors. Second conclusion is that it is still profitable to rethink computational problems of computer science as mathematical problems, because it sometimes leads to several orders of magnitude more efficient, fast, and accurate solution. Quantum algorithms (such as Grover and Shor algorithms), which replace a principle of one-by-one processing of elements by processing everything as a whole, looks like as promising candidates for expanding given approach (approach to count distinct calculation) even further. Such research can become very useful as fast as commercially available quantum computers become enough cheap and powerful.

References

1. Flajolet, P., Fusy, É., Gandouet, O., Meunier, F.: HyperLogLog: the analysis of a near-optimal cardinality estimation algorithm. In: Discrete Mathematics and Theoretical Computer Science Proceedings, Nancy, France, AH, pp. 127–146. CiteSeerX 10.1.1.76.4286
2. Wang, G., Zhang, X., Tang, S., Wilson, C., Zheng, H., Zhao, B.Y.: Clickstream user behavior models. ACM Trans. Web **11**(4), 1–37 (2017)
3. Banerjee, A., Ghosh, J.: Clickstream clustering using weighted longest common subsequences. In: Proceedings of Web Mining Workshop at the 1st SIAM Conference on Data Mining (2001)
4. Rönnbäck, L., Regardt, O., Bergholtz, M., Johannesson, P., Wohed, P.: Anchor modeling agile information modeling in evolving data environments. Data Knowl. Eng. **69**(12), 1229–1253 (2010)
5. Golov, N., Ronnback, L.: Big data normalization for massively parallel processing databases. Comput. Stand. Interfaces **54**(2), 86–93 (2017). https://doi.org/10.1016/j.csi.2017.01.009
6. Naspers takes full control of Russian classifieds site Avito in 1.16B dollars deal. https://techcrunch.com/2019/01/28/naspers-avito-1-16-billion/
7. Benchmarks of modern analytical databases for typical click stream analysis scenarious. https://clickhouse.yandex/benchmark.html
8. C++ Vertica extension implementing described algorithm. https://github.com/phil-88/vertica-udf

Implementing HuPf Algorithm
for the Inverse Kinematics of General
6R/P Manipulators

Jose Capco[1(✉)] and Saraleen Mae Manongsong[2]

[1] Research for Symbolic Computation, Johannes Kepler University, Linz, Austria
jcapco@risc.jku.at
[2] Institute of Mathematics, University of the Philippines Diliman,
1101 Quezon City, Philippines
smmanongsong@gmail.com

Abstract. We reformulate and extend the HuPf algorithm (see [7]), which was originally designed for a general 6R manipulator (i.e. 6 jointed open serial chain/robot with only rotational joints), to solve the inverse kinematic (IK) problem of 6R/P manipulators (6-jointed open serial robot with joints that are either rotational or prismatic/translational). For the algorithm we identify the kinematic images of 3R/P chains with a quasi-projective variety in $\mathbb{P}^7$ via dual quaternions. More specifically, these kinematic images are projections of the intersection of a Segre variety with a linear 3-space to an open subset of $\mathbb{P}^7$ (identified with the special Euclidean group SE(3)). We show an easy and efficient algorithm to obtain the linear varieties associated to 3R/P subchains of a 6R/P manipulator. We provide examples showing the linear spaces for different 3R/P chains (a full list of them is available in an upcoming paper). Accompanying the extended HuPf algorithm we provide numerical examples showing real IK solutions to some 6R/P manipulators.

Keywords: Inverse kinematics · Elimination theory ·
Serial manipulator

1 Introduction

The *Study quadric* is given by the $\sum_{i=0}^{3} x_i x_{i+4}$ for points

$$(x_0 : \cdots : x_7) \in \mathbb{P}^7(\mathbb{C})$$

J. Capco—Supported and funded by the Austrian Science Fund (FWF): Project P28349-N32 and W1214-N14 Project DK9.
S. M. Manongsong—Supported by the Office of the Chancellor of the University of the Philippines Diliman, through the Office of the Vice Chancellor for Research and Development, for funding support through the Outright Research Grant.

M. England et al. (Eds.): CASC 2019, LNCS 11661, pp. 78–90, 2019.
https://doi.org/10.1007/978-3-030-26831-2_6

There is a bijection between the special linear group SE(3) and a quasiprojective subset of the quadric via dual-quaternions. In this article we will give a rough idea what this means:

Let $\mathbb{H}$ be the classical (Hamiltonian) skew field of quaternions. One first shows a bijection between the quotient of multiplicative group $\mathbb{H}^*/\mathbb{R}^*$ and the special orthogonal group SO(3) (which is well-known, see [7,10]). The ring $\mathbb{H}[x]/\langle x^2 \rangle$ (indeterminate x commuting with coefficients in $\mathbb{H}$) is called the ring of *dual quaternions* and denoted $\mathbb{D}$. The elements in $\mathbb{D}$ are often written as $p + \epsilon q$ where ϵ is the equivalence class of x and p, q are quaternions in $\mathbb{H}$ (we compare this construction to construction of *dual numbers* in algebraic number theory). Clearly $\mathbb{H}$ is a subring of $\mathbb{D}$ and there is an injection from the elements of SE(3) to the quotient of multiplicative group $\mathbb{D}^*/\mathbb{R}^*$ and it is well-defined if we know how pure translations and pure rotations are mapped. Pure rotations are mapped to $\mathbb{H}$ which is in $\mathbb{D}$. Pure translations are mapped via

$$t \mapsto 0 + \epsilon \left(\frac{t}{2} \right)$$

where $t \in \mathbb{R}^3 \setminus \{\mathbf{0}\}$ is a non-zero translation vector and we regard elements in $\mathbb{H}$ as 4-tuples ($\mathbb{H}$ is a four-dimensional $\mathbb{R}$-algebra). One shows that this fully defines a group homomorphism SE(3) $\rightarrow \mathbb{D}^*$ and if we compose it with the canonical quotient $\mathbb{D}^* \rightarrow \mathbb{D}^*/\mathbb{R}^*$ we even have an injective group homomorphism. Since $\mathbb{D}$ is an eight-dimensional $\mathbb{R}$-algebra we can identify elements in $\mathbb{D}^*/\mathbb{R}^*$ with points in a subset of $\mathbb{P}^7(\mathbb{R})$. In fact the image of the composed map we just described SE(3) $\hookrightarrow \mathbb{D}^*/\mathbb{R}^*$ can be identified with points

$$\{(x_0 : \cdots : x_7) \in \mathbb{P}^7(\mathbb{R}) : x_0 \neq 0, \ldots, x_7 \neq 0 \text{ and } \sum_{i=0}^{3} x_i x_{i+4} = 0\}$$

which is a quasi-projective variety in $\mathbb{P}^7(\mathbb{R})$. For more details, we refer to [7,10] or our upcoming paper.

For engineering and applications, we deal with the real points of the Study quadric, but we also consider complex points because we discuss finiteness and existence of solutions to the inverse kinematic (IK) problem which involve some basic intersection theory where an algebraic closed base field (e.g. when using Hilbert's Nullstellensatz) is important. This allows us to give a general statement whether HuPf algorithm will work or not. We note that, we only assume that we are solving the inverse kinematic of regular values for a general 6R/P manipulator (finite solutions), complicated algorithms in real algebraic geometry like cylindrical algebraic decomposition (CAD, see [3]) to solve the real solutions is not necessary (there are at most 16 solutions, see [9]). Using CAD to describe real higher dimensional solutions (like the inverse kinematic of a redundant, e.g. 7 jointed, robot or the solutions within a kinematic singularity) is however very attractive but is beyond the scope of this work (we plan to investigate this in the future). Knowing if there are real solutions is usually done in the middle of the

HuPf algorithm e.g. when solving for the roots of a resultant to find the coordinates of the middle link of a 6R/P-chain. Thus, our objective in this manuscript is threefold:

1. We describe the algorithm to compute all the parameterized hyperplanes needed for the HuPf algorithm for different permutations of 3R/P joint-types and for different parameterization of joints in the 3R/P sub-chains (as described in [7–9]).
2. We discuss efficient choice of parameterized linear spaces when computing IK. We also discuss preprocessing (e.g. the linear spaces for the left 3-chain can be fully preprocessed) to make real-time IK computation possible.
3. We discuss special cases, i.e. cases when, for a given 3R/P chain, all the parameterized linear spaces defined by each of the parameterizing joints are inside the Study quadric (i.e. the HuPf algorithm may fail because there are infinite solution to the IK problem).

We shall use the Denavit-Hartenberg (DH) convention when describing relationship between two frames. More precisely, the transformation between the frames (of joints) is given by the following rule:

– The z-axis of the reference frame will be the axis of rotation if the joint is revolute or the translational direction if the joint is prismatic
– To obtain the next frame, one starts with a rotation about the z-axis of the reference frame, called the *rotation*, followed by
– a translation along the z-axis of the reference frame, called the *offset*, followed by
– a translation along the x-axis, called the *distance*, followed by
– a rotation about the x-axis, called the *twist*.

It is worth mentioning that, in order to solve IK from a system of polynomial equations, all rotations are parameterized by tangent of half-angles. In short, the transformation between frame i to frame $i + 1$ is given by

$$R_z(v_i)T_z(d_i)T_x(a_i)R_x(l_i)$$

where R_z, T_z, T_x, R_x are rotations or translations with respect to z- or x-axis parameterized by tangent of half-angle rotation v_i, offset d_i, distance a_i and tangent of half-angle twist l_i of the i-th frame (i-th joint). More thorough discussion on DH-parameters and the DH-convention is given in [11].

In the first step of the HuPf algorithm, one wants to compute a set of at least four hyperplanes which describes a parameterized linear space whose projection to $\mathbb{P}^7$ contains the image of the kinematic map. The way to compute this for a

6R manipulator is discussed in [9]. We give this algorithm when prismatic joints are also involved:

Algorithm 1. Computing 3-chain hyperplane parameterized by joint $k \in \{1, 3\}$

Input: DH-parameter of a 3-chain, $k \in \{1, 3\}$, **boolean** *parameterized*
Output: Hyperplane of 2-chains (non-parameterized) or hyperplane pencils
    – parameterized by joint k in $\mathbb{P}^7$ (3-chain)

1 Let ν_i, ν_j, ν_k be the three joint parameters with $i, j \in \{1, 2, 3\}$
2 Compute the forward kinematic image $\sigma = (s_0 : \cdots : s_7) \in \mathbb{P}^7$ of the 2-chain not
    containing k, with coordinates parameterized by ν_i, ν_j
3 Write each coordinate of σ as a 2-variate polynomial in $\mathbb{R}[\nu_i, \nu_j]$ with
    multi-degree at most $(1, 1)$
4 Create a 4×8 matrix M with l-th column having entries which are coefficients
    of $1, \nu_i, \nu_j, \nu_i \nu_j$ respectively in s_l
5 Compute the null space of the matrix M spanned by linearly independent
    vectors $w_1, \ldots, w_m$ each of length 8
6 **if not** *parameterized* **then**
7 | **return** the linear forms $H_l : \sum_{n=0}^{7} w_{ln} x_n \in \mathbb{R}[x_0, \ldots, x_7]$, where w_{ln} is the
    | n-th coordinate of a representative of w_l, with $l = 1, \ldots, m$
8 **end**
9 **else**
10 | **if** $k = 1$ **then**
11 | | Regard $(x_0 : \cdots : x_7)$ as a dual quaternion and pre-multiply it with a
    | | dual quaternion defined by motion of ν_1 to get $(x'_0 : \cdots : x'_7)$
12 | **end**
13 | **else**
14 | | Regard $(x_0 : \cdots : x_7)$ as a dual quaternion and post-multiply it with a
    | | dual quaternion defined by motion of ν_3 to get $(x'_0 : \cdots : x'_7)$
15 | **end**
16 | **return** the linear forms $H_l : \sum_{n=0}^{7} w_{ln} x'_n (x_0, \ldots, x_7) \in \mathbb{R}[x_0, \ldots, x_7]$, with
    | $l = 1, \ldots, m$
17 **end**

In Algorithm 1 (and further algorithms), the joint parameter ν_i is d_i if the i-th joint is prismatic otherwise it is v_i (i.e. if the i-th joint is revolute).

Algorithm 2. Computing 3-chain hyperplane parameterized by second joint

---

Input: DH-parameter of a 3-chain
Output: Hyperplane parameterized by 2nd joint in $\mathbb{P}^7$ corresponding to 3-chain motion

1 Let $\nu := (\nu_1, \nu_2, \nu_3)$ be the three joint parameters
2 Compute the forward kinematic $\sigma = (s_0 : \cdots : s_7) \in \mathbb{P}^7$ of the 3-chain with coordinates parameterized by all the joint parameters
3 Write each coordinate of σ as a 3-variate polynomial $\mathbb{R}[\nu_1, \nu_2, \nu_3]$ each of multi-degree at most $(1, 1, 1)$
4 Create an empty 12×16 matrix M
5 Set the $l \leq 8$ column of M have entries which are coefficients of ν^α, where $\alpha \leq (1, 1, 2)$, of s_l
6 Set the $l > 8$ columns of M have entries which are coefficients of ν^α, where $\alpha \leq (1, 1, 2)$, of $\nu_2 s_l$
7 Compute the null space of the matrix M spanned by vectors $w_1, \ldots, w_m$ each of length 16
8 Choose independent linear forms from the linear forms
$$H_l : \sum_{n=0}^{7} (w_{l,n} + w_{l,n+8}\nu_2)x_n \text{ with } l = 1, \ldots, m$$
9 **return** these independent linear forms

To compute the special cases, i.e. case for which hyperplane parameterized by any of the joints lie inside the Study quadric, one follows almost the same procedure as Algorithms 1 and 2. However, since now the hyperplane may not lie in general position (i.e. the DH-parameters may be very specific for them to all lie in the Study quadric), care must be taken when computing the null space of the coefficient matrices.

2   The Hyperplanes

For an efficient algorithm it is vital to choose the linear space that is described by linear forms with least complexity. Usually the linear space parameterized by the second joint is the most complex case. So one first computes linear space parameterized by the first joint and look for (DH parameter) conditions for which this linear space may lie in the Study quadric (say for RRR, this condition is when the twist half-angle tangent satisfies $l_2 = 0$ or the offset satisfies $a_2 = 0$). In the case that the linear space parameterized by ν_1 (see Algorithm 1) cannot be chosen, we look at the third joint and immediately apply the condition (so in our example if $l_2 = 0$ or $a_2 = 0$ then we immediately apply this in the set of equations), we then use Algorithm 1 to find the hyperplanes parameterized by ν_3. Finally if for this case the linear space lies in the Study quadric (in our example, this would be either $a_1 = 0$ or $l_1 = 0$ with the additional condition from the first investigation that $a_2 = 0$ or $l_2 = 0$), we immediately apply these conditions and look at Algorithm 1 to get hyperplane parameterized by ν_2. To summarize, the hyperplanes we provide cannot be used in general case but in cases of increasing level of complexity. One should use these steps in order when chosing the computed linear spaces in our work:

1. Check if $T(\nu_1)$ is applicable (for RRR we check if $a_2 \neq 0$ and $l_2 \neq 0$)
2. If not, check if $T(\nu_3)$ is applicable (for RRR, we check if $(a_2 = 0 \vee l_2 = 0) \wedge (a_1 \neq 0 \wedge l_2 \neq 0)$)
3. If not, supposing we are not in a *special case* (i.e. kinematic image of the 3-chain does not describe a planar, pure translation or spherical motion) then we use $T(\nu_2)$.

Because of the level of complexity of the equations describing the general $T(\nu_2)$, inverse kinematic computation using the algorithm will be much slower. So one tries to avoid this if either $T(\nu_1)$ or $T(\nu_3)$ is possible. Due to the length of the equations, we do not show all of the simplified $T(\nu_2)$ for all 3R/P chains. But we can show for all the subcases in the 3R-chain (for $T(\nu_1)$ and $T(\nu_3)$ one can look at other literatures which will have them in detail) (Table 1).

For the 3R/P chain these are the conditions to not choose $T(\nu_1)$ (in Step 1. above) respectively $T(\nu_3)$ (in Step 2. above, assuming we cannot choose $T(\nu_1)$):

Table 1. DH conditions for not choosing the linear space. For a given 3R/P if the condition in column 2 is satisfied, we have to look at column 3 and if that is also satisfied we may possibly choose $T(\nu_2)$

3R/P	$\not{X}T(\nu_1)$	$\not{X}T(\nu_3)$
RRP	$l_2 = \pm 1$	$a_1 = 0$ or $l_1 = 0$
RPR	$l_2 = \pm 1$	$l_1 = \pm 1$
RPP[a]	All	$l_1 = \pm 1$
PRR	$a_2 = 0$ or $l_2 = 0$	$l_1 = \pm 1$
PRP	$l_2 = \pm 1$	$l_1 = \pm 1$
PPR[a]	$l_2 = \pm 1$	All
RRR	$a_2 = 0$ or $l_2 = 0$	$a_1 = 0$ or $l_1 = 0$

[a] we exclude the case that the two consuctive prismatic joints allow movement in the same direction (i.e. twist angle is 0). This is a degenerate case that is not interesting and in this case a solution to the IK problem imply infinite solutions so that classical HuPf algorithm is not applicable.

Notice in the above table we disregarded PPP because this 3-chain describes a purely translational motion so its kinematic image lies in a 3-space living in the Study quadric. In fact, increasing prismatic joints should theoretically make it easier for us to compute inverse kinematics.

Here we show simplified parameterized (by a selected joint parameter) linear spaces of some 3R/P chains. By *simplified* we mean the following: the coordinates

$(x_0 : \cdots : y_3)$ are given up to post- or pre- multiplication by some fixed element in SE(3). For instance with the DH-parameter the kinematic image of an 3R chain parameterized by v_1 is given by:

$$R_z(v_1)T_z(d_1)T_x(a_1)R_x(l_1)M$$

where M is the kinematic image of a 2R-chain. But we can simply commute R_z and T_z and consider the image of

$$R_z(v_1)T_x(a_1)R_x(l_1)M$$

reducing the number of variables so we are able to display simpler linear forms (the actual hyperplane can be obtained by an easy transformation, in this case by a premultiplication of $T_z(d_1)$). Though, in all our cases, we assume for brevity that d_1 is always 0 (otherwise another simple transformation will yield the inverse kinematic).

Often $T(v_2)$ is very complicated and involves many subcases (in order to improve efficiency in the C++ implementation and the algorithm). So we will only show this in RRR and RRP case. For RRR the linear spaces $T(v_1)$ and $T(v_3)$ are well-studied (see [7–9]) so we will not show this.

2.1 RRR Hyperplanes, $T(v_2)$

For $T(v_2)$ with $[a_1, a_2] = [0, 0]$:

$H_1 : d_2 l_1 l_2 v_2 x_3 + d_2 l_1 l_2 x_0 - d_2 v_2 x_3 + d_2 x_0 - 2 l_1 l_2 v_2 x_4 +_2 l_1 l_2 x_7 - 2 v_2 x_4 - 2 x_7$

$H_2 : d_2 l_1 v_2 x_2 + d_2 l_1 x_1 + d_2 v_2 x_2 l_2 - d_2 x_1 l_2 - 2 l_1 v_2 x_5 +_2 l_1 x_6 +_2 v_2 l_2 x_5 +_2 l_2 x_6$

$H_3 : - d_2 l_1 v_2 x_1 + d_2 l_1 x_2 - d_2 v_2 x_1 l_2 - d_2 x_2 l_2 - 2 l_1 v_2 x_6 - 2 l_1 x_5 +_2 v_2 l_2 x_6$
  $- 2 l_2 x_5$

$H_4 : - d_2 l_1 l_2 v_2 x_0 + d_2 l_1 l_2 x_3 + d_2 v_2 x_0 + d_2 x_3 - 2 l_1 l_2 v_2 x_7 - 2 l_1 l_2 x_4 - 2 v_2 x_7$
  $+_2 x_4$

For $T(v_2)$ with $[a_1, l_2] = [0, 0]$:

$$H_1 : a_2 l_1 v_2 x_0 - a_2 l_1 x_3 - v_2 x_3 d_2 - 2 v_2 x_4 + x_0 d_2 - 2 x_7$$
$$H_2 : - a_2 v_2 x_1 - a_2 x_2 + v_2 x_2 d_2 l_1 - 2 v_2 l_1 x_5 + x_1 d_2 l_1 +_2 l_1 x_6$$
$$H_3 : - a_2 v_2 x_2 + a_2 x_1 - v_2 x_1 d_2 l_1 - 2 v_2 l_1 x_6 + x_2 d_2 l_1 - 2 l_1 x_5$$
$$H_4 : a_2 l_1 v_2 x_3 + a_2 l_1 x_0 + v_2 x_0 d_2 - 2 v_2 x_7 + x_3 d_2 +_2 x_4$$

For $T(v_2)$ with $[l_1, a_2] = [0, 0]$:

$$H_1 : a_1 l_2 v_2 x_0 - a_1 l_2 x_3 - v_2 x_3 d_2 - 2 v_2 x_4 + x_0 d_2 - 2 x_7$$
$$H_2 : - a_1 v_2 x_1 + a_1 x_2 - v_2 x_2 d_2 l_2 - 2 v_2 l_2 x_5 + x_1 d_2 l_2 - 2 l_2 x_6$$
$$H_3 : - a_1 v_2 x_2 - a_1 x_1 + v_2 x_1 d_2 l_2 - 2 v_2 l_2 x_6 + x_2 d_2 l_2 +_2 l_2 x_5$$
$$H_4 : a_1 l_2 v_2 x_3 + a_1 l_2 x_0 + v_2 x_0 d_2 - 2 v_2 x_7 + x_3 d_2 +_2 x_4$$

For $T(v_2)$ with $[l_1, l_2] = [0, 0]$:

$$H_1 : -x_1$$
$$H_2 : -x_2$$
$$H_3 : -d_2 x_3 - 2x_4$$
$$H_4 : d_2 x_0 - 2x_7$$

2.2   RRP Hyperplanes

For $T(v_1)$:

$$H_1 : l_2 x_0 - x_1$$
$$H_2 : -l_2 x_3 + x_2$$
$$H_3 : a_2 l_2^2 x_0 - a_2 x_0 - 2l_2 x_4 - 2x_5$$
$$H_4 : a_2 l_2^2 x_2 - a_2 x_2 - 2l_2^2 x_7 - 2l_2 x_6$$

For $T(d_3)$:

$$H_1 : a_1 l_1 x_0 - 2x_4$$
$$H_2 : -a_1 x_1 - 2l_1 x_5$$
$$H_3 : -a_1 x_2 - 2l_1 x_6$$
$$H_4 : a_1 l_1 x_3 - 2x_7$$

For $T(v_2)$ with $[a_1, l_2] = [0, 1]$:

$$H_1 : -l_1 v_2 x_0 - l_1 v_2 x_1 + l_1 x_2 + l_1 x_3 + v_2 x_0 - v_2 x_1 - x_2 + x_3$$
$$H_2 : -l_1 v_2 x_2 - l_1 v_2 x_3 - l_1 x_0 - l_1 x_1 - v_2 x_2 + v_2 x_3 - x_0 + x_1$$
$$\begin{aligned} H_3 : &2a_2 l_1^2 v_2 x_0 - 2a_2 l_1^2 x_2 - 2a_2 l_1 v_2 x_0 + 2a_2 l_1 x_2 + l_1^3 v_2 x_2 d_2 + l_1^3 v_2 x_4 - l_1^3 v_2 x_5 \\ &+ l_1^3 d_2 x_1 + l_1^3 x_6 - l_1^3 x_7 + l_1^2 v_2 x_2 d_2 - l_1^2 v_2 x_4 - l_1^2 v_2 x_5 + 2l_1^2 x_0 d_2 + l_1^2 d_2 x_1 \\ &- l_1^2 x_6 - l_1^2 x_7 + l_1 v_2 x_2 d_2 - l_1 v_2 x_4 + l_1 v_2 x_5 + 2l_1 x_0 d_2 - l_1 d_2 x_1 - l_1 x_6 \\ &+ l_1 x_7 + v_2 x_2 d_2 + v_2 x_4 + v_2 x_5 - d_2 x_1 + x_6 + x_7 \end{aligned}$$
$$\begin{aligned} H_4 : &2a_2 l_1^2 v_2 x_2 + 2a_2 l_1^2 x_0 + 2a_2 l_1 v_2 x_2 + 2a_2 l_1 x_0 - l_1^3 v_2 x_0 d_2 + l_1^3 v_2 x_6 - l_1^3 v_2 x_7 \\ &+ l_1^3 d_2 x_3 - l_1^3 x_4 + l_1^3 x_5 + l_1^2 v_2 x_0 d_2 + l_1^2 v_2 x_6 + l_1^2 v_2 x_7 - 2l_1^2 x_2 d_2 - l_1^2 d_2 x_3 \\ &- l_1^2 x_4 - l_1^2 x_5 - l_1 v_2 x_0 d_2 - l_1 v_2 x_6 + l_1 v_2 x_7 + 2l_1 x_2 d_2 - l_1 d_2 x_3 + l_1 x_4 \\ &- l_1 x_5 + v_2 x_0 d_2 - v_2 x_6 - v_2 x_7 + d_2 x_3 + x_4 + x_5 \end{aligned}$$

For $T(v_2)$ with $[a_1, l_2] = [0, -1]$:

$H_1 : l_1 v_2 x_0 - l_1 v_2 x_1 + l_1 x_2 - l_1 x_3 + v_2 x_0 + v_2 x_1 + x_2 + x_3$

$H_2 : l_1 v_2 x_2 - l_1 v_2 x_3 - l_1 x_0 + l_1 x_1 - v_2 x_2 - v_2 x_3 + x_0 + x_1$

$H_3 : - 2 a_2 l_1^2 v_2 x_0 - 2 a_2 l_1^2 x_2 - 2 a_2 l_1 v_2 x_0 - 2 a_2 l_1 x_2 + l_1^3 v_2 x_2 d_2 - l_1^3 v_2 x_4$
$\quad - l_1^3 v_2 x_5 + l_1^3 d_2 x_1 + l_1^3 x_6 + l_1^3 x_7 - l_1^2 v_2 x_2 d_2 - l_1^2 v_2 x_4 + l_1^2 v_2 x_5$
$\quad + 2 l_1^2 x_0 d_2 - l_1^2 d_2 x_1 + l_1^2 x_6 - l_1^2 x_7 + l_1 v_2 x_2 d_2 + l_1 v_2 x_4 + l_1 v_2 x_5$
$\quad - 2 l_1 x_0 d_2 - l_1 d_2 x_1 - l_1 x_6 - l_1 x_7 - v_2 x_2 d_2 + v_2 x_4 - v_2 x_5 + d_2 x_1 - x_6 + x_7$

$H_4 : - 2 a_2 l_1^2 v_2 x_2 + 2 a_2 l_1^2 x_0 + 2 a_2 l_1 v_2 x_2 - 2 a_2 l_1 x_0 - l_1^3 v_2 x_0 d_2 - l_1^3 v_2 x_6$
$\quad - l_1^3 v_2 x_7 + l_1^3 d_2 x_3 - l_1^3 x_4 - l_1^3 x_5 - l_1^2 v_2 x_0 d_2 + l_1^2 v_2 x_6 - l_1^2 v_2 x_7$
$\quad - 2 l_1^2 x_2 d_2 + l_1^2 d_2 x_3 + l_1^2 x_4 - l_1^2 x_5 - l_1 v_2 x_0 d_2 + l_1 v_2 x_6 + l_1 v_2 x_7$
$\quad - 2 l_1 x_2 d_2 - l_1 d_2 x_3 + l_1 x_4 + l_1 x_5 - v_2 x_0 d_2 - v_2 x_6 + v_2 x_7 - d_2 x_3 - x_4 + x_5$

For $T(v_2)$ with $[l_1, l_2] = [0, 1]$:

$$H_1 : x_0 - x_1$$
$$H_2 : x_2 - x_3$$
$$H_3 : - d_2 x_2 - x_4 - x_5$$
$$H_4 : d_2 x_0 - x_6 - x_7$$

For $T(v_2)$ with $[l_1, l_2] = [0, -1]$:

$$H_1 : - x_0 - x_1$$
$$H_2 : - x_2 - x_3$$
$$H_3 : - d_2 x_2 + x_4 - x_5$$
$$H_4 : d_2 x_0 + x_6 - x_7$$

The other linear forms for other 3-chain joint types can be similarly computed. They (esp. RPR and PRR chains) can found in [1] and in an upcoming paper.

2.3 The Right Chain

We have so far displayed linear forms describing the linear space for the left chain. Our point of reference is the base frame. Hyperplane from parameters of right chain can also be computed using the following algorithm:

Due to lack of space, we will not show the linear forms for the right-chain. This is available in the dataset [1]. In [2] we also include a Giac implementation of the algorithms (we us the giacpy python wrapper of Giac, see [5,6]).

Clearly, in the algorithms presented the computation of resultant is the 'bottleneck' (the resultant that one compute is that of two bivariate polynomials,

---

Algorithm 3. Computing hyperplanes parameterized by a joint in the right reversed 3-chain

---

Input: DH-parameter of the right 3-chain, end-effector transformation τ, $T(\nu_i)$
Output: $T(\nu_{7-i})$ described by DH-parameter of the right-chain
1 Make the following substitutions in the equations of $T(\nu_i)$ (say for RRP):

$$[a_1, a_2, a_3, l_1, l_2, l_3, v_1, v_2, v_3, d_1, d_2, d_3] \rightarrow$$

$$[-a_5, -a_4, 0, -l_5, -l_4, 0, -v_6, -v_5, -v_4, -d_6, -d_5, -d_4]$$

2 Replace $\sigma := (x_0 : \cdots : x_7)$ with $\tau\sigma$
3 **return** new linear forms describing $T(\nu_{7-i})$ having reversed joint-types (say for PRR)

---

with maximum total degree 14). However, we do not focus on complexity analysis because neither the degree (at most 14) nor the number of variables (two) will vary when computing the inverse kinematics of a general 6R/P manipulator using the HuPf algorithm. Moreover, for the magnitude of our problem even a naive resultant computation (e.g. using Sylvester matrix) would suffice and be fast enough. In fact, HuPf runtime is fast. For instance, the maximum runtime for solving one inverse kinematic query (C++ parallelized implementation) using HuPf in an Intel Core i5-6200U processor is 35ms. This is a tolerable number even by industrial standards. However in the future, we plan to study an extended version of HuPf algorithm devised to solve inverse kinematics of redundant manipulators where we will need to focus on time complexity as joint number varies.

3 The Inverse Kinematic Algorithm with an Example

Finally we can show HuPf algorithm for solving inverse kinematics of general 6R/P manipulators. We assume that the end-effector pose is reachable (i.e. the inverse kinematics has a real solution) and that the IK solutions are finite (for 6-jointed manipulators we only want solutions of regular values).

One can show, with assumption that the input is a reachable end-effector pose with finite IK solution, the algorithm ends successfully. This reasoning is also used to prove that the for loop in Line 2 will break successfully with one of the x_0, x_1, x_3, x_4 non-zero. Finally the finiteness of the IK solution will also guarantee us that f and g in Lines 7 and 8 are not identically 0 (it is not a constant because we have a solution to the IK problem).

We now show an example of an IK problem that we solve using Algorithm 4. Consider a 2R2P2R manipulator (i.e. a serial manipulator consisting of first two joints that are revloute, third and fourth joints that are prismatic and last two joints that are revolute) with DH-parameters given in the table below.

Algorithm 4. (HuPf) Computing IK of a general 6R/P manipulator for a general pose

---

Input: A reachable end-effector pose $\sigma \in$ SE(3) whose IK solution is finite. Parameterized linear spaces for the left-chain (see Algorithms 1 and 2) and right-chain (dependent on σ, see Algorithm 3) by joint-parameters μ and ν.

Output: IK solutions to σ

1 Let $\mathcal{L} := \{l_i\}_{i=1}^8$ be the linear forms describing the two linear spaces in the input

2 **for** $j = 1, \ldots, 8$ **do**

3   Solve for $(x_0 : \cdots : x_7) \in \mathbb{P}^7(\mathbb{C}(\mu, \nu))$ satisfying all linear forms in $\mathcal{L} \backslash \{l_j\}$

4   **if** $x_0 \neq 0$ *or* $x_1 \neq 0$ *or* $x_2 \neq 0$ *or* $x_3 \neq 0$ **then** break

5 **end**

6 Without loss of generality we may assume $x_0, \ldots, x_7 \in \mathbb{C}[\mu, \nu]$

7 Substitute $x_0, \ldots, x_7$ into l_j to obtain a polynomial $f(\mu, \nu) \in \mathbb{C}[\mu, \nu] \backslash \mathbb{C}$

8 Set $g := \sum_{i=0}^3 x_i x_{i+3}$ which is generally a polynomial in $\mathbb{C}[\mu, \nu] \backslash \mathbb{C}$

9 Common zeros of f and g are computed via resultant and elimination theory

10 **foreach** *common zero* (μ', ν') *of* f *and* g **do**

11   Let $x_0', \ldots, x_7'$ be the evaluations of $x_0, \ldots, x_7$ (pose of the middle link)

12   **foreach** *joint* λ *not* μ *and* ν **do**

13   There is a unique solution of λ via backsubstition of $x_0', \ldots, x_7'$ into a linear form paramaterized by λ from Algorithms 1, 2 and 3

14   **end**

15 **end**

16 **return** all joint values

Table 2. DH parameters for a 2R2P2R manipulator in our example. Parameters involving twist or rotation are tangent of half-angles.

i	v_i	d_i	a_i	l_i
1	*	0	0.2	0.2035
2	*	0.3	0.2	0.2035
3	-0.4142	*	0.3	0.4142
4	0.7133	*	0.4	0.3153
5	*	0.3	0	0.1763
6	*	0	0	0

If we apply Algorithm 4 we obtain 12 solutions to the IK problem for a generic pose of the end-effector. In our case we chose a pose given by joint values $10, 30, 0.1, -0.1, 31, 55$ (revolute joint values given in degree) or $0.0875, 0.1763, 0.1, -0.1, 0.2773, 0.5206$ (revolute joint values given as tangent of half-angles). Generally only 4 of these solutions are real, the real solutions are given in the table (Table 3) below and it is illustrated in the figure (Fig. 1) below

Table 3. Real inverse kinematics solutions to the given 2RP3R manipulator. Revolute joint values are given in tangent of half-angles.

	Solution 1	Solution 2	Solution 3	Solution 4
v_1	−0.0374	0.0875	1.2875	2.0551
v_2	0.7075	0.1763	−0.6262	−0.7273
d_3	−0.3786	0.1	0.0119	−0.3336
d_4	0.6391	−0.1	0.5796	1.03798
v_5	−1.4516	0.2773	0.0878	−0.3247
v_6	−9.3357	0.5206	1.3693	5.1233

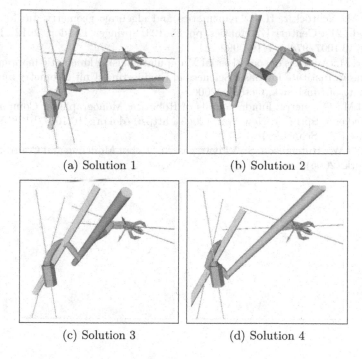

(a) Solution 1   (b) Solution 2

(c) Solution 3   (d) Solution 4

Fig. 1. Real inverse kinematic solutions to a certain end-effector pose of the 2R2P2R chain given in Table 2. The (light-blue) lines correspond to axis of rotations of the revolute joints and the (yellow) thick rods are the direction where the prismatic joint translates the links. (Color figure online)

References

1. Capco, J., Manongsong, S.M.: Linear Spaces Associated to 3R/P Kinematic Image [Data set]. Zenodo (2019). https://doi.org/10.5281/zenodo.3147394
2. Capco, J., Manongsong, S.M.: Code: Implementing HuPf Algorithm for the inverse Kinematics of General 6R/P Manipulators. Zenodo (2019). https://doi.org/10.5281/zenodo.3157441

3. Collins, G.E.: Quantifier elimination by cylindrical algebraic decomposition - twenty years of progress. In: Caviness, B.F., Johnson, J.R. (eds.) Quantifier Elimination and Cylindrical Algebraic Decomposition, pp. 8–23. Springer, Vienna (1998). https://doi.org/10.1007/978-3-7091-9459-1_2
4. Cox, D., Little, J., O'Shea, D.: Ideals, Varieties and Algorithms, 3rd edn. Springer, New York (2007). https://doi.org/10.1007/978-0-387-35651-8
5. Parisse B., De Graeve R.: Giac/Xcas. https://www-fourier.ujf-grenoble.fr/~parisse/giac.html. Accessed February 2019
6. Han F.: giacpy. https://gitlab.math.univ-paris-diderot.fr/han/giacpy. Accessed February 2019
7. Husty, M., Pfurner, M., Schröcker, H.-P.: A new and efficient algorithm for the inverse kinematics of a general serial 6R manipulator. Mech. Mach. Theory **42**, 66–81 (2007)
8. Husty, M., Schröcker, H.-P.: Kinematics and algebraic geometry. In: McCarthy, J.M. (ed.) 21st Century Kinematics, pp. 85–123. Springer, London (2012). https://doi.org/10.1007/978-1-4419-0999-2_4
9. Pfurner, M.: Analysis of spatial serial Manipulators using kinematic mapping. Doctoral thesis, Institute for Basic Sciences in Engineering, Unit Geometry and CAD, University of Innsbruck, October 2006
10. Selig, J.M.: Geometric Fundamentals of Robotics. Monographs in Computer Science, 2nd edn. Springer, New York (2005). https://doi.org/10.1007/b138859. (Ed.: D. Gries, F.B. Schneider)
11. Spong, M.W., Hutchinson, S., Vidyasagar, M.: Robot Modeling and Control. Wiley, New York (2005)

Symbolic-Numerical Algorithm for Large Scale Calculations the Orthonormal SU(3) BM Basis

A. Deveikis[1], A. A. Gusev[2]([✉]), V. P. Gerdt[2], S. I. Vinitsky[2,3], A. Góźdź[4], A. Pędrak[5], and Č. Burdik[6]

[1] Department of Applied Informatics, Vytautas Magnus University,
Kaunas, Lithuania
algirdas.deveikis@vdu.lt
[2] Joint Institute for Nuclear Research, Dubna, Russia
gooseff@jinr.ru
[3] RUDN University, 6 Miklukho-Maklaya, 117198 Moscow, Russia
[4] Institute of Physics, Maria Curie-Skłodowska University, Lublin, Poland
[5] National Centre for Nuclear Research, Warsaw, Poland
[6] Department of Mathematics, Faculty of Nuclear Sciences and Physical Engineering,
Czech Technical University, Prague, Czech Republic

Abstract. In this paper we proposed a new symbolic, non-standard recursive and fast orthonormalization procedure of linearly independent vectors but as in other approaches not orthonormal based on the Gram-Schmidt orthonormalization algorithm. Our adaptation of the Gram-Schmidt orthonormalization procedure provide simple analytic formulas for the SU(3) Bargmann-Moshinsky basis orthonormalization coefficients and do not involve any square root operation on the expressions coming from the previous iterative computation steps. This distinct features of the proposed orthonormalization algorithm may make the large scale symbolic calculations feasible. We demonstrate efficiency of our procedure by benchmark large-scale calculations of the non-canonical BM basis with the highest weight vectors of SO(3) irreducible representations.

1 Introduction

Despite the significant number of works on the development of algorithms and construction programs in both analytical and numerical form of the orthonormalized Bargman-Moshinsky basis, there are still no efficient and cost-effective algorithms and programs for its construction and calculation with its help, tensor operators necessary for constructing Hamiltonians by a collective of a model of a nucleus with tetrahedral symmetry under study by modern experiments [1]. Creation of such algorithms and programs is an actual problem in the field of Computer Algebra in Scientific Computing.

In our previous papers [2,3] noted below as I and II, we started to study optimal ways of building up fast versions of Gram-Schmidt orthonormalization

© Springer Nature Switzerland AG 2019
M. England et al. (Eds.): CASC 2019, LNCS 11661, pp. 91–106, 2019.
https://doi.org/10.1007/978-3-030-26831-2_7

procedure of the non-canonical BM basis in computer algebra systems and its application to scientific computing of a spectrum of the SU(3) collective nuclei models [4].

In this paper we developed the new symbolic, non-standard recursive and fast orthonormalization procedure of linearly independent vectors but as in other approaches not orthonormal, based on the Gram-Schmidt orthonormalization algorithm. Our adaptation of Gram-Schmidt orthonormalization procedure provide simple analytic formulas for BM basis orthonormalization coefficients and do not involve any square root operation on expressions coming from the previous iterative computation steps. This distinct features of the proposed orthonormalization algorithm may make the large scale symbolic calculations feasible. We demonstrate efficiency of our procedure by benchmark large-scale calculations of the non-canonical SU(3) BM basis [5–7] with the highest weight vectors of SO(3) irreducible representations(irreps). Note, the SU(3) irreps. presented in the form of expansions over the BM basis [6] have a wide range of applications in nuclei physics and quantum optics.

The structure of the paper is following. In the first section we present new *symbolic-numerical algorithm* of the Gram-Schmidt orthonormalization realized on example of non-canonical BM basis in a form of the program implemented in the computer algebra system Wolfram Mathematica 10.1. In the second section we present the best economical algorithm for generation of matrix of tensor operators and algebraic eigenvalue problem using calculated orthonormal BM basis as input and show final results of calculation of a spectrum of the SU(3) collective nuclear models. In conclusion we give a resume and point out some important problems for further applications of proposed algorithms.

2   Symbolic-Numerical Orthonormalization Algorithm

We start from the BM states constructed in the papers I and II:

$$|u_\alpha\rangle \equiv \left| \begin{matrix} (\lambda, \mu)_B \\ \alpha, L, L \end{matrix} \right\rangle, \tag{1}$$

which are linearly independent but as in other approaches not orthonormal. The quantum numbers λ, $\mu = 0, 1, 2, \ldots$ label irreducible representations of the SU(3) group. We assume that $\lambda \geq \mu$. The labels L, M are the quantum numbers of angular momentum and its projection (in our case $M = L$); α is the additional index required to distinguish equivalent irreducible representations of SO(3) appearing in a given irreducible representation of SU(3), the problem is not multiplicity free.

The orthogonalized BM states $|\psi_\alpha\rangle$ may be expressed in terms of the orthonormalized BM states $|\phi_\alpha\rangle$ as [7]:

$$|\psi_\alpha\rangle = -|u_\alpha\rangle + \sum_{\alpha'=\alpha+1}^{\alpha_{max}} c_{\alpha\alpha'}|\phi_{\alpha'}\rangle, \text{ when } 0 \leq \alpha < \alpha_{max}, \tag{2}$$

here $\alpha_{\max}$ is a number of linearly independent BM states $|u_\alpha\rangle$, and $c_{\alpha\alpha'}$ are linear coefficients. The orthonormalization process (2) starts (somewhat deliberately) by taking $|\psi_{\alpha_{\max}}\rangle = |u_{\alpha_{\max}}\rangle$.

The goal of this paper is to perform orthonormalization of the BM states $|u_\alpha\rangle$:

$$|\phi_i\rangle = \sum_{\alpha=0}^{\alpha_{\max}} A_{i,\alpha}|u_\alpha\rangle. \tag{3}$$

Here multiplicity index i is introduced to differentiate the orthonormalized BM states and takes the same range of values as α. The symbols $A_{i,\alpha}$ denotes matrix elements of the upper triangular matrix of the BM basis orthonormalization coefficients. These coefficients fulfill the following condition

$$A_{i,\alpha} = 0, \quad \text{if } i > \alpha. \tag{4}$$

In this paper we developed an analytical orthonormalization procedure based on the Gram-Schmidt orthonormalization algorithm (GSOA). For explicit construction of the orthonormalized BM basis let us consider step by step the *symbolic algorithm*.

Step 1. First step needs to perform initial setup and check the consistency of the input. The maximum possible value of α for a given μ is $\alpha_{\max}$. It is given by

$$\alpha_{\max} = \begin{cases} \frac{\mu}{2}, & \mu \text{ even,} \\ \frac{\mu-1}{2}, & \mu \text{ odd.} \end{cases} \tag{5}$$

The maximum value of L of the BM state is defined by the expression

$$L_{\max} = \mu - 2\alpha + \lambda - \beta, \tag{6}$$

where

$$\beta = \begin{cases} 0, & \lambda + \mu - L \text{ even,} \\ 1, & \lambda + \mu - L \text{ odd.} \end{cases} \tag{7}$$

To have a consistent input $L \leq L_{\max}$. From the expressions (6) and (7) it follows, that for some L values sufficiently close to the $L_{\max}$, the α values may be less than $\alpha_{\max}$ or even not exist. So, for every particular L value the expressions (6) and (7) allows to find the maximum value of α for which there exist the BM state i.e. $\alpha_{\max K}$.

At the same time there exists the lower boundary condition $L_{\min} \leq L$ that should be evaluated for every particular value of α. If $\alpha = 0$ and $\mu = 0$ the $L_{\min}$ is defined by

$$L_{\min} = \begin{cases} 0, & \lambda \text{ even,} \\ 1, & \lambda \text{ odd.} \end{cases} \tag{8}$$

In case of $\alpha = 0$ and $\mu \neq 0$ the minimum value of L will be $L_{\min} = \mu$. When $0 < \alpha < \alpha_{\max}$ then $L_{\min} = \mu - 2\alpha$. If $\alpha = \alpha_{\max}$ and μ is even then $L_{\min} = 1$. Finally, when $\alpha = \alpha_{\max}$ and μ is odd the $L_{\min}$ takes value given by expression (8). There may exist only the BM states for which the condition $L_{\min} \leq L$ is satisfied. So, the presented $L_{\min}$ calculation procedure allows to find for every particular L the minimum value of α for which there exists the BM state i.e. $\alpha_{\min K}$. So, for actual calculations of the BM basis orthonormalization coefficients $\alpha = \alpha_{\min K}, \alpha_{\min K} + 1, \ldots, \alpha_{\max K}$. An illustrative example for calculation of $\alpha_{\min K}$ and $\alpha_{\max K}$ for $\mu = 4$ when $\lambda = 0, \ldots, 6$ is presented in Table 1.

Table 1. The values of $\alpha_{\min K}$ and $\alpha_{\max K}$ for $\mu = 4$ when $\lambda = 0, \ldots, 6$.

L												
λ	α	0	1	2	3	4	5	6	7	8	9	10
0	$\alpha_{\min K}$	2		1		0						
	$\alpha_{\max K}$	2		1		0						
1	$\alpha_{\min K}$		2	1	1	0	0					
	$\alpha_{\max K}$		2	1	1	0	0					
2	$\alpha_{\min K}$	2	0	1	1	0	0	0				
	$\alpha_{\max K}$	2	1	2	1	1	0	0				
3	$\alpha_{\min K}$		2	1	1	0	0	0	0			
	$\alpha_{\max K}$		2	1	2	1	1	0	0			
4	$\alpha_{\min K}$	2		1	1	0	0	0	0	0		
	$\alpha_{\max K}$	2		2	1	2	1	1	0	0		
5	$\alpha_{\min K}$		2	1	1	0	0	0	0	0	0	
	$\alpha_{\max K}$		2	1	2	1	2	1	1	0	0	
6	$\alpha_{\min K}$	2	0	1	1	0	0	0	0	0	0	0
	$\alpha_{\max K}$	2	1	2	1	2	1	2	1	1	0	0

For practical calculation there may be useful the condition that allows to find such minimal value of L ($L_{\min}^{\text{total}}$) for which all the BM basis orthonormalization coefficients that forms the matrix A and as consequence the matrix A itself do not exist. For definition of this condition one should start with calculation of the quantity $K = \mu - 2\alpha_{\max K}$. If $K = 0$ then

$$L_{\min}^{\text{total}} = \begin{cases} 0, & \lambda \text{ even}, \\ 1, & \lambda \text{ odd}. \end{cases} \tag{9}$$

If $K \neq 0$ then $L_{\min}^{\text{total}} = K$.

Step 2. Second step needs to introduce and iteratively calculate the convenient intermediate quantities $f_{\alpha,\alpha'}^{(n)}$, here $n = 0, 1, \ldots, \alpha_{\max K} - 1$ indicates a number of iteration. The iteration starts at $n = 0$ by calculation of all $f_{\alpha,\alpha'}^{(0)}$ that are defined by the overlap integrals $\langle u_\alpha | u_{\alpha'} \rangle$ given in the paper I:

$$f_{\alpha,\alpha'}^{(0)} = \langle u_\alpha | u_{\alpha'} \rangle, \tag{10}$$

where $\alpha = \alpha_{\min K}, \alpha_{\min K} + 1, \ldots, \alpha_{\max K} - 1$ and $\alpha < \alpha' \leq \alpha_{\max K}$.

The above overlap was applied in symbolic calculations to test our procedure in paper I with analytical results of [7].

At the next iteration step $n = 1$ the calculation of $f_{\alpha,\alpha'}^{(1)}$ is defined by the formula

$$f_{\alpha,\alpha'}^{(1)} = -f_{\alpha,\alpha'}^{(0)} + \frac{f_{\alpha,\alpha_{\max K}}^{(0)} f_{\alpha',\alpha_{\max K}}^{(0)}}{\langle u_{\alpha_{\max K}} | u_{\alpha_{\max K}} \rangle}, \tag{11}$$

where $\alpha = \alpha_{\min K}, \alpha_{\min K} + 1, \ldots, \alpha_{\max K} - 2$ and $\alpha < \alpha' \le \alpha_{\max K} - 1$. For all next iteration steps $n > 1$, the $f_{\alpha,\alpha'}^{(n)}$ are defined by the formula

$$f_{\alpha,\alpha'}^{(n)} = f_{\alpha,\alpha'}^{(n-1)} + \frac{f_{\alpha,\alpha_{\max K}-n+1}^{(n-1)} f_{\alpha',\alpha_{\max K}-n+1}^{(n-1)}}{\langle \psi_{\alpha_{\max K}-n+1} | \psi_{\alpha_{\max K}-n+1} \rangle}, \tag{12}$$

where $\alpha = \alpha_{\min K}, \alpha_{\min K} + 1, \ldots, \alpha_{\max K} - n - 1$ and $\alpha < \alpha' \le \alpha_{\max K} - n$. Here the normalization integral is defined as

$$\langle \psi_\alpha | \psi_\alpha \rangle = \langle u_\alpha | u_\alpha \rangle - \sum_{\alpha'=\alpha+1}^{\alpha_{\max}} \frac{\left(f_{\alpha,\alpha'}^{(\alpha_{\max K}-\alpha')} \right)^2}{\langle \psi_{\alpha'} | \psi_{\alpha'} \rangle}. \tag{13}$$

It should be noted, that all quantities $f_{\alpha,\alpha'}^{(n)}$ at iteration step n may be calculated solely from the quantities $f_{\alpha,\alpha'}^{(n-1)}$ and normalization integrals $\langle \psi_\alpha | \psi_\alpha \rangle$ obtained at the previous iteration step $n - 1$. So, at the every iteration step (except the $n = 0$) the corresponding quantities: $f_{\alpha,\alpha'}^{(n)}$ and the normalization integrals $\langle \psi_\alpha | \psi_\alpha \rangle$, are calculated and put into the storage.

For the linear storage of the quantities $f_{\alpha,\alpha'}^{(n)}$ the corresponding sequence number s may be introduced. It depends on the quantities $n, \alpha, \alpha', \alpha_{\max}$ by the formula

$$s = \frac{1}{6} \left(\left(2 + 6\alpha_{\max} + 3\alpha_{\max}^2 \right) n - 3 \left(1 + \alpha_{\max} \right) n^2 + n^3 \right)$$
$$+ \left(\alpha_{\max} - n \right) \alpha + \frac{1}{2} \left(1 - \alpha \right) \alpha - \alpha + \alpha'.$$

Step 3. Finally, having calculated the quantities $f_{\alpha,\alpha'}^{(n)}$ and the normalization integrals $\langle \psi_\alpha | \psi_\alpha \rangle$, one may straightforwardly compute the required orthonormalization coefficients $A_{\alpha,\alpha'}$ of the expansion (3). In the case when $\alpha = \alpha' = \alpha_{\max K}$ the formula for calculation of $A_{\alpha,\alpha'}$ is

$$A_{\alpha_{\max K},\alpha_{\max K}} = \langle u_{\alpha_{\max K}} | u_{\alpha_{\max K}} \rangle^{-1/2}. \tag{14}$$

In case when $\alpha = \alpha'$ and $\alpha = \alpha_{\min K}, \alpha_{\min K} + 1, \ldots, \alpha_{\max K} - 1$ the formula for calculation of $A_{\alpha,\alpha'}$ is

$$A_{\alpha,\alpha} = -\langle \psi_\alpha | \psi_\alpha \rangle^{-1/2}. \tag{15}$$

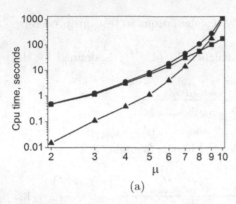 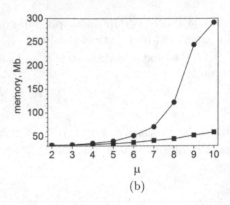

(a) (b)

Fig. 1. The CPU time versus parameter μ (a) and MaxMemoryUsed versus parameter μ (b): maximum number of Megabytes (Mb) used to store all data for the current Wolfram System session during the calculations of the orthogonal BM basis (circles) consisted of calculation of the overlap integrals (10) (squares) and execution of the othonormalization Gram–Schmidt procedure (11)–(17) (triangles).

In the case when $\alpha < \alpha' < \alpha_{\max K}$ the formula for calculation of $A_{\alpha,\alpha'}$ is

$$A_{\alpha,\alpha'} = \frac{1}{\langle \psi_{\alpha'} | \psi_{\alpha'} \rangle} \sum_{\alpha''=\alpha}^{\alpha'-1} A_{\alpha,\alpha''} f_{\alpha'',\alpha'}^{(\alpha_{\max K} - \alpha')}. \tag{16}$$

In the case when $\alpha = \alpha_{\min K}, \alpha_{\min K} + 1, \ldots, \alpha_{\max K} - 1$ and $\alpha' = \alpha_{\max K}$ the formula for calculation of $A_{\alpha,\alpha'}$ is

$$A_{\alpha,\alpha_{\max K}} = -\frac{1}{\langle \psi_{\alpha_{\max K}} | \psi_{\alpha_{\max K}} \rangle} \sum_{\alpha''=\alpha}^{\alpha_{\max K}-1} A_{\alpha,\alpha''} f_{\alpha'',\alpha_{\max K}}^{(0)}. \tag{17}$$

The above *algorithm* was realized in the form of the program implemented in the computer algebra system Wolfram Mathematica 10.1.

Remark 1. The two advantages of the proposed *algorithm*. First of all its simplicity: at any iterative step n the quantities $f_{\alpha,\alpha'}^{(n)}$ are composed of fragments that are not more complicated than that defined in the right hand side of Eq. (12) and the normalization integrals (13). Secondly, iterative calculation of the quantities $f_{\alpha,\alpha'}^{(n)}$ (12) and the normalization integrals (13) do not involve any square root operation in contradistinction to the conventional one [14]. This distinct features of the proposed orthonormalization algorithm make the large scale symbolic calculations *in principle* feasible.

In the case of the subset of three independent BM vectors (1) indicated by the displayed values of labels, expansion (3) demonstrated execution of the othonormalization Gram–Schmidt procedure (OGSP) (11)–(17) takes the form ($\mu = 4$, $\lambda - L$ even))

$$
\left| \begin{matrix} (\lambda, \mu) \\ f_2, L, L \end{matrix} \right\rangle = A_{2,2}^{(\lambda, \mu)}(L) \left| \begin{matrix} (\lambda, \mu)_B \\ 2, L, L \end{matrix} \right\rangle,
$$

$$
\left| \begin{matrix} (\lambda, \mu) \\ f_1, L, L \end{matrix} \right\rangle = A_{1,1}^{(\lambda, \mu)}(L) \left| \begin{matrix} (\lambda, \mu)_B \\ 1, L, L \end{matrix} \right\rangle + A_{1,2}^{(\lambda, \mu)}(L) \left| \begin{matrix} (\lambda, \mu)_B \\ 2, L, L \end{matrix} \right\rangle,
$$

$$
\left| \begin{matrix} (\lambda, \mu) \\ f_0, L, L \end{matrix} \right\rangle = A_{0,0}^{(\lambda, \mu)}(L) \left| \begin{matrix} (\lambda, \mu)_B \\ 0, L, L \end{matrix} \right\rangle + A_{0,1}^{(\lambda, \mu)}(L) \left| \begin{matrix} (\lambda, \mu)_B \\ 1, L, L \end{matrix} \right\rangle + A_{0,2}^{(\lambda, \mu)}(L) \left| \begin{matrix} (\lambda, \mu)_B \\ 2, L, L \end{matrix} \right\rangle,
$$

$$
A_{2,2}^{(\lambda, 4)}(L) = (\langle u_2|u_2 \rangle)^{-1/2},
$$

$$
A_{1,1}^{(\lambda, 4)}(L) = -\langle \psi_1|\psi_1 \rangle^{-1/2}, \qquad A_{1,2}^{(\lambda, 4)}(L) = \langle \psi_1|\psi_1 \rangle^{-1/2} \frac{\langle u_2|u_1 \rangle}{\langle u_2|u_2 \rangle},
$$

$$
A_{0,0}^{(\lambda, 4)}(L) = -\langle \psi_0|\psi_0 \rangle^{-1/2},
$$

$$
A_{0,1}^{(\lambda, 4)}(L) = -\frac{\langle \psi_0|\psi_0 \rangle^{-1/2}}{\langle \psi_1|\psi_1 \rangle} \left(-\langle u_1|u_0 \rangle + \frac{\langle u_2|u_1 \rangle \langle u_2|u_0 \rangle}{\langle u_2|u_2 \rangle} \right),
$$

$$
A_{0,2}^{(\lambda, 4)}(L) = \langle \psi_0|\psi_0 \rangle^{-1/2} \left[\frac{\langle u_2|u_0 \rangle}{\langle u_2|u_2 \rangle} \right.
$$

$$
\left. + \frac{1}{\langle \psi_1|\psi_1 \rangle} \left(-\langle u_1|u_0 \rangle + \frac{\langle u_2|u_1 \rangle \langle u_2|u_0 \rangle}{\langle u_2|u_2 \rangle} \right) \frac{\langle u_2|u_1 \rangle}{\langle u_2|u_2 \rangle} \right],
$$

$$
\langle \psi_0|\psi_0 \rangle = \langle u_0|u_0 \rangle - \frac{\langle u_2|u_0 \rangle^2}{\langle u_2|u_2 \rangle} - \frac{1}{\langle \psi_1|\psi_1 \rangle} \left(-\langle u_1|u_0 \rangle + \frac{\langle u_2|u_1 \rangle \langle u_2|u_0 \rangle}{\langle u_2|u_2 \rangle} \right)^2,
$$

$$
\langle \psi_1|\psi_1 \rangle = \langle u_1|u_1 \rangle - \frac{\langle u_2|u_1 \rangle^2}{\langle u_2|u_2 \rangle}.
$$

As an example, in Fig. 1 we show the CPU time and MaxMemoryUsed during of calculations of overlap integrals (13) and execution of the othonormalization Gram–Schmidt procedure (OGSP) (11)–(17) by the above *symbolic algorithm* versus parameter μ using the PC Intel Pentium CPU 1.50 GHz 4 GB 64bit Windows 8. One can see that the CPU time (in double logarithmic scale) of execution of the overlap integrals is linearly growing in contradistinction to the OGSP, whose execution time is growing faster than linearly due to manipulations with rational expressions.

3   Benchmark for Symbolic Numerical Algorithm

Because the BM vectors $|u_\alpha \rangle$ are linearly independent, one can require the orthonormalization properties for the vectors $|\phi_i \rangle$

$$
\langle \phi_i|\phi_j \rangle = \delta_{ij}. \tag{18}
$$

From Eq. (18) there may be derived the orthonormalization property of the orthonormalization coefficients $A_{\alpha, \alpha'}$ matrix $\mathcal{A}$

$$
\mathcal{A} \, \mathcal{U} \, \tilde{\mathcal{A}} = \mathcal{I}. \tag{19}
$$

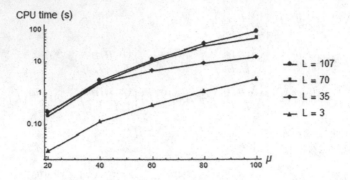

Fig. 2. The CPU time versus parameter μ for calculations of the A matrix with $\lambda = 119$ for $L = 3, 35, 70, 107$.

Here matrices $\boldsymbol{A}, \boldsymbol{U}$ and $\boldsymbol{I}$ have dimension $\alpha_{\max} + 1$. The matrix $\tilde{\boldsymbol{A}}$ is transposed of $\boldsymbol{A}$. In general case, when $\alpha_{\max K} < \alpha_{\max}$ and $\alpha_{\min K} > 0$, these matrices have the following block structure

$$\boldsymbol{A} = \begin{pmatrix} 0 & 0 & 0 \\ 0 & A & 0 \\ 0 & 0 & 0 \end{pmatrix}, \qquad \boldsymbol{U} = \begin{pmatrix} 0 & 0 & 0 \\ 0 & U & 0 \\ 0 & 0 & 0 \end{pmatrix}, \qquad \boldsymbol{I} = \begin{pmatrix} 0 & 0 & 0 \\ 0 & I & 0 \\ 0 & 0 & 0 \end{pmatrix}.$$

Here the matrices A, U and I have the dimension $\alpha_{\max K} - \alpha_{\min K} + 1$. The zeroes represents the sub-blocks with the appropriate dimensions that are filled with zeroes. The I is the unity matrix.

Let as display a structure of indices of the matrix A

$$A = \begin{pmatrix} A_{\alpha_{\min K}, \alpha_{\min K}} & \cdots & A_{\alpha_{\min K}, \alpha_{\max K}} \\ \vdots & \ddots & \vdots \\ A_{\alpha_{\max K}, \alpha_{\min K}} & \cdots & A_{\alpha_{\max K}, \alpha_{\max K}} \end{pmatrix}. \tag{20}$$

Finally, the entries of the matrix U are the overlap integrals $\langle u_\alpha | u_{\alpha'} \rangle$

$$U = \begin{pmatrix} \langle u_{\alpha_{\min K}} | u_{\alpha_{\min K}} \rangle & \cdots & \langle u_{\alpha_{\min K}} | u_{\alpha_{\max K}} \rangle \\ \vdots & \ddots & \vdots \\ \langle u_{\alpha_{\max K}} | u_{\alpha_{\min K}} \rangle & \cdots & \langle u_{\alpha_{\max K}} | u_{\alpha_{\max K}} \rangle \end{pmatrix}. \tag{21}$$

The Eq. (19) may be used for control of consistency and accuracy of calculations.

The efficiency of $\mathbf{A}$ matrix calculations for different values of the quantum number L is illustrated in Fig. 2. The computations were evaluated numerically to 150-digit precision. Such high precision was taken in order to compare these calculations with fuhrer calculations of $q_{ijk}^{(\lambda\mu)}(L)$ presented in Sect. 4.

Remark 2. From a conventional point of view the proposed symbolic orthonormalisation algorithm can be called as a non-standard recursive, or actually iterative, since it traverses the computation graph not from top to bottom, but from bottom to top. It has been organized in such a way, to have one to one correspondence of obtained results with definition of the orthonormalisation coefficients A assumed in Ref. [7]. This algorithm allows one to find the analytical expressions of the orthonormalized basis, if it is implemented in any computer algebra system, in particular, Wolfram Mathematica 10.1. It is given in the present paper and tested in the paper I with analytical results at small values λ and μ obtained in [7]. However, to perform the really fast large scale calculations with characteristics of computer time shown in Fig. 2, it has been implemented in the multi-precision arithmetics as a *symbolic-numerical algorithm*.

Table 2. The values of $(\alpha_{minK}, \alpha_{maxK})\dim(A)$ and α_{max} for $\mu = 60, 85, 100, 115$ and $\lambda = 120, 125$ when $L = 2, 31, 70, 120, 180$.

μ	λ	α_{max}	L 2	31	70	120	180
60	120	30	(29,30)2	(15,29)15	(0,30)31	(0,30)31	(0,0)1
	125		(29,29)1	(15,30)16	(0,29)30	(0,29)30	(0,2)3
85	120	42	(42,42)1	(27,42)16	(8,42)35	(0,42)43	(0,12)13
	125		(42,42)1	(27,42)16	(8,42)35	(0,42)43	(0,15)16
100	120	50	(49,50)2	(35,49)15	(15,50)36	(0,50)51	(0,20)21
	125		(49,49)1	(35,50)16	(15,49)35	(0,49)50	(0,22)23
115	120	57	(57,57)1	(42,57)16	(23,57)35	(0,57)58	(0,27)28
	125		(57,57)1	(42,57)16	(23,57)35	(0,57)58	(0,30)31

In numerical benchmark calculations given below we will demonstrate also the results of execution of the same OGSP (11)–(17) but with the normalized nonorthogonal eigenvectors $|\breve{u}_\alpha\rangle = N_{\alpha\alpha}^{-1}|u_\alpha\rangle$ and normalized overlap $\langle \breve{u}_\alpha | \breve{u}_{\alpha'} \rangle = \langle u_\alpha | N_{\alpha\alpha}^{-1} N_{\alpha'\alpha'}^{-1} | u_{\alpha'} \rangle$, $N_{\alpha\alpha} = (\langle u_\alpha | u_\alpha \rangle)^{1/2}$, respectively, i.e. $\langle \breve{u}_\alpha | \breve{u}_\alpha \rangle = 1$.

The examples of the output $(\alpha_{minK}, \alpha_{maxK}, \dim(A))$ of the program Abound.nb for some values of μ, λ, and L are presented in Table 2.

The orthonormalization of the BM basis, i.e. the calculation of the matrix A for given values of μ, λ, L and precision is provided by the program BMOrthonorm.nb. The calculation of the orthonormalized BM basis is based on the overlap integrals $\langle u_\alpha | u_{\alpha'} \rangle$. In case if that quantities are needed one may call the function overlapIntegral$[\mu, \alpha, \alpha', L, \lambda]$. As an example, we consider a case with $\mu = 10$, $\lambda = 11$ and $L = 6$ for $\alpha = 2$ and $\alpha' = 3$. In this case calling the function overlapIntegral produces the output $\langle u_\alpha | u_{\alpha'} \rangle$ = 595664650148853841920000000, i.e. the function overlapIntegral computes the *exact numerical* value of the overlap integral $\langle u_\alpha | u_{\alpha'} \rangle$.

Let us consider an example of the orthonormalization of the BM basis and take for it a case with $\mu = 10$, $\lambda = 11$ and $L = 6$. In this case $\alpha_{max} = 5$, $\alpha_{minK} = 2$, $\alpha_{maxK} = 4$, and precision was taken to be equal to prec $= 15$. For calculation of A matrix one may call the function Amatrix$[\mu, L, \lambda,$ prec$]$.

In this case the matrix $\mathcal{A}$ acting on unnormalized vector u prints as

$$\mathcal{A}^{\mu\lambda}(L) = \begin{pmatrix} 0\,0 & 0 & 0 & 0 & 0 \\ 0\,0 & 0 & 0 & 0 & 0 \\ 0\,0 & -5.268 \times 10^{-14} & 4.324 \times 10^{-14} & -5.671 \times 10^{-15} & 0 \\ 0\,0 & 0 & -1.271 \times 10^{-13} & 4.782 \times 10^{-14} & 0 \\ 0\,0 & 0 & 0 & 9.304 \times 10^{-14} & 0 \\ 0\,0 & 0 & 0 & 0 & 0 \end{pmatrix},$$

while $\check{\mathcal{A}} = \mathcal{A}\mathbf{N}$ acting on normalized vector $\check{u}$ prints as

$$\check{\mathcal{A}}^{\mu\lambda}(L) = \begin{pmatrix} 0\,0 & 0 & 0 & 0 & 0 \\ 0\,0 & 0 & 0 & 0 & 0 \\ 0\,0 & -1.06241748771592 & 0.382535950822453 & -0.060953283607289 & 0 \\ 0\,0 & 0 & -1.12436060747693 & 0.513991026814558 & 0 \\ 0\,0 & 0 & 0 & 1.00000000000000 & 0 \\ 0\,0 & 0 & 0 & 0 & 0 \end{pmatrix}.$$

The accuracy of calculations of the matrix $\mathcal{A}$ may be evaluated by the function `TestOrthonormalization`$[\mu,$ `L`$, \lambda,$ `base, prec`$]$ that use for this purpose the orthonormalization property Eq. (19). At first, the function `TestOrthonormalization`$[\mu,$ `L`$, \lambda,$ `base, prec`$]$ calculates the left hand side of the Eq. (19), i.e. the product of three matices $\mathcal{A}\,\mathcal{U}\check{\mathcal{A}}$. The result may be printed as the matrix `test` to ensure that its diagonal elements in the submatrix $\alpha_{\min K} \cdots \alpha_{\max K}$ are actually equal to one and other elements are equal to zero. Finally this submatrix is taken off (as matrix `testK`), printed and used to evaluate the accuracy of orthonormalization coefficients using the condition:

$$|10^{-\mathrm{prec}} - \|\mathrm{testK} - I\|| < \mathrm{base}^{-\mathrm{prec}}, \tag{22}$$

here the norm $\|...\|$ is defined as giving the maximum singular value of a matrix, and parameter `base` defines the accuracy of calculations of the matrix $\mathcal{A}^{\mu\lambda}(L)$ - in this case `base` is taken equal to 9.5. In the case under consideration the function `TestOrthonormalization`$[\mu,$ `L`$, \lambda,$ `base, prec`$]$ prints the following matrix for unnormalized $\mathcal{U}^{\mu\lambda}(L)$:

$$\mathcal{U}^{\mu\lambda}(L) = \begin{pmatrix} 0\,0 & 0 & 0 & 0 & 0 \\ 0\,0 & 0 & 0 & 0 & 0 \\ 0\,0 & 4.068 \times 10^{26} & 5.957 \times 10^{25} & 2.325 \times 10^{25} & 0 \\ 0\,0 & 5.957 \times 10^{25} & 7.826 \times 10^{25} & 4.347 \times 10^{25} & 0 \\ 0\,0 & 2.325 \times 10^{25} & 4.347 \times 10^{25} & 1.155 \times 10^{26} & 0 \\ 0\,0 & 0 & 0 & 0 & 0 \end{pmatrix},$$

while for normalized $\check{\mathcal{U}}^{\mu\lambda}(L)$

$$\check{\mathcal{U}}^{\mu\lambda}(L) = \begin{pmatrix} 0\,0 & 0 & 0 & 0 & 0 \\ 0\,0 & 0 & 0 & 0 & 0 \\ 0\,0 & 1.00000000000000 & 0.333834605013026 & 0.107226660720115 & 0 \\ 0\,0 & 0.333834605013026 & 1.00000000000000 & 0.457140728158342 & 0 \\ 0\,0 & 0.107226660720115 & 0.457140728158342 & 1.00000000000000 & 0 \\ 0\,0 & 0 & 0 & 0 & 0 \end{pmatrix}.$$

In this case the matrix product $\mathcal{A}\,\mathcal{U}\tilde{\mathcal{A}}$ will be printed as the matrix test: for the unnormalized overlap $\mathcal{U}$

$$
\left(\mathcal{A}\,\mathcal{U}\tilde{\mathcal{A}}\right)^{\mu\lambda}(L) =
\begin{pmatrix}
0\,0 & 0 & 0 & 0 & 0 \\
0\,0 & 0 & 0 & 0 & 0 \\
0\,0 & 1.0000000000 & 1.\times 10^{-15} & 1.\times 10^{-16} & 0 \\
0\,0 & 1.\times 10^{-15} & 1.0000000000 & 1.\times 10^{-15} & 0 \\
0\,0 & 1.\times 10^{-15} & 1.\times 10^{-15} & 1.0000000000 & 0 \\
0\,0 & 0 & 0 & 0 & 0
\end{pmatrix},
$$

and the matrix product $\check{\mathcal{A}}\check{\mathcal{U}}\tilde{\mathcal{A}}$ for the normalized overlap $\check{\mathcal{U}}$

$$
\left(\check{\mathcal{A}}\check{\mathcal{U}}\tilde{\mathcal{A}}\right)^{\mu\lambda}(L) =
\begin{pmatrix}
0\,0 & 0 & 0 & 0 & 0 \\
0\,0 & 0 & 0 & 0 & 0 \\
0\,0 & 1.0000000000 & 1.\times 10^{-15} & 1.\times 10^{-15} & 0 \\
0\,0 & 1.\times 10^{-15} & 1.0000000000 & 1.\times 10^{-15} & 0 \\
0\,0 & 1.\times 10^{-15} & 1.\times 10^{-15} & 1.0000000000 & 0 \\
0\,0 & 0 & 0 & 0 & 0
\end{pmatrix},
$$

here for saving the space we do not present last 4 zeroes after the decimal point for the diagonal matrix elements.

The efficiency of A matrix calculations for different values of parameter L is illustrated in Fig. 2. The computations were evaluated numerically to 15-digit precision. It should be pointed out that the CPU time for calculations of the A matrix less depend on the values of λ, L and the taken precision but more on the dimension of the matrix A. The apparent dependency of CPU time on the value of L reflects actually the changing of dimension of the matrix A depending on the value of L but not the change of the time for calculation of overlap integrals.

Remark 3. As shown by our benchmark calculation it would be appropriate in the future numerical calculation to provide scaling: the use of non-orthogonal normalized basis similar to Ref. [13] and the corresponding input matrix elements of scalar products – overlap integrals and intermediate output coefficients of orthogonalization and intermediate input matrix elements of tensor operators. Naturally, with such scaling, the result of calculating the orthonormal basis and the final values of the matrix elements of the tensor operators do not change. In this case, the desired numerical values coincide with the analytical values, but the intermediate values will remain within 16 significant figures, which corresponds to the accepted accuracy of the final results of $2 \cdot 10^{-16}$. Meanwhile, the principal problem of calculation exact numerical value of overlap integral in nonnormalized nonorthogonal BM basis at extremely large value of λ and μ will be solved using Wolfram Mathematica. The corresponding study of an efficiency of such adaptation of our code implemented in Wolfram Mathematica 10.1 and comparison with code implemented in Fortran are subject of a separate paper published elsewhere.

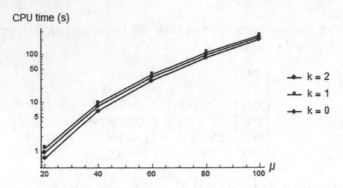

Fig. 3. The CPU time versus parameter μ for calculations of the $q_{ijk}^{(\lambda\mu)}(L)$ with $\lambda = 119$ and $L = 107$ for $k = 0, 1, 2$.

4   Generation and Solution of SU(3) algebraic problem

The Casimir operator of $SO(3)$ irreducible representations corresponding to the group chain $SU(3) \supset O(3) \supset O(2)$ have the form:

$$C_2(SU(3)) = Q \cdot Q + 3L \cdot L = 4(\lambda^2 + \mu^2 + \lambda\mu + 3\lambda + 3\mu). \qquad (23)$$

The dimension of the subspace for given λ, μ can be calculated by using the following formula:

$$D_{\lambda\mu} = \frac{1}{2}(\lambda + 1)(\mu + 1)(\lambda + \mu + 2). \qquad (24)$$

As a benchmark example we consider a perturbation operator announced in Ref. [4]

$$H_{4Q}'' = \sqrt{\frac{14}{5}}(\bar{Q} \otimes \bar{Q})_0^4 + (\bar{Q} \otimes \bar{Q})_{-4}^4 + (\bar{Q} \otimes \bar{Q})_4^4, \qquad (25)$$

where $(T_{\lambda'} \otimes T_\lambda)_M^L$ denotes the tensor product of two spherical tensors [8]. The matrix elements of the Hamiltonian (25) can be calculated by using the following formula:

$$(H_{4Q}'')_{\alpha L M, \alpha' L' M'}^{(\lambda\mu)} = \frac{1}{\sqrt{(2L+1)}} \sum_{L''=|L'-2|}^{L'+2} \frac{1}{\sqrt{(2L''+1)}} \sum_{\alpha''=0}^{\alpha_{max}}$$

$$\sum_{M''=-L''}^{L''} \left(H''^{(1)}_{LL'L''M''} + H''^{(2)}_{LL'L''M''} \right) R_{\alpha L, \alpha'' L''}^{(\lambda\mu)} R_{\alpha'' L'', \alpha' L'}^{(\lambda\mu)}. \qquad (26)$$

Matrix elements $H''^{(1)}_{LL'L''M''}$ and $H''^{(2)}_{LL'L''M''}$ read as

$$H''^{(1)}_{LL'L''M''} = \sqrt{\frac{14}{5}} \sum_{\eta=-2}^{2} \begin{bmatrix} 2 & 2 & 4 \\ \eta & -\eta & 0 \end{bmatrix} \begin{bmatrix} L'' & 2 & L \\ M'' & \eta & M \end{bmatrix} \begin{bmatrix} L' & 2 & L'' \\ M' & -\eta & M'' \end{bmatrix}, \qquad (27)$$

Table 3. The example of spectrum E of Hamiltonian (32) for $\gamma = 1.5$ and $h_{4Q} = 10$. The pair (λ, μ) labels the irreducible representations of the group SU(3) and the label ν denote degeneration of eigenvalues due to the intrinsic tetrahedral/octahedral symmetry.

(λ, μ)	$D_{\lambda\mu}$	$\gamma C_2(SU(3))$	E	ν	CPU time, s.
$(0,0)$	1	0	0	1	0.016
$(1,0)$	3	24	24	3	0.656
$(2,0)$	6	60	61.44	2	3.547
			60	1	
			59.04	3	
$(2,1)$	15	96	98.5042	3	28.531
			97.6949	3	
			96.96	1	
			95.52	2	
			93.9758	3	
			93.8251	3	

$$H''^{(2)}_{LL'L''M''} = \begin{bmatrix} L'' & 2 & L \\ M'' & -2 & M \end{bmatrix} \begin{bmatrix} L' & 2 & L'' \\ M' & -2 & M'' \end{bmatrix} + \begin{bmatrix} L'' & 2 & L \\ M'' & 2 & M \end{bmatrix} \begin{bmatrix} L' & 2 & L'' \\ M' & 2 & M'' \end{bmatrix}. \tag{28}$$

Here the notation of the Clebsh-Gordan coefficients [8] by the square brackets is introduced. The reduced matrix elements of the quadrupole operator have the form

$$R^{(\lambda\mu)}_{\alpha L, \alpha' L'} = \sqrt{2L'+1} \begin{bmatrix} L & 2 & L' \\ -L' & 0 & -L' \end{bmatrix}^{(-1)} q^{(\lambda\mu)}_{\alpha\alpha'(L-L')}(L). \tag{29}$$

If $L < L'$ then primed parameters should be interchanged with not primed parameters on the right hand side of the formula (29) and the overall sign should be changed as well if the $L - L'$ is the odd number. Matrix elements $q^{(\lambda\mu)}_{ijk}(L)$ read as

$$q^{(\lambda\mu)}_{ijk}(L) = \sum_{\substack{\alpha=0,\ldots,\alpha_{max} \\ s=0,\pm 1}} A^{(\lambda\mu)}_{i,\alpha}(L) a^{(k)}_s \tilde{A}^{(\lambda\mu)}_{j,(\alpha+s)}(L+k), \tag{30}$$

where coefficients $a^{(k)}_s$ are given in II: and $\tilde{A}^{(\lambda,\mu)}_{i,\alpha}(L)$ are elements of the inverse and the transpose of the matrix $\mathbf{A}$

$$\tilde{A}^{(\lambda,\mu)}_{i,\alpha}(L) = (A^{-1})^{(\lambda,\mu)}_{\alpha,i}(L). \tag{31}$$

The above formula was applied in symbolic calculations to test our procedure in paper II with analytical results of [9]. In present paper the efficiency of $q^{(\lambda\mu)}_{ijk}(L)$ calculations for different values of parameter k is illustrated in Fig. 3. The computations were evaluated numerically to 150-digit precision. Such high precision is necessary for accurate calculation of inverse matrix $\mathbf{A}^{-1}$.

Remark 4. If we wish to calculate $q_{ijk}^{(\lambda\mu)}(L)$ with help of the normalized matrix $\breve{A}$ then we will scale matrix **a** by such a way $\breve{a} = N^{-1}aN$ which corresponds to action of $\breve{a}$ on normalized vector $\breve{u}$.

Let us calculate for example the low lying energy levels $E_n \equiv E_{\lambda,\mu,\nu}$ of the Hamiltonian:

$$H \equiv H/h_{4Q} = \gamma C_2(SU(3)) + H_{4Q}''/h_{4Q}, \ H|\lambda,\mu,\nu> = E_n|\lambda,\mu,\nu>. \quad (32)$$

The computational results for an example of spectrum of the Hamiltonian (32) are presented in Table 3. The columns of the table are: (λ,μ) labels the irreducible representations of the group SU(3); $D_{\lambda\mu}$ is the dimension of the (λ,μ) irrep from Eq. (24) determining a complexity of the above algorithm; $C_2(SU(3))$ marks the eigenvalues of the second order Casimir operator (23); E presents the energy levels that results after diagonalization of the Hamiltonian (32); ν is the degeneration of the corresponding energy spectrum E; CPU time is the H_{4Q} matrix calculation time in seconds. The computations were evaluated numerically to 10-digit precision.

The computations was performed on Intel i7-3630QM 2.40 GHz CPU with 8 GB RAM running 64-bit Windows 8.

5   Conclusion

We present the effective and fast symbolic algorithm for constructing of the non-canonical Bargmann–Moshinsky (BM) basis with the highest weight vectors of SO(3) irreps which can be implemented in any computer algebra system. This kind of basis is widely used for calculating spectra and electromagnetic transitions in molecular and nuclear physics. The new symbolic algorithm for orthonormalisation of the obtained BM basis based on the Gram-Schmidt orthonormalisation procedure is developed.

To avoid misunderstanding we recall that from a conventional view point the proposed symbolic orthonormalisation algorithm can be called as a non-standard recursive, or 'actually iterative', since it traverses the computation graph not from top to bottom, but from bottom to top. It has been organized in such a way, to have one to one correspondence of obtained results with definition of the orthonormalisation coefficients of matrix **A** from Eqs. (3) and (20) assumed in Ref. [7].

This algorithm allows one to find the analytical expressions of the orthonormalized basis, if it is implemented in any computer algebra system, in particular, Wolfram Mathematica 10.1. It has been given in the present paper and tested explicitly on analytical results at small values λ and μ Refs. [7,9] in our previous papers I and II. However, to realized the really fast large scale calculations with characteristics of computer time shown in Figs. 2 and 3, it has been implemented in the multi-precision arithmetics as a symbolic-numerical algorithm. It can be also implemented in Fortran to apply in the fast large scale calculations like in Ref. [10].

The distinct advantage of this method is that it does not involve any square root operation on the expressions coming from the previous steps for computation of the orthonormalisation coefficients for this basis. This makes the proposed method very suitable for calculations on computer algebra systems. The symbolic nature of the developed algorithms allows one to avoid the numerical round-off errors in calculation of spectral characteristics (especially close to resonances) of quantum systems under consideration and to study their analytical properties for understanding the dominant symmetries [4]. The program in the Mathematica language for orthonormalisation of the non-canonical BM basis using the overlap integrals in Eq. (21) given by the analytical formula [2,7] is now prepared and will be published as an open code elsewhere. The great advantage of the program is the possibility to specify an arbitrary precision of calculations which is especially important for large scale calculations of physical quantities that involve procedures of matrices inversion. The high efficiency of the developed program was illustrated by orthonormalisation of BM basis up to extremely high quantum numbers (λ, μ), which is not given by other symbolic algorithms known in the literature [11,12].

Acknowledgements. The work was partially supported by the Bogoliubov-Infeld program, Votruba-Blokhintsev program, the RUDN University Program 5-100 and grant of Plenipotentiary of the Republic of Kazakhstan in JINR. AD is grateful to Prof. A. Góźdź for hospitality during visits in Institute of Physics, Maria Curie-Skłodowska University (UMCS).

The authors thank the both referees for their useful comments, remarks and suggestions.

References

1. Saha, A., et al.: Spectroscopy of a tetrahedral doubly magic candidate nucleus $^{160}_{70}\text{Yb}^{90}$. J. Phys. G Nucl. Part. Phys. **46**, 055102 (2019)
2. Deveikis, A., Gusev, A.A., Gerdt, V.P., Vinitsky, S.I., Góźdź, A., Pędrak, A.: Symbolic algorithm for generating the orthonormal Bargmann–Moshinsky basis for SU(3) group. In: Gerdt, V.P., Koepf, W., Seiler, W.M., Vorozhtsov, E.V. (eds.) CASC 2018. LNCS, vol. 11077, pp. 131–145. Springer, Cham (2018). https://doi.org/10.1007/978-3-319-99639-4_9
3. Vinitsky, S., et al.: On generation of the Bargmann-Moshinsky basis of SU(3) group. J. Phys. Conf. Ser. **1194**, 012109 (2019)
4. Gozdz, A., Pedrak, A., Gusev, A.A., Vinitsky, S.I.: Point symmetries in the nuclear SU(3) partner groups model. Acta Phys. Polonica B Proc. Suppl. **11**, 19–28 (2018)
5. Bargmann, V., Moshinsky, M.: Group theory of harmonic oscillators (II). Nucl. Phys. **23**, 177–199 (1961)
6. Moshinsky, M., Patera, J., Sharp, R.T., Winternitz, P.: Everything you always wanted to know about $SU(3) \supset O(3)$. Ann. Phys. (N.Y.) **95**, 139–169 (1975)
7. Alisauskas, S., Raychev, P., Roussev, R.: Analytical form of the orthonormal basis of the decomposition $SU(3) \supset O(3) \supset O(2)$ for some (λ, μ) multiplets. J. Phys. G Nucl. Phys. **7**, 1213–1226 (1981)
8. Varshalovitch, D.A., Moskalev, A.N., Hersonsky, V.K.: Quantum Theory of Angular Momentum. Nauka, Leningrad (1975). (also World Scientific (1988))

9. Raychev, P., Roussev, R.: Matrix elements of the generators of SU(3) and of the basic O(3) scalars in the enveloping algebra of SU(3). J. Phys. G Nucl. Phys. **7**, 1227–1238 (1981)

10. Cseh, J.: Algebraic models for shell-like quarteting of nucleons. Phys. Lett. B **743**, 213–217 (2015)

11. Dytrych, T., et al.: Efficacy of the SU(3) scheme for ab initio large-scale calculations beyond the lightest nuclei. Comput. Phys. Commun. **207**, 202–210 (2016)

12. Pan, F., Yuan, S., Launey, K.D., Draayer, J.P.: A new procedure for constructing basis vectors of SU(3)⊃SO(3). Nucl. Phys. A **743**, 70–99 (2016)

13. Asherova, R.M., Smirnov, Y.F.: On asymptotic properties of a quantum number Ω in a system with SU(3) symmetry. Repts. Math. Phys. **4**, 83–95 (1973)

14. Draayer, J.P., Akiyama, Y.: Wigner and Racah coefficients for SU3. J. Math. Phys. **14**, 1904–1912 (1973)

The Implementation of the Symbolic-Numerical Method for Finding the Adiabatic Waveguide Modes of Integrated Optical Waveguides in CAS Maple

D. V. Divakov[(✉)] and A. L. Sevastianov

Department of Applied Probability and Informatics,
Peoples' Friendship University of Russia (RUDN University),
6 Miklukho-Maklaya Street, Moscow 117198, Russia
{divakov-dv,sevastianov-al}@rudn.ru

Abstract. Computational problems of electrodynamics require an approximate solution of the system of Maxwell's vector equations for regions with different geometries. The main methods for solving problems with the Maxwell equations are either finite difference methods or methods based on the Galerkin and Kantorovich expansions, or the finite element method. Each of the classes of methods is characterised by a wide range of permissible objects, but in each of the methods, the solution contains a large number of quantities known only in numerical form.

We have chosen a different approach, in which to describe the waveguide propagation of electromagnetic radiation we propose using the model of adiabatic waveguide modes. This model allows reducing Maxwell equations to a system of ordinary differential equations, which allows analysis of its solutions at the symbolic level.

A fundamental system of solutions of the system is constructed in symbolic form. A numerical method for computing the guided modes of a planar three-layer open waveguide is formulated and implemented using a vector model of the adiabatic waveguide modes. Phase constants calculated in the framework of the model of adiabatic waveguide modes were verified by comparison with those calculated in the framework of the scalar model.

Keywords: Model of adiabatic waveguide modes ·
Integrated optical waveguide · Symbolic-numerical method ·
Waveguide propagation of electromagnetic radiation

The publication has been prepared with the support of the "RUDN University Program 5-100" and funded by RFBR according to the research projects No. 18-07-00567 and 19-01-00645.

M. England et al. (Eds.): CASC 2019, LNCS 11661, pp. 107–121, 2019.
https://doi.org/10.1007/978-3-030-26831-2_8

1 Introduction

Over the past 50–70 years, a tremendous development of numerical methods and computational capacities has occurred, which today allow solving the tasks of huge computation volume. Computational problems of electrodynamics require an approximate solution of the system of Maxwell vector equations for regions that differ in both linear dimensions and geometry. The approach to solving vector problems with the Maxwell equations, based on modern numerical methods, requires either the introduction of grids to construct difference schemes in the space or space-time domains, or the series expansion of solutions in some complete systems of functions. The series are further truncated to partial sums and Galerkin and Kantorovich methods [2, 3] are used for the final formulation of the computational problem. However, there are also methods that occupy an intermediate place between grid and projection methods, namely, the finite element method [4, 6–11].

In the authors' opinion, the methods for solving problems with the Maxwell equations mentioned above possess a high degree of arbitrariness in setting the parameters of the objects under consideration. As a payment for this arbitrariness all the described methods suffer from the corresponding main drawback: the solutions obtained are partially or completely specified as grid functions, for which the analysis at the symbolic level is inconvenient and sometimes impossible. Note also that for vector problems, the conditions at the interface of dielectric media that join the components of the electromagnetic field must also be satisfied. Such boundary conditions, as a rule, can be presented only approximately within the framework of the described methods, which is also a disadvantage in the case of applying these methods to waveguide problems.

For the physical interpretation of the solutions obtained, especially for vector problems, in which the desired quantities are vector fields, it is much more convenient to have a basic symbolic form of solution, only some selected characteristics of which will be found in numerical form. Moreover, according to the authors' idea, the form of solution should essentially reproduce the physical features of the waveguide propagation, the conditions at the interfaces between dielectric media, asymptotic conditions at infinity, and at the same time be a vector field at the symbolic level. At least the validity of the divergent equations should follow automatically from the validity of Maxwell equations. The price for the required characteristics will be the reduced universality of the structures considered within the framework of the method, namely, we will further consider a narrower class of waveguide structures for which the form of the solution is specified more precisely.

As an alternative to the numerical methods described in the introduction, it is proposed to introduce a physically more understandable type of solution, liked by physicists because of the possibility of constructing several approximations that specify the solution, namely, an asymptotic expansion. A limitation of the validity of such a decomposition is the presence of a small parameter, the smallness of which will determine the class of waveguide structures suitable for consideration in the framework of the method and the range of electromagnetic radiation.

In this paper, we have obtained for the first time the representation of the waveguide mode in a symbolic form (using computer algebra system), which is universal for all adiabatic waveguide structures in the zero approximation. Using this representation, it is possible to approximately calculate the waveguide modes of arbitrary layered waveguide structures.

2   Setting of the Problem

In this paper, the waveguide mode of propagation of monochromatic electromagnetic radiation in an integrated optical structure is considered [12–14]. In this case, the integrated optical structure itself, as a rule, is formed by applying additional dielectric layers on the waveguide base. A regular planar waveguide, e.g., a three-layer one, is often used as such a base. Suppose in this case that the coordinate system is set in such a way that the planes separating the dielectric layers of a three-layer waveguide are parallel to the plane (yOz). With this method of specifying the coordinate system, the wave vector of electromagnetic radiation propagation will also lie in the plane (yOz). The key role in obtaining integrated optical effects in the propagation of electromagnetic radiation is played by the profiles of additional dielectric layers, which will generally be specified through the equations of the surfaces $F_j(x, y, z) = x - h_j(y, z)$. Given that the dielectric layers are isotropic, the distributions of the permittivity and permeability parameters are real functions of the coordinates $\varepsilon = \varepsilon(x; y, z)$, $\mu = \mu(x; y, z)$ and for each fixed point (y^*, z^*) these functions will be piecewise constant. Note that the paper considers the material medium consisting of dielectric subdomains, filling together the entire three-dimensional space. Important parameters for specifying the waveguide propagation of electromagnetic radiation are the phase distribution $\varphi(y, z)$ and the distribution of the effective refractive index $n_{eff}(y, z)$. The phase velocity of the waveguide propagation of electromagnetic radiation is given by the following expression:

$$v = \frac{c}{n_{eff}(y, z)}. \tag{1}$$

Solving the problem of waveguide propagation of electromagnetic radiation for Maxwell equations without additional assumptions is an extremely difficult task. In our case, we limit the scope of the study to smoothly irregular integrated optical structures that satisfy the following constraints:

$$\left\| \frac{\partial h}{\partial y} \right\|, \left\| \frac{\partial h}{\partial y} \right\| \ll \frac{h k_0}{2\pi} = \frac{h}{\lambda}, \left\| \frac{\Delta \varphi}{\nabla \varphi} \right\| \ll k_0. \tag{2}$$

From the description of the class of integrated optical structures under consideration, it follows that in the absence of external currents and charges, the induced currents and charges are zero. In this case, the material coupling equations are assumed to be linear. Thus, the electromagnetic field in a space filled with dielectrics in the Gaussian system of units in the Cartesian coordinate system is described by the equations [12–14]:

$$\frac{\partial H_z}{\partial y} - \frac{\partial H_y}{\partial z} = \frac{\varepsilon}{c} \frac{\partial E_x}{\partial t}, \quad \frac{\partial E_z}{\partial y} - \frac{\partial E_y}{\partial z} = -\frac{\mu}{c} \frac{\partial H_x}{\partial t},$$

$$\frac{\partial H_x}{\partial z} - \frac{\partial H_z}{\partial x} = \frac{\varepsilon}{c} \frac{\partial E_y}{\partial t}, \quad \frac{\partial E_x}{\partial z} - \frac{\partial E_z}{\partial x} = -\frac{\mu}{c} \frac{\partial H_y}{\partial t}, \quad (3)$$

$$\frac{\partial H_y}{\partial x} - \frac{\partial H_x}{\partial y} = \frac{\varepsilon}{c} \frac{\partial E_z}{\partial t}, \quad \frac{\partial E_y}{\partial x} - \frac{\partial E_x}{\partial y} = -\frac{\mu}{c} \frac{\partial H_z}{\partial t}.$$

Here $\boldsymbol{E} = (E_x, E_y, E_z)^T$ and $\boldsymbol{H} = (H_x, H_y, H_z)^T$ are the vectors of electric and magnetic fields, and c is the speed of light in vacuum. At the interface between dielectric media, the following conditions are satisfied [12–14]:

$$\left[\boldsymbol{n} \times \boldsymbol{H}^1\right]\big|_{x=h(y,z)} = \left[\boldsymbol{n} \times \boldsymbol{H}^2\right]\big|_{x=h(y,z)},$$

$$\left[\boldsymbol{n} \times \boldsymbol{E}^1\right]\big|_{x=h(y,z)} = \left[\boldsymbol{n} \times \boldsymbol{E}^2\right]\big|_{x=h(y,z)}. \quad (4)$$

Here $\boldsymbol{n}$ is the normal vector to the interface between the first and the second dielectric medium, $\boldsymbol{E}^1$, $\boldsymbol{E}^2$, $\boldsymbol{H}^1$, and $\boldsymbol{H}^2$ are the vectors of the electric and magnetic field strength in the first and the second medium, respectively. Since an open integrated optical dielectric structure is considered, it is necessary to specify asymptotic boundary conditions at infinity for guided waveguide modes [12–14]:

$$\|\boldsymbol{E}\| \xrightarrow[x\to\pm\infty]{} 0, \quad \|\boldsymbol{H}\| \xrightarrow[x\to\pm\infty]{} 0. \quad (5)$$

In weakly inhomogeneous three-dimensional media, the propagation of electromagnetic radiation is described using locally plane waves or adiabatic approximations of the Maxwell equations obtained by the asymptotic method [1]. The propagation of electromagnetic radiation in the integrated optical structures can be described using the model of adiabatic waveguide modes (AWM).

3   Asymptotic Method for the Derivation of AWM Equations in Symbolic Form

To formulate the model of adiabatic waveguide modes, we will use the asymptotic expansions of the electric and magnetic field strength vectors. We will investigate asymptotic expansions, particularly used in the study of semiclassical approximations, in symbolic form involving no numerical methods. To construct the AWM model [5], we use the asymptotic method [1], in which the desired solutions of Maxwell equations (3) are presented in the form:

$$\boldsymbol{E}(x,y,z,t) = \sum_{s=0}^{\infty} \frac{\boldsymbol{E}_s(x;y,z)}{(-i\omega)^{\gamma+s}} \exp\left\{i\omega t - ik_0\varphi(y,z)\right\}, \quad (6)$$

$$H\left(x,y,z,t\right) = \sum_{s=0}^{\infty} \frac{H_s\left(x;y,z\right)}{\left(-i\omega\right)^{\gamma+s}} \exp\left\{i\omega t - ik_0\varphi\left(y,z\right)\right\}. \tag{7}$$

In this representation, the following notation and variables are used: ω is the angular frequency of the propagating monochromatic electromagnetic radiation, s is the asymptotic expansion index, E_s, H_s are the corresponding electric and magnetic strength vectors for the asymptotic expansion term of the order s, k_0 is the wave number, $\varphi\left(y,z\right)$ is the phase distribution. In this case, the angular frequency of propagating electromagnetic radiation should be large enough, so that the value $\frac{1}{\omega}$ is a small parameter of the asymptotic expansion (6)–(7) in the optical range of electromagnetic radiation.

In the notation $E_s\left(x;y,z\right)$, $H_s\left(x;y,z\right)$, the semicolon separator for x means the following assumption: $\partial E_s/\partial y, \partial E_s/\partial z, \partial H_s/\partial y$, and $\partial H_s/\partial z$ are small quantities, therefore, the following expressions for the derivatives of E and H are valid:

$$\frac{\partial E}{\partial y} = -ik_0\frac{\partial\varphi}{\partial y}E, \quad \frac{\partial E}{\partial z} = -ik_0\frac{\partial\varphi}{\partial z}E, \tag{8}$$

$$\frac{\partial H}{\partial y} = -ik_0\frac{\partial\varphi}{\partial y}H, \quad \frac{\partial H}{\partial z} = -ik_0\frac{\partial\varphi}{\partial z}H. \tag{9}$$

Let us substitute the expressions (6)–(7) into the system of Eqs. (3) and equate the coefficients for the same powers of $\frac{1}{\omega}$. As a result, in the zero approximation of the method, taking Eqs. (8) and (9) into account, we arrive at the system of homogeneous equations:

$$-ik_0\frac{\partial\varphi}{\partial y}H_0^z + ik_0\frac{\partial\varphi}{\partial z}H_0^y = ik_0\varepsilon E_0^x, \tag{10}$$

$$-ik_0\frac{\partial\varphi}{\partial z}H_0^x - \frac{\partial H_0^z}{\partial x} = ik_0\varepsilon E_0^y, \tag{11}$$

$$\frac{\partial H_0^y}{\partial x} + ik_0\frac{\partial\varphi}{\partial y}H_0^x = ik_0\varepsilon E_0^z, \tag{12}$$

$$-ik_0\frac{\partial\varphi}{\partial y}E_0^z + ik_0\frac{\partial\varphi}{\partial z}E_0^y = -ik_0\mu H_0^x, \tag{13}$$

$$-ik_0\frac{\partial\varphi}{\partial z}E_0^x - \frac{\partial E_0^z}{\partial x} = -ik_0\mu H_0^y, \tag{14}$$

$$\frac{\partial E_0^y}{\partial x} + ik_0\frac{\partial\varphi}{\partial y}E_0^x = -ik_0\mu H_0^z. \tag{15}$$

Comment. It is important to note that as a result of applying the asymptotic expansions (6)–(7), Maxwell's equations reduce to two algebraic equations and four first-order differential equations. The algebraic equations

$$E_0^x = \frac{1}{\varepsilon}\left(\frac{\partial\varphi}{\partial z}H_0^y - \frac{\partial\varphi}{\partial y}H_0^z\right), \tag{16}$$

$$H_0^x = \frac{1}{\mu}\left(\frac{\partial\varphi}{\partial y}E_0^z - \frac{\partial\varphi}{\partial z}E_0^y\right) \tag{17}$$

will be used below to formulate the model under study.

Substituting algebraic equations (16) and (17) into differential equations (11), (12), (14) and (15), we obtain a system of first-order differential equations; hereinafter, we will deal with the zero approximation of the asymptotic expansion for a small parameter, so the index of the approximation order is no longer preserved:

$$\frac{\partial H^z}{\partial x} + \frac{ik_0}{\mu}\frac{\partial\varphi}{\partial z}\left(\frac{\partial\varphi}{\partial y}E^z - \frac{\partial\varphi}{\partial z}E^y\right) + ik_0\varepsilon E^y = 0, \tag{18}$$

$$\frac{\partial H^y}{\partial x} + \frac{ik_0}{\mu}\frac{\partial\varphi}{\partial y}\left(\frac{\partial\varphi}{\partial y}E^z - \frac{\partial\varphi}{\partial z}E^y\right) - ik_0\varepsilon E^z = 0, \tag{19}$$

$$\frac{\partial E^z}{\partial x} + \frac{ik_0}{\varepsilon}\frac{\partial\varphi}{\partial z}\left(\frac{\partial\varphi}{\partial z}H^y - \frac{\partial\varphi}{\partial y}H^z\right) - ik_0\mu H^y = 0, \tag{20}$$

$$\frac{\partial E^y}{\partial x} + \frac{ik_0}{\varepsilon}\frac{\partial\varphi}{\partial y}\left(\frac{\partial\varphi}{\partial z}H^y - \frac{\partial\varphi}{\partial y}H^z\right) + ik_0\mu H^z = 0. \tag{21}$$

4   Construction of Fundamental System of Solutions (FSS) for the System of Ordinary Differential Equations (SODE) in Symbolic Form

We consider the problem of finding eigenvectors (guided modes) and eigenvalues using the model of adiabatic waveguide modes within the framework of the zero approximation of the asymptotic expansion for a planar regular three-layer optical waveguide. In this case, the problem will be solved at a fixed point (y, z). Allowing for the fact that permittivity and permeability coefficients are piecewise-constant functions, we solve the problem in each subdomain with constant values of ε, μ, and then join the solutions at the interface between dielectric media. In this case, in each layer ε, μ are fixed constants and $\frac{\partial\varphi}{\partial y}, \frac{\partial\varphi}{\partial z}$ are real numbers at fixed (y, z). To construct the total fields $\boldsymbol{E}$, $\boldsymbol{H}$ it is necessary to set and solve the problem of determining $\varphi(y, z)$, which is left beyond the scope of the present paper.

Thus, in each subdomain with constant values ε and μ, we obtain a system of ordinary differential equations (SODE) with constant coefficients of the first order having the form:

$$\boldsymbol{u}' = A\boldsymbol{u}, \tag{22}$$

where the vector of the desired quantities $\boldsymbol{u} = \boldsymbol{u}(x; y, z)$ and the matrix of coefficients $A = A(y, z)$ are defined as follows:

$$u = \begin{pmatrix} H^z(x; y, z) \\ E^z(x; y, z) \\ H^y(x; y, z) \\ E^y(x; y, z) \end{pmatrix}, \tag{23}$$

$$
A = \begin{bmatrix}
0 & \frac{ik_0}{\mu}\varphi_y\varphi_z & 0 & -\frac{ik_0}{\mu}\varphi_z{}^2 + ik_0\varepsilon \\
-\frac{ik_0}{\varepsilon}\varphi_y\varphi_z & 0 & \frac{ik_0}{\varepsilon}\varphi_z{}^2 - ik_0\mu & 0 \\
0 & \frac{ik_0}{\mu}\varphi_y{}^2 - ik_0\varepsilon & 0 & -\frac{ik_0}{\mu} \\
-\frac{ik_0}{\varepsilon}\varphi_y{}^2 + ik_0\mu & 0 & \frac{ik_0}{\varepsilon}\varphi_y\varphi_z & 0
\end{bmatrix}, \quad (24)
$$

where $\varphi_y = \frac{\partial\varphi}{\partial y}, \varphi_z = \frac{\partial\varphi}{\partial z}$.

It is important to note that the matrix of the system under consideration is sparse; in fact, half of the elements of the matrix are zero. In order to build a fundamental system of solutions (FSS) [17–19] for Eq. (22) in symbolic form, it is necessary to derive symbolic expressions for the eigenvalues and eigenvectors of the matrix A. Moreover, for efficient qualitative analysis of the solutions of the system, symbolic expressions of eigenvalues and eigenvectors must be compact and applicable for such an analysis.

Using the Maple computer algebra system [20], we calculate eigenvalues γ_j and eigenvectors ξ_j, $j = \overline{1,4}$ of the matrix A in a symbolic form. Let us denote by γ and X the vector composed of eigenvalues and the matrix of eigenvectors [15,16], respectively:

$$
\gamma = \begin{bmatrix}
k_0\sqrt{-\varepsilon\mu + \left(\frac{\partial\varphi}{\partial y}\right)^2 + \left(\frac{\partial\varphi}{\partial z}\right)^2} \\
-k_0\sqrt{-\varepsilon\mu + \left(\frac{\partial\varphi}{\partial y}\right)^2 + \left(\frac{\partial\varphi}{\partial z}\right)^2} \\
k_0\sqrt{-\varepsilon\mu + \left(\frac{\partial\varphi}{\partial y}\right)^2 + \left(\frac{\partial\varphi}{\partial z}\right)^2} \\
-k_0\sqrt{-\varepsilon\mu + \left(\frac{\partial\varphi}{\partial y}\right)^2 + \left(\frac{\partial\varphi}{\partial z}\right)^2}
\end{bmatrix}. \quad (25)
$$

As seen from the symbolic expressions for eigenvalues (25), there are two eigenvalues $\gamma_\pm = \pm k_0\sqrt{-\varepsilon\mu + \left(\frac{\partial\varphi}{\partial y}\right)^2 + \left(\frac{\partial\varphi}{\partial z}\right)^2}$, each having a multiplicity of 2, which may complicate the form of the FSS of the system under study. Using the computer algebra system, we obtain the matrix of eigenvectors X:

$$
\begin{bmatrix}
\frac{-\varepsilon\mu + \varphi_y{}^2}{\varphi_y\varphi_z} & -\frac{i\mu\sqrt{-\varepsilon\mu + \varphi_y{}^2 + \varphi_z{}^2}}{\varphi_y\varphi_z} & \frac{-\varepsilon\mu + \varphi_y{}^2}{\varphi_y\varphi_z} & \frac{i\mu\sqrt{-\varepsilon\mu + \varphi_y{}^2 + \varphi_z{}^2}}{\varphi_y\varphi_z} \\
-\frac{i\varepsilon\sqrt{-\varepsilon\mu + \varphi_y{}^2 + \varphi_z{}^2}}{\varphi_y\varphi_z} & \frac{-\varepsilon\mu + \varphi_z{}^2}{\varphi_y\varphi_z} & \frac{i\varepsilon\sqrt{-\varepsilon\mu + \varphi_y{}^2 + \varphi_z{}^2}}{\varphi_y\varphi_z} & \frac{-\varepsilon\mu + \varphi_z{}^2}{\varphi_y\varphi_z} \\
0 & 1 & 0 & 1 \\
1 & 0 & 1 & 0
\end{bmatrix}.
$$

From the form of the matrix X, it can be noted that despite the presence of multiple eigenvalues, the eigenvectors are linearly independent. Therefore, the matrix A can be diagonalised, which can be illustrated by constructing the Jordan form J of the matrix A using the Maple computer algebra system:

$$
J = \begin{bmatrix}
\gamma_+ & 0 & 0 & 0 \\
0 & \gamma_- & 0 & 0 \\
0 & 0 & \gamma_+ & 0 \\
0 & 0 & 0 & \gamma_-
\end{bmatrix}.
$$

Using the obtained symbolic expressions for eigenvalues and eigenvectors, we construct the FSS for SODE (22):

$$U\left(x; y, z\right) = C_1 \boldsymbol{\xi}_1 e^{\gamma + x} + C_2 \boldsymbol{\xi}_2 e^{\gamma - x} + C_3 \boldsymbol{\xi}_3 e^{\gamma + x} + C_4 \boldsymbol{\xi}_4 e^{\gamma - x} \qquad (26)$$

The behaviour of the solution of the system under study substantially depends on the sign of the radicand $\left(\frac{\partial \varphi}{\partial y}\right)^2 + \left(\frac{\partial \varphi}{\partial z}\right)^2 - \varepsilon\mu$. If $\left(\frac{\partial \varphi}{\partial y}\right)^2 + \left(\frac{\partial \varphi}{\partial z}\right)^2 \geq \varepsilon\mu$, then the solution of the system under study is a sum of the decreasing and increasing exponentials. Otherwise, the solution is a sum of the oscillating functions. The described properties of the solutions will be used below in setting the problem in terms of the formulation of asymptotic conditions.

Taking into account the fact that the FSS is known in symbolic form and is amenable to qualitative analysis, we will use it to state the problems of finding the modes of multilayer dielectric waveguides.

5   Statement of the Problem of Finding Guided Modes in a Symbolic Form

The configuration of an open multilayer planar isotropic waveguide is shown in Fig. 1.

In this configuration, there are always two cladding layers of semi-infinite thickness $(x > h_1(y, z)$ and $x < h_N(y, z))$. Between these layers, there is a finite number of layers of varying thickness. In this case, the interfaces between dielectric media can be arbitrary smooth surfaces. Thus, having FSS for each layer (26), the problem of finding guided modes of the waveguide structure is reduced

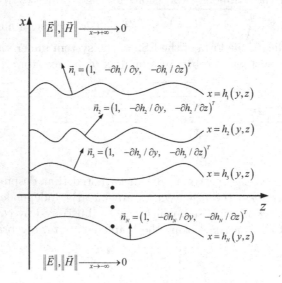

Fig. 1. Plot of a few smooth boundaries multilayered waveguide

to the following operations: (1) for layers of semi-infinite thickness, solutions are selected that satisfy the asymptotic conditions (5), i.e., the coefficients at increasing exponentials that enter the solution are set to be zero; (2) solutions for each layer are substituted into the boundary conditions. As a result of the substitution of solutions in the boundary conditions, we obtain a homogeneous system of linear algebraic equations with respect to the unknown coefficients, which determine the expansion of the solution in FSS for each layer. Note that the condition of solvability of a homogeneous system of linear algebraic equations is the equality to zero of the determinant of the matrix of its coefficients.

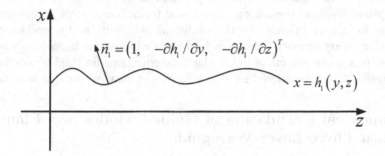

Fig. 2. Plot of a smooth boundary multilayered waveguide

With the above considerations taken into account, let us proceed to the formulation of two types of conditions in symbolic form.

Asymptotic conditions are imposed on solutions in semi-infinite layers, while relying on the analysis of FSSs, it is obvious that for the existence of a nontrivial solution, it is necessary that $\left(\frac{\partial\varphi}{\partial y}\right)^2 + \left(\frac{\partial\varphi}{\partial z}\right)^2 \geq \varepsilon\mu$, and that the coefficients of the growing exponents are zero. Consider conditions on an arbitrary smooth interface between two dielectric media given by $F(x,y,z) = x - h(y,z) = 0$ (Fig. 2).

$$[\boldsymbol{n} \times \boldsymbol{H}] = \begin{pmatrix} H_y\frac{\partial h}{\partial z} - H_z\frac{\partial h}{\partial y} \\ -H_z - H_x\frac{\partial h}{\partial z} \\ H_y - H_x\frac{\partial h}{\partial y} \end{pmatrix}. \tag{27}$$

When writing (27), the denominator $\sqrt{1 + \left(\frac{\partial h}{\partial y}\right)^2 + \left(\frac{\partial h}{\partial z}\right)^2}$ in the expression for the normal was omitted, since it does not vanish and coincides for both sides of (4). It should be noted that only two of the three components of the obtained vector (27) are linearly independent. Thus, to write the boundary conditions we will use the following expressions:

$$\left(A_z^1 + A_x^1\frac{\partial h}{\partial z}\right)\bigg|_{x=h(y,z)} = \left(A_z^2 + A_x^2\frac{\partial h}{\partial z}\right)\bigg|_{x=h(y,z)}, \tag{28}$$

$$\left(A_y^1 - A_x^1 \frac{\partial h}{\partial y}\right)\Bigg|_{x=h(y,z)} = \left(A_y^2 - A_x^2 \frac{\partial h}{\partial y}\right)\Bigg|_{x=h(y,z)}, \qquad (29)$$

where $A = \{E, H\}$. In addition, for the case of a flat boundary $F(x, y, z) = x - \text{const}$, these expressions are simplified to the following form:

$$A_y^1\big|_{x=const} = A_y^2\big|_{x=\text{const}}, \qquad (30)$$

$$A_z^1\big|_{x=const} = A_z^2\big|_{x=\text{const}const}. \qquad (31)$$

Summarising the intermediate results, we resume that a complete set of tools is obtained for the formulation of the problem of finding the guided modes of arbitrary multilayer waveguides in a symbolic form. Namely, we constructed an algorithm for generating asymptotic conditions at infinity and boundary conditions for arbitrary smooth interfaces between dielectric media, which together form a homogeneous system of linear algebraic equations for the FSS coefficients for all media of the considered integrated optical planar waveguide structure.

6   Numerical Calculation of Guided Modes of a Planar Open Three-Layer Waveguide

As an example, consider an open three-layer regular planar dielectric waveguide schematically presented below (Fig. 3). Note that such planar structures are well described in the framework of the scalar model, therefore, for verification, we compare the phase constants obtained in the model of adiabatic waveguide modes with appropriate coefficients obtained in the scalar model [21, 22].

Fig. 3. Open regular planar three-layer waveguide

The described structure occupies the entire space, the covering layer occupies a semi-infinite region $x > h$ and is characterised by a refractive index n_c, the substrate occupies a semi-infinite region $x < 0$ and is characterised by a refractive index n_s. The waveguide layer is enclosed in the region $0 < x < h$, has a finite thickness h and a refractive index n_f. Note that the refractive indices in the framework of the model under consideration are real values. We will consider electromagnetic radiation propagating along the axis Oz. In this case, the phase $\varphi(y, z)$ varies only along the axis Oz, and $\partial\varphi/\partial y = 0$. Below we present the formulation of the problem of finding guided waveguide modes of the structure described above in symbolic form.

7   Symbolic Formulation of the Problem with Symbolic Generation of System of Linear Algebraic Equations (SLAE)

Let us construct a fundamental system of solutions of SODE (22) for each of the layers in the case $\partial\varphi/\partial y = 0$ in the Maple computer algebra system by analogy with the more general case discussed in the above section "Asymptotic method for the derivation of AWM equations in symbolic form." The eigenvalues for each of the layers for the case $\partial\varphi/\partial y = 0$ are simplified to the form $\gamma_{\pm}^{\alpha} = \pm k_0\sqrt{-\varepsilon_\alpha\mu_\alpha + (\partial\varphi/\partial z)^2}$, where the index $\alpha = \{c, f, s\}$ indicates the corresponding dielectric layer, and the corresponding matrix of eigenvectors X_α takes the following form:

$$\begin{bmatrix} 0 & \dfrac{-i\mu_\alpha}{\sqrt{-\varepsilon_\alpha\mu_\alpha+(\partial\varphi/\partial z)^2}} & 0 & \dfrac{i\mu_\alpha}{\sqrt{-\varepsilon_\alpha\mu_\alpha+(\partial\varphi/\partial z)^2}} \\ 0 & 1 & 0 & 1 \\ \dfrac{i\varepsilon_\alpha}{\sqrt{-\varepsilon_\alpha\mu_\alpha+(\partial\varphi/\partial z)^2}} & 0 & \dfrac{-i\varepsilon_\alpha}{\sqrt{-\varepsilon_\alpha\mu_\alpha+(\partial\varphi/\partial z)^2}} & 0 \\ 1 & 0 & 1 & 0 \end{bmatrix}.$$

In this particular case, for each of the three layers considered ($\alpha = \{c, f, s\}$), the fundamental system of solutions takes the following form:

$$U_\alpha(x; y, z) = \begin{bmatrix} -\dfrac{iA_2^\alpha\mu_\alpha}{\sqrt{-\varepsilon_\alpha\mu_\alpha+(\partial\varphi/\partial z)^2}}e^{\gamma_+^\alpha x} + \dfrac{iA_4^\alpha\mu_\alpha}{\sqrt{-\varepsilon_\alpha\mu_\alpha+(\partial\varphi/\partial z)^2}}e^{\gamma_-^\alpha x} \\ A_2^\alpha e^{\gamma_+^\alpha x} + A_4^\alpha e^{\gamma_-^\alpha x} \\ \dfrac{iA_1^\alpha\varepsilon_\alpha}{\sqrt{-\varepsilon_\alpha\mu_\alpha+(\partial\varphi/\partial z)^2}}e^{\gamma_+^\alpha x} - \dfrac{iA_3^\alpha\varepsilon_\alpha}{\sqrt{-\varepsilon_\alpha\mu_\alpha+(\partial\varphi/\partial z)^2}}e^{\gamma_-^\alpha x} \\ A_1^\alpha e^{\gamma_+^\alpha x} + A_3^\alpha e^{\gamma_-^\alpha x} \end{bmatrix}.$$

Comment. Note also that for each of the semi-infinite layers corresponding to the substrate and the covering layer, the pair of coefficients with increasing exponentials is zero: for the covering layer $A_1^c = A_2^c = 0$, for the substrate $A_3^s = A_4^s = 0$.

We now write the boundary conditions for $x = 0$, $x = h$ and obtain a homogeneous system of linear algebraic equations (SLAE) for the vector of unknown quantities $A = \left(A_3^c, A_4^c, A_1^f, A_2^f, A_3^f, A_4^f, A_1^s, A_2^s\right)^T$. The structure of the resulting matrix is as follows (Fig. 4):

The matrix is seen to be sparse and has a checkerboard-like structure of blocks. In this case, the determinant of the matrix can be obtained in symbolic form; however, we do not present this cumbersome expression here for brevity.

Matrix M [8 × 8] structure

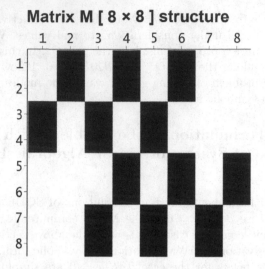

Fig. 4. The structure of the resulting matrix

A homogeneous system of linear algebraic equations has a non-zero solution if and only if the determinant of the matrix of its coefficients is zero. From this property, we obtain an equation for finding the phase constant, for which we introduce the following definition $\beta^2 = \left(\frac{\partial \varphi}{\partial z}\right)^2$, where β is determined from the equation $\det\left(M\left(\beta\right)\right) = 0$.

8   Numerical Calculations

Now let us proceed to numerical calculations. For this purpose we specify the required values as follows:

$$\lambda = 0.55; \ k_0 = \frac{2\pi}{\lambda}; \ n_c = 1.0; \ n_f = 1.565; \ n_s = 1.47; \ h = 2\lambda. \qquad (32)$$

For the given parameters, the plot of the real part of the determinant is as follows (Fig. 5):

The real part of the determinant shows four roots, while the imaginary part of the determinant is computer zero. Next, we will numerically find the zeros of the determinant of the matrix by bisectional search with a given accuracy. First, we localise the segments, at the ends of which the determinant changes sign and then in each such segment we apply the method of bisectional search to refine the root. The described algorithm is implemented in the Maple computer algebra system and used below to verify the formulated method.

9   Verification

To verify the described method, we compare the calculated phase constants with those obtained within the scalar model [21,22]. The phase constants in

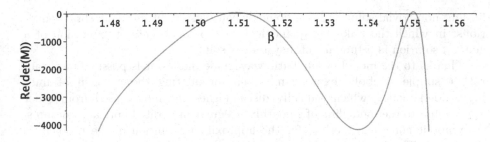

Fig. 5. Plot of the real part of the determinant

the framework of the described method are denoted by β^{vector}, and the phase constants obtained in the scalar model are denoted by β^{scalar}. In the framework of the algorithm described above, we will find the phase constants (the zeros of the determinant function) with the accuracy $\varepsilon = 4.44 \times 10^{-18}$; with the same accuracy, we calculate the phase constants using the scalar model. Let us compare the calculated phase constants and determine the absolute $\delta_{abs} = |\beta^{scalar} - \beta^{vector}|$ and relative $\delta_{rel} = |\beta^{scalar} - \beta^{vector}| / |\beta^{scalar}|$ errors. The verification results are presented in the table below.

β^{scalar}	β^{vector}	δ_{abs}	δ_{rel}
1.551492738069290 1290	1.5514927380692901222	6.8×10^{-18}	4.38×10^{-18}
1.5501811158901042864	1.5501811158901042882	1.8×10^{-18}	2.38×10^{-18}
1.5117506145374407068	1.5117506145374407104	3.6×10^{-18}	1.16×10^{-18}
1.5072749512764207624	1.5072749512764207660	3.6×10^{-18}	2.39×10^{-18}

Given the magnitude of the error $\varepsilon = 4.44 \times 10^{-18}$, we can conclude that the absolute error δ_{abs} has the order of the error of calculations, and the relative error does not exceed the error of the calculations. From the numerical results, it follows that the phase constants calculated in the framework of the method under study and in the framework of the scalar model coincide within the accuracy of the calculations performed.

10   Conclusion

In the paper, the vector problem of finding the guided modes of an open dielectric waveguide with an arbitrary number of dielectric layers having smooth boundaries is posed in symbolic form. The formulation of the problem is preceded by the symbolic derivation of a system of differential equations and its general solution, which is also studied in symbolic form. Thanks to the combination of symbolic transformations, it is possible to formulate a computational problem of finding non-trivial solutions of a homogeneous system of linear algebraic equations, in which the unknown quantities are the coefficients of expansion of a general solution in a fundamental system of solutions.

Thanks to the model of adiabatic waveguide modes, it is possible to obtain rather simple symbolic expressions when formulating the problem of finding guided modes, which radically distinguishes the model used from other approaches to the modelling of smoothly irregular integrated optical structures. A symbolic-numerical method for the approximate solution of the problem is developed. The method is implemented in the Maple computer algebra system. Verification was performed by comparing the phase constants calculated in the framework of the method under study with those obtained using the scalar model. The results coincide with an accuracy close to the accuracy of calculations, which demonstrates the consistency of the developed method.

In this paper, we obtained for the first time the representation of the waveguide mode in a symbolic form (using computer algebra system), which is universal for all adiabatic waveguide structures in the zero approximation. Using this representation, it is possible to approximately calculate the waveguide modes of arbitrary layered waveguide structures.

All results were obtained in the Maple computer algebra system using the LinearAlgebra and ArrayTools packages.

References

1. Babich, V.M., Buldyrev, V.S.: Asymptotic Methods in Short-Wave Diffraction Problems. Nauka, Moscow (1972). [English translation: Springer Series on Wave Phenomena 4. Springer, Berlin Heidelberg New York 1991]
2. Kantorovich, L.V., Krylov, V.I.: Approximate Methods of Higher Analysis. Wiley, New York (1964)
3. Fletcher, C.A.J.: Computational Galerkin Methods. Springer, Heidelberg (1984). https://doi.org/10.1007/978-3-642-85949-6
4. Gusev, A.A., et al.: Symbolic-numerical algorithms for solving the parametric self-adjoint 2D elliptic boundary-value problem using high-accuracy finite element method. In: Gerdt, V.P., Koepf, W., Seiler, W.M., Vorozhtsov, E.V. (eds.) CASC 2017. LNCS, vol. 10490, pp. 151–166. Springer, Cham (2017). https://doi.org/10.1007/978-3-319-66320-3_12
5. Sevastyanov, L.A., Sevastyanov, A.L., Tyutyunnik, A.A.: Analytical calculations in maple to implement the method of adiabatic modes for modelling smoothly irregular integrated optical waveguide structures. In: Gerdt, V.P., Koepf, W., Seiler, W.M., Vorozhtsov, E.V. (eds.) CASC 2014. LNCS, vol. 8660, pp. 419–431. Springer, Cham (2014). https://doi.org/10.1007/978-3-319-10515-4_30
6. Bathe, K.J.: Finite Element Procedures in Engineering Analysis. Prentice Hall, Englewood Cliffs (1982)
7. Ciarlet, P.: The Finite Element Method for Elliptic Problems. North Holland Publishing Company, Amsterdam (1978)
8. Strang, G., Fix, G.J.: An Analysis of the Finite Element Method. Prentice-Hall, Englewood Cliffs (1973)

9. Bogolyubov, A.N., Mukhartova, Yu.V., Gao, J., Bogolyubov, N.A.: Mathematical modeling of plane chiral waveguide using mixed finite elements. In: Progress in Electromagnetics Research Symposium, pp. 1216–1219 (2012)
10. Bogolyubov, A.N., Mukhartova, Y.V., Gao, T.: Calculation of a parallel-plate waveguide with a chiral insert by the mixed finite element method. Math. Models Comput. Simul. 5(5), 416–428 (2013)
11. Mukhartova, Y.V., Mongush, O.O., Bogolyubov, A.N.: Application of the finite-element method for solving a spectral problem in a waveguide with piecewise constant bi-isotropic filling. J. Commun. Technol. Electronics 62(1), 1–13 (2017)
12. Adams, M.J.: An Introduction to Optical Waveguides. Wiley, New York (1981)
13. Marcuse, D.: Light Transmission Optics. Van Nostrand, New York (1974)
14. Tamir, T.: Guided-Wave Optoelectronics. Springer, Berlin (1990). https://doi.org/10.1007/978-3-642-97074-0
15. Coddington, E.A., Levinson, N.: Theory of Ordinary Differential Equations. McGraw-Hill, New York (1955)
16. Hartman, P.: Ordinary Differential Equations, Classics in Applied Mathematics, 38. Society for Industrial and Applied Mathematics, Philadelphia (2002). [1964]
17. Johnson, W.: A Treatise on Ordinary and Partial Differential Equations. Wiley, New York (1913). In University of Michigan Historical Math Collection
18. Polyanin, A.D., Zaitsev, V.F., Moussiaux, A.: Handbook of First Order Partial Differential Equations. Taylor & Francis, London (2002)
19. Zwillinger, D.: Handbook of Differential Equations, 3rd edn. Academic Press, Boston (1997)
20. Mathematics-based software and services for education, engineering, and research. https://www.maplesoft.com/
21. Lovetskiy, K.P., Gevorkyan, M.N., Kulyabov, D.S., Sevastyanov, A.L., Sevastyanov, L.A.: Waveguide modes of a planar optical waveguide. Math. Model. Geom. 3(01), 43–63 (2015)
22. Ayryan, E.A., Egorov, A.A., Michuk, E.N., Sevastyanov, A.L., Sevastianov, L.A., Stavtsev, A.B.: Representations of Guided Modes of Integrated-Optical Multilayer Thin-Film Waveguides. E11–2011-31, LIT preprints (2011)

On Characteristic Decomposition
and Quasi-characteristic Decomposition

Rina Dong[1] and Chenqi Mou[1,2(✉)]

[1] LMIB–School of Mathematics and Systems Science,
Beihang University, Beijing 100191, China
{rina.dong,chenqi.mou}@buaa.edu.cn
[2] Beijing Advanced Innovation Center for Big Data and Brain Computing,
Beihang University, Beijing 100191, China

Abstract. In this paper, the concepts of quasi-characteristic pair and quasi-characteristic decomposition are introduced. The former is a pair $(\mathcal{G}, \mathcal{C})$ of a reduced lexicographic Gröbner basis $\mathcal{G}$ and the W-characteristic set $\mathcal{C}$ which is regular and extracted from $\mathcal{G}$; the latter is the decomposition of a polynomial set into finitely many quasi-characteristic pairs with associated zero relations. We show that the quasi-characteristic decomposition of any polynomial set can be obtained algorithmically regardless of the variable ordering condition. A new algorithm is presented for computing characteristic decomposition when the variable ordering condition is always satisfied, otherwise it degenerates to compute the quasi one. Some properties of quasi-characteristic pairs and decomposition are proved, and examples are given to illustrate the algorithm.

Keywords: Quasi-characteristic decomposition ·
Quasi-characteristic pair · W-characteristic set · Regular set ·
Gröbner basis

1 Introduction

Solving polynomial system is a fundamental problem in many areas of science and engineering such as commutative algebra [11], algebraic geometry [10,12], geometric reasoning [33], algebraic cryptanalysis [16], and biological analysis [5,31]. Compared with numeric methods such as homotopy continuation [26] for solving polynomial systems, symbolic methods produce rigorous and reliable solutions which are convenient for later manipulations. There are two main symbolic methods based respectively on triangular sets [24,28,33] and Gröbner bases [7,11] for solving polynomial systems, and both of them can be considered as polynomial generalizations of Gaussian elimination for linear systems.

This work was partially supported by the National Natural Science Foundation of China (NSFC 11771034) and the Fundamental Research Funds for the Central Universities in China (YWF-19-BJ-J-324).

M. England et al. (Eds.): CASC 2019, LNCS 11661, pp. 122–139, 2019.
https://doi.org/10.1007/978-3-030-26831-2_9

Triangular decomposition decomposes an arbitrary polynomial set into triangular sets which are special polynomial sets with a certain triangular structure, such that the zeros of the polynomial set is equal to the union of those of the triangular sets. By means of triangular decomposition one can effectively study the algebraic varieties defined by polynomial sets. Along with continuous research and development on its theory, methods, and algorithms, triangular decomposition has become a standard approach to polynomial elimination and polynomial system solving. Currently effective algorithms are available for decomposing polynomial sets into different kinds of triangular sets [8,9,18,21,28], e.g., characteristic sets, regular sets and normal sets.

The Gröbner basis is an important tool in computational ideal theory, in particular for polynomial system solving. For a given ideal and a term ordering, the Gröbner basis of this ideal is a set of special generators with good properties, such that this ideal could be handled algorithmically via these generators. Since its introduction by Buchberger in his Ph.D. thesis [6], Gröbner basis has gained extensive study [4,15,17,19,25,32], which makes it a powerful tool for computational commutative algebra and algebraic geometry. In particular, Gröbner bases with respect to the lexicographic term ordering (LEX) have rich properties, for example the elimination property for elimination ideals, such that they are convenient for polynomial system solving.

The structures of LEX Gröbner bases were studied first for bivariate ideals [20] and then extended to general zero-dimensional polynomial ideals [13,22]. For the case of arbitrary dimensions, the connections between Ritt characteristic sets and LEX Gröbner bases are exploited in [1,29] by extracting triangular sets from the LEX Gröbner bases. In particular, the W-characteristic sets which are the smallest triangular sets contained in the reduced LEX Gröbner bases are used to study the relationships between normal triangular sets and reduced LEX Gröbner bases in [29], and it is also shown there that when the W-characteristic set is abnormal and satisfies the variable ordering condition, there exist some pseudo-divisibility relationships among certain polynomials in it. By using such relationships, an algorithm is proposed in [30] for characteristic decomposition which decomposes an arbitrary polynomial set simultaneously into reduced LEX Gröbner bases and normal sets. In [14], the failure of the assumption on the variable ordering for the pseudo-divisibility to occur is further handled by temporarily changing the variable orderings.

In this paper, we introduce the concepts of quasi-characteristic pair and quasi-characteristic decomposition. The former is a pair $(\mathcal{G}, \mathcal{C})$ of reduced LEX Gröbner basis $\mathcal{G}$ and the W-characteristic set $\mathcal{C}$ which is regular and extracted from $\mathcal{G}$; the latter is the decomposition of a polynomial set into finitely many quasi-characteristic pairs with associated zero relations. We present a splitting strategy based on the properties of regularity without the assumption on the variable ordering. With this splitting strategy, we propose a new algorithm for computing characteristic decomposition of an arbitrary polynomial set when the variable ordering condition is always satisfied. When the condition is not always satisfied, instead of temporarily changing the variable orderings to handle the failure of the assumption on variable ordering as in [14], the proposed algorithm degenerates to compute quasi-characteristic decomposition. Our main contributions include: (1) the concepts of the quasi-characteristic pair and quasi-characteristic decomposition are introduced with some properties proved, (2) a new algorithm, which does not depend on pseudo-divisibility between polynomials ensured by satisfaction of the variable ordering condition as previous existing algorithms in [14,30] do, is proposed for computing (quasi-) characteristic decomposition of an arbitrary polynomial set, and (3) experimental results with our implementation are presented.

After a brief review of Gröbner bases, triangular decomposition and characteristic decomposition in Sect. 2, we present a new splitting strategy and another algorithm for characteristic decomposition in detail in Sect. 3. Then we define the quasi-characteristic pair and quasi-characteristic decomposition and prove some of their properties in Sect. 4, followed by two illustrative examples in Sect. 5. Then the experimental results with our implementation for (quasi-) characteristic decomposition are reported in Sect. 6.

2   Preliminaries

In this section some basic notions and notations used in the sequel are recalled. The reader is referred to [1,3,11,28] and the references therein for more details on the theories of Gröbner bases and triangular sets and to [30] for the definitions and properties of characteristic decomposition.

2.1   Gröbner Basis, Triangular Set and Triangular Decomposition

Let $\mathbb{K}$ be a field and $\mathbb{K}[x_1, \ldots, x_n]$ be the ring of polynomials in n ordered variables $x_1 < \cdots < x_n$ with coefficients in $\mathbb{K}$. In this paper, we write $\mathbb{K}[x_1, \ldots, x_n]$ as $\mathbb{K}[\boldsymbol{x}]$ for simplicity. With a fixed term ordering $<$ which is a total and well ordering on all the terms in $\mathbb{K}[\boldsymbol{x}]$, the greatest term in a polynomial $F \in \mathbb{K}[\boldsymbol{x}]$ with respect to $<$ is called the *leading term* of F and denoted by $\mathrm{lt}(F)$.

Here in this paper the LEX term ordering $<_{\mathrm{LEX}}$ is of our main concerns. For two different terms $\boldsymbol{x}^{\alpha}$ and $\boldsymbol{x}^{\beta}$ in $\mathbb{K}[\boldsymbol{x}]$ with $\boldsymbol{\alpha} = (\alpha_1, \ldots, \alpha_n)$ and $\boldsymbol{\beta} = (\beta_1, \ldots, \beta_n)$, we say that $\boldsymbol{x}^{\alpha} <_{\mathrm{LEX}} \boldsymbol{x}^{\beta}$ if there exists an integer i $(1 \leq i \leq n)$ such that $\alpha_i < \beta_i$ and for $j = i + 1, \ldots, n, \alpha_j = \beta_j$.

Definition 1. Let $\mathfrak{I} \subseteq \mathbb{K}[\boldsymbol{x}]$ be an ideal and $<$ be a term ordering. Denote by $\langle \mathrm{lt}(\mathfrak{I}) \rangle$ the ideal generated by the leading terms of all the polynomials in $\mathfrak{I}$. A finite set $\{G_1, \ldots, G_s\}$ of polynomials in $\mathfrak{I}$ is called a *Gröbner basis* of $\mathfrak{I}$ with respect to $<$ if $\langle \mathrm{lt}(G_1), \ldots, \mathrm{lt}(G_s) \rangle = \langle \mathrm{lt}(\mathfrak{I}) \rangle$.

Let $\mathcal{G} = \{G_1, \ldots, G_s\}$ be a Gröbner basis of an ideal $\mathfrak{I} \subseteq \mathbb{K}[\boldsymbol{x}]$ with respect to a fixed term ordering $<$. For an arbitrary polynomial $F \in \mathbb{K}[\boldsymbol{x}]$, there exists a unique polynomial $R \in \mathbb{K}[\boldsymbol{x}]$ corresponding to F such that $F - R \in \mathfrak{I}$ and no term of R is divisible by any of $\{\mathrm{lt}(G_1), \ldots, \mathrm{lt}(G_s)\}$. F is said to be *B-reduced* with respect to $\mathcal{G}$ if $F = R$. The Gröbner basis $\mathcal{G}$ is said to be *reduced* if each

G_i is monic and no term of G_i is divisible by any of $\{\mathrm{lt}(G_j) \mid j \neq i\}$ for all $i = 1, \ldots, s$. The reduced Gröbner basis with respect to a fixed term ordering is unique.

Let F be a polynomial in $\mathbb{K}[\boldsymbol{x}] \setminus \mathbb{K}$. With respect to the variable ordering, the greatest variable appearing in F is called the *leading variable* of F and denoted by $\mathrm{lv}(F)$. Let $\mathrm{lv}(F) = x_i$. Then F can be written as $F = I x_i^k + R$, where $I \in \mathbb{K}[x_1, \ldots, x_{i-1}]$, $R \in \mathbb{K}[x_1, \ldots, x_i]$, and $\deg(R, x_i) < k = \deg(F, x_i)$. The polynomial I is called the *initial* of F, denoted by $\mathrm{ini}(F)$. For any polynomial set $\mathcal{F} \subseteq \mathbb{K}[\boldsymbol{x}]$, $\mathrm{ini}(\mathcal{F})$ denotes the set $\{\mathrm{ini}(F) \mid F \in \mathcal{F}\}$.

Definition 2. A finite, nonempty, ordered set $\mathcal{T} = [T_1, \ldots, T_r]$ of polynomials in $\mathbb{K}[\boldsymbol{x}] \setminus \mathbb{K}$ is called a *triangular set* if $\mathrm{lv}(T_1) < \cdots < \mathrm{lv}(T_r)$.

We denote by $\mathrm{prem}(P, Q)$ the pseudo-reminder of $P \in \mathbb{K}[\boldsymbol{x}]$ with respect to $Q \in \mathbb{K}[\boldsymbol{x}] \setminus \mathbb{K}$ in $\mathrm{lv}(Q)$, and for any triangular set $\mathcal{T} = [T_1, \ldots, T_r] \subseteq \mathbb{K}[\boldsymbol{x}]$, the pseudo-reminder of P with respect to $\mathcal{T}$ is defined as

$$\mathrm{prem}(P, \mathcal{T}) := \mathrm{prem}(\cdots \mathrm{prem}(\mathrm{prem}(P, T_r), T_{r-1}), \ldots, T_1).$$

For any two polynomial sets $\mathcal{F}, \mathcal{G} \subset \mathbb{K}[\boldsymbol{x}]$, $\mathsf{Z}(\mathcal{F}/\mathcal{G})$ denotes the set

$$\mathsf{Z}(\mathcal{F}/\mathcal{G}) := \{\bar{\boldsymbol{x}} \in \bar{\mathbb{K}}^n \mid F(\bar{\boldsymbol{x}}) = 0, G(\bar{\boldsymbol{x}}) \neq 0, \text{ for any } F \in \mathcal{F}, G \in \mathcal{G}\},$$

where $\bar{\mathbb{K}}$ is the algebraic closure of $\mathbb{K}$. In particular, $\mathsf{Z}(\mathcal{F}) := \mathsf{Z}(\mathcal{F}/\{\ \})$.

Definition 3. Let $\mathcal{F} \subset \mathbb{K}[\boldsymbol{x}]$ be a polynomial set. A *triangular decomposition* of $\mathcal{F}$ is a finite number of triangular sets $\mathcal{T}_1, \mathcal{T}_2, \ldots, \mathcal{T}_t \subset \mathbb{K}[\boldsymbol{x}]$ such that $\mathsf{Z}(\mathcal{F}) = \bigcup_{i=1}^{t} \mathsf{Z}(\mathcal{T}_i / \mathrm{ini}(\mathcal{T}_i))$.

Let $\mathcal{F} \subseteq \mathbb{K}[\boldsymbol{x}]$ be a polynomial set. We denote by $\langle \mathcal{F} \rangle$ the ideal of $\mathbb{K}[\boldsymbol{x}]$ generated by $\mathcal{F}$ and by $\sqrt{\langle \mathcal{F} \rangle}$ the radical of $\langle \mathcal{F} \rangle$. For $H \in \mathbb{K}[\boldsymbol{x}]$ we use $\langle \mathcal{F} \rangle : H$ to denote the ideal quotient of $\langle \mathcal{F} \rangle$ by H, which is the set of the polynomials $P \in \mathbb{K}[\boldsymbol{x}]$ such that $PH \in \langle \mathcal{F} \rangle$. The *saturated ideal* of a triangular set $\mathcal{T} = [T_1, \ldots, T_r]$ is defined as $\mathrm{sat}(\mathcal{T}) := \{P \in \mathbb{K}[\boldsymbol{x}] \mid \exists\, i \geq 0 \text{ such that } PJ^i \in \langle \mathcal{T} \rangle\}$, where $J = \mathrm{ini}(T_1) \cdots \mathrm{ini}(T_r)$.

Definition 4. Let $\mathfrak{I}$ be an ideal of a polynomial ring $\mathbb{R}' \subseteq \mathbb{R} \subseteq \mathbb{K}[\boldsymbol{x}]$. An element $P \in \mathbb{R}$ is said to be *zero* (or a *zero-divisor*) *modulo* $\mathfrak{I}$ if P is zero (or a zero-divisor respectively) in $\mathbb{R}/\mathfrak{I}_{\mathbb{R}}$, where $\mathfrak{I}_{\mathbb{R}}$ denotes the ideal generated by $\mathfrak{I}$ in $\mathbb{R}$. Furthermore, we say that P is *regular* modulo $\mathfrak{I}$ if it is neither zero nor a zero-divisor modulo $\mathfrak{I}$.

Definition 5. Let $\mathcal{T} = [T_1, \ldots, T_r]$ be a triangular set in $\mathbb{K}[\boldsymbol{x}]$ and $x_{p_i} = \mathrm{lv}(T_i)$ for $i = 1, \ldots, r$. Then $\mathcal{T}$ is said to be *regular* (or called a *regular set*) if for $i = 2, \ldots, r$, we have $\mathrm{ini}(T_i) \in \mathbb{K}[x_1, \ldots, x_{p_i-1}]$ to be regular modulo $\mathrm{sat}(\mathcal{T}_{i-1})_{\mathbb{K}[x_1,\ldots,x_{p_i-1}]}$.

Regular sets [27] or regular chains [1] are one of the most commonly used triangular sets. It is proved in [27,28] that a triangular set T is regular if and only if $\mathrm{sat}(T) = \{P \in \mathbb{K}[\boldsymbol{x}] \mid \mathrm{prem}(P, T) = 0\}$.

Let $T \subset \mathbb{K}[\boldsymbol{x}]$ be a triangular set. The variables in $\{x_1, \ldots, x_n\} \setminus \{\mathrm{lv}(T_1), \ldots, \mathrm{lv}(T_r)\}$ are called its *parameters*, and T is said to be *normal* (or called a *normal set*) if $\mathrm{ini}(T)$ involves only the parameters of T. By definition any normal set is obviously regular. T is called an *ascending set* if $r = 1$ and $T_1 \in \mathbb{K} \setminus \{0\}$, or $\deg(T_j, \mathrm{lv}(T_i)) < \deg(T_i, \mathrm{lv}(T_i))$ for every pair $0 < i < j \leq r$.

Definition 6. Let $\mathcal{P}$ be a polynomial set in $\mathbb{K}[\boldsymbol{x}]$. An ascending set $\mathcal{C} \subseteq \mathbb{K}[\boldsymbol{x}]$ is called a *Ritt characteristic set* of the ideal $\langle \mathcal{P} \rangle$ if $\mathcal{C} \subseteq \langle \mathcal{P} \rangle$ and $\mathrm{prem}(P, \mathcal{C}) = 0$ for any $P \in \langle \mathcal{P} \rangle$.

The concepts of median set (see [1, Def. 3.4]) and W-characteristic set (see Definition 7 below) are introduced and applied to compute Ritt characteristic sets of given polynomial ideals.

2.2 W-Characteristic Set, Characteristic Pair and Characteristic Decomposition

For any polynomial ideal there exists a reduced Gröbner basis which is unique with respect to a fixed term ordering. In particular, from the reduced LEX Gröbner basis of an ideal, one can extract the W-characteristic set of this ideal as defined below.

Definition 7 ([29, **Def. 3.1**]). Let $\mathcal{P}$ be a polynomial set in $\mathbb{K}[\boldsymbol{x}]$ and $\mathcal{G}$ be the reduced LEX Gröbner basis of the ideal $\langle \mathcal{P} \rangle$. Denote $\mathcal{G}^{(i)} := \{G \in \mathcal{G} \mid \mathrm{lv}(G) = x_i\}$. Then the ordered set of all the smallest polynomials with respect to $<_{\mathrm{LEX}}$ in every set $\mathcal{G}^{(i)}$ for $i = 1, \ldots, n$ is called the *W-characteristic set* of $\langle \mathcal{P} \rangle$ (or of $\mathcal{P}$).

Given a minimal LEX Gröbner basis $\mathcal{G}$ which is not necessarily reduced, the minimal triangular set $T_{\mathcal{G}}$ [1, Notation 3.3] extracted from $\mathcal{G}$ in the same way as in Definition 7 is investigated in [1], with its basic properties proved when $\langle \mathcal{G} \rangle$ is a prime ideal. Note that the W-characteristic set is extracted from a reduced LEX Gröbner basis, and thus it is uniquely defined with respect to the polynomial ideal and may possess stronger properties.

Proposition 1 ([29, **Prop. 3.1**]). *Let $\mathcal{C}$ be the W-characteristic set of $\langle \mathcal{P} \rangle \subseteq \mathbb{K}[\boldsymbol{x}]$. Then*

(a) For any $P \in \langle \mathcal{P} \rangle$, $\mathrm{prem}(P, \mathcal{C}) = 0$;
(b) $\langle \mathcal{C} \rangle \subseteq \langle \mathcal{P} \rangle \subseteq \mathrm{sat}(\mathcal{C})$;
(c) $\mathsf{Z}(\mathcal{C}/\mathrm{ini}(\mathcal{C})) \subseteq \mathsf{Z}(\mathcal{P}) \subseteq \mathsf{Z}(\mathcal{C})$.

Note that the W-characteristic set $\mathcal{C}$ of $\langle \mathcal{P} \rangle$ is extracted from the LEX Gröbner basis $\mathcal{G}$ of $\langle \mathcal{P} \rangle$, thus $\mathcal{C} \subseteq \langle \mathcal{G} \rangle = \langle \mathcal{P} \rangle$. By Proposition 1(a) we also have $\mathrm{prem}(P, \mathcal{C}) = 0$ for any polynomial $P \in \langle \mathcal{P} \rangle$. Therefore, the W-characteristic set

C is quite close to the Ritt characteristic set of $\langle P \rangle$ by Definition 6. While for an arbitrary polynomial ideal, the W-characteristic set and Ritt characteristic set are not necessarily the same, since a W-characteristic set may not be an ascending set.

Let $P \subseteq \mathbb{K}[x]$ be a polynomial set and C be the W-characteristic set of $\langle P \rangle$. For a fixed variable ordering $<$, we say that the *variable ordering condition is satisfied* for C if all the parameters of C are ordered before the leading variables of polynomials in C with respect to $<$. When the variable ordering condition is satisfied for a W-characteristic set C, we have the following property on the regularity and normality of C.

Theorem 1 ([29, **Thm. 3.9**]). *Let $C = [C_1, \ldots, C_r]$ be the W-characteristic set of $\langle P \rangle \subseteq \mathbb{K}[x]$. If the variable ordering condition is satisfied for C and C is not normal, then there exists an integer k $(1 \le k < r)$ such that $[C_1, \ldots, C_k]$ is normal and $[C_1, \ldots, C_{k+1}]$ is not regular.*

Remark 1. When the variable ordering condition is not satisfied for the W-characteristic set, Theorem 1 does not hold in general. See [29, Ex. 3.1(b)] for an example where the W-characteristic set is not normal but regular.

There are more pseudo-divisibility relationships between polynomials in W-characteristic sets proved in [29]. Based on those relationships, Proposition 1, and Theorem 1, an effective algorithm for normal triangular decomposition of polynomial sets is proposed in [30].

Definition 8 ([30, **Def. 3.1**]). A pair $(\mathcal{G}, C)$ of polynomial sets $\mathcal{G}, C \subseteq \mathbb{K}[x]$ is called a *characteristic pair* if $\mathcal{G}$ is a reduced LEX Gröbner basis, C is the W-characteristic set of $\langle \mathcal{G} \rangle$, and C is normal.

For any polynomial set $\mathcal{F} \subseteq \mathbb{K}[x]$, A finite number of characteristic pairs $(\mathcal{G}_1, C_1), \ldots, (\mathcal{G}_t, C_t)$ is said to be a *characteristic decomposition* of $\mathcal{F}$ if the following zero relationship holds.

$$\mathsf{Z}(\mathcal{F}) = \bigcup_{i=1}^{t} \mathsf{Z}(\mathcal{G}_i) = \bigcup_{i=1}^{t} \mathsf{Z}(C_i / \mathrm{ini}(C_i)) = \bigcup_{i=1}^{t} \mathsf{Z}(\mathrm{sat}(C_i)). \tag{1}$$

As can be seen from the definition of characteristic decomposition above, from a characteristic decomposition $(\mathcal{G}_1, C_1), \ldots, (\mathcal{G}_t, C_t)$ of a polynomial set $\mathcal{F}$ one can easily extract a normal decomposition $C_1, \ldots, C_t$ of $\mathcal{F}$.

Definition 9 ([30, **Def. 3.8**]). A reduced LEX Gröbner basis $\mathcal{G}$ is said to be *characterizable* if $\langle \mathcal{G} \rangle = \mathrm{sat}(C)$, where C is the W-characteristic set of $\mathcal{G}$.

Obviously, the W-characteristic set is uniquely extracted from the reduced LEX Gröbner basis. Furthermore, it is proved that the W-characteristic set of a characterizable Gröbner basis is also normal [30, Prop. 3.9], and thus a characterizable Gröbner basis and its W-characteristic set can determine each other.

3 A New Algorithm for Characteristic Decomposition

In this section we first propose a new algorithm for characteristic decomposition under the assumption on the variable ordering. It is also the base of the algorithm for quasi-characteristic decomposition presented in Sect. 4.

By Theorem 1 we know that for the W-characteristic set C of any polynomial set $\mathcal{P}$, if the variable ordering condition is satisfied, the regularity and normality of C are equivalent. The property of irregularity allows us to split the Gröbner basis $\mathcal{G}$ of $\langle \mathcal{P} \rangle$ to compute a regular decomposition which is nicely also a normal decomposition. We assume here that the variables are properly ordered such that the variable ordering condition is always satisfied. Next let us explain the splitting process.

Let Φ be a set of polynomial sets, initialized as $\{\mathcal{F}\}$ with $\mathcal{F}$ being the input polynomial set in $\mathbb{K}[\boldsymbol{x}]$ for characteristic decomposition, and Ψ be a set which contains the characteristic pairs already computed. Now we pick a polynomial set $\mathcal{P}$ and remove it from Φ, compute the reduced LEX Gröbner basis $\mathcal{G}$ of the ideal $\langle \mathcal{P} \rangle$, and extract the W-characteristic set $C = [C_1, \ldots, C_r]$ of $\langle \mathcal{P} \rangle$ from $\mathcal{G}$. Let $I_i = \mathrm{ini}(C_i)$ for $i = 1, \ldots, r$.

(1) If C is normal, then from Proposition 1 we know that

$$\mathsf{Z}(C/\mathrm{ini}(C)) \subseteq \mathsf{Z}(\mathcal{P}) \subseteq \mathsf{Z}(C). \tag{2}$$

In view of this zero relation, we put the characteristic pair $(\mathcal{G}, C)$ into Ψ and adjoin the polynomial sets $\mathcal{G} \cup \{I_1\}, \ldots, \mathcal{G} \cup \{I_r\}$ to Φ for further processing.

(2) If C is not normal, let $k > 1$ be the smallest integer such that $C_k = [C_1, \ldots, C_k]$ is abnormal. Then by Theorem 1 we know that $[C_1, \ldots, C_{k-1}]$ is regular.

   (a) If I_k is zero modulo $\mathrm{sat}(C_{k-1})$ which means $I_k \in \mathrm{sat}(C_{k-1})$, then one can see that $(I_1 \cdots I_{k-1})^q I_k \in \langle C_{k-1} \rangle \subset \langle \mathcal{G} \rangle$ for some integer $q \geq 0$. Note that $I_1, \ldots, I_k$ are all B-reduced with respect to $\mathcal{G}$. In this case, we adjoin the polynomial sets $\mathcal{G} \cup \{I_1\}, \ldots, \mathcal{G} \cup \{I_k\}$ to Φ.

   (b) Otherwise I_k is a zero divisor modulo $\mathrm{sat}(C_{k-1})$. Then there exists a polynomial F in the Gröbner basis of $(\mathrm{sat}(C_{k-1}) : I_k)$ and $F \notin \mathrm{sat}(C_{k-1})$, meaning $FI_k \in \mathrm{sat}(C_{k-1})$ and $F \notin \mathrm{sat}(C_{k-1})$. It follows that

$$(I_k \cdots I_{k-1})^{\bar{q}} F I_k \in \langle C_{k-1} \rangle \subset \langle \mathcal{G} \rangle$$

   for some integer $\bar{q} \geq 0$. Since $F \notin \langle \mathcal{G} \rangle$ (the reason will be detailed in the proof of Theorem 2), we adjoin $\mathcal{G} \cup \{I_1\}, \ldots, \mathcal{G} \cup \{I_k\}, \mathcal{G} \cup \{F\}$ to Φ.

After the splitting of $\mathcal{G}$, we continue picking another polynomial set $\mathcal{P}'$ (and meanwhile remove it) from Φ, and split the Gröbner basis $\mathcal{G}'$ of $\mathcal{P}'$ with the same strategy above. In any case above, the new polynomial set $\mathcal{P}'$ after the splitting is obtained by adjoining a polynomial which does not belong to $\langle \mathcal{G} \rangle$. Hence $\langle \mathcal{G} \rangle \subsetneq \langle \mathcal{G}' \rangle$. Then we could consider further the reduced LEX Gröbner basis and the W-characteristic set of each $\langle \mathcal{P}' \rangle$ and continue the splitting process. This process is repeated until Φ becomes empty. In view of the Ascending Chain Condition of polynomial ideals, this process must terminate in a finite number of steps. Finally, we shall obtain finitely many reduced LEX Gröbner bases $\mathcal{G}_1, \ldots, \mathcal{G}_s$ such that $\mathsf{Z}(\mathcal{P}) = \mathsf{Z}(\mathcal{G}_1) \cup \cdots \cup \mathsf{Z}(\mathcal{G}_s)$, or equivalently $\sqrt{\langle \mathcal{P} \rangle} = \sqrt{\langle \mathcal{G}_1 \rangle} \cap \cdots \cap \sqrt{\langle \mathcal{G}_s \rangle}$, and the W-characteristic set $\mathcal{C}_i$ of each $\langle \mathcal{G}_i \rangle$ is normal for $i = 1, \ldots, s$. This method for characteristic decomposition, whose main steps are outlined above, is described formally as Algorithm 1. Note that $F \notin \mathrm{sat}(\mathcal{C}_{k-1})$ in Line 19 can be effectively tested with $\mathrm{prem}(F, \mathcal{C}_{k-1}) \neq 0$, since $\mathcal{C}_{k-1}$ is normal and thus regular.

Algorithm 1. $\Psi := \mathsf{CharDec}(\mathcal{F}, <)$ (algorithm for characteristic decomposition)

Input: a finite, nonempty set $\mathcal{F}$ of nonzero polynomials in $\mathbb{K}[\boldsymbol{x}]$.
Output: a characteristic decomposition Ψ of $\mathcal{F}$ such that
$\mathsf{Z}(\mathcal{F}) = \bigcup_{(\mathcal{G}, \mathcal{C}) \in \Psi} \mathsf{Z}(\mathcal{G}) = \bigcup_{(\mathcal{G}, \mathcal{C}) \in \Psi} \mathsf{Z}(\mathcal{C}/\mathrm{ini}(\mathcal{C}))$ if the variable ordering condition is satisfied; "Fail" otherwise.

```
1  Ψ := ∅, Φ := {F};
2  while Φ ≠ ∅ do
3      Choose P from Φ and set Φ := Φ \ {P};
4      G := reduced LEX Gröbner basis of ⟨P⟩;
5      if G ≠ {1} then
6          Extract the W-characteristic set C of ⟨P⟩ from G;
7          if The variable ordering condition is not satisfied for C then
8              return "Fail";
9          if C is normal then
10             Ψ := Ψ ∪ {(G, C)};
11             Φ := Φ ∪ {G ∪ {ini(C)} | ini(C) ∉ 𝕂, C ∈ C};
12         else
13             Let C_k be the first polynomial such that C_k is abnormal;
14             I_k := ini(C_k);
15             R := prem(I_k, C_{k-1});
16             if R = 0 then
17                 Φ := Φ ∪ {G ∪ {ini(C_j)} | j ≤ k};
18             else
19                 F := any polynomial in Gröbner basis of (sat(C_{k-1}) : I_k) such
                       that F ∉ sat(C_{k-1});
20                 Φ := Φ ∪ {G ∪ {ini(C_j)} | j ≤ k} ∪ {G ∪ {F}};
21 return Ψ
```

Theorem 2. *If the variable ordering condition is always satisfied, then Algorithm 1 terminates in a finite number of steps with correct output.*

Proof. (Termination) We can see that every time a polynomial set $\mathcal{P}$ is picked from Φ, splitting occurs only in Lines 11, 17 and 20. To prove the termination of

Algorithm 1, we only need to show that for all the three splitting cases, each polynomial set $\mathcal{G}' = \mathcal{G} \cup \{H\}$ adjoined to Φ generates a polynomial ideal $\langle \mathcal{G}' \rangle$ strictly larger than $\langle \mathcal{G} \rangle$. Let $\mathcal{C} = [C_1, \ldots, C_r]$ be the W-characteristic set extracted from $\mathcal{G}$ with $x_{p_i} = \mathrm{lv}(C_i)$ and $I_i = \mathrm{ini}(C_i)$ for $i = 1, \ldots, r$. One can see that I_i is B-reduced with respect to $\mathcal{G}$, for otherwise it contradicts with the fact that $\mathcal{G}$ is the reduced LEX Gröbner basis of $\langle \mathcal{P} \rangle$. Then $I_i \notin \langle \mathcal{G} \rangle$ and thus $\langle \mathcal{G} \cup I_i \rangle$ is strictly larger than $\langle \mathcal{G} \rangle$. To complete the proof of termination, it remains to show that F in Line 20 is not in $\langle \mathcal{G} \rangle$. Suppose that $F \in \langle \mathcal{G} \rangle$. Since $\mathrm{lv}(F) < \mathrm{lv}(C_k) = x_{p_k}$, by Proposition 1(b) we have $F \in \langle \mathcal{G} \rangle \cap \mathbb{K}[x_1, \ldots, x_{p_k-1}] = \langle \mathcal{G}_{k-1} \rangle \subseteq \mathrm{sat}(\mathcal{C}_{k-1})$, which leads to a contradiction.

(Correctness) When $\mathsf{Z}(\mathcal{F}) = \emptyset$, $\mathcal{G} = \{1\}$ in Line 5 and $\Psi = \emptyset$ is returned. Therefore, we need to show that when $\Psi \neq \emptyset$ and the variable ordering condition is always satisfied, all the pairs $(\mathcal{G}, \mathcal{C}) \in \Psi$ are characteristic pairs and the zero relation (1) holds. The property that $(\mathcal{G}, \mathcal{C})$ are all characteristic pairs is obvious, for only in Line 10 an element, which is a characteristic pair, is adjoined to Ψ. Next we prove the zero relation (1).

(1) When the W-characteristic set $\mathcal{C}$ in Line 6 is normal, let $J = I_1 \cdots I_r$. Then $\mathsf{Z}(\mathcal{P}) = (\mathsf{Z}(\mathcal{P}) \setminus \mathsf{Z}(J)) \cup \mathsf{Z}(\mathcal{P} \cup \{J\})$. By Theorem 1(c) we know that $\mathsf{Z}(\mathcal{P}) \setminus \mathsf{Z}(J) = \mathsf{Z}(\mathcal{C}/\mathrm{ini}(\mathcal{C}))$. Moreover,

$$\mathsf{Z}(\mathcal{P} \cup \{J\}) = \bigcup_{i=1}^{r} \mathsf{Z}(\mathcal{P} \cup \{I_i\}) = \bigcup_{i=1}^{r} \mathsf{Z}(\mathcal{G} \cup \{I_i\}).$$

Therefore,

$$\mathsf{Z}(\mathcal{P}) = \mathsf{Z}(\mathcal{C}/\mathrm{ini}(\mathcal{C})) \cup \bigcup_{i=1}^{r} \mathsf{Z}(\mathcal{G} \cup \{I_i\}). \tag{3}$$

(2) When $\mathcal{C}$ is abnormal. (a) If I_k in Line 14 is zero modulo $\mathrm{sat}(\mathcal{C}_{k-1})$, then by Theorem 1 we have $\mathrm{prem}(I_k, \mathcal{C}_{k-1}) = 0$. Then there exist $Q_1, \ldots, Q_{k-1} \in \mathbb{K}[\boldsymbol{x}]$ and $q_1, \ldots, q_{k-1} \in \mathbb{Z}_{\geq 0}$ such that $I_1^{q_1} \ldots I_{k-1}^{q_{k-1}} I_k = Q_1 C_1 + \cdots + Q_{k-1} C_{k-1} \in \langle \mathcal{C}_{k-1} \rangle \subseteq \langle \mathcal{G} \rangle$. Therefore,

$$\mathsf{Z}(\mathcal{P}) = \mathsf{Z}(\mathcal{G}) = \mathsf{Z}(\mathcal{G} \cup \{I_1^{q_1} \ldots I_{k-1}^{q_{k-1}} I_k\}) = \bigcup_{i=1}^{k} \mathsf{Z}(\mathcal{G} \cup \{I_i\}). \tag{4}$$

(b) Otherwise I_k is a zero-divisor modulo $\mathrm{sat}(\mathcal{C}_{k-1})$, and thus $\mathrm{prem}(I_k, \mathcal{C}_{k-1}) \neq 0$. Then there exists $F \in \mathbb{K}[\boldsymbol{x}]$ such that $F I_k \in \mathrm{sat}(\mathcal{C}_{k-1})$, and it follows that there exist $q_1, \ldots, q_{k-1} \in \mathbb{Z}_{\geq 0}$ such that $I_1^{q_1} \ldots I_{k-1}^{q_{k-1}} F I_k \in \langle \mathcal{C}_{k-1} \rangle \subseteq \langle \mathcal{G} \rangle$, and therefore

$$\mathsf{Z}(\mathcal{P}) = \mathsf{Z}(\mathcal{G}) = \mathsf{Z}(\mathcal{G} \cup \{I_1^{q_1} \ldots I_{k-1}^{q_{k-1}} F I_k\}) = \bigcup_{i=1}^{k} \mathsf{Z}(\mathcal{G} \cup \{I_i\}) \cup \mathsf{Z}(\mathcal{G} \cup \{F\}). \tag{5}$$

The zero relations in (3), (4) and (5) show that for each polynomial set $\mathcal{P} \in \Phi$, any zero of $\mathcal{P}$ is either in $\mathsf{Z}(\mathcal{C}/\mathrm{ini}(\mathcal{C}))$ if the W-characteristic set $\mathcal{C}$ of $\langle \mathcal{P} \rangle$ is normal or in $\mathsf{Z}(\mathcal{P}')$, where $\mathcal{P}'$ is a new polynomial set adjoined to Φ for later computation. This proves the zero relation $Z(\mathcal{F}) = \bigcup_{(\mathcal{G},\mathcal{C})\in\Psi} Z(\mathcal{C}/\operatorname{ini}(\mathcal{C}))$, for Algorithm 1 terminates when Φ becomes empty.

On one hand, by the zero relation (2), we have

$$Z(\mathcal{F}) = \bigcup_{(\mathcal{G},\mathcal{C})\in\Psi} Z(\mathcal{C}/\operatorname{ini}(\mathcal{C})) \subseteq \bigcup_{(\mathcal{G},\mathcal{C})\in\Psi} Z(\mathcal{G}).$$

On the other hand, $Z(\mathcal{G}) \subseteq Z(\mathcal{F})$ holds for all $(\mathcal{G},\mathcal{C}) \in \Psi$ according to the zero relations (3), (4) and (5) for the splitting. This proves the equality $Z(\mathcal{F}) = \bigcup_{(\mathcal{G},\mathcal{C})\in\Psi} Z(\mathcal{G})$. For each $(\mathcal{G},\mathcal{C}) \in \Psi$, we have $Z(\mathcal{C}/\operatorname{ini}(\mathcal{C})) \subseteq Z(\operatorname{sat}(\mathcal{C})) \subseteq Z(\mathcal{G})$, and thus

$$Z(\mathcal{F}) = \bigcup_{(\mathcal{G},\mathcal{C})\in\Psi} Z(\mathcal{C}/\operatorname{ini}(\mathcal{C})) \subseteq \bigcup_{(\mathcal{G},\mathcal{C})\in\Psi} Z(\operatorname{sat}(\mathcal{C})) \subseteq \bigcup_{(\mathcal{G},\mathcal{C})\in\Psi} Z(\mathcal{G}) = Z(\mathcal{F}).$$

This completes the proof of the zero relation (1).

For an arbitrary polynomial set, if the variable ordering condition is always satisfied, the output of Algorithm 1 is definitely a characteristic decomposition. Otherwise, "Fail" will be returned. It needs to be emphasized that the satisfaction of the variable ordering condition is only sufficient for characteristic decomposition here, but not necessary. In fact, we can carry out splitting in the algorithm regardless of the assumption on the variable ordering. In Sect. 4, instead of simply returning "Fail" in the algorithm, we propose quasi-characteristic decomposition to handle the case where the variable ordering condition is not always satisfied.

4   Decomposition into Quasi-characteristic Pairs

In this section we discuss the decomposition of an arbitrary polynomial set into quasi-characteristic pairs with associated zero relations and prove some properties about the decomposition. A decomposition algorithm is also presented.

To get rid of the assumption on the variable ordering, we first introduce the concept of quasi-characteristic pair which is close to but weaker than the characteristic pair.

Definition 10. A pair $(\mathcal{G},\mathcal{C})$ with $\mathcal{G},\mathcal{C} \subseteq \mathbb{K}[\boldsymbol{x}]$ is called a *quasi-characteristic pair* if $\mathcal{G}$ is a reduced LEX Gröbner basis, $\mathcal{C}$ is the W-characteristic set of $\langle\mathcal{G}\rangle$, and $\mathcal{C}$ is regular.

By definition any characteristic pair is obviously a quasi one, but a quasi-characteristic pair is not necessarily a characteristic pair. They are equivalent when the variable ordering condition is satisfied for the W-characteristic set in the pair. The following results exhibit some of the nice properties of characteristic pairs which the quasi ones also possess.

Proposition 2. Let $\mathcal{C}$ be the regular W-characteristic set of $\langle\mathcal{P}\rangle \subseteq \mathbb{K}[\boldsymbol{x}]$. If $\operatorname{sat}(\mathcal{C}) = \langle\mathcal{C}\rangle$, then $\operatorname{sat}(\mathcal{C}) = \langle\mathcal{P}\rangle$.

Proof. The proposition follows immediately from $\langle C \rangle \subseteq \langle P \rangle \subseteq \mathrm{sat}(C)$.

Proposition 3. *The W-characteristic set of any characterizable Gröbner basis is regular.*

Proof. Let G be any characterizable Gröbner basis and $C = [C_1, \ldots, C_r]$ be the W-characteristic set of $\langle G \rangle$. Denote p-sat$(C) := \{P \in \mathbb{K}[x] \mid \mathrm{prem}(P, C) = 0\}$. On one hand, for any $P \in$ p-sat(C), we have $\mathrm{prem}(P, C) = 0$, and thus there exist $q_1, \ldots, q_r \in \mathbb{Z}_{\geq 0}$ and $Q_1, \ldots, Q_r \in \mathbb{K}[x]$ such that $\mathrm{ini}(C_1)^{q_1} \cdots \mathrm{ini}(C_r)^{q_r} P = \sum_{i=1}^r Q_i C_i \in \langle C \rangle$. It follows that $P \in \mathrm{sat}(C)$, which means p-sat$(C) \subseteq \mathrm{sat}(C)$. On the other hand, for any $P \in \mathrm{sat}(C)$, we have $P \in \langle G \rangle$ since $\langle G \rangle = \mathrm{sat}(C)$. By Proposition 1(a) one can see that $\mathrm{prem}(P, C) = 0$, which means $\mathrm{sat}(C) \subseteq$ p-sat(C). Hence $\mathrm{sat}(C) = $ p-sat(C), and by [1, Thm. 6.1] C is regular.

Remark 2. Proposition 3 above holds regardless of the assumption on the variable ordering. Under this additional assumption, [30, Prop. 3.9] shows that the W-characteristic set C extracted from any characterizable Gröbner basis G is normal. In fact, the proof of [30, Prop. 3.9] also implicitly depends on the assumption on the variable ordering, which makes the pseudo-divisibility relationships among the polynomials in C happen. Our proof of Proposition 3, however, does not require this assumption and is much simpler. Furthermore, when the variable ordering condition is satisfied for C, one can also see that C is normal due to the equivalence of regularity and normality of C.

Let $\mathcal{F}$ be a finite, nonempty set of polynomials in $\mathbb{K}[x]$. We call a finite number of quasi-characteristic pairs $(G_1, C_1), \ldots, (G_t, C_t)$ a *quasi-characteristic decomposition* of $\mathcal{F}$ if the zero relation (1) holds. From any quasi-characteristic decomposition of $\mathcal{F}$, one can extract a regular decomposition $C_1, \ldots, C_t$ of $\mathcal{F}$ such that $\mathsf{Z}(\mathcal{F}) = \bigcup_{i=1}^t \mathsf{Z}(C_i / \mathrm{ini}(C_i))$.

According to [2, Prop. 4.1.3], we know that $\mathrm{sat}(\mathcal{T})$ is purely equidimensional for any regular set $\mathcal{T} \subseteq \mathbb{K}[x]$. It follows that for any quasi-characteristic pair (G, C) in the quasi-characteristic decomposition of $\mathcal{F}$, by Proposition 2 $\langle G \rangle$ is purely equidimensional if $\mathrm{sat}(C) = \langle C \rangle$ is verified, and thus we have the following.

Proposition 4. *Let Φ be a quasi-characteristic decomposition of $\mathcal{F} \subseteq \mathbb{K}[x]$ and assume that $\mathrm{sat}(C) = \langle C \rangle$ for every quasi-characteristic pair $(G, C) \in \Psi$. Then $\sqrt{\langle \mathcal{F} \rangle} = \bigcap_{(G,C) \in \Psi} \sqrt{\langle G \rangle}$ and each $\langle G \rangle$ is purely equidimensional.*

Proposition 5. *Let $C = [C_1, \ldots, C_r]$ be the W-characteristic set of $\langle P \rangle \subseteq \mathbb{K}[x]$ and*

$$C^* = [C_1, \mathrm{prem}(C_2, [C_1]), \ldots, \mathrm{prem}(C_r, [C_1, \ldots, C_{r-1}])]. \tag{6}$$

If C is regular, then

(1) C^* is regular,
(2) C^* is a Ritt characteristic set of $\langle P \rangle$,
(3) $\mathsf{Z}(C^* / \mathrm{ini}(C^*)) = \mathsf{Z}(C / \mathrm{ini}(C))$.

Proof. (1–2) According to [29, Thm. 3.4], C^* is regular and C^* is a Ritt characteristic set of $\langle \mathcal{P} \rangle$. (3) This can be easily obtained from [30, Thm. 3.14].

Let Ψ be a quasi-characteristic decomposition of $\mathcal{F} \subseteq \mathbb{K}[x]$ and C^* be computed from C according to (6) for each quasi-characteristic pair $(\mathcal{G}, \mathcal{C}) \in \Psi$. Then obviously we have

$$\mathsf{Z}(\mathcal{F}) = \bigcup_{(\mathcal{G},\mathcal{C}) \in \Psi} \mathsf{Z}(C^*/\operatorname{ini}(C^*)) = \bigcup_{(\mathcal{G},\mathcal{C}) \in \Psi} \mathsf{Z}(\operatorname{sat}(C^*)), \qquad (7)$$

and C^* is the Ritt characteristic set of $\langle \mathcal{G} \rangle$ for each $(\mathcal{G}, \mathcal{C}) \in \Psi$.

Theorem 3. *From any finite, nonempty polynomial set $\mathcal{F} \subseteq \mathbb{K}[x]$, one can compute a quasi-characteristic decomposition of $\mathcal{F}$ in a finite number of steps.*

To give an algorithm for computing the quasi-characteristic decomposition, we just need to change the condition of normality in Line 9 of Algorithm 1 to regularity, and the algorithm for quasi-characteristic decomposition will be embedded in Algorithm 2 as one degenerate case. The proof of Theorem 3 is very similar to that of Theorem 2.

Next Algorithm 2 is proposed for computing a characteristic decomposition of a polynomial set when the variable ordering condition is always satisfied, and a quasi-characteristic decomposition if not. Obviously, the correctness and termination of Algorithm 2 could be proved by Theorems 2 and 3. Note that regularity of the W-characteristic set C in Line 7 of Algorithm 2 can be effectively tested with iterated resultant [28, Lem. 4.3.2].

Remark 3. Based on the pseudo-divisibility relationships among polynomials in the reduced LEX Gröbner basis $\mathcal{G}$ proved in [29], an algorithm for computing characteristic decomposition is proposed in [30, Sec. 4], where the assumption on variable ordering is necessary for splitting $\mathcal{G}$. Otherwise the algorithm may fail to properly split in the decomposition process. It worths mentioning that Algorithm 2 can always correctly split $\mathcal{G}$ without the assumption on the variable ordering, and the output is either a characteristic decomposition or a quasi one in the worse case, where the variable ordering condition is not always satisfied. Actually, even in the worse case, the quasi-characteristic decomposition computed with Algorithm 2 is still possibly a characteristic decomposition (see Sect. 5 for some examples).

5   Two Illustrative Examples

5.1   Characteristic Decomposition

Consider $\mathcal{F} = \{ay - x - 1, -xyz + az, xz^2 - az + y\} \subseteq \mathbb{K}[a, x, y, z]$ with $a < x < y < z$. The procedure to compute a characteristic decomposition of $\mathcal{F}$ using Algorithm 1 is illustrated below.

Algorithm 2. $\Psi := \mathsf{QCharDec}(\mathcal{F}, <)$ (algorithm for (quasi-) characteristic decomposition)

Input: a finite, nonempty set $\mathcal{F}$ of nonzero polynomials in $\mathbb{K}[\boldsymbol{x}]$.
Output: either a characteristic decomposition (if output $flag = 1$) or a
   quasi-characteristic decomposition (if output $flag = 2$) Ψ of $\mathcal{F}$ such
   that $\mathsf{Z}(\mathcal{F}) = \bigcup_{(\mathcal{G},\mathcal{C})\in\Psi} \mathsf{Z}(\mathcal{G}) = \bigcup_{(\mathcal{G},\mathcal{C})\in\Psi} \mathsf{Z}(\mathcal{C}/\operatorname{ini}(\mathcal{C}))$, and the type $flag$
   of decomposition.

1 $\Psi := \emptyset, \Phi := \{\mathcal{F}\}, flag := 1;$
2 **while** $\Phi \neq \emptyset$ **do**
3 | Choose $\mathcal{P}$ from Φ and set $\Phi := \Phi \setminus \{\mathcal{P}\};$
4 | $\mathcal{G} :=$ reduced LEX Gröbner basis of $\langle\mathcal{P}\rangle;$
5 | **if** $\mathcal{G} \neq \{1\}$ **then**
6 | | $\mathcal{C} :=$ W-characteristic set of $\langle\mathcal{G}\rangle;$
7 | | **if** $\mathcal{C}$ *is regular* **then**
8 | | | **if** *the variable ordering condition is not satisfied for* $\mathcal{C}$ **then**
9 | | | | $flag := 2;$
10 | | | $\Psi := \Psi \cup \{(\mathcal{G},\mathcal{C})\};$
11 | | | $\Phi := \Phi \cup \{\mathcal{G} \cup \{\operatorname{ini}(C)\} \mid \operatorname{ini}(C) \notin \mathbb{K}, C \in \mathcal{C}\};$
12 | | **else**
13 | | | **if** *the variable ordering condition is not satisfied for* $\mathcal{C}$ **then**
14 | | | | $C_k :=$ the first polynomial that makes $\mathcal{C}_k$ irregular;
15 | | | **else**
16 | | | | $C_k :=$ the first polynomial that makes $\mathcal{C}_k$ abnormal;
17 | | | $I_k := \operatorname{ini}(C_k);$
18 | | | $R := \operatorname{prem}(I_k, \mathcal{C}_k);$
19 | | | **if** $R = 0$ **then**
20 | | | | $\Phi := \Phi \cup \{\mathcal{G} \cup \{\operatorname{ini}(C_j)\} \mid j \leq k\};$
21 | | | **else**
22 | | | | $F :=$ any polynomial in Gröbner basis of $(\operatorname{sat}(\mathcal{C}_{k-1}) : I_k)$ such
   | | | | that $F \notin \operatorname{sat}(\mathcal{C}_{k-1});$
23 | | | | $\Phi := \Phi \cup \{\mathcal{G} \cup \{\operatorname{ini}(C_j)\} \mid j \leq k\} \cup \{\mathcal{G} \cup \{F\}\};$

24 **return** $\Psi, flag$

With the reduced LEX Gröbner basis of $\langle\mathcal{F}\rangle$ computed as

$$\mathcal{G} = [-a^2x + x^3 - a^2 + 2x^2 + x, ay - x - 1, x^2y - ax + xy - a, xy^2 - x - 1,$$
$$- a^2z + x^2z + xz, xyz - az, y^3 - yz + z^2 - y],$$

and the W-characteristic set as $\mathcal{C} = [-a^2x + x^3 - a^2 + 2x^2 + x, ay - x - 1, -a^2z + x^2z + xz]$, one can check that the W-characteristic set $\mathcal{C}$ is abnormal, and it satisfies the variable ordering condition. One can also check that $\operatorname{prem}(\operatorname{ini}(C_3), \mathcal{C}_2) = -a^2 + x^2 + x$, which means $\operatorname{ini}(\mathcal{C}_3) \notin \operatorname{sat}(\mathcal{C}_2)$. Then we can find a polynomial $F = x + 1 \notin \operatorname{sat}(\mathcal{C}_2)$ but $F \cdot \operatorname{ini}(C_3) \in \operatorname{sat}(\mathcal{C}_2)$, which leads the splitting in Line 20: $\mathcal{F}_1 := \mathcal{G} \cup \{a\}$, $\mathcal{F}_2 := \mathcal{G} \cup \{-a^2 + x^2 + x\}$, and $\mathcal{F}_3 := \mathcal{G} \cup \{x+1\}$ are adjoined to Φ for further computation.

For $\mathcal{F}_1 \in \Phi$, by computing the reduced LEX Gröbner basis $\mathcal{G}_1 = \{a, x + 1, y^2, yz, z^2 - y\}$ of $\langle \mathcal{F}_1 \rangle$ and the W-characteristic set $\mathcal{C}_1 = [a, x+1, y^2, yz]$ of $\mathcal{G}_1$, one finds that $\mathcal{C}_1$ is abnormal, then some quick calculation leads the splitting in Line 20: $\mathcal{F}_4 := \mathcal{G}_1 \cup \{y\}$ is adjoined to Φ.

For $\mathcal{F}_2 \in \Phi$, after the computation of the reduced LEX Gröbner basis $\mathcal{G}_2 = \{x + 1, ay, y^2, a^2z, az + yz, az + z^2 - y\}$ of $\langle \mathcal{F}_2 \rangle$ and the W-characteristic set $\mathcal{C}_2 = [x + 1, ay, a^2z]$ of $\mathcal{G}_2$, one finds that $\mathcal{C}_2$ is normal and thus $(\mathcal{G}_2, \mathcal{C}_2)$ is one characteristic pair. Meanwhile, $\mathcal{F}_5 := \mathcal{G}_2 \cup \{a\}$ is adjoined to Φ.

For $\mathcal{F}_3 \in \Phi$, the reduced LEX Gröbner basis of $\mathcal{F}_3$ is computed as $\mathcal{G}_3 = \{-a^2 + x^2 + x, ay - x - 1, xy^2 - x - 1, xyz - az, y^3 - yz + z^2 - y\}$ and from it the W-characteristic set $\mathcal{C}_3 = [-a^2 + x^2 + x, ay - x - 1, xyz - az]$ of $\mathcal{G}_3$ is extracted. Clearly $\mathcal{C}_3$ is abnormal, then some quick calculation leads the splitting in Line 17: $\mathcal{F}_6 := \mathcal{G}_3 \cup \{a\}$ and $\mathcal{F}_7 := \mathcal{G}_3 \cup \{xy - a\}$ are adjoined to Φ.

Then $\mathcal{F}_4, \mathcal{F}_6 \in \Phi$ furnish two characteristic pairs $(\mathcal{G}_4, \mathcal{C}_4) = (\{a, x+1, y, z^2\}, [a, x + 1, y, z^2])$ and $(\mathcal{G}_6, \mathcal{C}_6) = (\{-a^2 + x^2 + x, ay - x - 1, xy - a, y^3 - yz + z^2 - y\}, [-a^2 + x^2 + x, ay - x - 1, y^3 - yz + z^2 - y])$, and only $\mathcal{F}_8 := \mathcal{G}_6 \cup \{a\}$ is adjoined to Φ. After handling $\mathcal{F}_7$ and $\mathcal{F}_8$, no more new characteristic pairs are

Table 1. Illustration for Algorithm 2

i	Φ	$\mathcal{P}$	$\mathcal{G}_i$	$\mathcal{C}_i$	VOC	Line	Ψ
1	$\{\mathcal{F}\}$	$\mathcal{F}$	$\{G_1, G_2, G_3,$ $G_4, G_5, G_6,$ $G_7, G_8\}$	$[G_1, G_2, G_5]$	Yes	11	$\{(\mathcal{G}_1, \mathcal{C}_1)\}$
2	$\{\mathcal{F}_1, \mathcal{F}_2\}$	$\mathcal{F}_1$	$\{y, x^2, tx, t^2z,$ $G_6, G_7, G_8\}$	$[y, x^2, tx, t^2z, G_6]$	Yes	23	$\{(\mathcal{G}_1, \mathcal{C}_1)\}$
3	$\{\mathcal{F}_2, \mathcal{F}_3\}$	$\mathcal{F}_2$	$\{z, y^2, xy,$ $x^2, G_3, ry,$ $rx, G_7, r^2\}$	$[z, y^2, xy, G_3, ry]$	Yes	20	$\{(\mathcal{G}_1, \mathcal{C}_1)\}$
4	$\{\mathcal{F}_3, \mathcal{F}_4\}$	$\mathcal{F}_3$	$\{y, x, tz, rt, G_8\}$	$[y, x, tz, rt]$	Yes	20	$\{(\mathcal{G}_1, \mathcal{C}_1)\}$
5	$\{\mathcal{F}_4, \mathcal{F}_5, \mathcal{F}_6\}$	$\mathcal{F}_4$	$\{z, y, x^2, tx,$ $rx, G_7, r^2\}$	$[z, y, x^2, tx, rx]$	Yes	23	$\{(\mathcal{G}_1, \mathcal{C}_1)\}$
6	$\{\mathcal{F}_5, \mathcal{F}_6, \mathcal{F}_7\}$	$\mathcal{F}_5$	$\{y, x, t, G_8\}$	$[y, x, t, G_8]$	Yes	11	$\{(\mathcal{G}_1, \mathcal{C}_1),$ $(\mathcal{G}_6, \mathcal{C}_6)\}$
7	$\{\mathcal{F}_6, \mathcal{F}_7\}$	$\mathcal{F}_6(\mathcal{F}_7)$	$\{z, y, x, rt, r^2\}$	$[z, y, x, rt]$	No	11	$\{(\mathcal{G}_1, \mathcal{C}_1),$ $(\mathcal{G}_6, \mathcal{C}_6),$ $(\mathcal{G}_7, \mathcal{C}_7)\}$
8	$\{\mathcal{F}_8\}$	$\mathcal{F}_8$	$\{z, y, x, t, r^2\}$	$[z, y, x, t, r^2]$	Yes	11	$\{(\mathcal{G}_1, \mathcal{C}_1),$ $(\mathcal{G}_6, \mathcal{C}_6),$ $(\mathcal{G}_7, \mathcal{C}_7),$ $(\mathcal{G}_8, \mathcal{C}_8)\}$

computed and $\varPhi$ becomes empty. The computation in Algorithm 1 ends, and the characteristic decomposition of $\mathcal{F}$ is output as $(\mathcal{G}_2, \mathcal{C}_2), (\mathcal{G}_4, \mathcal{C}_4), (\mathcal{G}_6, \mathcal{C}_6)$.

5.2   Quasi-characteristic Decomposition

Consider $\mathcal{F} = \{-rt + x, -rt^2 + y, -r^2 + z\} \subseteq \mathbb{K}[z, y, x, t, r]$ with $z < y < x < t < r$. The procedure to compute a quasi-characteristic decomposition of $\mathcal{F}$ using Algorithm 2 is shown in Table 1, where $\mathcal{G}_i$ is the computed reduced LEX Gröbner basis and $\mathcal{C}_i$ is its W-characteristic set in the ith loop. "VOC" records if the variable ordering condition is satisfied for $\mathcal{C}_i$, "Line" records the splitting case by the number of line in Algorithm 2.

The polynomial sets $\mathcal{F}_i$ and polynomials G_j in Table 1 are listed below:

$$\mathcal{F}_1 = \mathcal{G}_1 \cup \{y\},\ \mathcal{F}_2 = \mathcal{G}_1 \cup \{y\},\ \mathcal{F}_3 = \mathcal{G}_2 \cup \{x\},\ \mathcal{F}_4 = \mathcal{G}_3 \cup \{y\},$$
$$\mathcal{F}_5 = \mathcal{G}_4 \cup \{t\},\ \ \mathcal{F}_6 = \mathcal{G}_4 \cup \{z\},\ \mathcal{F}_7 = \mathcal{G}_5 \cup \{x\},\ \mathcal{F}_8 = \mathcal{G}_7 \cup \{t\};$$
$$G_1 = x^4 - y^2 z,\ G_2 = tyz - x^3,\ G_3 = tx - y,\ \ \ G_4 = t^2 z - x^2,$$
$$G_5 = ry - x^2,\ \ G_6 = rx - tz,\ \ G_7 = rt - x,\ \ G_8 = r^2 - z.$$

Table 2. Timings for (quasi-) characteristic decomposition (in second)

Ex	Label	Var	Eqs	Deg	Type	GB	SAT	Total	Branches
1	E1	10	10	9	char	0.608	0.001	0.651	5
2	E5†	15	17	3	quasi	2.645	0.001	2.976	7
3	E23†	9	5	2	quasi	0.119	0	0.239	11
4	E32†	8	6	3	quasi	0.077	0.001	0.112	3
5	E33†	13	11	4	quasi	14.335	0	14.678	6
6	E35†	8	8	4	quasi	0.265	0.001	0.324	11
7	S8	4	3	12	char	0.015	0	0.020	2
8	S12	8	4	2	char	0.088	0	0.093	1
9	S13†	5	2	4	quasi	0.104	0	0.139	13
10	Maclane†	10	6	2	quasi	155.682	0.117	165.686	345
11	Leykin_1†	8	6	4	quasi	385.091	1.139	392.746	128
12	F663†	10	9	2	quasi	1.536	0	1.624	5
13	Wang16	4	4	4	char	0.049	0	0.051	1
14	Cyclic5	5	5	5	char	0.208	0.002	0.305	11
15	Filter9	9	9	5	char	0.585	0	0.593	1
16	Cyclic6	6	6	6	char	2.948	0.004	3.234	25
17	N16	5	5	13	char	76.975	0	76.977	1
18	4-body-homog	3	3	8	char	1.258	0	1.530	5
19	Circles	2	2	10	char	3.753	0	3.754	1
20	Katsura-4	5	5	2	char	184.373	0	184.376	1

The output quasi-characteristic decomposition of $\mathcal{P}$, as shown in Table 1, is $(\mathcal{G}_1, \mathcal{C}_1), (\mathcal{G}_6, \mathcal{C}_6), (\mathcal{G}_7, \mathcal{C}_7), (\mathcal{G}_8, \mathcal{C}_8)$. One can check that it is indeed also a characteristic decomposition.

6 Implementation and Experimental Results

We have implemented Algorithms 1 and 2 in MAPLE 18 and carried out experiments with the implementation on an Intel(R) Core(TM) i5-4210U CPU at 1.70 GHz×4 with 7.7 GB RAM under Ubuntu 16.04 LTS. The implementation is based on the functions for Gröbner basis computation available in the FGb library and MAPLE's built-in packages. Selected results of the experiments on the benchmark polynomial sets are presented in Table 2: Ex 1–6 are taken from the Epsilon package, Ex 7–9 from [23], Ex 10–11 from [8], Ex 12–16 from the FGb library, Ex 17 from [28], and Ex 18–20 can be found with this link[1].

In Table 2, "Label" indicates the label in the above-cited references, "Var", "Eqs" and "Deg" indicate the number of variables, the number of polynomials, and the maximal degree of the polynomials in the examples respectively. "Type" indicates the output is a characteristic decomposition (char) or quasi one (quasi). "GB" and "SAT" record the time for computing all the reduced LEX Gröbner bases and saturated ideals, respectively, and "Total" records the total time for (quasi-) characteristic decomposition using Algorithm 2, and "Branches" indicates the number of the (quasi-) charactcristic pairs computed.

The variable ordering condition is not always satisfied for 9 of the test examples (marked with † in Table 2). Quasi-characteristic decompositions of them are computed by Algorithm 2, but in fact 7 of the computed quasi-characteristic decompositions are also characteristic ones. This verifies that the variable ordering condition is only a sufficient condition for computing characteristic decomposition but not a necessary one.

Acknowledgments. The authors would like to thank the anonymous reviewers for their detailed and helpful comments on an earlier version of this paper.

References

1. Aubry, P., Lazard, D., Moreno Maza, M.: On the theories of triangular sets. J. Symbolic Comput. **28**(1–2), 105–124 (1999)
2. Aubry, P.: Ensembles Triangulaires de Polynômes et Résolution de Systemes Algébriques. Implantation en Axiom. Ph.D. thesis, Université Pierre et Marie Curie, France (1999)
3. Becker, T., Weispfenning, V., Kredel, H.: Gröbner Bases: A Computational Approach to Commutative Algebra. Graduate Texts in Mathematics. Springer, New York (1993)

---

[1] http://www.lifl.fr/~lemaire/BCLM09/BCLM09-systems.txt.

4. Bender, M.R., Faugère, J.C., Tsigaridas, E.: Towards mixed Gröbner basis algorithms: The multihomogeneous and sparse case. In: Proceedings of ISSAC 2018, pp. 71–78. ACM Press (2018). https://doi.org/10.1145/3208976.3209018, https://hal.inria.fr/hal-01787423
5. Boulier, F., Han, M., Lemaire, F., Romanovski, V.G.: Qualitative investigation of a gene model using computer algebra algorithms. Prog. Comput. Softw. 41(2), 105–111 (2015)
6. Buchberger, B.: Ein Algorithmus zum Auffinden der Basiselemente des Restklassenrings nach einem nulldimensionalen Polynomideal. Ph.D. thesis, Universität Innsbruck, Austria (1965)
7. Buchberger, B.: Gröbner bases: an algorithmic method in polynomial ideal theory. In: Bose, N. (ed.) Multidimensional Systems Theory, pp. 184–232. Springer, Dordrecht (1985)
8. Chen, C., Golubitsky, O., Lemaire, F., Maza, M.M., Pan, W.: Comprehensive triangular decomposition. In: Ganzha, V.G., Mayr, E.W., Vorozhtsov, E.V. (eds.) CASC 2007. LNCS, vol. 4770, pp. 73–101. Springer, Heidelberg (2007). https://doi.org/10.1007/978-3-540-75187-8_7
9. Chen, C., Moreno Maza, M.: Algorithms for computing triangular decompositions of polynomial systems. J. Symbolic Comput. 47(6), 610–642 (2012)
10. Cheng, J.S., Jin, K., Lazard, D.: Certified rational parametric approximationof real algebraic space curves with local generic position method. J. Symbolic Comput. 58, 18–40 (2013). https://doi.org/10.1016/j.jsc.2013.06.004. http://www.sciencedirect.com/science/article/pii/S0747717113000953
11. Cox, D., Little, J., O'Shea, D.: Ideals, Varieties, and Algorithms: An Introduction to Computational Algebraic Geometry and Commutative Algebra. Undergraduate Texts in Mathematics. Springer, New York (1997). https://doi.org/10.1007/978-3-662-41154-4
12. Cox, D., Little, J., O'Shea, D.: Using Algebraic Geometry. Graduate Texts in Mathematics. Springer, New York (1998). https://doi.org/10.1007/978-1-4757-6911-1
13. Dahan, X.: On lexicographic Gröbner bases of radical ideals in dimension zero: Interpolation and structure, preprint at arXiv:1207.3887 (2012)
14. Dong, R., Mou, C.: Decomposing polynomial sets simultaneously into Gröbner bases and normal triangular sets. In: Gerdt, V.P., Koepf, W., Seiler, W.M., Vorozhtsov, E.V. (eds.) CASC 2017. LNCS, vol. 10490, pp. 77–92. Springer, Cham (2017). https://doi.org/10.1007/978-3-319-66320-3_7
15. Faugère, J.C., Gianni, P., Lazard, D., Mora, T.: Efficient computation of zero-dimensional Gröbner bases by change of ordering. J. Symbolic Comput. 16(4), 329–344 (1993)
16. Faugère, J.-C., Joux, A.: Algebraic cryptanalysis of hidden field equation (HFE) cryptosystems using Gröbner bases. In: Boneh, D. (ed.) CRYPTO 2003. LNCS, vol. 2729, pp. 44–60. Springer, Heidelberg (2003). https://doi.org/10.1007/978-3-540-45146-4_3
17. Gao, S., Volny, F., Wang, M.: A new framework for computing Gröbner bases. Math. Comput. 85(297), 449–465 (2016)
18. Kalkbrener, M.: A generalized Euclidean algorithm for computing triangular representations of algebraic varieties. J. Symbolic Comput. 15(2), 143–167 (1993)
19. Kapur, D., Lu, D., Monagan, M.B., Sun, Y., Wang, D.: An efficient algorithm for computing parametric multivariate polynomial GCD. In: Proceedings of ISSAC 2018, pp. 239–246. ACM Press (2018). https://doi.org/10.1145/3208976.3208980
20. Lazard, D.: Ideal bases and primary decomposition: case of two variables. J. Symbolic Comput. 1(3), 261–270 (1985)

21. Li, B., Wang, D.: An algorithm for transforming regular chain into normal chain. In: Kapur, D. (ed.) ASCM 2007. LNCS (LNAI), vol. 5081, pp. 236–245. Springer, Heidelberg (2008). https://doi.org/10.1007/978-3-540-87827-8_20
22. Marinari, M.G., Mora, T.: A remark on a remark by Macaulay or enhancing Lazard structural theorem. Bull. Iran. Math. Soc. **29**(1), 1–45 (2003)
23. Mou, C., Wang, D., Li, X.: Decomposing polynomial sets into simple sets over finite fields: the positive-dimensional case. Theoret. Comput. Sci. **468**, 102–113 (2013)
24. Ritt, J.F.: Differential Algebra. American Mathematical Society, New York (1950)
25. Shimoyama, T., Yokoyama, K.: Localization and primary decomposition of polynomial ideals. J. Symbolic Comput. **22**(3), 247–277 (1996)
26. Verschelde, J., Verlinden, P., Cools, R.: Homotopies exploiting Newton polytopes for solving sparse polynomial systems. SIAM J. Numer. Anal. **31**(3), 915–930 (1994)
27. Wang, D.: Computing triangular systems and regular systems. J. Symbolic Comput. **30**(2), 221–236 (2000)
28. Wang, D.: Elimination Methods. Springer-Verlag, Wien (2001). https://doi.org/10.1007/978-3-7091-6202-6
29. Wang, D.: On the connection between Ritt characteristic sets and Buchberger-Gröbner bases. Math. Comput. Sci. **10**, 479–492 (2016)
30. Wang, D., Dong, R., Mou, C.: Decomposition of polynomial sets into characteristic pairs. arXiv:1702.08664 (2017)
31. Weber, A., Sturm, T., Abdel-Rahman, E.O.: Algorithmic global criteria for excluding oscillations. Bull. Math. Biol. **73**(4), 899–917 (2011). https://doi.org/10.1007/s11538-010-9618-0
32. Weispfenning, V.: Comprehensive Gröbner bases. J. Symbolic Comput. **14**(1), 1–29 (1992)
33. Wu, W.T.: Basic principles of mechanical theorem proving in elementary geometries. J. Automated Reasoning **2**(3), 221–252 (1986)

About Integrability of the Degenerate System

Victor F. Edneral[1,2]([⊠])

[1] Skobeltsyn Institute of Nuclear Physics, Lomonosov Moscow State University,
Leninskie Gory 1(2), Moscow 119991, Russian Federation
edneral@theory.sinp.msu.ru
[2] Peoples' Friendship University of Russia (RUDN University),
6 Miklukho-Maklaya street, Moscow 117198, Russian Federation
edneral-vf@rudn.ru

Abstract. We study integrability of an autonomous planar polynomial system of ODEs with a degenerate singular point at the origin depending on five parameters. By mean of the Power Geometry Method, this degenerated system is reduced to a non-degenerate form by the blow-up process. After, we search for the necessary conditions of local integrability by the normal form method. We look for the set of necessary conditions on parameters under which the original system is locally integrable near the degenerate stationary point. We found seven two-parametric families in the five-parameter space. Then first integrals of motion were found for six families. For the seventh family, we found the formal first integral. So, at least six of these families in parameters space are manifolds where the global integrability of the original system takes place.

Keywords: Ordinary differential equations · Integrability ·
Resonant normal form · Power geometry · Computer algebra

1 Introduction

At the investigation of nonlinear ODEs systems, it is very important to find the exact solutions of such systems. If such solutions are opened, it is possible to study the system near such an exact solution and to create approximate solutions in these domains. Unfortunately, most systems of ODEs cannot be solved exactly. But sometimes the systems depend on parameters and we may try to find families of the values of the parameters at which such a system has an exact solution. Here we propose and check a new approach for searching such families.

For this checking, we consider a degenerate planar system of autonomous ordinary differential equations resolved with respect to derivatives and with a polynomial right-hand side of the fifth order. The system depends on five free parameters.

The publication has been prepared with the support of the "RUDN University Program 5–100".

Choosing this model system was defined by several factors. First, it belongs to the ODEs class with a non-nilpotent matrix of the linear part which almost was not investigated before. Second, this system has many domains for an investigation and all of them do via studying nilpotent systems, so we check our approach for nilpotent systems also. This is a very rich system.

This paper closes the series of papers [9–13]. Here we study the case of a last non-studied domain in the parametrical space. This task was very hard from the computational side. We opened a new family of parameters at which we have integrability and calculated the corresponding first integral of motion. So we have 7 domains of integrability and seven corresponding integrals.

The structure of the actual paper reflects the steps of the discussed approach. First, we transfer the original degenerate system to the nilpotent form [5,9,12].

Then, we split investigation in two cases with respect to the value of the parameter b. Postponing the case of the again degenerate system we apply the normal form method for the nilpotent case. This method allows for creating a necessary and sufficient condition for local integrability for each stationary point of the system. For each such point, the condition **A** is an infinite set of algebraic equations in system parameters. A solution of a finite subset of these equations which includes equations from each stationary point is a necessary condition of local integrability of the original degenerate system. Such subsets can be calculated by the computer algebra program. For the first case above, we found four two-dimensional families in the parameter space [9].

After that, we transfer the new degenerated system to the nilpotent form by power transformation and split it into two subcases. The first one leads to a couple new two dimension families in the parameter space [7,11,12]. So, we get six families. This paper studies the last subcase. Here we calculate the seventh family.

Finally, we have obtained seven two-dimensional families in the parameter space which give the necessary condition of local integrability. We do not pretend that we have found all such families. But for each such family, we have calculated the corresponding first integral of motion. So, we got families parameters along which we have global integrability [14].

2   The Model System

We consider an autonomous system of ordinary differential equations of the form

$$
\begin{aligned}
dx/dt &= -y^3 - b\,x^3 y + a_0\,x^5 + a_1\,x^2 y^2, \\
dy/dt &= (1/b)\,x^2 y^2 + x^5 + b_0\,x^4 y + b_1\,x\,y^3.
\end{aligned}
\tag{1}
$$

Thus, we consider the system with five arbitrary parameters a_i, b_i, $(i = 0, 1)$ and b. The coefficient $1/b$ provides the integrability of the first quasi-homogeneous approximation of the system [9]. A similar system was originally studied in [1].

3   Transferring the System to Non-degenerate Form

It can be done using the power transformation by blow-up process (see Chap. 1, Par. 1.8 in [4, 13])

$$x = u\,v^2, \quad y = u\,v^3 \tag{2}$$

with the time rescaling $u^2 v^7 dt = d\tau$. As the result, we obtain system (1) in the form

$$
\begin{aligned}
du/d\tau &= -3\,u - [3\,b + (2/b)]u^2 - 2\,u^3 + (3\,a_1 - 2\,b_1)u^2 v + \\
&\quad (3\,a_0 - 2\,b_0)u^3 v, \\
dv/d\tau &= v + [b + (1/b)]u\,v + u^2 v + (b_1 - a_1)u\,v^2 + (b_0 - a_0)u^2 v^2.
\end{aligned}
\tag{3}
$$

We should study this system along the invariant lines $u = 0$, $v = 0$. More about this transformation see [13].

4   About the Normal Form and the Condition A

The formal pseudo-identical change of coordinates

$$Y = Z + \Xi(Z), \tag{4}$$

where $\Xi = (\xi_1, \ldots, \xi_n)$ and $\xi_j(Z)$ are formal power series, transforms a system with diagonal matrix of linear part to the normal form [3, 4, 9]

$$\dot{z}_j = z_j g_i(Z) = z_j \sum g_{jQ} Z^Q \text{ over } Q \in \mathbb{N}_j, \, j = 1, \ldots, n, \tag{5}$$

where $Q = (q_1, \ldots, q_n)$, $Z^Q = z_1^{q_1} \ldots z_n^{q_n}$,

$$\mathbb{N}_j = \{Q : Q \in \mathbb{Z}^n, \; Q + E_j \geq 0\}, \, j = 1, \ldots, n,$$

E_j means the unit vector. Denote

$$\mathbb{N} = \mathbb{N}_1 \cup \ldots \cup \mathbb{N}_n. \tag{6}$$

The diagonal $\Lambda = (\lambda_1, \ldots, \lambda_n)$ of J consists of eigenvalues of the matrix A.

System (5) is called the *resonant normal form* if:

(a) J is the Jordan matrix,

(b) in writing (5), there are only the *resonant terms*, for which the scalar product

$$\langle Q, \Lambda \rangle \overset{\text{def}}{=} q_1 \lambda_1 + \ldots + q_n \lambda_n = 0. \tag{7}$$

Theorem 1. ([3]) *There exists a formal change (4) reducing a system to its normal form (5).*

In [3,4], condition **A** was formulated on the coefficients of normal form (5), which guarantees the convergence of the normalizing transformation (4).

Condition **A**. *In the normal form (5)*

$$g_j = \lambda_j \alpha(Z) + \bar{\lambda}_j \beta(Z), \quad j = 1, \ldots, n, \tag{8}$$

where $\alpha(Z)$ and $\beta(Z)$ are some power series.

Let

$$\omega_k = \min |\langle Q, \Lambda \rangle| \text{ over } Q \in \mathbb{N}, \ \langle Q, \Lambda \rangle \neq 0, \ \sum_{j=1}^{n} q_j < 2^k, \ k = 1, 2, \ldots$$

There is also the condition on small divisors ω_k. It is satisfied in the planar case.

Let us see the condition **A** in a two-dimensional case in more detail. Let

$$\frac{\lambda_1}{\lambda_2} = -\frac{m}{n} \quad \text{or} \quad n\lambda_1 = -m\lambda_2, \tag{9}$$

where m and n are natural numbers. With respect of (5) and (9), we can write the condition **A** as a system

$$\begin{aligned} n\frac{d \log(z_1)}{dt} &= n(\lambda_1 \cdot \alpha(Z) + \bar{\lambda}_1 \cdot \beta(Z)) \\ m\frac{d \log(z_2)}{dt} &= m(\lambda_2 \cdot \alpha(Z) + \bar{\lambda}_2 \cdot \beta(Z)). \end{aligned} \tag{10}$$

Summing these two equalities with respect to (9) we have

$$\frac{d \log(z_1^n z_2^m)}{dt} = 0 \quad \text{or} \quad z_1^n z_2^m = \text{const}(t). \tag{11}$$

So, $z_1^n z_2^m$ is the first integral of motion. It corresponds to the local integrability of the original system because the condition **A** is enough for convergence of the normalizing transformation near the stationary point. If some $\lambda_i = 0$ then the condition **A** has the form $g_i(Z) = 0$ and z_1 is the first integral of motion.

5   Necessary Condition of Local Integrability

The satisfaction of the condition **A** is the necessary and sufficient condition for local integrability of a planar system near an elementary stationary point [3,4,9]. The condition **A** is a strong algebraic condition on coefficients of the normal form. For local integrability of original system (1) near a degenerate (non-elementary) stationary point, it is necessary to have local integrability near each of elementary stationary points, which are produced by the blowing up process described above. The condition **A** is usually an infinite series of algebraic equations. Each of them will be an equation on parameters of the original system and each of these equations gives a necessary condition of local integrability of the system near the corresponding stationary point. So, the necessary condition will be a satisfaction of some equations from infinite sets of (8) at the same values of the system parameters at all stationary points of the system simultaneously. Let us demonstrate the satisfaction of this condition for system (1).

Under the power transformation (2), the point $x = y = 0$ blows up into two straight invariant lines $u = 0$ and $v = 0$. Along the line $u = 0$, system (3) has a single stationary point $u = v = 0$. Along the second line $v = 0$, this system has four elementary stationary points

$$u = 0, \quad u = -\frac{1}{b}, \quad u = -\frac{3b}{2}, \quad u = \infty. \tag{12}$$

So, the necessary condition of local integrability of system (1) near the point $x = y = 0$ is local integrability near all stationary points of system (3) simultaneously.

Lemma 1. ([9,11]). *Near the points $u = v = 0$ and $u = \infty, v = 0$, system (3) is locally integrable.*

Let us consider two other stationary points. Firstly we restrict ourselves to the case $b \neq 0$, $b^2 \neq 2/3$ when the linear part of system (3) has non-vanishing eigenvalues after the both shifts $u = w - 1/b$ and $u = w - 3/b$. At $b^2 = 2/3$, the matrix of the linear part of the shifted system (14) is degenerate again. This case is studied by means of one more power transformation in Sect. 7.

The algorithm for calculating the normal form and the normalizing transformation together with the corresponding computer program is described in [6,8]. We calculated the lowest orders of the condition **A** with these programs. There are four of two sets of parameters (in a_1 and b) that satisfy these conditions for $b \neq 0$ [11]. They correspond to simultaneous satisfaction of **A** at stationary points $u = -1\,b/2, v = 0$ and $u = -3\,b/2, v = 0$

$$
\begin{aligned}
&1) \quad a_0 = a_1\,b, \quad\quad b_0 = b_1\,b, \quad b^2 \neq 2/3, \\
&2) \quad b_1 = -2\,a_1, \quad a_0 = a_1 b, \quad b_0 = b_1 b, \quad b^2 \neq 2/3, \\
&3) \quad b_1 = (3/2)\,a_1, a_0 = a_1 b, \quad b_0 = b_1 b, \quad b^2 \neq 2/3, \\
&4) \quad b_1 = (8/3)\,a_1, a_0 = a_1 b, \quad b_0 = b_1 b, \quad b^2 \neq 2/3.
\end{aligned}
\tag{13}
$$

Thus, we have proved

Theorem 2. ([9,11]) *Conditions (13) form the set of necessary conditions of local integrability of system (3) near all its stationary points and the local integrability of system (1) at the stationary point $x = y = 0$.*

6   Sufficient Conditions of Integrability

The conditions presented in Theorem 2 as the necessary conditions for the local integrability of system (1) at the stationary point at the origin can be considered as good candidates for sufficient conditions of global integrability. However, it is necessary to prove the sufficiency of these conditions by independent methods. It is necessary to do it for each of four conditions (13) at each of stationary points $u = -3b/2, v = 0$ and $u = -1/b, v = 0$, for $b^2 \neq 2/3$.

In [7], we found first integrals of system (3) for all cases (13) (mainly by the Darboux method, see e.g. [16]).

We found four families of such solutions which exhausted all cases mentioned above:

1. At $a_0 = 0$, $a_1 = -b_0\, b$, $b_1 = 0$

$$I_{1uv} = u^2(3\,b + 2\,u)v^6,$$
$$I_{1xy} = 2\,x^3 + 3\,b\,y^2.$$

2. At $b_1 = -2a_1$, $a_0 = a_1 b$, $b_0 = b_1 b$

$$I_{2uv} = u^2\, v^6\, (3\,b + u\,(2 - 6\,a_1\,b\,v)),$$
$$I_{2xy} = 2\,x^3 - 6\,a_1\,b\,x^2\,y + 3\,b\,y^2.$$

3. At $b_1 = 3a_1/2$, $a_0 = a_1 b$, $b_0 = b_1 b$

$$I_{3uv} = [4 - 4a_1\,u\,v + 3^{5/6}a_1 \times {}_2F_1\,(2/3, 1/6; 5/3; -2u/(3b)) \times$$
$$u\,v\,(3 + 2u/b)^{1/6}]/[u^{1/3}v\,(3b + 2u)^{1/6}],$$

$$I_{3xy} = [a_1 x^2(-4 + 3^{5/6}\,{}_2F_1\,(2/3, 1/6; 5/3; -2\,x^3/(3\,b\,y^2)) \times$$
$$(3 + 2x^3/(b\,y^2))^{1/6}) + 4y]/[y^{4/3}(3\,b + 2\,x^3/y^2)^{1/6}],$$

4. At $b_1 = 8a_1/3$, $a_0 = a_1 b$, $b_0 = b_1 b$

$$I_{4u,v} = [u\,(3 + 2\,a_1^2 bu) + 6\,a_1\,b\,v]/$$
$$[3\,u\,[u^3(6 + a_1^2 b\,u) + 6\,a_1^2 b\,u^2 v + 9\,b\,v^2]^{1/6}]-$$
$$8\,a_1\sqrt{-b}/3^{5/3}B_{6+a_1\,\sqrt{-6\,b\,u}+3\,v\,\sqrt{-6\,b/u^3}}(5/6, 5/6),$$

where $B_t(a, b)$ is the incomplete beta function and ${}_2F_1(a, b; c; z)$ is the hypergeometric function [2].

The first integrals and solutions do not have any singularities for the values $b^2 = 2/3$, but the approach using which these solutions were found, has the limitation $b^2 \neq 2/3$, so there are possible additional integrals at this values. Thus, we need to study the case $b^2 = 2/3$ separately.

7   Case $b^2 = 2/3$

Remark that a choice of b sign does not matter because of linear automorphism $\{x, y, b, a_0, a_1, b_0, b_1\} \rightarrow \{-x, -y, -b, a_0, -a_1, b_0, -b_1\}$ of system (1). We suppose below $b = +\sqrt{2/3}$.

7.1   Subcase $3a_0 - 2b_0 = b(3a_1 - 2b_1)$

This subcase is an extension of conditions $a_0 = a_1 b$, $b_0 = b_1 b$ from (13). Let us consider the case $b = \sqrt{2/3}$. At these values of b, both stationary points

$u = -3b/2, v = 0$ and $u = -1/b, v = 0$ are collapsing and after the shift $u \to w - 1/b$ we have instead of (3) the degenerate system

$$
\begin{aligned}
\frac{dw}{d\tau} &= v(-\tfrac{9}{2}\sqrt{\tfrac{3}{2}}\,a_0 + \tfrac{9}{2}\,a_1 + 3\sqrt{\tfrac{3}{2}}\,b_0 - 3\,b_1) + \\
&\quad wv(\tfrac{27}{2}\,a_0 - 3\sqrt{6}\,a_1 - 9\,b_0 + 2\sqrt{6}\,b_1) + \\
&\quad \sqrt{6}\,w^2 + w^2 v(-9\sqrt{\tfrac{3}{2}}\,a_0 + 3\,a_1 + 3\sqrt{6}\,b_0 - 2\,b_1) - \\
&\quad 2\,w^3 + w^3 v(3\,a_0 - 2\,b_0), \\
\frac{dv}{d\tau} &= -\tfrac{\sqrt{6}}{6}\,wv + v^2(-\tfrac{3}{2}\,a_0 + \sqrt{\tfrac{3}{2}}\,a_1 + \tfrac{3}{2}\,b_0 - \sqrt{\tfrac{3}{2}}\,b_1) + \\
&\quad w^2 v + wv^2(\sqrt{6}\,a_0 - a_1 - \sqrt{6}\,b_0 + b_1) + \\
&\quad + w^2 v^2(-a_0 + b_0).
\end{aligned}
\tag{14}
$$

This system has zero eigenvalues at the stationary point $w = v = 0$, so we should apply a power transformation once again.

If $b^2 = 2/3$, Eq. (14) can be rewritten as

$$
\begin{aligned}
\frac{dw}{d\tau} &= -3v/(2b)[(3a_0 - 2b_0) - b(3a_1 - 2b_1)] + \\
&\quad wv(\tfrac{27}{2}\,a_0 - 3\sqrt{6}\,a_1 - 9\,b_0 + 2\sqrt{6}\,b_1) + \\
&\quad \sqrt{6}\,w^2 + w^2 v(-9\sqrt{\tfrac{3}{2}}\,a_0 + 3\,a_1 + 3\sqrt{6}\,b_0 - 2\,b_1) - \\
&\quad 2\,w^3 + w^3 v(3\,a_0 - 2\,b_0), \\
\frac{dv}{d\tau} &= -\tfrac{\sqrt{6}}{6}\,wv + v^2(-\tfrac{3}{2}\,a_0 + \sqrt{\tfrac{3}{2}}\,a_1 + \tfrac{3}{2}\,b_0 - \sqrt{\tfrac{3}{2}}\,b_1) + \\
&\quad w^2 v + wv^2(\sqrt{6}\,a_0 - a_1 - \sqrt{6}\,b_0 + b_1) + \\
&\quad + w^2 v^2(-a_0 + b_0).
\end{aligned}
\tag{15}
$$

We see that in systems (14) and (15), the coefficient of v in the linear part of the first equation is zero if $3a_0 - 2b_0 = b(3a_1 - 2b_1)$. So we have the special subcase [11]. For this subcase, we use the transformation

$$
u \to w - 1/b, \quad v \to rw, \quad \dot{v} \to \dot{r}w + r\dot{w}, \tag{16}
$$

with the time scaling by division of the equations by $w/\sqrt{6}$, so $\tilde{\tau} = w\tau/\sqrt{6}$. Then, from (3) we have

$$
\begin{aligned}
\frac{dw}{d\tilde{\tau}} &= 6w + 3(3a_1 - 2b_1)rw - 2\sqrt{6}w^2 - 2\sqrt{6}(3a_1 - 2b_1)rw^2 + \\
&\quad 2(3a_1 - 2b_1)rw^3, \\
\frac{dr}{d\tilde{\tau}} &= -7r - (9a_1 - \sqrt{\tfrac{3}{2}}b_0 - 5b_1)r^2 + 3\sqrt{6}rw + \\
&\quad (7\sqrt{6}a_1 - 2b_0 - 13\sqrt{\tfrac{2}{3}}b_1)r^2 w - (8a_1 - \sqrt{\tfrac{3}{2}}b_0 - \tfrac{16}{3}b_1)r^2 w^2.
\end{aligned}
\tag{17}
$$

This is a three-parameter system with the resonance of the 13th order at the stationary point $w = 0, r = 0$ on the invariant line $w = 0$. Along this line, there is also another stationary point. It is possible to prove integrability of the system there and this point does not supply any additional restriction on the parameters.

We calculated the normal form for (17) till the 26th order and got two equations for the condition **A**. They are $A_{13} = 0$ and $A_{26} = 0$ where A_{13} and A_{26}

are given in [17]. Each of these equations is homogeneous in parameters a_1, b_0, and b_1 of system (1) of sixth and twelfth orders. Both A_{13} and A_{26} are equal to zero under conditions (13).

Homogeneous algebraic equations in three variables can be rewritten as non-homogeneous equations in two variables. If we suppose that $a_1 = 0$ we get only one- and zero-dimensional solutions in the parameter space. In general case, $a_1 \neq 0$. In this case, we substitute $b_0 = c_0 a_1, b_1 = c_1 a_1$ and obtain the system of two equations in two variables $A_{13}(c_0, c_1) = 0$, $A_{26}(c_0, c_1) = 0$. The resultant of two corresponding polynomials in each of two variables is identically equal to zero. So it is enough to solve only equation $A_{13}(c_0, c_1) = 0$. This equation can be factorized as the product of four factors including a_1^6:

$$
\begin{aligned}
A_{13} = {} & 48(c_1 - 3/2) \times \\
& (c_0 - 1/12\sqrt{6}c_1 + 1/2\sqrt{6})^2 \times \\
& [409790784c_0^3 - 104\sqrt{6}c_0^2(-9152256 + 3385633c_1) - \\
& 208c_0(-10917702 + c_1(-360720 + 3319927c_1)) + \\
& \sqrt{6}(-718439040 + c_1(2461047528 + \\
& c_1(-1944898681 + 441207868c_1)))] \times \\
& a_1^6.
\end{aligned}
\tag{18}
$$

From the first two factors, we get a couple of two-parametric solutions $c_1 = 3/2$ and $c_1 = 6 + 2\sqrt{6}c_0$ or

$$
\begin{aligned}
&5)\ b_1 = 3a_1/2, && a_0 = (2b_0 + b(3a_1 - 2b_1))/3, && b = \sqrt{2/3}, \\
&6)\ b_1 = 6a_1 + 2\sqrt{6}b_0, && a_0 = (2b_0 + b(3a_1 - 2b_1))/3, && b = \sqrt{2/3}.
\end{aligned}
\tag{19}
$$

For solutions (19) we calculate the normal form of (17) till the 36th order and obtain that for each solution, it is a diagonal linear system. So, it is integrable.

The use of general roots of the polynomial, which is a cubic factor in (18), is out of our consideration. We restrict possible sets of parameters by a two-dimensional case with coefficients over algebraic extension of rational numbers with the algebraic number $\sqrt{6}$. The last factor of (18) has no such roots.

For each set of parameters (19), one can find the Darboux integrating factor [7, 11, 16].

A couple of the first integrals of (17) with respect to found parameters (19) is

$$
\begin{aligned}
I_{5rw} = {} & w^{-7/6}(1 - \sqrt{\tfrac{2}{3}}w)^{-1/3}[-9a_1 + 3\sqrt{6}b_0 - \tfrac{42}{r} - 6(\sqrt{6}a_1 + 5b_0)w + \\
& 2(9a_1 + 4\sqrt{6}b_0)w^2 - 2^{1/6}(9\sqrt{2}a_1 + 8\sqrt{3}b_0)w^{5/3}(-\sqrt{6} + 2w)^{1/3} \times \\
& {}_2F_1(-1/2, 1/3; 1/2; \sqrt{2/3}/w)],
\end{aligned}
\tag{20}
$$

$$
\begin{aligned}
I_{6rw} = {} & r^{3\frac{3a_1}{3a_1 + \sqrt{6}b_0}} \cdot w^{7/3 + \frac{7b_0}{3\sqrt{6}a_1 + 6b_0}} \cdot (1 - \sqrt{2/3}w)^{\frac{-a_1}{3a_1 + \sqrt{6}b_0}} \times \\
& \{\tfrac{-6 + 2\sqrt{6}w}{6a_1 + 3\sqrt{6}b_0} + r[3 + 2w(-\sqrt{6} + w)]\}.
\end{aligned}
$$

In the original variables x, y of Eq. (1), these integrals up to a numeric factor have the form

$$
\begin{aligned}
I_{5xy} = {} & (y/x^2)(\sqrt{6} + 2x^3/y^2)^{-7/6}(x^3/y^2)^{2/3} \cdot \{42\sqrt{6} + \\
& 1/(xy^3)[-36a_1x^6 - 16\sqrt{6}b_0x^6 + 84x^4y24\sqrt{6}a_1x^3y^2 - \\
& 36b_0x^3y^2 + 2^{1/3}(x^3/y^2)^{1/3}y^2(\sqrt{6} + (x^3/y^2)^{2/3} \times \\
& (2(9a_1 + 4\sqrt{6}b_0)x^3 + 3(3\sqrt{6}a_1 + 8b_0)y^2) \times \\
& {}_2F_1(-1/2, 1/3; 1/2; \tfrac{3y^2}{3y^2 + \sqrt{6}x^3})]\},
\end{aligned}
\tag{21}
$$

$$
\begin{aligned}
I_{6xy} = {} & y(\sqrt{2/3} + x^3/y^2)^{-1/2 + \frac{a_1}{-6a_1 - 2\sqrt{6}b_0}}(x^2/y)^{-\frac{a_1}{3a_1 + \sqrt{6}b_0}} \times \\
& \{3 + (x^2/y^2)[\sqrt{6}x + 3(2a_1 + \sqrt{6}b_0)y]\}.
\end{aligned}
$$

In the case of $b = -\sqrt{2/3}$, we obtain an analogous formula. It seems that cases (3) and (5) above are the same but they are defined in different manifolds of parameters.

Analytical properties of first integrals in cases (1)–(6) were discussed in [7, 12].

7.2   Subcase $3a_0 - 2b_0 \neq b(3a_1 - 2b_1)$

If we use the transformation

$$
u \to w - 1/b, \quad v \to w^2 p, \quad \dot{v} \to 2\dot{w}wp + w^2\dot{p} \quad \text{and} \quad \tilde{\tau} = w\tau/\sqrt{6}, \tag{22}
$$

then at $b^2 = 2/3$ we get from (14) the systems

$$
\begin{aligned}
\frac{dp}{d\tilde{\tau}} = {} & -13p - 7\sqrt{6}a_0p^2w^3 + 60a_0p^2w^2 - 57\sqrt{\tfrac{3}{2}}a_0p^2w \\
& + 27a_0p^2 - 7\sqrt{6}a_1p^2w^2 + 39a_1p^2w - 9\sqrt{6}a_1p^2 + 5\sqrt{6}b_0p^2w^3 \\
& - 42b_0p^2w^2 + 39\sqrt{\tfrac{3}{2}}b_0p^2w - 18b_0p^2 + 5\sqrt{6}b_1p^2w^2 - 27b_1p^2w \\
& + 6\sqrt{6}b_1p^2 + 5\sqrt{6}pw, \\
\frac{dw}{d\tilde{\tau}} = {} & +6w + 3\sqrt{6}a_0pw^4 - 27a_0pw^3 + 27\sqrt{\tfrac{3}{2}}a_0pw^2 - \tfrac{27}{2}a_0pw \\
& + 3\sqrt{6}a_1pw^3 - 18a_1pw^2 + 9\sqrt{\tfrac{3}{2}}a_1pw - 2\sqrt{6}b_0pw^4 \\
& + 18b_0pw^3 - 9\sqrt{6}b_0pw^2 + 9b_0pw - 2\sqrt{6}b_1pw^3 \\
& + 12b_1pw^2 - 3\sqrt{6}b_1pw - 2\sqrt{6}w^2.
\end{aligned}
\tag{23}
$$

Along the invariant line $w = 0$, this system has two stationary points

$$
(w = 0, p = 0) \quad \text{and} \quad (w = 0, p = \frac{13}{3(9a_0 - 3\sqrt{6}a_1 - 6b_0 + 2\sqrt{6}b_1)}). \tag{24}
$$

At the origin, it has a resonance of the 19th order and at the second stationary point, we have equation with the resonance of the 27th order.

The denominator in the coordinates of the second point is non-zero because of $3a_0 - 2b_0 \neq b(3a_1 - 2b_1)$. At the 19th order resonance, the condition **A** is satisfied at zeroing a homogeneous polynomial in 4 parameters of 6th order over algebraic extension of integer numbers [18]. It consists of 84 terms and it can be factorized by system MATHEMATICA over an algebraic extension with the number $\sqrt{6}$ into two factors. The first factor is linear

$$20a_0 + 2\sqrt{6}a_1 + 4b_0 + 3\sqrt{6}b_1. \tag{25}$$

The second factor is a homogeneous polynomial of the 5th order. It includes 5th order of each of the parameters in isolation and its roots can be calculated in elliptic functions only. So, a usable solution of the condition **A** is a root of polynomial (25).

At the second stationary point, we have a resonance of the 27th order. So, we should compute series in 6 variables $p, w, a0, a1, b0, b1$ till this order with rational coefficients. Non-numeric denominators arise from the rational shift in (24). This is a very hard task. For simplification, we use the solution of condition **A** for the 19th order, i.e., equality of (25) to zero. Thus, we put $a_0 \rightarrow \frac{1}{20}(-2\sqrt{6}a_1 - 4b_0 - 3\sqrt{6}b_1)$ and we get 3 parameters system in rational functions with non-zero denominators. For simplification, we introduce a new parameter $r = 6\sqrt{6}a_1 + 12b_0 - \sqrt{6}b_1$, which is the denominator of the shift $\frac{13}{3(9a_0 - 3\sqrt{6}a_1 - 6b_0 + 2\sqrt{6}b_1)}$ at corresponding a_0 and put $b_0 \rightarrow \frac{1}{12}(-6\sqrt{6}a_1 + \sqrt{6}b_1 + r)$. After that we should diagonalize the linear part matrix at the second stationary point. During this process, we get one more denominator which we denote as $h = 130a_1 - 75b_1 - 9\sqrt{6}r$. We put $b_1 \rightarrow 26a_1/15 - h/75 - 3\sqrt{6}r/25$. First, we suppose that $h \neq 0$. Then we calculated the normal form till the 27th order (all smaller orders are zero in definition of the resonance normal form) in 3 parameters r, h, and a_1 [19]. It took with MATHEMATICA-11 about 700000 seconds or 8 days at a scalar 3.60 GHz processor. The MATHEMATICA used only a half GBt of RAM for this calculation. We get 8248088 terms of the normalizing transformation and 754 terms of the normal form. The numerator of this normal form is a homogeneous polynomial of the 54th order in 3 parameters. Coefficients of this polynomial have more than hundred digits. This polynomial can be factorized. Numerator consists of non-zero factor r^{27}, homogeneous factor of the 24th order, and a square of a linear polynomial $2a_1 + 3b_1$. Before we calculate this result, we try to use lazy calculations, modular arithmetic etc.

So the last 7th candidate for integrability is

$$a_1 = -3/2b_1, \quad b = \sqrt{2/3}, \quad 20a_0 + 2\sqrt{6}a_1 + 4b_0 + 3\sqrt{6}b_1 = 0,$$
$$3a_0 - 2b_0 \neq b(3a_1 - 2b_1). \tag{26}$$

The case of $h = 0$ is simpler for calculation and it gives a condition

$$a_1 = -\tfrac{3}{2}b_1, \quad b = \sqrt{2/3}, \quad 20a_0 + 2\sqrt{6}a_1 + 4b_0 + 3\sqrt{6}b_1 = 0,$$
$$b_0 = -\tfrac{5a_1}{3\sqrt{6}}. \tag{27}$$

This is a particular case of (26).

The seventh condition looks like cases (3) and (5) above, but a location of its parameters is not intersecting with any of cases (1)–(6). We calculated the corresponding formal first integral for the 7th case

$$
\begin{aligned}
I_{7pw} = (&77760(5\sqrt{6}a_1 + 9b_1)^2 \times \\
&(-12\log(4(-5 + (5\sqrt{6}a_1 + 9b_1)w)) - \\
&\log(856485660833532609561432309910464b_1^{13}(3(50\sqrt{6}a_1^2 + 180a_1b_1 + \\
&27\sqrt{6}b_1^2)w - 12b_1p^3(-5 + (5\sqrt{6}a_1 + 9b_1)w) - \\
&2p(300a_1^2w + 25\sqrt{6}a_1(-1 + 9b_1w) + 9b_1(-10 + 27b_1w)) \\
&+2p^2(50\sqrt{6}a_1^2w + 9\sqrt{6}b_1(-5 + 9b_1w) + 10a_1(-5 + 36b_1w)))) \\
&-((26\log(\sqrt{6} - 2p) - 26\log(p) - 29\log(1/29(-\sqrt{6} + 2p))) \\
&(\sqrt{6} - 2p)(-50a_1^2(39\sqrt{6} + p(-165 + 29\sqrt{6}p))w + \\
&3b_1(-351\sqrt{6}b_1w + p(-780 + 2187b_1w + 2p(-42\sqrt{6} + 29p) \\
&(-5 + 9b_1w))) + 10a_1(-702b_1w + p(\sqrt{6}(-65 + 612b_1w) + \\
&p(145 + 3b_1(-339 + 29\sqrt{6}p)w))))))/ \\
&((13\sqrt{6} - 29p)(50a_1^2(3\sqrt{6} + 2p(-6 + \sqrt{6}p))w + \\
&3b_1(27\sqrt{6}b_1w + p(60 - 162b_1w + 2(3\sqrt{6} - 2p)p(-5 + 9b_1w))) \\
&-10a_1(-54b_1w + p(5\sqrt{6}(-1 + 9b_1w) + \\
&2p(5 + 3b_1(-12 + \sqrt{6}p)w)))))))/(16000\sqrt{6}a_1^3 + 2400a_1^2(1 + 36b_1) + \\
&60a_1(-27 + \sqrt{6} + 24\sqrt{6}b_1(1 + 18b_1)) + 9(1 + 12b_1(1 - 9\sqrt{6} + \\
&12b_1(1 + 12b_1)))).
\end{aligned}
$$

We believe that it can be rewritten in analytic form. Existence of this integral is very important.

8   Conclusions

For a five-parameter non-Hamiltonian degenerate planar system (1), we have studied possible necessary conditions of integrability. We found 7 families in parameter space which satisfy these conditions. These families were found as two-dimension manifolds in the five parameter space. For six from seven of them, we have calculated the analytic first integrals of motion that is the original system is globally integrable at these values of parameters. For the seventh case, we calculated the formal first integral.

Acknowledgements. The author is very grateful to Profs. A.D. Bruno, V.G. Romanovski, and A.B. Batkhin for important advices, discussions, and assistance.

References

1. Algaba, A., Gamero, E., Garcia, C.: The integrability problem for a class of planar systems. Nonlinearity **22**, 395–420 (2009)
2. Bateman, H., Erdêlyi, A.: Higher Transcendental Functions, vol. 1. Mc Graw-Hill Book Company Inc., New York (1953)
3. Bruno, A.D.: Analytical form of differential equations (I, II). Trudy Moskov. Mat. Obsc. 25, 119–262 (1971), 26, 199–239 (1972) (in Russian). Trans. Moscow Math. Soc. 25, 131–288 (1971), 26, 199–239 (1972) (in English)
4. Bruno, A.D.: Local Methods in Nonlinear Differential Equations. Nauka, Moscow 1979 (in Russian). Springer, Berlin (1989) (in English)

5. Bruno, A.D.: Power Geometry in Algebraic and Differential Equations. Fizmatlit, Moscow (1998) (in Russian). Elsevier Science, Amsterdam (2000) (in English)
6. Edneral, V.F., Khanin, R.: Application of the resonant normal form to high-order nonlinear ODEs using mathematica. Nucl. Instrum. Methods Phys. Res. Sect. A **502**(2–3), 643–645 (2003)
7. Edneral, V., Romanovski, V.G.: On sufficient conditions for integrability of a planar system of odes near a degenerate stationary point. In: Gerdt, V.P., Koepf, W., Mayr, E.W., Vorozhtsov, E.V. (eds.) CASC 2010. LNCS, vol. 6244, pp. 97–105. Springer, Heidelberg (2010). https://doi.org/10.1007/978-3-642-15274-0_9
8. Edneral, V.F.: An algorithm for construction of normal forms. In: Ganzha, V.G., Mayr, E.W., Vorozhtsov, E.V. (eds.) CASC 2007. LNCS, vol. 4770, pp. 134–142. Springer, Heidelberg (2007). https://doi.org/10.1007/978-3-540-75187-8_10
9. Bruno, A.D., Edneral, V.: On Integrability of a Planar ODE System Near a Degenerate Stationary Point. In: Gerdt, V.P., Mayr, E.W., Vorozhtsov, E.V. (eds.) CASC 2009. LNCS, vol. 5743, pp. 45–53. Springer, Heidelberg (2009). https://doi.org/10.1007/978-3-642-04103-7_4
10. Bruno, A.D., Edneral, V.F.: On integrability of a planar system of ODEs near a degenerate stationary point. J. Math. Sci. **166**(3), 326–333 (2010)
11. Bruno, A.D., Edneral, V.F.: On possibility of additional solutions of the degenerate system near double degeneration at the special value of the parameter. In: Gerdt, V.P., Koepf, W., Mayr, E.W., Vorozhtsov, E.V. (eds.) CASC 2013. LNCS, vol. 8136, pp. 75–87. Springer, Cham (2013). https://doi.org/10.1007/978-3-319-02297-0_6
12. Bruno, A.D., Edneral, V.F., Romanovski, V.G.: On new integrals of the algabagamero-garcia system. In: Gerdt, V.P., Koepf, W., Seiler, W.M., Vorozhtsov, E.V. (eds.) CASC 2017. LNCS, vol. 10490, pp. 40–50. Springer, Cham (2017). https://doi.org/10.1007/978-3-319-66320-3_4
13. Edneral, V.F.: Application of power geometry and normal form methods to the study of nonlinear ODEs. EPJ Web. Conf. **173**, 01004 (2018)
14. Edneral, V., Romanovski, V.: Local and global odes properties. In: Proceedings of 24th Conference on Applications of Computer Algebra - ACA, Santiago de Compostela, Spain (2018). https://doi.org/10.15304/9788416954872
15. Christopher, C., Mardešić, P., Rousseau, C.: Normalizable, integrable, and linearizable saddle points for complex quadratic systems in C^2. J. Dyn. Control Sys. **9**, 311–363 (2003)
16. Romanovski, V.G., Shafer, D.S.: The Center And Cyclicity Problems: A Computational Algebra Approach. Birkhüser, Boston (2009)
17. http://theory.sinp.msu.ru/~edneral/CASC2017/a13-26.txt
18. http://theory.sinp.msu.ru/~edneral/CASC2019/a19factorized.nb
19. http://theory.sinp.msu.ru/~edneral/CASC2019/a27factorized.nb

Construction of a New Implicit Difference Scheme for 2D Boussinesq Paradigm Equation

Yu. A. Blinkov[1], V. P. Gerdt[2,3,4(✉)], I. A. Pankratov[1], and E. A. Kotkova[2,4]

[1] Saratov State University, Saratov 413100, Russian Federation
{BlinkovUA,PankratovIA}@info.sgu.ru
[2] Joint Institute for Nuclear Research, Dubna 141980, Russian Federation
gerdt@jinr.ru, ekaterina.a.kotkova@gmail.com
[3] Peoples' Friendship University of Russia, Moscow 117198, Russian Federation
[4] Dubna State University, Dubna 141982, Russian Federation

Abstract. Given an orthogonal and uniform solution grid with equal spatial grid sizes, we construct a new second-order implicit conservative finite difference scheme for the fourth-order 2D Boussinesq paradigm equation with quadratic nonlinear part. We apply the algebraic approach to the construction of difference schemes suggested by the first two authors and based on a combination of the finite volume method, difference elimination, and numerical integration. For the difference elimination, we make use of the techniques of Gröbner bases; in so doing, we introduce an extra difference indeterminate to reduce the nonlinear elimination problem to the pure linear one. It allows us to apply the Gröbner bases algorithm and software designed for linear generating sets of difference polynomials. Additionally, for the obtained difference scheme and also for another scheme known in the literature, we compute the modified differential equations and compare them.

Keywords: Computer algebra · Difference elimination ·
Finite difference approximation · Gröbner basis · Modified equation ·
Consistency · Conservativity

1 Introduction

The aim of this paper is to apply the algebraic approach devised in [1] to the generation of a finite difference scheme for the two-dimensional Boussinesq Paradigm Equation (BPE) [2] with the quadratic nonlinearity

$$u_{tt} = \Delta\left[u - \alpha f(u) + \beta_1 u_{tt} - \beta_2 \Delta u\right], \quad f(u) := u^2, \tag{1}$$

where $u = u(t, x, y)$ is the surface elevation of the wave, $\alpha > 0$ is an amplitude parameter, $\beta_1, \beta_2 > 0$ are two dispersion coefficients (parameters) and $\Delta := \partial_{xx} + \partial_{yy}$ is the Laplace operator.

© Springer Nature Switzerland AG 2019
M. England et al. (Eds.): CASC 2019, LNCS 11661, pp. 152–163, 2019.
https://doi.org/10.1007/978-3-030-26831-2_11

For the numerical solution of the fourth-order Eq. (1), it is convenient to rewrite it [3] as the system of two second-order partial differential equations (PDE)

$$\begin{cases} v - u + \beta_1 \triangle u = 0, \\ v_{tt} - \dfrac{\beta_2}{\beta_1} \triangle v - \dfrac{\beta_1 - \beta_2}{\beta_1^2}(u - v) + \triangle F = 0, \quad F := \alpha\, u^2, \end{cases} \tag{2}$$

consisting of the elliptic equation in u and the hyperbolic equation in v.

Equations (1) with various nonlinear parts $f(u)$ and their numerical solutions were intensively studied in the literature (see book [4], Ch. 1 and references therein). In doing so, the most of numerical results were obtained by means of finite difference methods applied to system (2).

In the present paper, we construct a new implicit difference scheme for the governing Eqs. (2) by the method proposed in paper [1] and based on combination of the finite volume method, numerical integration, and difference elimination. Taking into account the symmetry of Eqs. (2) under the permutation $x \leftrightarrow y$ of the spatial variables, we choose a uniform and orthogonal solution grid with equal grid spacings in x and y. Such choice is natural for preservation of the permutational symmetry at the discrete level. As shown by the research results presented in [5], *mimetic discretizations*, i.e., such discrete approximations to PDE that mimic their basic algebraic properties, are more likely to produce highly accurate and stable numerical results.

In addition to the permutational symmetry, among the algebraic properties of the Eqs. (2) to be preserved at the discrete level, we make sure of the *conservativity* and *strong consistency* or *s-consistency* of the discretization. Conservativity means inheritance at the discrete level of the underlying conservation laws of Eqs. (2). The conventional notion of consistency (cf. [6], Ch. 7) provides reduction of the finite difference approximation (FDA) to the original PDE when the grid spacings tend to zero. In other words, a consistent discretization is a FDA to PDE. Strong consistency is the novel concept introduced in [7,8]. For Eqs. (2), such consistency means approximation of elements in the *differential ideal* generated by the polynomials in Eqs. (2) by elements in the *difference ideal* (cf. [9], Ch. 2) generated by the polynomials in the FDA to Eqs. (2).

Generally, non-linearity of differential equations is a major obstacle for their discretization by the method of paper [1] and also for the verification of s-consistency, since a difference Gröbner basis algorithm providing the elimination of grid functions for partial derivatives in constructing of a difference scheme may not terminate in view of non-Noetherianity of difference polynomial rings [9,10]. Moreover, even though the algorithm terminates, no software implementing non-linear difference Gröbner basis algorithms exists. However, for Eqs. (2), one can "hide" the nonlinear terms into an extra difference indeterminate F, as we have done in the lower equation of (2), and use a linear difference Gröbner basis algorithm if no nonlinear terms occur among the leaders of intermediate polynomials. We exploit this fact in our construction of difference scheme for Eqs. (2), and perform computations with implementation [11] in Python of the difference version of the classical Buchberger's algorithm [12].

For the constructed difference scheme and for the scheme devised in [3], under the assumption of equal spatial grid spacings, we compute their *modified equations* and compare the both schemes via these equations. Nowadays, the method of modified equations suggested in [16] is widely used (see [17], Ch. 8 and [18], Sect. 5.5) in studying difference schemes. This method provides a natural and unified platform to study such basic properties of the scheme as order of approximation, consistency, stability, convergence, dissipativity, dispersion, and invariance.

The present paper is organized as follows. In Sect. 2, we generate for Eqs. (2) a difference scheme by applying the approach of paper [1]. The consistency and conservativity analysis of the constructed scheme is done in Sect. 3. Computation of modified equation for our scheme and for the scheme of paper [3] is presented in Sect. 4. Some concluding remarks are given in Sect. 5.

2   Generation of a Difference Scheme for BPE

We consider the orthogonal and uniform solution grid with the spacing h for the spatial independent variables x, y and with the time spacing τ. Then we apply the approach of paper [1] to generation of a difference scheme for Eqs. (2).

Step 1. Completion to involution (we refer to [19] and to the references therein for the theory of involution). It is easy to see that system (2) is involutive.

Remark 1. For the lexicographical orderly ranking such that

$$t \succ x \succ y \quad \text{and} \quad v \succ F \succ u,$$

the differential polynomials occurring in the left-hand sides of (2) form a Gröbner basis of the differential ideal they generate in the ring $\mathcal{R} := \mathbb{Q}(\alpha, \beta_1, \beta_2)\{u, v\}$ of differential polynomials over the field $\mathbb{Q}(\alpha, \beta_1, \beta_2)$ of rational functions in parameters α, β_1, β_2.

Step 2. Conversion into the integral form. To provide *conservativity* of the scheme to be generated, we choose a rectangular parallelepiped as a "control volume" (stencil) shown in Fig. 1. Then we can rewrite equations differential system (2) into the following equivalent *integral conservation (balance) law form*

$$\begin{cases} \iint_\Omega (v - u)\, dxdy + \beta_1 \oint_{\partial\Omega} -u_y\, dx + u_x\, dy = 0, \\[2mm] \iint_\Omega v_t\, dxdy \Big|_{n\tau}^{(n+2)\tau} - \int_{n\tau}^{(n+2)\tau} \left(\frac{\beta_2}{\beta_1} \oint_{\partial\Omega} -v_y\, dx + v_x\, dy + \right. \\[2mm] \left. \frac{\beta_1 - \beta_2}{\beta_1^2} \int_\Omega (u - v)\, dxdy - \oint_{\partial\Omega} -F_y\, dx + F_x\, dy \right) dt = 0. \end{cases} \quad (3)$$

Here Ω is the square domain $jh \leq x \leq (j+2)h$, $kh \leq y \leq (k+2)h$ and $\partial\Omega$ is the boundary of Ω.

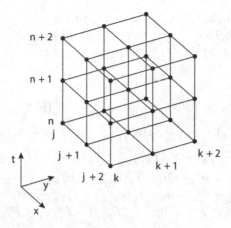

Fig. 1. Stencil $3 \times 3 \times 3$.

Step 3. Addition of integral relations for derivatives. We add to system (3) the *exact integral relations* between the partial derivatives of the independent variables occurring in (3) and the dependent variables themselves:

$$
\begin{cases}
\int_{x_j}^{x_{j+1}} u_{x\,j\,k}^{n}\,dx = u_{j+1\,k}^{n} - u_{j\,k}^{n}, & \int_{y_k}^{y_{k+1}} u_{y\,j\,k}^{n}\,dy = u_{j\,k+1}^{n} - u_{j\,k}^{n}, \\
\int_{x_j}^{x_{j+1}} v_{x\,j\,k}^{n}\,dx = v_{j+1\,k}^{n} - v_{j\,k}^{n}, & \int_{y_k}^{y_{k+1}} v_{y\,j\,k}^{n}\,dy = v_{j\,k+1}^{n} - v_{j\,k}^{n}, \\
\int_{x_j}^{x_{j+1}} F_{x\,j\,k}^{n}\,dx = F_{j+1\,k}^{n} - F_{j\,k}^{n}, & \int_{y_k}^{y_{k+1}} F_{y\,j\,k}^{n}\,dy = F_{j\,k+1}^{n} - F_{j\,k}^{n}, \\
\int_{t_n}^{t_{n+1}} v_{t\,j\,k}^{n}\,dt = v_{j\,k}^{n+1} - v_{j\,k}^{n}.
\end{cases}
\tag{4}
$$

Step 4. Numerical evaluation of integrals. We apply the Simpson's rule to the integrals in Eqs. (3) and the trapezoidal rule to the integrals in Eqs. (4). These rules are given in the table:

Trapezoidal rule	Simpson's rule
$T_x(f_{j\,k}^{n}) = \dfrac{f_{j\,k}^{n} + f_{j+1\,k}^{n}}{2}\,h$ $T_y(f_{j\,k}^{n}) = \dfrac{f_{j\,k}^{n} + f_{j\,k+1}^{n}}{2}\,h$	$S_x(f_{j\,k}^{n}) = \dfrac{f_{j-1\,k}^{n} + 4f_{j\,k}^{n} + f_{j+1\,k}^{n}}{3}\,h \xrightarrow[h\to 0]{} 2hf(t,x,y)$ $S_y(f_{j\,k}^{n}) = \dfrac{f_{j\,k-1}^{n} + 4f_{j\,k}^{n} + f_{j\,k+1}^{n}}{3}\,h \xrightarrow[h\to 0]{} 2hf(t,x,y)$ $S_t(f_{j\,k}^{n}) = \dfrac{f_{j\,k}^{n-1} + 4f_{j\,k}^{n} + f_{j\,k}^{n+1}}{3}\,\tau \xrightarrow[h\to 0]{} 2\tau f(t,x,y)$

As a result, we obtain the system (2) of difference equations for the grid functions $u_{j,\,k}^{n} \approx u(n\tau, jh, kh)$, $v_{j,\,k}^{n} \approx v(n\tau, jh, kh)$, $F_{j,k}^{n} \approx F(n\tau, jh, kh)$, approximating functions $u(t,x,y), v(t,x,y), F(t,x,y)$, and the grid functions approximating partial derivatives at the grid nodes

$$\begin{cases} u_{xj,k}^n \approx u_x(n\tau, jh, kh)\,, & u_{yj,k}^n \approx u_y(n\tau, jh, kh)\,, \\ v_{xj,k}^n \approx v_x(n\tau, jh, kh)\,, & v_{yj,k}^n \approx v_y(n\tau, jh, kh)\,, & v_{tj,k}^n \approx v_t(n\tau, jh, kh)\,, \\ F_{xj,k}^n \approx F_x(n\tau, jh, kh)\,, & F_{yj,k}^n \approx F_y(n\tau, jh, kh)\,, \end{cases}$$

where $j, k, n \in \mathbb{Z}$.

$$\begin{cases} S_x S_y(v_{jk}^n - u_{jk}^n) + \beta_1 \left(S_x(u_{yjk+2}^n - u_{yjk}^n) + S_y(u_{xj+2k}^n - u_{xjk}^n) \right) = 0\,, \\ S_x S_y(v_{tjk}^{n+2} - v_{tjk}^n) - S_t \left(\frac{\beta_2}{\beta_1} S_x(v_{yjk+2}^n - v_{yjk}^n) + \frac{\beta_2}{\beta_1} S_y(v_{xj+2k}^n - v_{xjk}^n) - \right. \\ \left. \frac{\beta_1 - \beta_2}{\beta_1^2} S_x S_y(u_{jk}^n - v_{jk}^n) + S_x(F_{yjk+2}^n - F_{yjk}^n) + S_y(F_{xj+2k}^n - F_{xjk}^n) \right) = 0\,, \\ T_x(u_{xjk}^n) - u_{j+1k}^n + u_{jk}^n = 0\,, \quad T_y(u_{yjk}^n) - u_{jk+1}^n + u_{jk}^n = 0\,, \\ T_x(v_{xjk}^n) - v_{j+1k}^n + v_{jk}^n = 0\,, \quad T_y(v_{yjk}^n) - v_{jk+1}^n + v_{jk}^n = 0\,, \\ T_x(F_{xjk}^n) - F_{j+1k}^n + F_{jk}^n = 0\,, \quad T_y(F_{yjk}^n) - F_{jk+1}^n + F_{jk}^n = 0\,, \\ T_t(v_{tjk}^n) - v_{jk}^{n+1} + v_{jk}^n = 0\,. \end{cases}$$

Step 5. Difference elimination of derivatives. To eliminate the grid functions of the partial derivatives,

$$v_t, v_x, v_y, u_x, u_y, F_x, F_y\,, \tag{5}$$

we construct a difference Gröbner basis form of the set of difference polynomials in the left-hand sides of (2) with the Python package [11] and for the lexicographic POT (Position Over Term) ranking (cf. [13], Def. 3.5.3) such that:

$$n \succ j \succ k\,, \quad v_t \succ v_x \succ v_y \succ F_x \succ F_y \succ u_x \succ u_y \succ v \succ F \succ u\,. \tag{6}$$

The output Gröbner basis includes two difference polynomials not containing the grid functions (5). These polynomials comprise the difference scheme:

$$\begin{cases} \frac{1}{4h^2} S_x S_y(v_{jk}^n - u_{jk}^n) + \frac{1}{2h} \beta_1 \left(D_{xx} S_y(u_{jk}^n) + D_{yy} S_x(u_{jk}^n) \right) = 0\,, \\ \frac{1}{4h^2} D_{tt} S_x S_y(v_{jk}^n) - \frac{1}{4\tau h} \frac{\beta_2}{\beta_1} S_t \left(D_{xx} S_y(v_{jk}^n) + D_{yy} S_x(v_{jk}^n) \right) - \\ \frac{\beta_1 - \beta_2}{8h^2 \beta_1^2} S_t S_x S_y(u_{jk}^n - v_{jk}^n) + \frac{1}{4\tau h} S_t \left(D_{xx} S_y(F_{jk}^n) + D_{yy} S_x(F_{jk}^n) \right) = 0\,, \end{cases} \tag{7}$$

where

$$D_{xx}(f_{jk}^n) = \frac{f_{j-1k}^n - 2f_{jk}^n + f_{j+1k}^n}{h^2} \xrightarrow[h \to 0]{} f_{xx}\,,$$

$$D_{yy}(f_{jk}^n) = \frac{f_{jk-1}^n - 2f_{jk}^n + f_{jk+1}^n}{h^2} \xrightarrow[h \to 0]{} f_{yy}\,, \tag{8}$$

$$D_{tt}(f_{jk}^n) = \frac{f_{jk}^{n-1} - 2f_{jk}^{n+1} + f_{jk}^{n+1}}{\tau^2} \xrightarrow[\tau \to 0]{} f_{tt}\,.$$

3   Consistency and Conservativity

In Remark 1, we introduced the ring of *differential polynomials*

$$\mathcal{R} := \mathbb{Q}(\alpha, \beta_1, \beta_2)\{u, v\}$$

over the field of rational functions in *parameters* α, β_1, β_2, with *derivations* $\partial_t, \partial_x, \partial_y$ and with *differential indeterminates* u, v.

In a similar manner, we consider the *difference polynomial ring*

$$\tilde{\mathcal{R}} := \mathbb{Q}(\alpha, \beta_1, \beta_2, \tau, h)\{u_{jk}^n, v_{jk}^n\}$$

over the extended coefficient field, with the grid functions u_{jk}^n, v_{jk}^n as *indeterminates* and the right-shift operators σ_t, σ_x and σ_y acting as translations, for example,

$$\sigma_t(v_{j,k}^n) = v_{j,k}^{n+1}, \quad \sigma_x(v_{j,k}^n) = v_{j+1,k}^n, \quad \sigma_y(v_{j,k}^n) = v_{j,k+1}^n.$$

We denote by $\mathcal{I} \subset \mathcal{R}$ the *differential ideal* generated by the polynomials in left-hand sides in (2) and by $\tilde{\mathcal{I}} \subset \tilde{\mathcal{R}}$ the *difference ideal* generated by the left-hand sides of scheme (7). The elements in $\mathcal{I}$ vanish on solutions of (2) and those in $\tilde{\mathcal{I}}$ vanish on solutions of (7). We refer to an element in $\mathcal{I}$ (respectively, in $\tilde{\mathcal{I}}$) as to a *consequence* of the governing Eqs. (2) (respectively, of Eqs. (7)). Apparently, the both ideals are radical. Among consequences of Eqs. (2), there are the following ones (cf. [15]).

Definition 1. *A consequence of Eqs. (2) of the form*

$$f = 0, \quad f := p_t + q_x + r_y + s \in \mathcal{I}, \quad p, q, r, s \in \mathcal{R}, \qquad (9)$$

is called conservation law *in the differential form with* density p, *fluxes* q, r *and source* s.

The difference polynomials in Eqs. (7) approximate those in Eqs. (2). It follows from the Taylor expansion about a grid point $t = nh, x = jh, y = kh$ of the difference operators shown in the above table and in formulae (8). Hence, scheme (7) is *consistent* with Eqs. (2). If one denotes by g_1 and g_2 the left-hand side of elliptic and hyperbolic equation in (2), respectively, and denote by $\tilde{g}_1$ and $\tilde{g}_2$ the discrete version of these polynomials from Eqs. (7), then the consistency conditions read

$$\tilde{g}_1 \xrightarrow[\tau, h \to 0]{} g_1 + \mathcal{O}(\tau, h), \quad \tilde{g}_2 \xrightarrow[\tau, h \to 0]{} g_2 + \mathcal{O}(\tau, h), \qquad (10)$$

where $O(\tau, h)$ denotes terms that reduce to zero when $\tau, h \to 0$.

Proposition 1. *For each differential polynomial* $p \in \mathcal{I}$, *there is a difference polynomial* $\tilde{p} \in \tilde{\mathcal{I}}$ *approximating* p, *i.e.,*

$$\tilde{p} \xrightarrow[\tau, h \to 0]{} p + \mathcal{O}(\tau, h). \qquad (11)$$

Proof. Since $\{g_1, g_2\}$ is a Gröbner basis of $\mathcal{I}$, the polynomial p can be represented as (cf. [14], Eq. (2.44))

$$p = \sum_{i=1}^{2} \left(\sum_{j=1}^{k_j} c_{i,j} \Theta_{i,j} \right) g_i, \quad \Theta_{i,j} \in \mathbf{Mon}(\partial), \quad c_{i,j} \in \mathcal{R}, \quad k_j \in \mathbb{N}_{>0},$$

where
$$\mathbf{Mon}(\partial) := \{\partial_t^{i_1} \partial_x^{i_2} \partial_y^{i_2} \mid i_1, i_2, i_3 \in \mathbb{N}_{\geq 0}\}. \tag{12}$$

We approximate the first-order partial derivatives occurring in p by the central differences

$$D_t = \frac{\sigma_t - \sigma_t^{-1}}{2\tau}, \quad D_x = \frac{\sigma_x - \sigma_x^{-1}}{2h}, \quad D_y = \frac{\sigma_y - \sigma_y^{-1}}{2h}. \tag{13}$$

Accordingly, we approximate the derivation operators (12) as

$$\partial_t^{i_1} \partial_x^{i_2} \partial_y^{i_2} \Longrightarrow D_t^{i_1} D_x^{i_2} D_y^{i_2}.$$

Then the polynomial $\tilde{p}$ given by

$$\tilde{p} := \sum_{i=1}^{2} \left(\sum_{j=1}^{k_j} \tilde{c}_{i,j} \tilde{D}_{i,j} \right) \tilde{g}_i, \quad \tilde{D}_{i,j} \in \mathbf{Mon}(D), \quad \tilde{c}_{i,j} \in \tilde{\mathcal{R}}, \quad k_j \in \mathbb{N}_{>0}, \tag{14}$$

where
$$\mathbf{Mon}(D) := \{D_t^{i_1} D_x^{i_2} D_y^{i_2} \mid i_1, i_2, i_3 \in \mathbb{N}_{\geq 0}\}.$$

satisfies (11). □

Remark 2. The consistency conditions (10) imply the *strong consistency* or *s-consistency* of $\{\tilde{g}_1, \tilde{g}_2\}$ with $\{g_1, g_2\}$ (see [8], Thr. 3). In the proof of Proposition 1, we explicitly demonstrate how to construct a difference consequence of scheme (7) which approximates a given consequence of the differential system (2).

Corollary 1. *For a conservation law of the form* (9), *there is a consequence of scheme* (7) *which approximates that law.*

By the construction of the difference scheme (7) as a discrete version of the integral conservation law form (3) and by Corollary 1, the smooth solutions of Eqs. (7) in the limit $\tau, h \longrightarrow 0$ satisfy all the polynomial conservation laws (9). Thus, the scheme is conservative.

4 Modified Equation

Given a finite difference scheme, the main idea of *modified equation method* (see [16] and its bibliography) is to construct another PDE which is approximated better by the difference scheme. In the case of Boussinesq Paradigm Equation (BPE) (2) and its finite difference approximation (FDA) (7), the modified equation method (cf. [18], Sect. 5.5) is based on the consideration of a solution to (7) as a set of smooth functions $\{u, v\}$ whose values at the grid points satisfy the difference scheme (7). Since the difference equations (7) describe BPE only approximately, one cannot expect that a solution interpolating the grid values exactly satisfies BPE. In fact, it satisfies another set of differential equations which we shall call the *modified BPE*.

To obtain a modified BPE, the shift operators occurring in (7) are replaced by their Taylor expansions about a grid point

$$\sigma_t = \sum_{k \geq 0} \frac{\tau^k}{k!} \partial_t^k, \quad \sigma_x = \sum_{k \geq 0} \frac{h^k}{k!} \partial_x^k, \quad \sigma_y = \sum_{k \geq 0} \frac{h^k}{k!} \partial_y^k.$$

This replacement converts the difference equations comprising the scheme into infinite order differential equations of the following form

$$
\begin{cases}
\beta_1(\underline{u_{xx}} + u_{yy}) + v - u + h^2 \left(\frac{\beta_1}{12}(u_{xxxx} + 4u_{xxyy} + u_{yyyy}) \right. \\
\left. - \frac{u_{xx}+u_{yy}}{6} + \frac{v_{xx}+v_{yy}}{6} \right) + \mathcal{O}(h^4) = 0, \\
\underline{v_{tt}} - \frac{\beta_2}{\beta_1}(v_{xx} + v_{yy}) - \frac{\beta_1-\beta_2}{\beta_1^2}(u-v) - (F_{xx} + F_{yy}) \\
+ \tau^2 \left(\frac{F_{ttxx}-F_{ttyy}}{6} + \frac{v_{tttt}}{12} - \frac{\beta_2}{6\beta_1}(v_{ttxx} + v_{ttyy}) \right. \\
\left. - \frac{\beta_1-\beta_2}{3\beta_1^2}(u_{tt} - v_{tt}) \right) + h^2 \left(\frac{F_{xxxx}+4F_{xxyy}+F_{yyyy}}{12} \right. \\
\left. - \frac{v_{ttxx}+v_{ttyy}}{6} - \frac{\beta_2}{12\beta_1}(v_{xxxx} + 4v_{xxyy} + v_{yyyy}) \right. \\
\left. - \frac{\beta_1-\beta_2}{3\beta_1^2}(u_{xx} + u_{yy}) + \frac{\beta_1-\beta_2}{3\beta_1^2}(v_{xx} + v_{yy}) \right) + \mathcal{O}(\tau^4, \tau^2 h^2, h^4) = 0.
\end{cases}
\tag{15}
$$

where the terms of order τ^2 and h^2 are written explicitly. We underlined the highest ranked terms (leaders) in BPE (2) for ranking (4). The calculation of the right-hand sides in Eq. (15) as well as the computation of the expressions given below was done with the use of freely available Python library [11].

Remark 3. The Taylor expansions of the s-consistent difference scheme (7) over the chosen grid points contain only the even powers of h. It follows immediately from the fact that all the finite differences occurring in the equations of both schemes are the central difference approximations of the partial derivatives occurring in (2).

Furthermore, we substitute $F \rightarrow \alpha u^2$ and reduce the terms of order τ^2 and h^2 in (15) modulo their left-hand sides which form the differential Gröbner basis (2) for the lexicographic orderly ranking

$$t \succ x \succ y, \quad v \succ u.$$

This reduction will give us a *canonical form* of the second order modified flow, since given a Gröbner basis, the *normal form* of a polynomial modulo this basis is uniquely defined (cf. [13], Sect. 2.1). Moreover, since FDA (7) does not have S-polynomials, it is sufficient to perform autoreduction of (15).

Thus, after the Taylor expansion of FDA (scheme) and its autoreduction yields the *modified BPE*:

$$\begin{cases}
\beta_1(u_{xx} + u_{yy}) + (v - u) + h^2\left(\frac{\beta_1 u_{yyyy}}{6} - \frac{2u_{yy} + v_{xx} - v_{yy}}{12} - \frac{v - u}{12\beta_1}\right) \\
\qquad\qquad + \mathcal{O}(h^4) = 0, \\
v_{tt} - \frac{\beta_2}{\beta_1}(v_{xx} + v_{yy}) - \frac{\beta_1 - \beta_2}{\beta_1^2}(u - v) + 2\alpha\left(u_x^2 + u_y^2 + \frac{u^2 - uv}{\beta_1}\right) \\
+ \tau^2\left(\frac{\alpha u_{ttx}u_x}{3} + \frac{\alpha u_{tty}u_y}{3} + \frac{\alpha u_{tx}^2}{3} + \frac{\alpha u_{ty}^2}{3} + \frac{2\alpha\beta_2 u_{xy}^2}{3\beta_1} + \frac{2\alpha\beta_2 u_{yy}^2}{3\beta_1} + \frac{\alpha u u_{tt}}{3\beta_1}\right. \\
\quad - \frac{\alpha u v_{tt}}{6\beta_1} - \frac{\alpha u_{tt}v}{6\beta_1} + \frac{\alpha u_t^2}{3\beta_1} - \frac{\alpha u_t v_t}{3\beta_1} + \frac{\alpha u_x^2}{2\beta_1} + \frac{\alpha u_y^2}{2\beta_1} + \frac{2\alpha\beta_2 u u_{yy}}{3\beta_1^2} - \frac{\alpha\beta_2 u v_{xx}}{6\beta_1^2} \\
\quad - \frac{\alpha\beta_2 u v_{yy}}{6\beta_1^2} + \frac{7\alpha\beta_2 u_x^2}{6\beta_1^2} - \frac{2\alpha\beta_2 u_x v_x}{3\beta_1^2} + \frac{2\alpha\beta_2 u_{yy}v}{3\beta_1^2} + \frac{7\alpha\beta_2 u_y^2}{6\beta_1^2} - \frac{2\alpha\beta_2 u_y v_y}{3\beta_1^2} \\
\quad - \frac{\alpha u^2}{2\beta_1^2} + \frac{\alpha u v}{2\beta_1^2} + \frac{7\alpha\beta_2 u^2}{6\beta_1^3} - \frac{5\alpha\beta_2 u v}{3\beta_1^3} + \frac{\alpha\beta_2 v^2}{2\beta_1^3} - \frac{u_{tt}}{4\beta_1} - \frac{\beta_2^2 v_{xxxx}}{12\beta_1^2} \\
\quad - \frac{\beta_2^2 v_{xxyy}}{6\beta_1^2} - \frac{\beta_2^2 v_{yyyy}}{12\beta_1^2} + \frac{\beta_2 u_{tt}}{4\beta_1^2} + \frac{\beta_2 v_{xx}}{3\beta_1^2} + \frac{\beta_2 v_{yy}}{3\beta_1^2} + \frac{u}{4\beta_1^2} - \frac{v}{4\beta_1^2} - \frac{\beta_2^2 v_{xx}}{3\beta_1^3} \\
\left.\quad - \frac{\beta_2^2 v_{yy}}{3\beta_1^3} - \frac{7\beta_2 u}{12\beta_1^3} + \frac{7\beta_2 v}{12\beta_1^3} + - \frac{\beta_2^2 u}{3\beta_1^4} - \frac{\beta_2^2 v}{3\beta_1^4}\right) + h^2\left(\frac{2\alpha u_{xyy}u_x}{3} - \frac{2\alpha u_{yyy}u_y}{3}\right. \\
\quad - \alpha u_{yy}^2 + \frac{\alpha u u_{yy}}{\beta_1} - \frac{2\alpha u_x^2}{3\beta_1} + \frac{2\alpha u_x v_x}{3\beta_1} - \frac{\alpha u_{yy}v}{\beta_1} - \frac{\alpha u^2}{2\beta_1^2} + \frac{\alpha u v}{\beta_1^2} \\
\left.\quad - \frac{\alpha v^2}{2\beta_1^2} + \frac{\beta_2 v_{xxxx}}{12\beta_1} + \frac{\beta_2 v_{yyyy}}{12\beta_1} + \frac{v_{xx}}{6\beta_1} + \frac{v_{yy}}{6\beta_1} - \frac{\beta_2 v_{xx}}{6\beta_1^2} - \frac{\beta_2 v_{yy}}{6\beta_1^2} - \frac{u}{6\beta_1^2}\right. \\
\left.\qquad\qquad + \frac{v}{6\beta_1^2} + \frac{\beta_2 u}{6\beta_1^3} - \frac{\beta_2 v}{6\beta_1^3}\right) + \mathcal{O}(\tau^4, \tau^2 h^2, h^4) = 0.
\end{cases} \tag{16}$$

Formulae (16) show that the scheme (7) has the second-order of accuracy in the grid spacings τ and h. Being implicit, the scheme is unconditionally stable. Therefore, in accordance to the Lax-Richtmyer equivalence theorem (see [6], Thm. 1.5.1) proved for linear scalar PDE and initial - value problem, it has been accepted that the convergence is ensured if a given FDA to the PDE is consistent and stable.

In paper [3], another finite difference scheme was designed. That scheme is implicit, unconditionally stable, also has the second-order of accuracy, and for the case of equal spacial grid spacings this scheme reads

$$\begin{cases}
(v_{jk}^{n+1} - u_{jk}^{n+1}) + \beta_1\left(D_{xx}(u_{jk}^{n+1}) + D_{yy}(u_{jk}^{n+1})\right) = 0, \\
D_{tt}(v_{jk}^n) - \frac{\beta_2}{2\beta_1}D_2\left(v_{jk}^{n+1} + v_{jk}^{n-1}\right) - \frac{\beta_1 - \beta_2}{2\beta_1^2}(u_{jk}^{n+1} + u_{jk}^{n-1}) + \\
\frac{\beta_1 - \beta_2}{2\beta_1^2}(v_{jk}^{n+1} + v_{jk}^{n-1}) - \frac{\beta_1}{3}D_2\left(u_{jk}^{n+1}{}^2 + u_{jk}^{n+1}u_{jk}^{n-1} + u_{jk}^{n-1}{}^2\right) = 0,
\end{cases} \tag{17}$$

where

$$D_2 := D_{xx} + D_{yy}$$

approximates the Laplace operator. The Taylor expansion of Eqs. (17) and autoreduction yield the following modified BPE

$$
\begin{cases}
\beta_1(u_{xx} + u_{yy}) + (v - u) + h^2 \left(\dfrac{\beta_1 u_{yyyy}}{6} - \dfrac{2u_{yy} + v_{xx} - v_{yy}}{12} - \dfrac{v - u}{12\beta_1} \right) \\
\qquad + \mathcal{O}(h^4) = 0 \\[4pt]
v_{tt} - \dfrac{\beta_2}{\beta_1}(v_{xx} + v_{yy}) - \dfrac{\beta_1 - \beta_2}{\beta_1^2}(u - v) + 2\alpha \left(u_x^2 + u_y^2 + \dfrac{u^2 - uv}{\beta_1} \right) + \\[4pt]
\tau^2 \left(\alpha u u_{ttyy} - \dfrac{\alpha u u_{tt}}{6} + 2\alpha u_{ttx} u_x + 2\alpha u_{tty} u_y + \dfrac{2\alpha u_{tx}^2}{3} + \dfrac{2\alpha u_{tyy} u_t}{3} + \dfrac{2\alpha u_{ty}^2}{3} \right. \\[4pt]
- \dfrac{\alpha u_t^2}{6} - \dfrac{\alpha v_t^2}{6} - \dfrac{\alpha v v_{tt}}{6} + \dfrac{5\alpha\beta_2 u_x^2}{6\beta_1} + \dfrac{5\alpha\beta_2 u_y^2}{6\beta_1} + \dfrac{5\alpha\beta_2 v_{xx} v}{6\beta_1} + \dfrac{5\alpha\beta_2 v_x^2}{6\beta_1} \\[4pt]
+ \dfrac{5\alpha\beta_2 v_{yy} v}{6\beta_1} + \dfrac{5\alpha\beta_2 v_y^2}{6\beta_1} + \dfrac{\alpha u u_{tt}}{\beta_1} + \dfrac{11\alpha u u_{yy}}{6\beta_1} - \dfrac{\alpha u_{tt} v}{\beta_1} - \dfrac{11\alpha u_x^2}{6\beta_1} \\[4pt]
- \dfrac{11\alpha v_{xx} v}{6\beta_1} - \dfrac{11\alpha v_x^2}{6\beta_1} + \dfrac{5\alpha\beta_2 u^2}{6\beta_1^2} - \dfrac{11\alpha\beta_2 u u_{yy}}{6\beta_1^2} - \dfrac{5\alpha\beta_2 u v}{6\beta_1^2} + \dfrac{11\alpha\beta_2 u_x^2}{6\beta_1^2} \\[4pt]
+ \dfrac{11\alpha\beta_2 v_{xx} v}{6\beta_1^2} + \dfrac{11\alpha\beta_2 v_x^2}{6\beta_1^2} - \dfrac{11\alpha u^2}{6\beta_1^2} + \dfrac{11\alpha u v}{6\beta_1^2} + \dfrac{11\alpha\beta_2 u^2}{6\beta_1^3} - \dfrac{11\alpha\beta_2 u v}{6\beta_1^3} \\[4pt]
+ \dfrac{11 u_{tt}}{12\beta_1} - \dfrac{5\beta_2^2 v_{xxxx}}{12\beta_1^2} - \dfrac{5\beta_2^2 v_{xxyy}}{6\beta_1^2} - \dfrac{5\beta_2^2 v_{yyyy}}{12\beta_1^2} + \dfrac{11\beta_2 u_{tt}}{12\beta_1^2} + \dfrac{4\beta_2 v_{xx}}{3\beta_1^2} \\[4pt]
+ \dfrac{4\beta_2 v_{yy}}{3\beta_1^2} + \dfrac{11 u}{12\beta_1^2} - \dfrac{11 v}{12\beta_1^2} - \dfrac{4\beta_2^2 v_{xx}}{3\beta_1^3} - \dfrac{4\beta_2^2 v_{yy}}{3\beta_1^3} - \dfrac{9\beta_2 u}{4\beta_1^3} + \dfrac{9\beta_2 v}{4\beta_1^3} + \dfrac{4\beta_2^2 u}{3\beta_1^4} \\[4pt]
\left. - \dfrac{4\beta_2^2 v}{3\beta_1^4} \right) + h^2 \left(\dfrac{\alpha u u_{yyyy}}{6} + \dfrac{2\alpha u_{yyy} u_y}{3} + \alpha u_{yy}^2 - \dfrac{\alpha u u_{yy}}{\beta_1} + \dfrac{\alpha u_{yy} v}{\beta_1} + \dfrac{\alpha u^2}{2\beta_1^2} \right. \\[4pt]
\left. - \dfrac{\alpha u v}{\beta_1^2} + \dfrac{\alpha v^2}{2\beta_1^2} - \dfrac{\beta_2 v_{xxxx}}{12\beta_1} - \dfrac{\beta_2 v_{yyyy}}{12\beta_1} \right) + \mathcal{O}(\tau^4, \tau^2 h^2, h^4) = 0.
\end{cases}
$$

$$(18)$$

We see that in the order h^2, the first equation in (18) coincides with that in (16) whereas the second equations of the schemes differ in the terms of order τ^2 and h^2. Thus, schemes (7) and (17) differ from each other.

5   Conclusion

We applied our computer algebra - aided approach [1] to the fourth order Boussinesq paradigm equation (1), rewritten in the form of two second order equations (2), and derived the finite difference scheme (7) defined on the uniform and orthogonal grid with spacings τ and h. Due to the use of integral conservation law form (3) for Eqs. (2), the obtained scheme is conservative. By its construction, scheme (7) approximates the differential system (2) (cf. (10)) and, hence, is consistent.

Furthermore, since the differential polynomials in the left-hand side of (2) and the difference polynomials in the left-hand side of (7) form a differential Gröbner basis and a difference Gröbner basis, respectively, of the ideal they generate, the consistency of (7) with (2) implies the s-consistency (cf. Remark 2). Therefore, any differential consequence of Eqs. (2) is approximated by a consequence of Eqs. (7) (Proposition 1). In particular, any smooth solution to Eqs. (7) in the limit $\tau, h \longrightarrow 0$ satisfies to all conservation law of the form (9) (Corollary 1), i.e., scheme (7) is conservative.

For scheme (7), we computed its modified equation (16) with accuracy $\mathcal{O}(\tau^2, h^2)$. It is the differential system (16), such that a smooth solution to the scheme satisfies this system. The canonical form of modified equation given by formulae (16) is obtained by the reduction of terms order τ^2 and h^2 modulo the ideal generated by the set of differential polynomials in the left-hand side of Eqs. (2). For the ranking (4) this set is a Gröbner basis.

Given two difference schemes for a PDE, one can compare them by computing their modified equations. In our case, we confront the scheme (7) with another scheme taken from [3]. The both schemes have the same order $\mathcal{O}(\tau^2, h^2)$ of accuracy and their first equations approximating $v - u + \beta_1 \triangle u = 0$ coincide if the terms of order $\mathcal{O}(\tau^4, \tau^2 h^2, h^4)$ are neglected. But the second equations of the schemes which approximate

$$v_{tt} - \frac{\beta_2}{\beta_1}\triangle v - \frac{\beta_1 - \beta_2}{\beta_1^2}(u - v) + \triangle F = 0, \quad F := \alpha\, u^2,$$

are significantly distinct. In our further comparative analysis of the schemes, we plan to confront them on numerical simulation.

In this paper, we use the Python-based library [11], as a tool for doing all computations with differential and difference polynomials we needed including construction of Gröbner bases and Taylor expansions. The denotation $\alpha u^2 \longrightarrow F$ reduces the nonlinear problem (1) to the linear one, and allows to apply the software packages oriented to computation of Gröbner bases. For example, to compute linear differential Gröbner bases one can apply the Maple package JANET [20], and to compute linear difference Gröbner bases one can apply the Maple package LDA [21].

Acknowledgements. This work has been partially supported by the Russian Foundation for Basic Research (grant No. 18-51-18005) and by the RUDN University Program (5-100).

References

1. Gerdt, V.P., Blinkov, Y.A., Mozzhilkin, V.V.: Gröbner bases and generation of difference schemes for partial differential equations. SIGMA **2**, 051 (2006)
2. Christov, C.I.: An energy - consistent Galilean - invariant dispersive shallow - water model. Wave Motion **34**, 161–174 (2001)
3. Christov, C.I., Kolkovska, N., Vasileva, D.: On the numerical simulation of unsteady solutions for the 2D Boussinesq paradigm equation. In: Dimov, I., Dimova, S., Kolkovska, N. (eds.) NMA 2010. LNCS, vol. 6046, pp. 386–394. Springer, Heidelberg (2011). https://doi.org/10.1007/978-3-642-18466-6_46

4. Todorov, M.D.: Nonlinear Waves: Theory, Computer Simulation, Experiment. Morgan & Claypool Publishers, San Rafael (2018)
5. Koren, B., Abgrall, R., Bochev, P., Frank, J.: Physics-compatible numerical methods. J. Comput. Phys. **257**, 1039–1526 (2014)
6. Strikwerda, J.C.: Finite Difference Schemes and Partial Differential Equations, 2nd edn. SIAM, Philadelphia (2004)
7. Gerdt, V.P., Robertz, D.: Consistency of finite difference approximations for linear PDE systems and its algorithmic verification. In: Watt, S.M. (ed.) ISSAC 2010, pp. 53–59. Association for Computing Machinery, New York (2010)
8. Gerdt, V.P.: Consistency analysis of finite difference approximations to PDE systems. In: Adam, G., Buša, J., Hnatič, M. (eds.) MMCP 2011. LNCS, vol. 7125, pp. 28–42. Springer, Heidelberg (2012). https://doi.org/10.1007/978-3-642-28212-6_3
9. Levin, A.: Difference Algebra. Algebra and Applications, vol. 8. Springer, Dordrecht (2008). https://doi.org/10.1007/978-1-4020-6947-5
10. Gerdt, V., La Scala, R.: Noetherian quotients of the algebra of partial difference polynomials and Gröbner bases of symmetric ideals. J. Algebra **423**, 1233–1261 (2015)
11. https://github.com/blinkovua/GInv/tree/master/pyginv
12. Buchberger, B.: Gröbner bases: an algorithmic method in polynomial ideal theory. In: Bose, N.K. (ed.) Recent Trends in Multidimensional System, pp. 184–232. Reidel, Dordrecht (1985)
13. Adams, W.W., Loustanau, P.: Introduction to Gröbner Bases. Graduate Studies in Mathematics, vol. 3. American Mathematical Society (1994)
14. Robertz, D.: Formal Algorithmic Elimination for PDEs. LNM, vol. 2121. Springer, Cham (2014). https://doi.org/10.1007/978-3-319-11445-3
15. Chertock, A., Christov, C.I., Kurganov, A: Central-upwind schemes for Boussinesq paradigm equations. In: Krause, E., Shokin, Yu., Resch, M., Kröner, D., Shokina, N. (eds.) Computational Science and High Performance Computing IV, vol. 115, pp. 267–281 (2011)
16. Shokin, Y.I.: The Method of Differential Approximation, 1st edn. Springer, Heidelberg (1983)
17. Ganzha, V.G., Vorozhtsov, E.V.: Computer-aided Analysis of Difference Schemes for Partial Differential Equations. Wiley, New York (1996)
18. Moin, P.: Fundamentals of Engineering Numerical Analysis, 2nd edn. Cambridge University Press, Cambridge (2010)
19. Seiler, W.M.: Involution: The Formal Theory of Differential Equations and its Applications in Computer Algebra. Algorithms and Computation in Mathematics, vol. 24, 1st edn. Springer, Heidelberg (2010). https://doi.org/10.1007/978-3-642-01287-7
20. Blinkov, Yu.A., Cid, C.F., Gerdt, V.P., Plesken, W., Robertz, D.: The MAPLE package janet: II. Linear partial differential equations. In: Ganzha, V.G., Mayr, E.W., Vorozhtsov, E.V. (eds.) Proceedings of the 6th International Workshop on Computer Algebra in Scientific Computing, CASC 2003, pp. 41–54. Technische Universität München (2003). Package Janet is freely available on the web page http://134.130.169.213/Janet/
21. Gerdt, V.P., Robertz, D.: Computation of difference Gröbner bases. Comput. Sc. J. Moldova. **20**(2), 203–226 (2012). Package LDA is freely available on the web page http://134.130.169.213/Janet/

Symbolic Investigation of the Dynamics of a System of Two Connected Bodies Moving Along a Circular Orbit

Sergey A. Gutnik[1,2]([⊠]) [iD] and Vasily A. Sarychev[3] [iD]

[1] Moscow State Institute of International Relations (MGIMO University),
76, Prospekt Vernadskogo, Moscow 119454, Russia
s.gutnik@inno.mgimo.ru
[2] Moscow Institute of Physics and Technology,
Institutskii per. 9, Dolgoprudny 141701, Russia
[3] Keldysh Institute of Applied Mathematics (Russian Academy of Sciences),
4, Miusskaya Square, Moscow 125047, Russia
vas31@rambler.ru

Abstract. The dynamics of the system of two bodies, connected by a spherical hinge, that moves along a circular orbit under the action of gravitational torque is investigated. Computer algebra method based on the resultant approach was applied to reduce the satellite stationary motion system of algebraic equations to a single algebraic equation in one variable that determines all planar equilibrium configurations of the two–body system. Classification of domains with equal numbers of equilibrium solutions is carried out using algebraic methods for constructing discriminant hypersurfaces. Bifurcation curves in the space of system parameters that determine boundaries of domains with a fixed number of equilibria of the two–body system were obtained symbolically. Depending on the parameters of the problem, the number of equilibria was found by analyzing the real roots of the algebraic equations.

Keywords: Satellite-stabilizer system · Gravitational torque ·
Circular orbit · Lagrange equations · Algebraic equations ·
Equilibrium orientation · Computer algebra ·
Discriminant hypersurface

1 Introduction

In this work, we investigate the dynamics of a system of two bodies (satellite and stabilizer) connected by a spherical hinge that moves in a central Newtonian force field on a circular orbit using computer algebra methods.

Determining the equilibria for the system of connected bodies on a circular orbit is of practical interest for designing composite gravitational orientation systems of satellites that can stay on the orbit for a long time without energy consumption. The dynamics of various composite schemes for satellite–stabilizer gravitational orientation systems was discussed in detail in [1].

© Springer Nature Switzerland AG 2019
M. England et al. (Eds.): CASC 2019, LNCS 11661, pp. 164–178, 2019.
https://doi.org/10.1007/978-3-030-26831-2_12

The study of the satellite–stabilizer dynamics under the influence of gravitational torque is an important topic for the practical implementation of attitude control systems of the artificial satellites. The dynamics of a satellite–stabilizer subjected to gravitational torque was considered in many papers indicated in [1]. In [2] and [3], planar equilibrium orientations were found in special cases, when the spherical hinge is located at the intersection of the satellite and stabilizer principal central axis of inertia. In [4], all equilibrium orientations were found in the case of axisymmetric satellite and stabilizer. In paper [5], some classes of spatial equilibrium orientations of the satellite–stabilizer system in the orbital coordinate system were analyzed, using computer algebra methods.

In this paper, we consider the planar equilibria (equilibrium orientations) of the satellite–stabilizer system in the orbital coordinate frame for certain values of the principal central moments of inertia of the bodies when the spherical hinge is located at the intersection of the satellite and stabilizer principal central planes of inertia. The action of the stabilizer on the satellite provides new equilibrium orientations for the two-body system, as well as introduces dissipation into the system. The investigation of satellite equilibria was performed by using the Computer Algebra resultant method. The regions with an equal number of equilibria were specified by using the Meiman theorem [13] for the construction of discriminant hypersurfaces.

The algebraic methods for determining the equilibrium orientations of the two-body system described in this work were successfully used to analyze the dynamics of a satellite–gyrostat system [6,7] as well as the dynamics of a satellite with an aerodynamic orientation system [8,9].

In mechanics, computer algebra is widely employed to analyze polynomial systems with the use of symbolic computations. Some computer algebra algorithms for solving these problems were described in [11,12,15]. The question of finding regions of parameter space with certain equilibria properties also occurred in relevance to a biology problem was presented at the CASC 2017 Workshop [16].

2   Equations of Motion

Let us consider the system of two bodies connected by a spherical hinge that moves along a circular orbit [1]. To write equations of motion for two bodies, we introduce the following right-handed Cartesian coordinate systems (Fig. 1). The absolute coordinate system $CX_aY_aZ_a$ with the origin at the Earth's center of mass C. The plane CX_aY_a coincides with the equatorial plane and the CZ_a axis coincides with the Earth axis of rotation, and $OXYZ$ is the orbital coordinate system. The OZ axis is directed along the radius vector that connects the Earth center of mass C with the center of mass of the two–body system O, the OX axis is directed along the linear velocity vector of the center of mass O. Then, the OY axis is directed along the normal to the orbital plane. The coordinate system for the ith body $(i = 1, 2)$ is $Ox_iy_iz_i$, where Ox_i, Oy_i, and Oz_i are the principal central axes of inertia for the ith body. The orientation of the coordinate system $Ox_iy_iz_i$ with respect to the orbital coordinate system is determined using the pitch (α_i), yaw (β_i), and roll (γ_i) angles, and the direction cosines in the transformation matrix between the orbital coordinate system $OXYZ$ and $Ox_iy_iz_i$ are expressed in terms of aircraft angles using the relations [1]:

$$
\begin{aligned}
a_{11}^{(i)} &= \cos\alpha_i \cos\beta_i, \\
a_{12}^{(i)} &= \sin\alpha_i \sin\gamma_i - \cos\alpha_i \sin\beta_i \cos\gamma_i, \\
a_{13}^{(i)} &= \sin\alpha_i \cos\gamma_i + \cos\alpha_i \sin\beta_i \sin\gamma_i, \\
a_{21}^{(i)} &= \sin\beta_i, \quad a_{22}^{(i)} = \cos\beta_i \cos\gamma_i, \\
a_{23}^{(i)} &= -\cos\beta_i \sin\gamma_i, \quad a_{31}^{(i)} = -\sin\alpha_i \cos\beta_i, \\
a_{32}^{(i)} &= \cos\alpha_i \sin\gamma_i + \sin\alpha_i \sin\beta_i \cos\gamma_i, \\
a_{33}^{(i)} &= \cos\alpha_i \cos\beta_i - \sin\alpha_i \sin\beta_i \sin\gamma_i.
\end{aligned} \tag{1}
$$

Suppose that (a_i, b_i, c_i) are the coordinates of the spherical hinge P in the body coordinate system $Ox_iy_iz_i$, A_i, B_i, C_i are the principal central moments of inertia; $M = M_1M_2/(M_1 + M_2)$; M_i is the mass of the ith body; p_i, q_i, and r_i are the projections of the absolute angular velocity of the ith body onto the axes Ox_i, Oy_i, and Oz_i; and ω_0 is the angular velocity for the center of mass of the two-body system moving along a circular orbit. Then, using expressions for kinetic energy and force function, which determines the effect of the Earth gravitational field on the system of two bodies connected by a hinge [1], the equations of motion for this system can be written as Lagrange equations of the second kind by symbolic differentiation in the Maple system [10] in the case when $b_1 = b_2 = 0$:

$$
\begin{aligned}
&(A_i + Mc_i^2)\dot{p}_i - Ma_ic_i\dot{r}_i - Mc_ic_j(a_{12}^{(i)}a_{12}^{(j)} + a_{22}^{(i)}a_{22}^{(j)} + a_{32}^{(i)}a_{32}^{(j)})\dot{p}_j \\
&-M\big(a_jc_i(a_{12}^{(i)}a_{13}^{(j)} + a_{22}^{(i)}a_{23}^{(j)} + a_{32}^{(i)}a_{33}^{(j)}) - c_ic_j(a_{12}^{(i)}a_{11}^{(j)} + a_{22}^{(i)}a_{21}^{(j)} \\
&\quad + a_{32}^{(i)}a_{31}^{(j)})\big)\dot{q}_j + Ma_jc_i(a_{12}^{(i)}a_{12}^{(j)} + a_{22}^{(i)}a_{22}^{(j)} + a_{32}^{(i)}a_{32}^{(j)})\dot{r}_j \\
&+Ma_jc_i\big(a_{12}^{(i)}(r_j(p_ja_{13}^{(j)} - r_ja_{11}^{(j)}) - q_j(q_ja_{11}^{(j)} - p_ja_{12}^{(j)})) \\
&\quad +a_{22}^{(i)}(r_j(p_ja_{23}^{(j)} - r_ja_{21}^{(j)}) - q_j(q_ja_{21}^{(j)} - p_ja_{22}^{(j)})) \\
&\quad +a_{32}^{(i)}(r_j(p_ja_{33}^{(j)} - r_ja_{31}^{(j)}) - q_j(q_ja_{31}^{(j)} - p_ja_{32}^{(j)}))\big) \\
&+Mc_ic_j\big((a_{12}^{(i)}(q_j(r_ja_{12}^{(j)} - q_ja_{13}^{(j)}) - p_j(p_ja_{13}^{(j)} - r_ja_{11}^{(j)})) \\
&\quad +a_{22}^{(i)}(q_j(r_ja_{22}^{(j)} - q_ja_{23}^{(j)}) - p_j(p_ja_{23}^{(j)} - r_ja_{21}^{(j)})) \\
&\quad +a_{32}^{(i)}(q_j(r_ja_{32}^{(j)} - q_ja_{33}^{(j)}) - p_j(p_ja_{33}^{(j)} - r_ja_{31}^{(j)}))\big) \\
&+((C_i - B_i) - Mc_i^2)q_ir_i - Ma_ip_ic_iq_i + 3\omega_0^2(C_i - B_i)a_{32}^{(i)}a_{33}^{(i)} \\
&\quad +M\omega_0^2c_i\big(a_j(a_{12}^{(i)}a_{11}^{(j)} + a_{22}^{(i)}a_{21}^{(j)} + a_{32}^{(i)}a_{31}^{(j)}) \\
&\quad +c_j(a_{12}^{(i)}a_{13}^{(j)} + a_{22}^{(i)}a_{23}^{(j)} + a_{32}^{(i)}a_{33}^{(j)})\big) \\
&+3M\omega_0^2c_ia_{32}^{(i)}(a_ia_{31}^{(i)} + c_ia_{33}^{(i)} - a_ja_{31}^{(j)} - c_ja_{33}^{(j)}) = 0, \quad (2)
\end{aligned}
$$

$$(B_i + M(a_i^2 + c_i^2))\dot{q}_i - M(a_i c_j(a_{13}^{(i)}a_{12}^{(j)} + a_{23}^{(i)}a_{22}^{(j)} + a_{33}^{(i)}a_{32}^{(j)})$$

$$-c_i c_j(a_{11}^{(i)}a_{12}^{(j)} + a_{21}^{(i)}a_{22}^{(j)} + a_{31}^{(i)}a_{32}^{(j)}))\dot{p}_j$$

$$-M\big(a_i a_j(a_{13}^{(i)}a_{13}^{(j)} + a_{23}^{(i)}a_{23}^{(j)} + a_{33}^{(i)}a_{33}^{(j)}) - a_i c_j(a_{13}^{(i)}a_{11}^{(j)} + a_{23}^{(i)}a_{21}^{(j)} + a_{33}^{(i)}a_{31}^{(j)})$$

$$-c_i a_j(a_{11}^{(i)}a_{13}^{(j)} + a_{21}^{(i)}a_{23}^{(j)} + a_{31}^{(i)}a_{33}^{(j)}) + c_i c_j(a_{11}^{(i)}a_{11}^{(j)} + a_{21}^{(i)}a_{21}^{(j)} + a_{31}^{(i)}a_{31}^{(j)}))\dot{q}_j$$

$$+M(a_i a_j(a_{13}^{(i)}a_{12}^{(j)} + a_{23}^{(i)}a_{22}^{(j)} + a_{33}^{(i)}a_{32}^{(j)})$$

$$-c_i a_j(a_{11}^{(i)}a_{12}^{(j)} + a_{21}^{(i)}a_{22}^{(j)} + a_{31}^{(i)}a_{32}^{(j)}))\dot{r}_j$$

$$+Ma_j\big((a_i a_{13}^{(i)} - c_1 a_{11}^{(i)})(r_j(p_j a_{13}^{(j)} - r_j a_{11}^{(j)}) - q_j(q_j a_{11}^{(j)} - p_j a_{12}^{(j)}))$$

$$+(a_i a_{23}^{(i)} - c_1 a_{21}^{(i)})(r_j(p_j a_{23}^{(j)} - r_j a_{21}^{(j)}) - q_j(q_j a_{11}^{(j)} - p_j a_{12}^{(j)}))$$

$$+(a_i a_{33}^{(i)} - c_1 a_{31}^{(i)})(r_j(p_j a_{33}^{(j)} - r_j a_{31}^{(j)}) - q_j(q_j a_{31}^{(j)} - p_j a_{32}^{(j)})))$$

$$+Mc_j\big((a_i a_{13}^{(i)} - c_i a_{11}^{(i)})(q_j(r_j a_{12}^{(j)} - q_j a_{13}^{(j)}) - p_j(p_j a_{13}^{(j)} - r_j a_{11}^{(j)}))$$

$$+(a_i a_{23}^{(i)} - c_i a_{21}^{(i)})(q_j(r_j a_{22}^{(j)} - q_j a_{23}^{(j)}) - p_j(p_j a_{23}^{(j)} - r_j a_{21}^{(j)}))$$

$$+(a_i a_{33}^{(i)} - c_i a_{31}^{(i)})(q_j(r_j a_{32}^{(j)} - q_j a_{33}^{(j)}) - p_j(p_j a_{33}^{(j)} - r_j a_{31}^{(j)})))$$

$$+((A_i - C_i) - M(a_i^2 - c_i^2))r_i p_i - Ma_i c_i(r_i^2 - p_i^2)$$

$$-3\omega_0^2(A_i - C_i)a_{31}^{(i)}a_{33}^{(i)} - M\omega_0^2(c_i c_j(a_{11}^{(i)}a_{13}^{(j)} + a_{21}^{(i)}a_{23}^{(j)} + a_{31}^{(i)}a_{33}^{(j)})$$

$$+c_i a_j(a_{11}^{(i)}a_{11}^{(j)} + a_{21}^{(i)}a_{21}^{(j)} + a_{31}^{(i)}a_{31}^{(j)}) - a_i c_j(a_{13}^{(i)}a_{13}^{(j)} + a_{23}^{(i)}a_{23}^{(j)} + a_{33}^{(i)}a_{33}^{(j)})$$

$$-a_i a_j(a_{13}^{(i)}a_{11}^{(j)} + a_{23}^{(i)}a_{21}^{(j)} + a_{33}^{(i)}a_{31}^{(j)}))$$

$$-3M\omega_0^2(c_i a_{31}^{(i)} - a_i a_{33}^{(i)})(a_i a_{31}^{(i)} + c_i a_{33}^{(i)} - a_j a_{31}^{(j)} - c_j a_{33}^{(j)}) = 0,$$

$$(C_i + Ma_i^2)\dot{r}_i - Ma_i c_i\dot{p}_i + Ma_i c_j(a_{12}^{(i)}a_{12}^{(j)} + a_{22}^{(i)}a_{22}^{(j)} + a_{32}^{(i)}a_{32}^{(j)})\dot{p}_j$$

$$+M(a_i a_j(a_{12}^{(i)}a_{13}^{(j)} + a_{22}^{(i)}a_{23}^{(j)} + a_{32}^{(i)}a_{33}^{(j)})$$

$$-a_i c_j(a_{12}^{(i)}a_{11}^{(j)} + a_{22}^{(i)}a_{21}^{(j)} + a_{32}^{(i)}a_{31}^{(j)}))\dot{q}_j$$

$$-Ma_i a_j(a_{12}^{(i)}a_{12}^{(j)} + a_{22}^{(i)}a_{22}^{(j)} + a_{32}^{(i)}a_{32}^{(j)})\dot{r}_j$$

$$-Ma_i a_j\big(a_{12}^{(i)}(r_j(p_j a_{13}^{(j)} - r_j a_{11}^{(j)}) - q_j(q_j a_{11}^{(j)} - p_j a_{12}^{(j)}))$$

$$+a_{22}^{(i)}(r_j(p_j a_{23}^{(j)} - r_j a_{21}^{(j)}) - q_j(q_j a_{21}^{(j)} - p_j a_{22}^{(j)}))$$

$$+a_{32}^{(i)}(r_j(p_j a_{33}^{(j)} - r_j a_{31}^{(j)}) - q_j(q_j a_{31}^{(j)} - p_j a_{32}^{(j)})))$$

$$-Ma_i c_j\big((a_{12}^{(i)}(q_j(r_j a_{12}^{(j)} - q_j a_{13}^{(j)}) - p_j(p_j a_{13}^{(j)} - r_j a_{11}^{(j)}))$$

$$+a_{22}^{(i)}(q_j(r_j a_{22}^{(j)} - q_j a_{23}^{(j)}) - p_j(p_j a_{23}^{(j)} - r_j a_{21}^{(j)}))$$

$$+a_{32}^{(i)}(q_j(r_j a_{32}^{(j)} - q_j a_{33}^{(j)}) - p_j(p_j a_{33}^{(j)} - r_j a_{31}^{(j)})))$$

$$+((B_i - A_i) - Ma_i^2)p_i q_i + Ma_i c_i r_i q_i$$

$$-3\omega_0^2(B_i - A_i)a_{31}^{(i)}a_{32}^{(i)} - M\omega_0^2 a_i(c_j(a_{12}^{(i)}a_{13}^{(j)} + a_{22}^{(i)}a_{23}^{(j)} + a_{32}^{(i)}a_{33}^{(j)})$$

$$+a_j(a_{12}^{(i)}a_{11}^{(j)} + a_{22}^{(i)}a_{21}^{(j)} + a_{32}^{(i)}a_{31}^{(j)}))$$

$$+3M\omega_0^2 a_i a_{32}^{(i)}(a_i a_{31}^{(i)} + c_i a_{33}^{(i)} - a_j a_{31}^{(j)} - c_j a_{33}^{(j)}) = 0.$$

Here

$$p_i = (\dot\alpha_i + \omega_0)a_{21}^{(i)} + \dot\gamma_i,$$
$$q_i = (\dot\alpha_i + \omega_0)a_{22}^{(i)} + \dot\beta_i \sin\gamma_i, \qquad (3)$$
$$r_i = (\dot\alpha_i + \omega_0)a_{23}^{(i)} + \dot\beta_i \cos\gamma_i.$$

In the first three equations of (2), $i = 1$ and $j = 2$; in the next three equations of (2), $i = 2$ and $j = 1$. In (3), $i = 1, 2$. In (2) and (3), the dot denotes the differentiation with respect to time t.

3   Equilibrium Orientations of Satellite-Stabilizer System

Assuming the initial condition $(\alpha_i, \beta_i, \gamma_i) = (\alpha_{i0} = \text{const}, \beta_{i0} = \text{const}, \gamma_{i0} = \text{const})$, also $A_i \neq B_i \neq C_i$, and introducing the notations $a_{ij}^{(1)} = a_{ij}, a_{ij}^{(2)} = b_{ij}$, we obtain from (2) and (3) the equations

$$((C_1 - B_1) - Mc_1^2))(a_{22}a_{23} - 3a_{32}a_{33}) - Ma_1c_1(a_{21}a_{22} - 3a_{31}a_{32})$$
$$+ Mc_1a_{22}(a_2b_{21} + c_2b_{23}) - 3Mc_1a_{32}(a_2b_{31} + c_2b_{33}) = 0,$$
$$((A_1 - C_1) - M(a_1^2 - c_1^2))(a_{23}a_{21} - 3a_{33}a_{31}) - Mc_1a_1((a_{23}^2 - a_{21}^2)$$
$$- 3(a_{33}^2 - a_{31}^2)) - M(c_1a_{21} - a_1a_{23})(a_2b_{21} + c_2b_{23})$$
$$+ 3M(c_1a_{31} - a_1a_{33})(a_2b_{31} + c_2b_{33}) = 0,$$
$$((B_1 - A_1) + Ma_1^2))(a_{21}a_{22} - 3a_{31}a_{32}) + Ma_1c_1(a_{22}a_{23} - 3a_{32}a_{33})$$
$$- Ma_1a_{22}(a_2b_{21} + c_2b_{23}) + 3Ma_1a_{32}(a_2b_{31} + c_2b_{33}) = 0, \qquad (4)$$
$$((C_2 - B_2) - Mc_2^2))(b_{22}b_{23} - 3b_{32}b_{33}) - Ma_2c_2(b_{21}b_{22} - 3b_{31}b_{32})$$
$$+ Mc_2b_{22}(a_1a_{21} + c_1a_{23}) - 3Mc_2b_{32}(a_1a_{31} + c_1a_{33}) = 0,$$
$$((A_2 - C_2) - M(a_2^2 - c_2^2))(b_{23}b_{21} - 3b_{33}b_{31}) - Mc_2a_2((b_{23}^2 - b_{21}^2)$$
$$- 3(b_{33}^2 - b_{31}^2)) - M(c_2b_{21} - a_2b_{23})(a_1a_{21} + c_1a_{23})$$
$$+ 3M(c_2b_{31} - a_2b_{33})(a_1a_{31} + c_1a_{33}) = 0,$$
$$((B_2 - A_2) + Ma_2^2))(b_{21}b_{22} - 3b_{31}b_{32}) + + Ma_2c_2(b_{22}b_{23} - 3b_{32}b_{33})$$
$$- Ma_2b_{22}(a_1a_{21} + c_1a_{23}) + 3Ma_2b_{32}(a_1a_{31} + c_1a_{33}) = 0,$$

which allow us to determine the equilibrium orientation for the system of two bodies connected by a spherical hinge in the orbital coordinate system. Taking into account the expressions for the direction cosines from (1), system (4) can be considered as a system of six equations with six unknowns α_i, β_i, and γ_i ($i = 1, 2$).

Another way of closing Eq. (4) is to add six orthogonality conditions for the direction cosines:

$$a_{21}^2 + a_{22}^2 + a_{23}^2 - 1 = 0,$$
$$a_{31}^2 + a_{32}^2 + a_{33}^2 - 1 = 0,$$

$$a_{21}a_{31} + a_{22}a_{32} + a_{23}a_{33} = 0, \tag{5}$$
$$b_{21}^2 + b_{22}^2 + b_{23}^2 - 1 = 0,$$
$$b_{31}^2 + b_{32}^2 + b_{33}^2 - 1 = 0,$$
$$b_{21}a_{31} + b_{22}b_{32} + b_{23}b_{33} = 0.$$

For this system, the following problem is formulated: for given 11 parameters, determine all twelve direction cosines. The other six direction cosines (a_{11}, a_{12}, a_{13} and b_{11}, b_{12}, b_{13}) can be obtained from the orthogonality conditions.

The system of Eqs. (4) and (5) was solved only for the following case: $b_1 = b_2 = 0$, $c_1 = c_2 = 0$. Equilibrium solutions in this case for the system of two bodies in the orbital plane for $\beta_{i0} = \gamma_{i0} = 0$ and $\alpha_{i0} \neq 0$ were considered in [2,3]. In [3], planar oscillations of the two-body system were analyzed, all equilibrium orientations were determined, and sufficient conditions for the stability of the equilibrium orientations were obtained using the energy integral as a Lyapunov function. In [5], for this case the system of 12 algebraic Eqs. (4) and (5) was decomposed using linear algebra methods and algorithms for the Gröbner basis construction. Some classes of spatial equilibrium solutions were obtained from algebraic equations included in the Gröbner basis. The parameter values that cause the change in the number of equilibrium orientations for the satellite–stabilizer system were found.

Construction of the Gröbner basis for the system (4) and (5) of 12 second-order algebraic equations, whose coefficients depend on 11 parameters, is a very complicated algorithmic problem. In general case, the system of algebraic Eqs. (4) and (5) cannot be solved by direct application of the Gröbner basis construction methods. We will solve system (4) and (5) in the special case, when all equilibrium configurations of the two–body system are located in the plane of the circular orbit. In that case, $\alpha_{10} \neq 0$ and $\alpha_{20} \neq 0$, $\beta_{10} = \beta_{20} = \gamma_{10} = \gamma_{20} = 0$.

Substituting the expressions for the direction cosines from (1) in terms of the aircraft angles into Eq. (4) and taking into account the condition $\beta_{10} = \beta_{20} = \gamma_{10} = \gamma_{20} = 0$, we obtain two equations with two unknowns α_{10} and α_{20}

$$d_1 \sin \alpha_{10} \cos \alpha_{10} + a_1 c_1 (\cos \alpha_{10}^2 - \sin \alpha_{10}^2) + a_1 a_2 \cos \alpha_{10} \sin \alpha_{20}$$
$$-a_1 c_2 \cos \alpha_{10} \cos \alpha_{20} + a_2 c_1 \sin \alpha_{10} \sin \alpha_{20} - c_1 c_2 \sin \alpha_{10} \cos \alpha_{20} = 0, \tag{6}$$
$$d_2 \sin \alpha_{20} \cos \alpha_{20} + a_2 c_2 (\cos \alpha_{20}^2 - \sin \alpha_{20}^2) + a_1 a_2 \cos \alpha_{20} \sin \alpha_{10}$$
$$-a_2 c_1 \cos \alpha_{20} \cos \alpha_{10} + a_1 c_2 \sin \alpha_{20} \sin \alpha_{10} - c_1 c_2 \sin \alpha_{20} \cos \alpha_{10} = 0.$$

Equations (6) form a closed system of two equations with respect to the two aircraft angles α_{10} and α_{20}, that determines the flat satellite–stabilizer equilibrium orientations. In (6), we introduce the following designations: $d_1 = ((A_1 - C_1) - M(a_1^2 - c_1^2))/M$, $d_2 = ((A_2 - C_2) - M(a_2^2 - c_2^2))/M$.

Trigonometric system (6) in the α_{10} and α_{20} angles cannot be solved directly. Therefore, for this system, we used the universal change of sines and cosines through the half-angle tangent

$$\sin \alpha_{i0} = \frac{2 \tan(\frac{\alpha_{i0}}{2})}{1 + \tan^2(\frac{\alpha_{i0}}{2})} = \frac{2t_i}{1 + t_i^2}, \quad \cos \alpha_{i0} = \frac{1 - \tan^2(\frac{\alpha_{i0}}{2})}{1 + \tan^2(\frac{\alpha_{i0}}{2})} = \frac{1 - t_i^2}{1 + t_i^2}, \tag{7}$$

where $t_i = \tan(\frac{\alpha_{i0}}{2})$.

Substituting expressions (7) in terms of half-angle tangent into Eqs. (6) we obtain two algebraic equations with two unknowns t_1 and t_2

$$a_0 t_1^4 + a_1 t_1^3 + a_2 t_1^2 + a_4 = 0,$$
$$b_0 t_1^2 + b_1 t_1 + b_2 = 0, \tag{8}$$

where

$$a_0 = a_1(c_1 - c_2)t_2^2 - 2a_1a_2t_2 + a_1(c_1 + c_2)),$$
$$a_1 = 2(c_1c_2 - d_1)t_2^2 + 4a_2c_1t_2 - 2(c_1c_2 + d_1),$$
$$a_2 = -6a_1c_1(1 + t_2^2),$$
$$a_4 = a_1(c_1 + c_2)t_2^2 + 2a_1a_2t_2 + a_1(c_1 - c_2)),$$
$$b_0 = (1 - t_2^2)(a_2c_2(1 - t_2^2) + a_2c_1(1 + t_2^2) + 2d_2t_2)$$
$$+2c_1c_2t_2(1 + t_2^2) - 4a_2c_2t_2^2,$$
$$b_1 = 2a_1(1 + t_2^2)(a_2(1 - t_2^2) + 2c_2t_2),$$
$$b_2 = (1 - t_2^2)(a_2c_2(1 - t_2^2) - a_2c_1(1 + t_2^2) + 2d_2t_2)$$
$$-2c_1c_2t_2(1 + t_2^2) - 4a_2c_2t_2^2.$$

Using the resultant concept we eliminate the variable t_1 from Eq. (8). Expanding the determinant of resultant matrix of Eq. (8) with the help of Maple symbolic matrix function, we obtain the 16th order algebraic equation in t_2 variable

$$p_0 t_2^{16} + p_1 t_2^{15} + p_2 t_2^{14} + p_3 t_2^{13} + p_4 t_2^{12} + p_5 t_2^{11} + p_6 t_2^{10} + p_7 t_2^9 + p_8 t_2^8$$
$$+ p_9 t_2^7 + p_{10} t_2^6 + p_{11} t_2^5 + p_{12} t_2^4 + p_{13} t_2^3 + p_{14} t_2^2 + p_{15} t_2 + p_{16} = 0, \tag{9}$$

the coefficients of which depend on the parameters a_1, a_2, c_1, c_2, d_1, d_2 in the form

$$p_0 = p_{16} = a_2^4(a_1^2 - c_2^2)(c_1^2 - c_2^2)(a_1^2 - c_1^2 + d_1)^2,$$
$$p_1 = -p_{15} = -4a_2^3c_2(a_1^2 - c_1^2 + d_1)(a_1^4(a_2^2 + 2(c_1^2 - c_2^2))$$
$$+ a_1^2(2c_1^2(d_1 - c_1^2) + (c_2^2 - d_1)(c_2^2 + d_2) - a_2^2c_2^2) + a_2^2c_1^2(c_2^2 - c_1^2)$$
$$+ c_1^2c_2^2(2c_1^2 - c_2^2 - d_1 - d_2) + d_1d_2(c_1^2 + 2c_2^2)),$$
$$p_2 = p_{14} = -4a_2^2(a_1^6(a_2^4 - 9a_2^2c_2^2 - 6c_1^2c_2^2 + 6c_2^4)$$
$$+ a_1^4(12c_1^4c_2^2 + a_2^4(3c_1^2 - c_2^2) - c_2^2(c_2^4 + d_2(d_2 - 6d_1) + 2c_2^2(2d_2$$
$$- 3d_1)) - 2c_1^2(c_2^4 - 2d_2^2 + 2c_2^2(3d_1 + 2d_2)) + a_2^2(8c_2^4 - 2d_1d_2$$
$$+ 2c_2^2(2d_2 - 5d_1) + c_1^2(c_2^2 + 8d_2)))+,$$
$$+ a_1^2(a_2^4c_1^2(3c_1^2 - 2c_2^2) - 6c_1^6c_2^2 - 2c_1^2c_2^2(c_2^4 + 3d_1^2 - 8c_2^2d_2 + d_2^2)$$
$$+ d_1(c_2^4(d_1 - 6d_2) + 2c_2^2d_2(2d_1 - 3d_2) + d_1d_2^2)$$
$$- 2c_1^4(c_2^4 - 2d_2^2 + 2c_2^2(2d_2 - 3d_1)) + a_2^2(3c_2^2d_1(4c_2^2 - d_1 + 2d_2)$$
$$+ c_1^4(c_2^2 + 8d_2) - 8c_1^2(c_2^4 + 2c_2^2d_2))) + 6c_1^6c_2^4 - c_1^4c_2^6 + a_2^4(c_1^6 - c_1^4c_2^2)$$

$$- 6c_1^4c_2^4d_1 + c_1^2c_2^4d_1^2 - 4c_1^4c_2^4d_2 - 6c_1^4c_2^2d_1d2 + 6c_1^2c_2^4d_1d_2$$
$$+ 4c_1^2c_2^2d_1^2d_2 - c_1^4c_2^2d_2^2 + 6c_1^2c_2^2d_1d_2^2 + c_1^2d_1^2d_2^2 - 6c_2^2d_1^2d_2^2$$
$$+ a_2^2(6c_2^4d_1^2 - 9c_1^6c_2^2 - 3c_1^2c_2^2d_1(4c_2^2 + d_1 + 2d_2)$$
$$+ 2c_1^4(4c_2^4 + d_1d_2 + c_2^2(5d_1 + 2d_2)))), \tag{10}$$

$$p_3 = -p_{13} = 4a_2c_2(a_1^6(11a_2^4 - 8c_1^2c_2^2 + 8c_2^4 - 2a_2^2(c_1^2 + 15c_2^2))$$
$$+ a_1^4(a_2^4(5c_1^2 - 19c_2^2 + 7d_1 - 4d_2) + a_2^2(4c_1^4 + 19c_2^4$$
$$- 2c_1^2(c_2^2 + 2(d_1 - 8d_2)) + d_2(4d_2 - 15d_1) + 3c_2^2(9d_2 - 11d_1))$$
$$+ 4(4c_1^4c_2^2 + c_2^2(c_2^2(d_1 - d_2) + d_2(3d_1 - d_2)) + 2c_1^2(c_2^4 + 2d_2^2$$
$$- 2c_2^2(d_1 + 2d_2)))) + a_1^2(a_2^4(5c_1^4 - 19c_2^2d_1 + c_1^2(38c_2^2 - 8d_2))$$
$$- a_2^2(2c_1^6 + d_1(c_2^2(11d_1 - 57d_2) + (7d_1 - 12d_2)d_2) - 19c_2^4$$
$$+ 2c_1^2(19c_2^4 + d_1^2 + 11c_2^2d_2 + 20d_2^2) + 2c_1^4(c_2^2 - 2(d_1 + 8d_2)))$$
$$- 4(2c_1^6c_2^2 - 2c_1^4(c_2^4 + 2c_2^2(d_1 - 2d_2) + 2d_2^2)$$
$$+ d_1d_2(d_2(d_2 - d_1) - c_2^2(d_1 - 3d_2)) + 2c_1^2(c_2^4d_2 + 2d_2^3$$
$$+ c_2^2(d_1^2 - 5d_2^2)))) + a_2^4(11c_1^6 + 19c_1^2c_2^2d_1 - c_1^4(19c_2^2 + 7d_1 + 4d_2))$$
$$+ 4(2c_1^6c_2^2 - 2d_1^2d_2^3 - c_1^4c_2^2(c_2^2(d_1 + d_2) + d_2(3d_1 + d_2))$$
$$+ c_1^2d_1d_2(d_2(d_1 + d_2) + c_2^2(d_1 + 3d_2))) + a_2^2(38c_2^2d_1^2d_2 - 30c_1^6c_2^2$$
$$+ c_1^4(19c_2^4 + d_2(15d_1 + 4d_2) + 3c_2^2(11d_1 + 9d_2))$$
$$- c1^2d_1(19c_2^4 + d_2(7d_1 + 12d_2) + c_2^2(11d_1 + 57d_2))))),$$

$$p_4 = p_{12} = 4(a_1^6(2a_2^6 + 4c_2^4(c_1^2 - c_2^2) - a_2^4(c_1^2 + 41c_2^2)$$
$$+ 4a_2^2(3c_1^2c_2^2 + 11c_2^4)) + a_1^4(2a_2^6(3c_1^2 - 7c_2^2) + a_2^4(2c_1^4 + 83c_2^4$$
$$+ 4d_2(d_2 - d_1) + c_1^2(9c_2^2 - 2d_1 + 16d_2) + 14c_2^2(4d_2 - 3d_1))$$
$$- 4(2c_1^4c_2^4 + c_2^2(2d_1 - d_2)d_2 + c_1^2(3c_2^6 + 4c_2^2d_2^2 - 2c_2^4(d_1 + 4d_2)))$$
$$- 2a_2^2(12c_1^4c_2^2 + 2c_1^2(c_2^4 - 6c_2^2(d_1 - 2d_2) - 2d_2^2)$$
$$+ c_2^2(7c_2^4 + d_2(11d_2 - 18d_1) + 2c_2^2(14d_2 - 11d_1))))$$
$$+ a_1^2(a_2^6(6c_1^4 - 28c_1^2c_2^2) - a_2^4(c_1^6 + c_2^2d_1(9d_1 - 110c_2^2 - 84d_2)$$
$$- c_1^4(9c_2^2 + 2d_1 + 16d_2) + c_1^2(54c_2^4 + d_1^2 + 224c_2^2d_2 - 8d_2^2))$$
$$+ 4(c_1^6c_2^4 - c_2^2d_1(d_1 - 2d_2)d_2^2 - c_1^4(3c_2^6 + 2c_2^4(d_1 - 4d_2) + 4c_2^2d_2^2)$$
$$+ c_1^2(4d_2^4 - 8c_2^2d_2^3 + c_2^4(d_1^2 + 2d_2^2)))2a_2^2(6c_1^6c_2^2 + d_1(c_2^4(5d_1 - 42d_2)$$
$$+ 6c_2^2(2d_1 - 7d_2)d_2 + (d_1 - 4d_2)d_2^2) - 2c_1^4(c_2^4 - 2d_2^2$$
$$+ 6c_2^2(d_1 + 2d_2)) - 2c_1^2(7c_2^6 - 56c_2^4d_2 - 8d_2^3 + c_2^2(11d_2^2 - 3d_1^2))))$$
$$+ 2a_2^6c_1^4(c_1^2 - 7c_2^2) - 4(c_1^2c_2^2 - d_1d_2)^2(c_1^2c_2^2 - d_2^2)$$
$$+ a_2^4(55c_2^4d_1^2 - 41c_1^6c_2^2 - c_1^2c_2^2d_1(110c_2^2 + 9d_1 + 84d_2)$$
$$+ c_1^4(83c_2^4 + 4d_2(d_1 + d_2) + 14c_2^2(3d_1 + 4d_2))) + 2a_2^2(22c_1^6c_2^4$$
$$- 42c_2^2d_1^2d_2^2 - c_1^4c_2^2(7c_2^4 + d_2(18d_1 + 11d_2) + 2c_2^2(11d_1 + 14d_2))$$
$$+ c_1^2d_1(d_2^2(d_1 + 4d_2) + 6c_2^2d_2(2d_1 + 7d_2) + c_2^4(5d_1 + 42d_2)))),$$

$$p_5 = -p_{11} = -4a_2c_2\big(a_1^6(9a_2^4 + 24c_2^2(c_1^2 + c_2^2) - 2a_2^2(3c_1^2 + 29c_2^2))$$
$$+ a_1^4(a_2^4(7c_1^2 - 129c_2^2 + 5d_1 - 36d_2) + a_2^2(12c_1^4 + 129c_2^4$$
$$+ 2c_1^2(13c_2^2 - 6d_1 + 16d_2) + d_2(36d_2 - 13d_1) + c_2^2(201d_2 - 75d_1))$$
$$- 4(12c_1^4c_2^2 + 2c_1^2(c_2^4 - 2d_2^2 + 2c_2^2(2d_2 - 3d_1))$$
$$+ c_2^2(d_2(9d_2 - 7d_1) + c_2^2(9d_2 - 5d_1))))$$
$$+ a_1^2(a_2^4(7c_1^4 - 129c_2^2d_1 + 6c_1^2(43c_2^2 - 12d_2))$$
$$+ a_2^2(2c_1^4(13c_2^2 + 6d_1 + 16d_2) - 6c_1^6 - 6c_1^2(43c_2^4 + d_1^2 + 19c_2^2d_2$$
$$+ 60d_2^2) + d_1(129c_2^4 + d_2(108d_2 - 5d_1) + c_2^2(387d_2 - 25d_1)))$$
$$+ 4(6c_1^6c_2^2 + d_1d_2(c_2^2(5d_1 - 27d_2) + d_2(d_1 - 9d_2))$$
$$- 2c_1^4(c_2^4 - 2d_2^2 + c_2^2(6d_1 + 4d_2)) - 6c_1^2(3c_2^4d_2 + 6d_2^3$$
$$- c_2^2(d_1^2 + 15d_2^2)))) + a_2^4(9c_1^6 + 129c_1^2c_2^2d_1$$
$$- c_1^4(129c_2^2 + 5d_1 + 36d_2)) + 4(6c_1^6c_2^4 - 18d_1^2d_2^3$$
$$- c_1^4c_2^2(c_2^2(5d_1 + 9d_2) + d_2(7d_1 + 9d_2))$$
$$+ c_1^2d_1d_2(d_2(d_1 + 9d_2) + c_2^2(5d_1 + 27d_2)))$$
$$+ a_2^2(258c_2^2d_1^2d_2 - 58c_1^6c_2^2 + c_1^4(129c_2^4 + d_2(13d_1 + 36d_2)$$
$$+ 3c_2^2(25d_1 + 67d_2)) - c_1^2d_1(129c_2^4 + d_2(5d_1 + 108d_2)$$
$$+ c_2^2(25d_1 + 387d_2)))),$$

$$p_6 = p_{10} = 4\big(a_1^6(a_2^6 - 9a_2^4c_2^2 + 16c_1^2c_2^4 - 2a_2^2(3c_1^2c_2^2 + 5c_2^4))$$
$$+ a_1^4(3a_2^6(c_1^2 + 21c_2^2) - 16c_2^4(2c_1^4 - 2c_1^2d_1 + d_2^2) + a_2^4(c_1^2(c_2^2 + 8d_2)$$
$$- 2(180c_2^4 + d_2(d_1 + 8d_2) + c_2^2(5d_1 + 126d_2)))$$
$$+ a_2^2(12c_1^4c_2^2 + 2c_1^2(7c_2^4 + 2d_2^2 - 2c_2^2(3d_1 + 2d_2))$$
$$+ c_2^2(63c_2^4 + d2(6d_1 + 95d_2) + c_2^2(252d_2 - 26d_1))))$$
$$+ a_1^2(3a_2^6(c_1^4 + 42c_1^2c_2^2) + a_2^4(c_1^4(c_2^2 + 8d_2) - 3c_2^2d_1(156c_2^2 + d_1$$
$$+ 126d_2) + 8c_1^2(27c_2^4 + 126c_2^2d_2 - 4d_2^2)) + 16(c_1^6c_2^4 - 2c_1^4c_2^4d_1$$
$$- 2c_2^2d_1d_2^3 + c_1^2(8c_2^2d_2^3 - 4d_2^4 + c_2^4(d_1^2 - 2d_2^2)))$$
$$+ a_2^2(2c_1^4(7c_2^4 + c_2^2(6d_1 - 4d_2) + 2d_2^2) - 6c_1^6c_2^2$$
$$+ d_1(d_2^2(d_1 + 32d_2) + 2c_2^2d_2(2d_1 + 189d_2) + c_2^4(378d_2 - 15d_1))$$
$$+ 2c_1^2(63c_2^6 - 504c_2^4d_2 - 64d_2^3 + c_2^2(95d_2^2 - 3d_1^2)))))$$
$$+ a_2^6c_1^4(c_1^2 + 63c_2^2) - 16d_2^2(c_1^2c_2^2 - d_1d_2)^2$$
$$- a_2^4(9c_1^6c_2^2 + 234c_2^4d_1^2 + 3c_1^2c_2^2d_1(d_1 - 156c2^2 - 26d2)$$
$$+ 2c_1^4(180c_2^4 + d_2(8d_2 - d_1) + c_2^2(126d_2 - 5d_1)))$$
$$+ a_2^2(378c_2^2d_1^2d_2^2 - 10c_1^6c_2^4 + c_1^2d_1(2c_2^2(2d_1 - 189d_2)d_2$$
$$+ (d_1 - 32d_2)d_2^2 - 3c_2^4(5d_1 + 126d_2))$$
$$+ c_1^4c_2^2(63c_2^4 + d_2(95d_2 - 6d_1) + 2c_2^2(13d_1 + 126d_2)))),$$

$$p_7 = -p_9 = -4a_2c_2\big(a_1^6(21a_2^4 + 16c_2^2(c_1^2 + 2c_2^2) - 6a_2^2(c_1^2 + 15c_2^2))$$
$$+ a_1^4(a_2^4(11c_1^2 + 363c_2^2 + 13d_1 + 88d_2)$$
$$+ a_2^2(12c_1^4 - 363c_2^4 + 2c_1^2(13c_2^2 - 6d_1 + 32d_2)$$

$$- d_2(29d_1 + 88d_2) - c_2^2(111d_1 + 539d_2)) + 8(4c_1^2(c_2^2(d_1 - 2d_2)$$
$$+ d_2^2) - 4c_1^4 c_2^2 + c_2^2(c_2^2(3d_1 + 11d_2) + d_2(5d_1 + 11d_2))))$$
$$+ a_1^2(11a_2^4(c_1^4 + 33c_2^2 d_1 + c_1^2(16d_2 - 66c_2^2))$$
$$+ a_2^2(2c_1^4(13c_2^2 + 6d_1 + 32d_2) - 6c_1^6$$
$$+ c_1^2(726c_2^4 - 6d_1^2 + 374c_2^2 d_2 + 880d_2^2)$$
$$- d_1(363c_2^4 + d_2(13d_1 + 264d_2) + c_2^2(37d_1 + 1089d_2)))$$
$$+ 8(2c_1^6 c_2^2 + d_1 d_2(3c_2^2 + d_2)(d_1 + 11d_2) - 4c_1^4(c_2^2(d_1 + 2d_2)$$
$$- d_2^2) + 2c_1^2(11c_2^4 d_2 + 22d_2^3 + c_2^2(d_1^2 - 55d_2^2))))$$
$$+ a_2^4(21c_1^6 - 363c_1^2 c_2^2 d_1 + c_1^4(363c_2^2 - 13d_1 + 88d_2))$$
$$+ 8(4c_1^6 c_2^4 + 22d_1^2 d_2^3 + c_1^2 d_1(d_1 - 11d_2)d_2(3c_2^2 + d_2)$$
$$+ c_1^4 c_2^2(d_2(11d_2 - 5d_1) + c_2^2(11d_2 - 3d_1)))$$
$$- a_2^2(90c_1^6 c_2^2 + 726c_2^2 d_1^2 d_2 + c_1^2 d_1(c_2^2(37d_1 - 1089d_2)$$
$$+ d_2(13d_1 - 264d_2) - 363c_2^4)$$
$$+ c_1^4(363c_2^4 + d_2(88d_2 - 29d_1) + c_2^2(539d_2 - 111d_1)))),$$

$$p_8 = -2\big(a_1^6(8a_2^6 + 48a_2^2 c_2^2(c_1^2 + 5c_2^2) - 3a_2^4(c_1^2 + 55c_2^2)$$
$$- 16(3c_1^2 c_2^4 + c_2^6)) + a_1^4(8a_2^6(3c_1^2 + 25c_2^2) + a_2^4(6c_1^4 - 1267c_2^4$$
$$- 16d_2(d_1 + 3d_2) + c_1^2(37c_2^2 - 6d_1 + 64d_2)$$
$$- 10c_2^2(17d_1 + 80d_2)) + 16(6c_1^4 c_2^2 - c_2^4 d_2(2d_1 + 3d_2)$$
$$- c_1^2(3c_2^6 + c_2^4(6d_1 - 8d_2) + 4c_2^2 d_2^2))$$
$$- 8a_2^2(12c_1^4 c_2^2 + 2c_1^2(5c_2^4 - 6c_2^2(d_1 - 2d_2) - 2d_2^2)$$
$$- c_2^2(25c_2^4 + d_2(18d_1 + 37d_2) + 2c_2^2(19d_1 + 50d_2))))$$
$$+ a_1^2(8a_2^6(3c_1^4 + 50c_1^2 c_2^2) + a_2^4(c_1^4(37c_2^2 + 6d_1 + 64d_2)$$
$$- 3c_1^6 - c_2^2 d_1(1734c_2^2 + 37d_1 + 1200d_2)$$
$$+ c_1^2(934c_2^4 - 3d_1^2 + 3200c_2^2 d_2 - 96d_2^2))$$
$$- 16(3c_1^6 c_2^4 + c_2^2 d_1 d_2^2(d_1 + 6d_2)c_1^4(3c_2^6 + 4c_2^2 d_2^2$$
$$- 2c_2^4(3d_1 + 4d_2)) + 3c_1^2(4d_2^4 - 8c_2^2 d_2^3 + c_2^4(d_1^2 + 2d_2^2)))$$
$$+ 8a_2^2(6c_1^6 c_2^2 - 2c_1^4(5c_2^4 - 2d_2^2 + 6c_2^2(d_1 + 2d_2))$$
$$+ d_1(d_2^2(d_1 + 12d_2) + 6c_2^2 d_2(2d_1 + 25d_2)$$
$$+ c_2^4(13d_1 + 150d_2)) + c_1^2(50c_2^6 - 400c_2^4 d2 - 48d_2^3$$
$$+ c_2^2(6d_1^2 + 74d_2^2)))) + 8a_2^6(c_1^6 + 25c_1^4 c_2^2)$$
$$- 16(c_1^2 c_2^2 - d_1 d_2)^2(c_1^2 c_2^2 + 3d_2^2) + 8a_2^2(30c_1^6 c_2^4$$
$$+ 150c_2^2 d_1^2 d_2^2 + c_1^2 d_1(c_2^4(13d_1 - 150d_2) + 6c_2^2(2d_1 - 25d_2)d_2$$
$$+ (d_1 - 12d_2)d_2^2) + c_1^4 c_2^2(25c_2^4 + d_2(37d_2 - 18d_1)$$
$$+ c_2^2(100d_2 - 38d_1))) - a_2^4(165c_1^6 c_2^2 + 867c_2^4 d_1^2$$
$$+ c_1^2 c_2^2 d_1(37d_1 - 1734c_2^2 - 1200d_2)$$
$$+ c_1^4(1267c_2^4 + 16d_2(3d_2 - d_1) + c2^2(800d_2 - 170d_1)))).$$

By the definition of resultant, to every root t_2 of Eq. (9) there corresponds a common root t_1 of system (8). It can easily be shown that to every real root t_2 of Eq. (9), there corresponds one equilibrium solution of the original system (6). Since the number of real roots of Eq. (9) does not exceed 16, the two bodies system satellite–stabilizer in the plane of a circular orbit can have at most 16 equilibrium configurations in the orbital coordinate system.

From the form of the coefficients of the algebraic Eq. (9), it follows that this equation is recurrent. Then dividing Eq. (9) by t_2^8 we will get the equation

$$p_0(t_2^8 + \frac{1}{t_2^8}) + p_1(t_2^7 - \frac{1}{t_2^7}) + p_2(t_2^6 + \frac{1}{t_2^6}) + p_3(t_2^5 - \frac{1}{t_2^5}) + p_4(t_2^4 + \frac{1}{t_2^4})$$

$$+ p_5(t_2^3 - \frac{1}{t_2^3}) + p_6(t_2^2 + \frac{1}{t_2^2}) + p_7(t_2 - \frac{1}{t_2}) + p_8 = 0. \quad (11)$$

After replacing in (11) $x = (t_2 - \frac{1}{t_2}) = (2/\tan\alpha_{i0})$, $(t_2^2 + \frac{1}{t_2^2}) = x^2 + 2$, $(t_2^3 - \frac{1}{t_2^3}) = x^3 + 3x$ and so on, we will get the equation of the 8th degree

$$P(x) = \bar{p}_0 x^8 + \bar{p}_1 x^7 + \bar{p}_2 x^6 + \bar{p}_3 x^5 + \bar{p}_4 x^4 \bar{p}_5 x^3 + \bar{p}_6 x^2 + \bar{p}_7 x + \bar{p}_8 = 0. \quad (12)$$

Here

$$\bar{p}_0 = p_0, \quad \bar{p}_1 = p_1,$$
$$\bar{p}_2 = p_2 + 8p_0, \quad \bar{p}_3 = p_3 + 7p_1,$$
$$\bar{p}_4 = p_4 + 20p_0 + 6p_2,$$
$$\bar{p}_5 = p_5 + 14p_1 + 5p_3,$$
$$\bar{p}_6 = p_6 + 16p_0 + 9p_2 + 4p_4,$$
$$\bar{p}_7 = p_7 + 7p_1 + 5p_3 + 3p_5,$$
$$\bar{p}_8 = p_8 + 2(p_0 + p_2 + p_4 + p_6).$$

Using Eqs. (12) and (8), for each set of the system parameters, we can determine numerically the angles α_{20} and α_{10}, that is, all the planar equilibrium orientations of the satellite–stabilizer system.

4 Investigation of Equilibria

Equations (8) and (12) make it possible to determine all the plane equilibrium configurations of the satellite–stabilizer, due to the action of the gravity torque for the given values of system parameters a_1, a_2, c_1, c_2, and d_1, d_2 of the problem.

In studying the two–body system equilibrium orientations, we determine the domains with an equal number of real roots of Eq. (12) in the space of 6 parameters. To identify these domains, we use the Meiman theorem [13], which yields that the decomposition of the space of parameters into domains with an equal number of real roots is determined by the discriminant hypersurface. It is also possible to calculate the number of real roots of a polynomial by means of ith subdiscriminants using Jacobi theorem [14,15].

In our case, the discriminant hypersurface is given by the discriminant of polynomial (12). This hypersurface contains a component of codimension 1, which is the boundary of domains with an equal number of real roots. The set of singular points of the discriminant hypersurface in the space of parameters a_1, a_2, c_1, c_2, and d_1, d_2 is given by the following system of algebraic equations:

$$P(x) = 0, \quad P'(x) = 0. \tag{13}$$

Here the symbol "prime" denotes differentiation with respect to x.

We can eliminate the variable x from system (13) by calculating the determinant of the resultant matrix of Eq. (13) with the help of symbolic matrix functions in Maple. The form of the discriminant of the polynomial $P(x)$ is a very cumbersome expression.

Let us consider a simpler case when $a_1 = a_2 = c_1 = c_2 = a$. Then introducing the new parameters in (6) $d_{01} = (A_1 - C_1))/Ma^2$, $d_{02} = (A_2 - C_2)/Ma^2$, we obtain from (9) a simpler algebraic equation of the 8th degree, whose coefficients depend only on two parameters d_{01} and d_{02}

$$p_{00}t_2^8 + p_{01}t_2^7 + p_{02}t_2^6 + p_{03}t_2^5 + p_{04}t_2^4 + p_{05}t_2^3 + p_{06}t_2^2 + p_{07}t_2 + p_{08} = 0, \tag{14}$$

where

$$p_{00} = p_{08} = d_{01}^2(d_{02}^2 - 1)^2 - d_{02}^2,$$
$$p_{01} = -p_{07} = 2(d_{02} - 2)(d_{01}^2(d_{02}^2 + d_{02} - 2) + 2d_{02}^2),$$
$$p_{02} = p_{06} = d_{01}^2(d_{02}^4 - 20d_{02}^2 + 8d_{02} + 20) + 4d_{02}^2(d_{02}^2 - 5),$$
$$p_{03} = -p_{05} = 2(d_{01}^2(7d_{02}^3 + d_{02}^2 - 28d_{02} - 4) + 2d_{02}^2(7d_{02} + 2)),$$
$$p_{04} = -2(d_{01}^2(d_{02}^4 - 27d_{02}^2 - 10d_{02} + 13) + 4d_{02}^4 - 13d_{02}^2).$$

After replacing in (14) $x = (t_2 - \frac{1}{t_2})$, we will obtain the equation of the 4th degree

$$P_1(x) = p_{00}x^4 + p_{01}x^3 + (p_{02} + 4p_{00})x^2$$
$$+ (p_{03} + 3p_{01})x + p_{04} + 2(p_{02} + p_{00}) = 0. \tag{15}$$

Now we determine the conditions for the existence of real roots of Eq. (15). To identify these conditions, we use the Meiman theorem [13]. In our case, the discriminant hypersurface is given by the discriminant of polynomial $P_1(x)$. The boundary of domains with the equal number of real roots on the plane of parameters d_{01} and d_{02} is given by the following system of algebraic equations:

$$P_1(x) = 0, \quad P_1'(x) = 0. \tag{16}$$

We eliminate the variable x from system (16) by calculating the determinant of the resultant matrix of Eqs. (16) and obtain an algebraic equation of the discriminant hypersurface as

$$P_2(d_{01}, d_{02}) = 256 d_{02}^{12} P_3(d_{01}, d_{02}) P_4(d_{01}, d_{02}) P_5(d_{01}, d_{02}) = 0. \qquad (17)$$

Here

$$P_3(d_{01}, d_{02}) = (d_{01}^2 (d_{02} - 1)^2 - d_{02}^2),$$
$$P_4(d_{01}, d_{02}) = (d_{01}^4 (d_{02}^2 + 8)^2 + 16 d_{01}^2 d_{02}^2 (d_{02}^2 + 10) + 64 d_{02}^4 - 6912),$$
$$P_5(d_{01}, d_{02}) = ((4 d_{01}^2 + d_{01}^2 d_{02}^2 + 4 d_{02}^2)^2 - 64 d_{01}^2 d_{02}^2).$$

Now we should check the change in the number of equilibria when the curve (17) is intersected. This can be done numerically by determining the number of equilibria at a single point of each domain at the plane (d_{01}, d_{02}). This analysis showed that only the curve $P_4(d_{01}, d_{02}) = 0$ separates the domains with different number of equilibria.

Figure 2 presents an example of the properties and form of the discriminant hypersurface $P_2(d_{01}, d_{02}) = 0$, which are the set of curves on the plane (d_{01}, d_{02}). Fig. 2 shows the distributions of domains with equal number of real roots of Eq. (17) and indicates the domains where four and two real solutions exist (8 and 4 equilibrium orientations). In Fig. 2, four branches of two hyperbolas indicate the boundaries $P_3(d_{01}, d_{02}) = 0$, where the number of real roots of Eq. (17) does not change. Therefore, in the case when $a_1 = a_2 = c_1 = c_2 = a$, there exist only 8 and 4 planar equilibrium orientations for the satellite–stabilizer system.

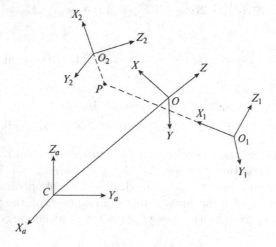

Fig. 1. Basic coordinate systems

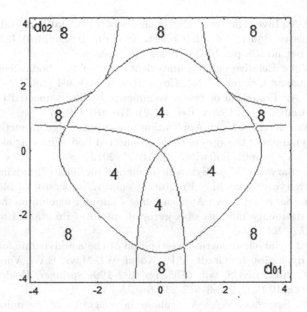

Fig. 2. The regions with the fixed number of equilibria

5 Conclusion

In this paper, we present the study of the dynamics of the rotational motion of the satellite–stabilizer system subject to the gravitational torque in the plane of the orbit. The computer algebra method (based on the resultant approach) of determining all equilibrium orientations of the satellite–stabilizer system in the orbital coordinate system in the plane of a circular orbit was presented. The conditions for the existence of these equilibria were obtained.

We have made an analysis of the evolution of domains of existence of equilibrium orientations in the plane of system parameters d_{01} and d_{02} for the special case when the coordinates of the spherical hinge in the satellite body coordinate system $Ox_1y_1z_1$ and stabilizer body coordinate system $Ox_2y_2z_2$ are equal. For this case, we have indicated the analytic equation of the discriminant hypersurfaces that limits regions with different number of equilibrium configurations of the satellite–stabilizer system. The hypersurface equation was computed symbolically using the resultant approach.

The obtained results can be used to design gravitational attitude control systems for the artificial Earth satellites.

References

1. Sarychev, V.A.: Problems of orientation of satellites, Itogi Nauki i Tekhniki. Ser. Space Research, vol. 11. VINITI, Moscow (1978). (in Russian)

2. Sarychev, V.A.: Investigation of the dynamics of a gravitational stabilization system, Sb. Iskusstv. Sputniki Zemli (Collect. Artif. Earth Satellites). Izd. Akad. Nauk SSSR, Moscow, no. 16, pp. 10–33 (1963). (in Russian)
3. Sarychev, V.A.: Relative equilibrium orientations of two bodies connected by a spherical hinge on a circular orbit. Cosm. Res. **5**, 360–364 (1967)
4. Sarychev, V.A.: Equilibria of two axisymmetric bodies connected by a spherical hinge in a circular orbit. Cosm. Res. **37**(2), 176–181 (1999)
5. Gutnik, S.A., Sarychev, V.A.: Application of computer algebra methods to investigate the dynamics of the system of two connected bodies moving along a circular orbit. Program. Comput. Softw. **45**(2), 51–57 (2019)
6. Gutnik, S.A., Sarychev, V.A.: Symbolic-numerical methods of studying equilibrium positions of a gyrostat satellite. Program. Comput. Softw. **40**(3), 143–150 (2014)
7. Gutnik, S.A., Sarychev, V.A.: Application of computer algebra methods for investigation of stationary motions of a gyrostat satellite. Program. Comput. Softw. **43**(2), 90–97 (2017)
8. Gutnik, S.A.: Symbolic-numeric investigation of the aerodynamic forces influence on satellite dynamics. In: Gerdt, V.P., Koepf, W., Mayr, E.W., Vorozhtsov, E.V. (eds.) CASC 2011. LNCS, vol. 6885, pp. 192–199. Springer, Heidelberg (2011). https://doi.org/10.1007/978-3-642-23568-9_15
9. Gutnik, S.A., Sarychev, V.A.: A symbolic investigation of the influence of aerodynamic forces on satellite equilibria. In: Gerdt, V.P., Koepf, W., Seiler, W.M., Vorozhtsov, E.V. (eds.) CASC 2016. LNCS, vol. 9890, pp. 243–254. Springer, Cham (2016). https://doi.org/10.1007/978-3-319-45641-6_16
10. Char, B.W., Geddes, K.O., Gonnet, G.H., Monagan, M.B., Watt, S.M.: Maple Reference Manual. Watcom Publications Limited, Waterloo (1992)
11. Michels, D.L., Lyakhov, D.A., Gerdt, V.P., Hossain, Z., Riedel-Kruse, I.H., Weber, A.G.: On the general analytical solution of the kinematic cosserat equations. In: Gerdt, V.P., Koepf, W., Seiler, W.M., Vorozhtsov, E.V. (eds.) CASC 2016. LNCS, vol. 9890, pp. 367–380. Springer, Cham (2016). https://doi.org/10.1007/978-3-319-45641-6_24
12. Chen, C., Maza, M.M.: Semi-algebraic description of the equilibria of dynamical systems. In: Gerdt, V.P., Koepf, W., Mayr, E.W., Vorozhtsov, E.V. (eds.) CASC 2011. LNCS, vol. 6885, pp. 101–125. Springer, Heidelberg (2011). https://doi.org/10.1007/978-3-642-23568-9_9
13. Meiman, N.N.: Some problems on the distribution of the zeros of polynomials. Uspekhi Mat. Nauk **34**, 154–188 (1949). (in Russian)
14. Gantmacher, F.R.: The Theory of Matrices. Chelsea Publishing Company, New York (1959)
15. Batkhin, A.B.: Parameterization of the discriminant set of a polynomial. Program. Comput. Softw. **42**(2), 65–76 (2016)
16. England, M., Errami, H., Grigoriev, D., Radulescu, O., Sturm, T., Weber, A.: Symbolic versus numerical computation and visualization of parameter regions for multistationarity of biological networks. In: Gerdt, V.P., Koepf, W., Seiler, W.M., Vorozhtsov, E.V. (eds.) CASC 2017. LNCS, vol. 10490, pp. 93–108. Springer, Cham (2017). https://doi.org/10.1007/978-3-319-66320-3_8

Parametric Standard Bases and Their Applications

Amir Hashemi[(✉)] and Mahsa Kazemi

Department of Mathematical Sciences, Isfahan University of Technology,
Isfahan 84156-83111, Iran
Amir.Hashemi@cc.iut.ac.ir, mahsa.kazemi@math.iut.ac.ir

Abstract. In this paper, by stating a local variant of stability criteria due to Kalkbrener [25] and based on the Kapur et al. algorithm [30] for computing comprehensive Gröbner systems, we present an algorithm for the computation of *comprehensive standard systems*. Although our algorithm is a straightforward extension of the mentioned algorithm, however the effectiveness of our approach can be seen in its applications. To this end, we study some applications of parametric standard bases in *catastrophe and singularity theories* as well as in *automated geometric theorem discovery*. In particular, in the last application, it is demonstrated that for a given geometric theorem (which is not always true), our algorithm is able to construct all possible conditions under which the geometric conclusion remains locally true.

1 Introduction

Gröbner bases, along with an algorithm for their computation, were introduced by Buchberger in his PhD thesis [6]. He then proposed two criteria to enhance the efficiency of his algorithm [5]. Relying on linear algebra techniques and Buchberger's ideas, Lazard [33] developed a computationally more efficient algorithm for this task. By exploiting syzygy methods and substituting Buchberger's Spolynomial test/completion with the test/completion based on Möller's lifting theorem [38], Gebauer and Möller [18] provided a variant of Buchberger's algorithm by applying Buchberger's criteria in a more efficient way. Finally, Faugère presented his F_4 and F_5 algorithms [15,16] to compute effectively Gröbner bases.

Standard bases, which are known as a local variant of Gröbner bases, originated in the works of Gordan [21]. The notion of standard bases was introduced by Hironaka [24] and Grauert (for special cases of local orderings) [22]. Standard bases have many applications in commutative algebra and algebraic geometry including geometry theorem proving, computing the multiplicity of points in the variety associated to a zero-dimensional ideal, and deriving Milnor/Tjurina numbers of an isolated singular point of a hypersurface. We refer the reader to [12] for more information on these topics. In addition to these applications, standard bases have been recently considered as a core routine in many of the algorithms implemented within the Singularity library [17] to analyze local bifurcations of a given scalar singularity.

© Springer Nature Switzerland AG 2019
M. England et al. (Eds.): CASC 2019, LNCS 11661, pp. 179–196, 2019.
https://doi.org/10.1007/978-3-030-26831-2_13

Another important topic considered in this paper is *Comprehensive Gröbner bases* (CGB) which was first presented and computed through an algorithm by Weispfenning [49]. A CGB forms a finite parametric basis for a given parametric polynomial ideal such that for any specialization of parameters, the specialization of the basis, leads to a Gröbner basis for the specialization of the parametric ideal.

Comprehensive Gröbner system (CGS) is encoded by a finite set of triples, called branches, where each triple contains null/non-null condition sets along with a finite set of parametric polynomials. For any specialization of parameters, there exists a branch corresponding to the specialization so that the null/non-null condition sets are satisfied and the corresponding set gives a Gröbner basis for the parametric ideal. Solving parametric polynomial systems was discussed through different approaches by Sit [44] and Kapur [27]. Afterwards, Montes [40] proposed a more efficient algorithm, so-called DISPGB, in order to compute CGS's. References related to the works on canonical CGB, improvement of CGB and minimal canonical CGS are given in [34,35,50], respectively. A significant improvement on computing CGB's and CGS's is due to Suzuki and Sato [46] by recursively computing reduced Gröbner bases. Combining the Suzuki-Sato and Weispfenning algorithms, Kapur et al. achieved an efficient algorithm for deriving CGS's [28,30]. The GRÖBNERCOVER algorithm described by Montes and Wibmer [42] decomposes the parameter space into locally closed subsets together with polynomial data from which the reduced Gröbner basis for each parameter point is determined. Finally, an algorithm for computing simultaneously CGB and CGS was proposed by Kapur et al. [29–31].

In this paper, we are interested in computing a comprehensive standard systems (CSS) for a given parametric polynomial ideal. To the best of our knowledge, this topic has been studied only in [1] where an approach close to the one presented by Montes [40] was developed to compute a CSS to study the local Hilbert-Samuel functions. Based on a local variant of stability criteria due to Kalkbrener [25] and the Kapur et al. algorithm [30], we propose an algorithm to compute a CSS for a given parametric ideal. Therefore our new algorithm may be simpler and more efficient than the one given in [1]. Furthermore, we study several applications of this algorithm in catastrophe and singularity theories and automated geometric theorem discovery. In the latter application, our aim here is to propose an effective approach using CSS's to fully analyze the correctness of a geometric statement. It is worth noting that automated theorem proving deals with the development of computer programs to decide whether or not a (geometrical) claim is a logical consequence of a set of hypotheses. This concept has been treated first by Gelernter et al. [19], Tarski [47], Seidenberg [43], and Collins [10]. Thereafter, two different yet essential approaches based on Wu's and Gröbner bases methods were proposed for this task, see e.g. [9,26,32].

The rest of this paper is organized as follows. In Sect. 2, we review some preliminaries about standard bases and discuss different existing methods for computing them. Section 3 deals with the local analogue of Kalkbrener's stability criteria to describe a new algorithm for computing CSS's. Section 4 is devoted to a discussion on some applications of our approach via three examples. Finally, in Sect. 5, we make some concluding remarks.

2   Standard Bases Computation

Throughout this paper, we let $\mathcal{R} = \mathbb{K}[\mathbf{x}]_{\langle \mathbf{x} \rangle}$ be the local ring in terms of the variables $\mathbf{x} = x_1, \ldots, x_n$ over the field $\mathbb{K}$, whose maximal ideal is generated by $\langle \mathbf{x} \rangle$. Recall that a *semigroup ordering* on $\mathcal{R}$ is a total order $\prec$ on the set of monomials in $\mathcal{R}$ such that for each two monomials m, m' if $m' \prec m$ then $m'u \prec mu$ for every monomial u. Notice that $\prec$ is not necessarily well-ordering on the set of all monomials of $\mathcal{R}$. A semigroup ordering is called *local* if for every monomial m we have $m \preceq 1$. As a well-known example of local monomial orderings that we use in this paper, we mention the *degree-anticompatible lexicographic* ordering (or simply alex($\mathbf{x}$)): For two given monomials $m, m' \in \mathcal{R}$ we write $m' \prec_{alex} m$ if either $\deg(m) < \deg(m')$ or $\deg(m) = \deg(m')$ and $m' \prec_{lex} m$. Suppose that $f \in \mathcal{R}$ is a polynomial, and let $\prec$ be a semigroup ordering on the set of all monomials in $\mathcal{R}$. The *leading monomial* of f, denoted by $\mathrm{LM}(f)$, is the greatest monomial with respect to $\prec$ in f and its coefficient, denoted by $\mathrm{LC}(f)$, is called the *leading coefficient* of f. The *leading term* of f is defined to be $\mathrm{LT}(f) = \mathrm{LC}(f)\mathrm{LM}(f)$. Let $\mathcal{I} \subset \mathcal{R}$ be an ideal. A finite subset $\{g_1, \ldots, g_s\}$ of $\mathcal{I}$ is called a *standard basis* for $\mathcal{I}$ with respect to $\prec$ if $\mathrm{LT}(\mathcal{I}) = \langle \mathrm{LT}(g_1), \ldots, \mathrm{LT}(g_s) \rangle$. For more details we refer the reader to [12]. Standard bases can be computed via a variant o Buchberger's algorithm relying on *Mora normal form* algorithm as follows in which ecart(g) stands for $\deg(g) - \deg(\mathrm{LT}(g))$.

Algorithm 1. MORANORMALFORM

1: **Input:** A polynomial $f \in \mathcal{R}$, a set of polynomials $G \subset \mathcal{R}$ and a semigroup ordering $\prec$
2: **Output:** A Mora normal form of f w.r.t. G
3: $h := f$
4: $T := G$
5: **while** ($h \neq 0$ and $T_h := \{g \in T \mid \mathrm{LM}(g) \mid \mathrm{LM}(h)\} \neq \emptyset$) **do**
6:    choose $g \in T_h$ with minimal ecart(g) w.r.t. $\prec$
7:    **if** ecart(g) > ecart(h) **then**
8:       $T := T \cup \{h\}$
9:    **end if**
10:    $h := h - \frac{\mathrm{LT}(h)}{\mathrm{LT}(g)}g$
11: **end while**
12: **return**(h)

Remark 2.1. *An important issue with this algorithm is the discussion of its arithmetic complexity. For example, dividing $zxy^3 - yz^4 - y^3 + x^3z^2y + y^2z^4$ by $\{y - xy^3 - yz^3, x^2z + yz^2 + x^2yz^2, y^2 - x^3z^2 - yz^4 + z\}$ by using the above algorithm, does not terminate after 72 h (the experimental results in this paper have been obtained* *on a Ubuntu desktop 2.2 GHz AMD Opteron, 96 GB). This simple observation illustrates that the computation of standard bases is an established hard problem in practice.*

Note that the classical approaches to compute Gröbner bases are applicable directly to compute standard bases. On the other hand, to the best of our knowledge, efficient computation of standard bases has not been addressed in the literature. For this purpose, we took into account the local version of UPDATE algorithm [2], MMT algorithm [39] and the recent BERKESCHREYER algorithm due to Berkesch and Schreyer [3] which have been applied widely to compute Gröbner bases. We have implemented five algorithms, namely UPDATE, UPDATE-MINIMAL, UPDATE+MMT, UPDATE-MINIMAL+MMT and BERKESCHREYER in the local setting. The second algorithm is referred to as the UPDATE algorithm without performing the final minimalization process. By the minimalization process we mean that in each iteration of computing standard basis, if in the computed basis, the leading term of a polynomial divides the leading term of another polynomial then the latter is removed. The next two are targeted the installation of the UPDATE structure on MMT algorithm with/without the minimalization process. Finally, the last one is based on the procedure proposed in [3]. Our experimental comparison of these algorithms on more than 50 benchmark polynomial systems shows that UPDATE-MINIMAL+MMT outperforms the others in terms of running time and used memory (see the following figure); thus, we use it in the rest of the paper to calculate standard bases. For the sake of shortness, the comparison results are not included in this paper. The MAPLE code of our implementations, together with the list of examples, are available at http://amirhashemi.iut.ac.ir/softwares.

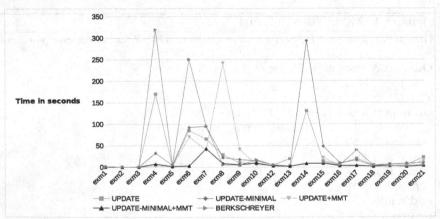

3   Comprehensive Standard Systems

In this section, we adapt the concept of CGS's to the local setting. For this purpose, after discussing a local variant of stability criteria of Kalkbrener [25], we give a local variant of the Kapur et al. algorithm [30] to compute CSS's.

Consider $F = \{f_1, \ldots, f_l\} \subset S = \mathbb{K}[\mathbf{a}][\mathbf{x}]_{\langle \mathbf{x} \rangle}$ where $\mathbf{a}$ represents a sequence $a_1, \ldots, a_m$ of parameters such that $\{\mathbf{x}\} \cap \{\mathbf{a}\} = \emptyset$. Let $\prec_{\mathbf{x}}$ be a semigroup ordering on the variables and $\prec_{\mathbf{a}}$ be a monomial ordering on the parameters. In order to define a CSS we shall need the following definitions:

- The *block ordering* $\prec_{\mathbf{x},\mathbf{a}} = (\prec_{\mathbf{x}}, \prec_{\mathbf{a}})$ on the monomials in terms of $\mathbf{a}$ and $\mathbf{x}$ is defined as follows: We write $\mathbf{x}^{\alpha}\mathbf{a}^{\beta} \prec_{\mathbf{x},\mathbf{a}} \mathbf{x}^{\alpha'}\mathbf{a}^{\beta'}$ if either $\mathbf{x}^{\alpha} \prec_{\mathbf{x}} \mathbf{x}^{\alpha'}$ or $\mathbf{x}^{\alpha} = \mathbf{x}^{\alpha'}$ and $\mathbf{a}^{\beta} \prec_{\mathbf{a}} \mathbf{a}^{\beta'}$.
- Let L be the algebraic closure of $\mathbb{K}$. A morphism $\sigma : \mathbb{K}[\mathbf{a}] \to L$ is called a *specialization* of parameters. For each $f \in S$ we define $\sigma(f) = f \mid_{\mathbf{a}=t_1,\ldots,t_m}$ where $\sigma(a_i) = t_i$ for each i.
- For a finite set $A \subset \mathbb{K}[\mathbf{a}]$ the *variety* of A, denoted by $\mathbb{V}(A)$, is the set of all common zeros of A.
- Given a pair $(N, W) \subset \mathbb{K}[\mathbf{a}] \times \mathbb{K}[\mathbf{a}]$, we say that a specialization σ satisfies (N, W) when $\sigma(p) = 0$ and $\sigma(q) \neq 0$ are valid for all $p \in N$ and for some $q \in W$, respectively. In other words, a specialization σ corresponding to $(t_1, \ldots, t_m) \in L^m$ satisfies (N, W) when $(t_1, \ldots, t_m) \in \mathbb{V}(N) \setminus \mathbb{V}(W)$. Moreover, the pair (N, W) is said to be *inconsistent* whenever $\mathbb{V}(N) \setminus \mathbb{V}(W) = \emptyset$. Here, (N, W) is referred to as a condition sets containing null-condition set N and non-null condition set W.

Definition 3.1. *Let* $F, S_i \subset S$ *and* $(N_i, W_i) \subset \mathbb{K}[\mathbf{a}] \times \mathbb{K}[\mathbf{a}]$ *for* $i = 1, \ldots, \ell$. *The triple set* $G = \{(N_i, W_i, S_i)\}_{i=1}^{\ell}$ *is called a CSS for* $\langle F \rangle$ *w.r.t.* $\prec_{\mathbf{x},\mathbf{a}}$ *over* $V \subset L^m$ *if for any* i *the following conditions hold.*

(a) *For any specialization* $\sigma : \mathbb{K}[\mathbf{a}] \to L$ *satisfying* (N_i, W_i), $\sigma(S_i) \subset \mathcal{R}$ *is a standard basis for* $\langle \sigma(F) \rangle$ *w.r.t.* $\prec_{\mathbf{x}}$.
(b) $V \subseteq \bigcup_{i=1}^{\ell} \mathbb{V}(N_i) \setminus \mathbb{V}(W_i)$.

The set G *is called a CSS for* $\langle F \rangle$, *if it is a CSS for* $\langle F \rangle$ *over whole* L^m.

It should be noted that the Kapur et al. algorithm [30] is based on several basic properties of Gröbner bases. All these properties are held in local setting; for instance, the local analogue of the elimination theorem is given by [12, Theorem 5.4, Page 184]. For some other useful properties of standard bases, we refer to [14, Theorem 1.7.3, Pages 54–55]. In particular, we shall need the local version of stability criteria of Kalkbrener [25, Theorem 3.1] and minimal Dickson basis [30] that will be restated here for the convenience of the reader. We give the proof of the first result by adapting the proof of the original result to the local setting.

Theorem 3.2. *Let* K *be a Noetherian commutative ring with identity and* $\sigma : K \to \mathbb{K}$ *be a ring homomorphism which is naturally extended to the homomorphism* $K[\mathbf{x}] \to \mathcal{R}$. *Let* $\mathcal{I}$ *be an ideal in* $K[\mathbf{x}]$ *and* $S = \{g_1, \ldots, g_r, g_{r+1}, \ldots, g_s\}$ *be a standard basis for* $\mathcal{I}$ *with respect to* $\prec_{\mathbf{x}}$. *Further, suppose that* $\sigma(\mathrm{LC}(g_i)) \neq 0$ *for* $i \leq r$ *and* $\sigma(\mathrm{LC}(g_i)) = 0$ *for* $i > r$. *Let* $S_r = \{g_1, \ldots, g_r\}$. *Then the following statements are equivalent.*

(a) $\langle \sigma(\mathrm{LT}(\mathcal{I})) \rangle = \mathrm{LT}(\langle \sigma(\mathcal{I}) \rangle)$.
(b) $\sigma(S_r)$ *is a standard basis for* $\langle \sigma(\mathcal{I}) \rangle \subset \mathcal{R}$ *with respect to* $\prec_{\mathbf{x}}$.
(c) *For each* $i \in \{r+1, \ldots, s\}$, $\sigma(g_i)$ *is reduced to zero modulo* $\sigma(S_r)$.

Proof. For the sake of simplicity, for an ideal $\mathcal{J}$, we denote $\langle \sigma(\mathcal{J}) \rangle$ by $\sigma(\mathcal{J})$.

$(a) \Rightarrow (b)$ From assumption, we have obviously $\sigma(\mathrm{LT}(\mathcal{I})) = \mathrm{LT}(\sigma(\mathcal{I}))$. Since $\sigma(\mathrm{LC}(g_i)) = 0$ for each $i \in \{r+1, \ldots, s\}$, we can then write

$$\mathrm{LT}(\sigma(\mathcal{I})) = \sigma(\mathrm{LT}(\mathcal{I})) = \langle \sigma(\mathrm{LT}(g_i)), i = 1, \ldots, r \rangle. \tag{1}$$

Now assume that $f \in \sigma(\mathcal{I})$ with $f \neq 0$. Thus, $\mathrm{LT}(f) \in \mathrm{LT}(\sigma(\mathcal{I}))$ and in turn $\mathrm{LT}(f) \in \sigma(\mathrm{LT}(\mathcal{I}))$. Therefore, there exists $g_i \in S$ such that $\sigma(\mathrm{LT}(g_i)) \mid \mathrm{LT}(f)$. From assumption, we have $i \leq r$ and in consequence $\sigma(S_r)$ is a standard basis for $\sigma(\mathcal{I})$ with respect to $\prec_{\mathbf{x}}$.

$(b) \Rightarrow (a)$ Since $\sigma(S_r)$ is a standard basis for $\sigma(\mathcal{I})$ with respect to $\prec_{\mathbf{x}}$, for each $\sigma(f) \in \sigma(I)$, there exists $i \in \{1, \ldots, r\}$ such that $\mathrm{LT}(\sigma(g_i)) \mid \mathrm{LT}(\sigma(f))$. The same holds when we take $f \in \{g_{r+1}, \ldots, g_s\}$. It follows that

$$\begin{aligned}
\mathrm{LT}(\sigma(\mathcal{I})) &= \langle \mathrm{LT}(\sigma(g_1)), \ldots, \mathrm{LT}(\sigma(g_r)) \rangle \\
&= \langle \sigma(\mathrm{LT}(g_1)), \ldots, \sigma(\mathrm{LT}(g_r)) \rangle \\
&= \langle \sigma(\mathrm{LT}(S)) \rangle = \sigma(\langle \mathrm{LT}(S) \rangle) = \sigma(\mathrm{LT}(\mathcal{I})).
\end{aligned}$$

$(b) \Rightarrow (c)$ This part is concluded by [14, Pages 54–55] and the fact that $\sigma(g_i) \in \sigma(\mathcal{I})$ for each $g_i \in \mathcal{S}$.

$(c) \Rightarrow (b)$ The hypothesis of part (c) implies that $\sigma(S \setminus S_r) \subseteq \langle \sigma(S_r) \rangle$ and thus $\sigma(S) \subseteq \langle \sigma(S_r) \rangle$. Therefore, $\sigma(S_r)$ generates $\langle \sigma(S) \rangle$ and it suffices to show that $\sigma(S_r)$ is a standard basis for this ideal. For this purpose, and using the local variant of Buchberger's criterion [14, Theorem 1.7.3, Pages 54–55] (see also [2]), we must show that for each $i, j \in \{1, \ldots, r\}$, $\mathrm{spoly}(\sigma(g_i), \sigma(g_j))$ reduces to zero modulo $\sigma(S_r)$. Since $\sigma(\mathrm{LC}(g_i)) \neq 0$ for each $i \in \{1, \ldots, r\}$, then we have

$$\mathrm{spoly}(\sigma(g_i), \sigma(g_j)) = c^{ij} \sigma(\mathrm{spoly}(g_i, g_j)), \tag{2}$$

where $c^{ij} \in \mathbb{K}$ is a non-zero constant. Furthermore, since S is a standard basis and $\mathrm{spoly}(g_i, g_j) \in \langle S \rangle$, by [14, Theorem 1.7.3, Pages 54–55], $\mathrm{spoly}(g_i, g_j) = \sum_{l=1}^{s} a_l^{ij} g_l$ where $a_l^{ij} \in K$ and

$$\mathrm{LM}(g_l) \preceq_{\mathbf{x}} \mathrm{LM}(\mathrm{spoly}(g_i, g_j)) \prec_{\mathbf{x}} \mathrm{lcm}(\mathrm{LM}(g_i), \mathrm{LM}(g_j)). \tag{3}$$

Substituting the last equality into Equality (2) gives rise to

$$\mathrm{spoly}(\sigma(g_i), \sigma(g_j)) = c^{ij} \sum_{l=1}^{s} \sigma(a_l^{ij}) \sigma(g_l),$$

Note that from assumption we conclude that for each $l \in \{r+1, \ldots, s\}$, if $\sigma(g_l)$ appears in this expression, then we can replace it by a combination of the elements of $\sigma(S_r)$. Hence, $\mathrm{spoly}(\sigma(g_i), \sigma(g_j))$ can be written as a combination of $\sigma(g_1), \ldots, \sigma(g_r)$ and in turn $\sigma(S_r)$ is a standard basis for the ideal it generates, completing the proof. $\qquad \square$

Definition 3.3. *Let $F \subset S$ and $\prec_{\mathbf{x},\mathbf{a}}$ be the block ordering of $\prec_{\mathbf{x}}$ and $\prec_{\mathbf{a}}$. A subset $F' \subset F$ is called a minimal Dickson basis for F if*

- $\langle \mathrm{LM}_{\prec_{\mathbf{x}}}(F') \rangle = \langle \mathrm{LM}_{\prec_{\mathbf{x}}}(F) \rangle$,
- *For any two distinct elements $f_1, f_2 \in F'$ we have $\mathrm{LM}_{\prec_{\mathbf{x}}}(f_1) \nmid \mathrm{LM}_{\prec_{\mathbf{x}}}(f_2)$.*

Definition 3.4. *Let $F = \{f_1, \ldots, f_k\} \subset \mathcal{R}$ and $\prec$ be a semigroup ordering on $\mathcal{R}$. We say that a polynomial $f \in \mathcal{R}$ has a standard representation w.r.t. F if f can be written as $f = q_1 f_1 + \cdots + q_k f_k$ with $\mathrm{LM}(q_i f_i) \preceq \mathrm{LM}(f)$.*

In [14, Theorem 1.7.3, Pages 54–55], it was shown that a finite set S of polynomials in $\mathcal{R}$ forms a standard basis iff every polynomial in the ideal generated by S has a standard representation w.r.t. S. In addition, it was proved that this is equivalent to the fact that the Spolynomial of any pair of polynomials in S has a standard representation w.r.t. S. We use the following local variant of [30, Lemma 4.5] to prove Theorem 3.6.

Lemma 3.5. *Let $\mathcal{I} \subset S$ be an ideal and S a standard basis for $\mathcal{I}$ w.r.t. the block ordering of the semigroup ordering $\prec_{\mathbf{x}}$ and the monomial ordering $\prec_{\mathbf{a}}$. Let $S_1 = \{g_1, \ldots, g_s\} \subset S$ and σ be a specialization from $\mathbb{K}[\mathbf{a}]$ to $L \supset \mathbb{K}$ so that $\sigma(\mathrm{LC}_{\mathbf{x}}(g_i)) \neq 0$ for each i. In addition, assume that for every $f \in S \setminus S_1$, $\sigma(f)$ has a standard representation w.r.t. $\sigma(S_1)$. Then $\sigma(S_1)$ standard basis for $\sigma(\mathcal{I})$ w.r.t. $\prec_{\mathbf{x}}$.*

Proof. Since S is a standard basis for $\mathcal{I}$ and for every $f \in S \setminus S_1$, $\sigma(f)$ has a standard representation w.r.t. $\sigma(S_1)$ then $\sigma(S_1)$ generates $\sigma(\mathcal{I})$. Now, by applying the local variant of Buchberger's criterion [12, Theorem 4.3, Page 175], we shall show that for each i, j, $\mathrm{spoly}(\sigma(g_i), \sigma(g_j))$ has a standard representation w.r.t. $\sigma(S_1)$. Since S is a standard basis for $\mathcal{I}$, then we know that $\mathrm{spoly}(g_i, g_j) = q_1 g_1 + \cdots + q_\ell g_\ell$ where $S \setminus S_1 = \{g_{s+1}, \ldots, g_\ell\}$ and $\mathrm{LM}(q_i g_i) \preceq \mathrm{LM}(\mathrm{spoly}(g_i, g_j))$. From $\sigma(\mathrm{LC}_{\mathbf{x}}(g_i)) \neq 0$ for each $i = 1, \ldots, s$ we obtain $\mathrm{spoly}(\sigma(g_i), \sigma(g_j)) = c\sigma(\mathrm{spoly}(g_i, g_j))$ where $0 \neq c \in L$. It follows that

$$\mathrm{spoly}(\sigma(g_i), \sigma(g_j)) = c(\sigma(q_1)\sigma(g_1) + \cdots + \sigma(q_\ell)\sigma(g_\ell)).$$

On the other hand, $\sigma(g_i)$ for $i = s+1, \ldots, \ell$ has a standard representation w.r.t. $\sigma(S_1)$, and this shows that $\mathrm{spoly}(\sigma(g_i), \sigma(g_j))$ has a standard representation w.r.t. $\sigma(S_1)$, completing the proof. $\qquad\square$

Theorem 3.6. *Let $\mathcal{I} \subset S$ be an ideal and S a standard basis for $\mathcal{I}$ w.r.t. the block ordering of the semigroup ordering $\prec_{\mathbf{x}}$ and the monomial ordering $\prec_{\mathbf{a}}$. Let $S_r = S \cap \mathbb{K}[\mathbf{a}]$ and $S_m = MinimalDicksonBasis(S \setminus S_r)$. Let σ be a specialization from $\mathbb{K}[\mathbf{a}]$ to $L \supset \mathbb{K}$ so that $\sigma(f) = 0$ for $f \in S_r$ and $\sigma(h) \neq 0$ with $h = \prod_{f \in S_m} \mathrm{LC}_{\mathbf{x}}(f)$. Then, $\sigma(S_m)$ is a minimal standard basis for $\sigma(\mathcal{I})$ w.r.t. $\prec_{\mathbf{x}}$.*

Proof. We follow the proof of [30, Theorem 4.3]. Assume that $f \in S \setminus (S_r \cup S_m)$. We know that $\mathrm{LM}_{\prec_{\mathbf{x}}}(f)$ is divisible by $\mathrm{LM}_{\prec_{\mathbf{x}}}(g)$ for some $g \in S_m$. Thus, applying the Mora normal form algorithm [12, Page 170], we can write

$$\mathrm{LC}_{\prec_{\mathbf{x}}}(g_1) \cdots \mathrm{LC}_{\prec_{\mathbf{x}}}(g_s) u f = q_1 g_1 + \cdots + q_s g_s + p$$

where $g_i \in S_m$, $q_i, p \in \mathbb{K}[\mathbf{a}, \mathbf{x}]$, u is a unit in $\mathcal{S}$ and the leading monomial of p w.r.t. $\mathbf{x}$ can not be divisible by $\mathrm{LM}_{\prec_{\mathbf{x}}}(g)$ for some $g \in S_m$. In addition, we have $\mathrm{LM}_{\prec_{\mathbf{x}}}(q_i g_i) \preceq \mathrm{LM}_{\prec_{\mathbf{x}}}(f)$ for each i. It is clear that $p \in \mathcal{I}$. Since S is a standard basis for $\mathcal{I}$ then p reduces to zero on division by S. From the above construction and assumptions, it follows that p reduces to zero by S_r and in turn $\sigma(p) = 0$. Thus, applying σ on the both side of the above equality, we derive that $\sigma(f)$ reduces to zero by $\sigma(S_m)$ using the Mora normal form algorithm. This proves that for every $f \in S \setminus (S_r \cup S_m)$, $\sigma(f)$ has a standard representation w.r.t. $\sigma(S_m)$. From Lemma 3.5, it follows that $\sigma(S_m)$ is a minimal standard basis for $\sigma(\mathcal{I})$ w.r.t. $\prec_{\mathbf{x}}$. □

Based on Theorem 3.6, we are now able to present a variant of the Kapur et al. algorithm [30] to compute CSS's. In the next algorithm, we assume that C is a global variable which is initially assumed to be the empty list [ ].

Algorithm 2. CSSMAIN

1: **Input:** A finite set $F \subset \mathcal{S}$, two finite sets $N, W \subset \mathbb{K}[\mathbf{a}]$ and orderings $\prec_{\mathbf{x}}$ and $\prec_{\mathbf{a}}$
2: **Output:** A CSS for $\langle F \rangle$ over $\mathbb{V}(N) \setminus \mathbb{V}(W)$
3: **if** (N, W) is inconsistent **then**
4:     $\mathrm{return}(\emptyset)$
5: **end if**
6: $S := \mathrm{STANDARDBASIS}(F \cup N, \prec_{\mathbf{x}, \mathbf{a}})$
7: **if** $1 \in S$ **then**
8:     $\mathrm{return}\{(N, W, \{1\})\}$
9: **end if**
10: $S_r := S \cap \mathbb{K}[\mathbf{a}]$
11: **if** (S_r, W) is inconsistent **then**
12:     $\mathrm{return}(C)$
13: **else**
14:     $S_m := \mathrm{MINIMALDICKSONBASIS}(S \setminus S_r)$
15:     $h = \mathrm{lcm}(h_1, \ldots, h_k)$ with $h_i = \mathrm{LC}_{\prec_{\mathbf{x}}}(g_i)$ for each $g_i \in S_m = \{g_1, \ldots, g_k\}$
16:     **if** $(S_r, W \times \{h\})$ is consistent **then**
17:         $C := C \cup \{S_r, W \times \{h\}, S_m\}$
18:     **end if**
19:     $\mathrm{return}(C \cup \bigcup_{i=1}^k \mathrm{CSSMAIN}(S_r \cup h_i, W \times \{h_1 \ldots h_{i-1}\}, S \setminus S_r) \cup \{$ (Other cases, $\{1\})\})$
20: **end if**

Remark 3.7. *Algorithm 2 can be extended when $\mathbb{K}[\mathbf{x}]_{\langle \mathbf{x} \rangle}$ is replaced with $\mathbb{K}[[\mathbf{x}]]$ or E_n (see Subsect. 4.2). To do so, the applied division algorithm should be replace by the corresponding division algorithm, see [2, Pages 251–252] and [17, Definition 2.3], respectively. This plays a key role in the development of* PARAMET-RICSINGULARITIES *module of the* Singular *library, see [17].*

4 Applications

This section is devoted to some applications of the CSSMAIN to compute comprehensive Milnor systems, comprehensive determinacy systems and automated geometric theorem discovery.

4.1 Comprehensive Milnor Systems

The *Milnor number*, amongst the most important concepts in the study of complex singularities, was developed by John Milnor as an invariant of a singular function [37]. Keeping the notations of the previous section, let us continue with the following definition where $\mathbb{K} = \mathbb{C}$ or $\mathbb{R}$ and $\dim_{\mathbb{K}}(A)$ denotes the dimension of A as a $\mathbb{K}$-vector space.

Definition 4.1. *Suppose $f \in \mathcal{R}$ has an isolated singularity at the origin and $J(f) = \langle \frac{\partial f}{\partial x_1}, \dots, \frac{\partial f}{\partial x_n} \rangle$. The Milnor number of f at the origin is defined by*

$$\mu = \dim_{\mathbb{K}}(\frac{\mathcal{R}}{J(f)}).$$

We refer the reader to [12, Page 155] for more details on Milnor number. One interesting application of CSS's arises in the study of the Milnor number of a given parametric polynomial, see the next example.

Example 4.2. *Let $f = ax^4 + bxz^3 + cy^2 + xy^2 + z^2 \in \mathbb{K}[a, b, c][x, y, z]_{\langle x,y,z \rangle}$. Table 1 illustrates the results obtained for the Milnor number of f through applying CSSMAIN to $J(f) = \langle 4ax^3 + bz^3 + y^2, 2cy + 2xy, 3bxz^2 + 2z \rangle$ w.r.t. the block ordering of the monomial orderings $\mathrm{alex}(x, y, z)$ and $\mathrm{lex}(a, b, c)$ to compute a CSS for $J(f)$. Note that in each branch, from the computed standard basis, one is able to compute easily the dimension of the corresponding factor ring as a $\mathbb{K}$-vector space (see the local variant of Macaulay's theorem [12, Theorem 4.3, Page 177]).*

Table 1. A comprehensive Milnor system for a paramtric polynomial

Null set	Non-null set	Standard basis	Milnor number
{}	{ac}	$\{\frac{\partial f}{\partial z}, \frac{\partial f}{\partial y}, -8acx^3 - 2bcz^3 + 2xy^2\}$	3
{a}	{c}	$\{\frac{\partial f}{\partial z}, \frac{\partial f}{\partial y}\}$	infinite
{c}	{a}	$\{\frac{\partial f}{\partial z}, \frac{\partial f}{\partial x}, -2xy, 8ax^4 + 2bxz^3\}$	5
{a, c}	{}	$\{\frac{\partial f}{\partial z}, \frac{\partial f}{\partial x}, -2xy\}$	infinite

4.2   Comprehensive Determinacy Systems

Let us first give some background on determinacy. Two smooth functions are equivalent as germ when they are identical in the vicinity of the origin. The space of all n-variate smooth germs is denoted by E_n which is a local ring with the unique maximal ideal $\mathcal{M} = \langle x_1, \ldots, x_n \rangle$; see [20, Page 56]. Developed by René Thom using the ideas in differential topology and dynamical system theory [48], *catastrophe theory* studies the local structure of critical points of real-valued smooth functions in E_n. As this theory is a local one, the elements of E_n are considered as of their germ equivalent class. Due to the important applications in science and engineering, catastrophe theory has come to the attention of many researchers for decades [23, 45]. In this subsection, we are concerned with one of the fundamental concepts involved in this theory, namely *determinacy*. We say that two smooth germs f and g are *right-equivalent* if there exists a diffeomorphism mapping from f to g. The germ f is called k-*determined* if for any germ g which is equal to f modulo degree $k+1$, f and g are right-equivalent. It is proved that f is k-determined if

$$\mathcal{M}^{k+1} \subset \mathcal{M}^2 J(f), \tag{4}$$

where $\mathcal{M}^k = \langle \mathbf{x}^\alpha : |\alpha| = k \rangle$; see [20, Page 101]. We remark that, E_n, containing flat germs and/or germs with infinite Taylor series, is a computationally expensive ring. Thus, we propose the following theorem (see [17, Theorem 2.16(1)]) so that computations are converted to the smaller ring $\mathcal{R}$.

Theorem 4.3. *Let* $f \in \mathcal{R} \subset E_n$. *Then,* $\mathcal{M}_{E_n}^{k+1} \subset \mathcal{M}_{E_n}^2 J(f)_{E_n}$ *iff* $\mathcal{M}_{\mathcal{R}}^{k+1} \subset \mathcal{M}_{\mathcal{R}}^2 J(f)_{\mathcal{R}}$.

In order to determine the value of k satisfying inclusion (4), we apply the *Hilbert series* methods [11] and propose a new and more efficient procedure (compared to the one given in [36]). Let us briefly review the notion of Hilbert series. Assume that $\mathcal{I}$ is a homogeneous ideal in the ordinary polynomial ring $\mathbb{K}[\mathbf{x}]$. For a subset $A \subset \mathbb{K}[\mathbf{x}]$, let A_i be the set of all homogeneous polynomials of A of total degree i. The *Hilbert series* of $\mathcal{I}$ is defined by

$$\mathrm{HS}_{\mathcal{I}}(t) = \Sigma_{i=0}^{\infty} \mathrm{HF}_{\mathcal{I}}(i) t^i,$$

where $\mathrm{HF}_{\mathcal{I}}(i) = \dim_{\mathbb{K}}(\frac{\mathbb{K}[\mathbf{x}]_i}{\mathcal{I}_i})$ is the *Hilbert function* of $\mathcal{I}$. One sees readily that the Hilbert series of $\mathcal{M}^k$ is equal to the polynomial $1 + 2t + \ldots + \binom{n+k-2}{k-1} t^{k-1}$ and, in turn, the value of k for which inclusion (4) is satisfied equals to the Hilbert series degree of $\mathcal{M}^2 J(f)$. It is worth noting that, in the case of isolated singularities, the procedure in [36] runs a for-loop until k is found. The pseudocode for our approach is stated in Algorithm 3. Note that we utilize the MAPLE's built-in command HILBERTSERIES for computing the Hilbert series of an ideal.

Algorithm 3. DETERMINACY

1: **Input:** A germ $f \in E_n$
2: **Output:** Determinacy of f
3: I := LT(S) where S is a standard basis for $\mathcal{M}^2 J(f)$
4: h := HILBERTSERIES(I)
5: k := deg(h)
6: **return**(k)

Example 4.4. *In this example, we discuss the determinacy of the polynomial f given in Example 4.2 for all possible values of parameters. Applying* CSSMAIN *to* $\mathcal{M}^2 J(f)$ *we get Table 2. Then, the next step consists of computing the leading term ideal for each branch and then the corresponding Hilbert series to get the desired determinacy. Here, we restrict our considerations only to the first and the third branches. Thus, we get*

$$\mathcal{I}_1 = \langle x^2 z, x^2 y, y^2 z, xy^2, y^3, z^3, xyz, xz^2, yz^2, x^5 \rangle,$$
$$\mathcal{I}_3 = \langle x^2 z, y^2 x^2, y^2 z, y^4, z^3, xyz, y^3 x, xz^2, yz^2, x^5, x^3 y \rangle,$$

with $\mathrm{HS}_{\mathcal{I}_1}(t) = t^4 + t^3 + 6t^2 + 3t + 1$ *and* $\mathrm{HS}_{\mathcal{I}_3}(t) := t^4 + 4t^3 + 6t^2 + 3t + 1$, *respectively. The degrees of these Hilbert series are the corresponding determinacies.*

Table 2. A comprehensive determinacy system for a parametric polynomial

Null set	Non-null set	Standard basis	Determinacy
{}	{ac}	$\{x^2 \frac{\partial f}{\partial z}, x^2 \frac{\partial f}{\partial y}, y^2 \frac{\partial f}{\partial z}, y^2 \frac{\partial f}{\partial y}, z^2 \frac{\partial f}{\partial z}, xy \frac{\partial f}{\partial z}, xy \frac{\partial f}{\partial y}, xz \frac{\partial f}{\partial z},$ $yz \frac{\partial f}{\partial z}, -8acx^5 - 2bcx^2 z^3 + 2x^3 y^2\}$	4
{a}	{c}	$\{x^2 \frac{\partial f}{\partial z}, x^2 \frac{\partial f}{\partial y}, y^2 \frac{\partial f}{\partial z}, y^2 \frac{\partial f}{\partial y}, z^2 \frac{\partial f}{\partial z}, xy \frac{\partial f}{\partial z}, xy \frac{\partial f}{\partial y}, xz \frac{\partial f}{\partial z},$ $yz \frac{\partial f}{\partial z}\}$	0
{c}	{a}	$\{x^2 \frac{\partial f}{\partial z}, x^2 \frac{\partial f}{\partial x}, y^2 \frac{\partial f}{\partial z}, y^2 \frac{\partial f}{\partial x}, z^2 \frac{\partial f}{\partial z}, xy \frac{\partial f}{\partial z}, xy \frac{\partial f}{\partial x}, xz \frac{\partial f}{\partial z},$ $yz \frac{\partial f}{\partial z}, -2x^3 y, 8ax^5 + 2bx^2 z^3\}$	4
{a, c}	{}	$\{x^2 \frac{\partial f}{\partial z}, x^2 \frac{\partial f}{\partial x}, y^2 \frac{\partial f}{\partial z}, y^2 \frac{\partial f}{\partial x}, z^2 \frac{\partial f}{\partial z}, xy \frac{\partial f}{\partial z}, xy \frac{\partial f}{\partial x}, xz \frac{\partial f}{\partial z},$ $yz \frac{\partial f}{\partial z}, -2x^3 y\}$	0

4.3 Automated Geometric Theorem Discovery

Automated geometric theorem discovery, which has been investigated using algebraic geometric tools for decades, aims at detecting complementary conditions under which a given geometric statement is true; see [4, 7, 8, 13, 26, 41, 51, 52]. Suppose that a given theorem can be described through the polynomial hypotheses $\{f_1, \ldots, f_k\}$ and the conclusion statement g. Indeed, to describe an admissible geometric theorem, we shall need a number of arbitrary coordinates (parameters), namely $u_1, \ldots, u_m$ and a number of dependent coordinates (variables), namely $x_1, \ldots, x_n$. Therefore, $f_1, \ldots, f_k, g \in \mathbb{K}[u_1, \ldots, u_m, x_1, \ldots, x_n]$. Let us first discuss three different types of correctness of a geometric theorem. The conclusion g follows *strictly* from the hypotheses $\{f_1, \ldots, f_k\}$ if it vanishes on $\mathbb{V}(f_1, \ldots, f_k)$. We know that if $g \in \sqrt{\langle f_1, \ldots, f_k \rangle}$ then g results strictly from the hypotheses, see e.g. [11, Proposition 5, Page 325]. In addition, the conclusion g follows *generically* from the hypotheses $f_1, \ldots, f_k$ if it vanishes on the variety of the f_i's as polynomials in $\mathbb{K}(u_1, \ldots, u_m)[x_1, \ldots, x_n]$. From [11, Corollary 9, Page 301], it follows that g results generically from $\{f_1, \ldots, f_k\}$ provided that $1 \in \mathcal{I} = \langle f_1, \ldots, f_k, 1 - yg \rangle \subset \mathbb{K}(u_1, \ldots, u_m)[x_1, \ldots, x_n]$. Now, assume that the triple set $G = \{(N_i, W_i, G_i)\}_{i=1}^{\ell}$ is a CGS for $\mathcal{I}$. Applying the Kapur et al. algorithm [30] on $\mathcal{I}$, we obtain a finite number of branches through which one can determine whether or not the theorem holds. It should be noted that the validity of $1 \in \mathcal{I}$ is equivalent to the fact that the Gröbner basis of the generic branch of G is $\{1\}$, i.e. for the branch (N_i, W_i, G_i) with $N_i = \{\}$ we have $G_i = \{1\}$. More generally, any branch (N_i, W_i, G_i) with $G_i = \{1\}$ provides the conditions for the correctness of the theorem, see [52] for more details on this approach. In this case, we say that the conclusion g is *generically true*. For other branches, let us say (N_i, W_i, G_i) with $G_i \neq \{1\}$, the theorem is not generically true, however it might be valid on some components. In [52], the authors propose a method to decide whether g is generically true on some components. This method uses a parametric radical membership which may be, in practice, a computational bottleneck. On the other hand, in order to avoid applying the radical computation, one can use standard bases to decide the validity of a statement. For this, we will apply the next proposition from [12, Proposition 5.2, Page 183].

Proposition 4.5. *Keeping the above notations, assume that the origin is contained in an irreducible component W of $\mathbb{V}(f_1, \ldots, f_k)$. Then g holds over W if the remainder on the division of g by a standard basis of $\langle f_1, \ldots, f_k \rangle$ is zero.*

In this case, we say that the conclusion g is *generically true on components* or equivalently *locally true*; that is g vanishes on some but not all non-degenerate components of $\mathbb{V}(f_1, \ldots, f_k)$. Using CSS's we derive necessary and sufficient conditions on the parameters under which the conclusion is true on components. Let us consider a CGS containing a branch (N_i, W_i, G_i) with $G_i \neq \{1\}$. Following the idea of [12, Page 182], we first apply the change of coordinates $u_i \mapsto U_i + a_i$ leading to $\tilde{f}_1, \ldots, \tilde{f}_k, \tilde{g} \in \mathcal{S} = \mathbb{K}[a_1, \ldots, a_m][U_1, \ldots, U_m]_{\langle U_1, \ldots, U_m \rangle}[x_1, \ldots, x_n]$ which translates $(a_1, \ldots, a_m)$ to the origin. Passing $\{\tilde{f}_1, \ldots, \tilde{f}_k\}$ together with the condition sets $\{N_i, W_i\}$ to the procedure CSSMAIN, produces a number of branches with this property so that for each branch if the remainder of $\tilde{g}$ by the corresponding standard basis using the Mora normal form algorithm is zero then the theorem is locally true, and incorrect otherwise. Let us introduce two notations that we use in the next theorem. Assume that $V \subset \mathbb{K}^n$ is a variety, i.e. there exist a finite sequence $f_1, \ldots, f_k \in \mathbb{K}[\mathbf{x}]$ so that $V = \mathbb{V}(f_1, \ldots, f_k)$. Then, the ideal of V, denoted by $\mathbb{I}(V)$, is defined to be the ideal $\{f \in \mathbb{K}[\mathbf{x}] \mid f(a_1, \ldots, a_n) = 0 \text{ for all } (a_1, \ldots, a_n) \in V\}$. Also, the sequence $U_1, \ldots, U_m$ is denoted by $\mathbf{U}$.

Theorem 4.6. *With the above notations, suppose that $(t_1, \ldots, t_m) \in \mathbb{K}^m$. Then, the conclusion g is locally true at this point iff there exists a branch $\{N, W, S\}$ in a CSS of $\tilde{\mathcal{I}} = \langle \tilde{f}_1, \ldots, \tilde{f}_k \rangle$ so that $(t_1, \ldots, t_m) \in \mathbb{V}(N) \backslash \mathbb{V}(W)$ and the Mora normal form of $\tilde{g}$ by S considering the condition sets (N, W) is zero. In particular, a CSS of $\tilde{\mathcal{I}}$ determines all the conditions (on the parameters) under which g is locally true.*

Proof. Assume that the Mora normal form of $\tilde{g}$ by S is zero. Then, by replacing the values of parameters by $t_1, \ldots, t_m$ in this division, we can see easily that there exists $u \in \mathbb{K}[\mathbf{U}, \mathbf{x}]$ so that $u \in \mathbb{K}[\mathbf{U}, \mathbf{x}]$ is a unit in $\mathbb{K}[\mathbf{U}]_{\langle \mathbf{U} \rangle}$ and $u\tilde{g} = h_1\tilde{f}_1 + \cdots + h_k\tilde{f}_k$ where $h_i \in \mathbb{K}[\mathbf{U}, \mathbf{x}]$, see [12, Corollary 3.13, Page 170]. Note that since all the polynomials have been translated to the origin by the map $u_i \mapsto U_i + t_i$, we then have $u(0) \neq 0$. Now, assume that V' be an irreducible component of $\mathbb{V}(f_1, \ldots, f_k)$ containing $(t_1, \ldots, t_m)$. Then, $\tilde{V}'$ is an irreducible component of $\mathbb{V}(\tilde{f}_1, \ldots, \tilde{f}_k)$ containing the origin and $u\tilde{g} \in \mathbb{I}(\mathbb{V}(\tilde{f}_1, \ldots, \tilde{f}_k)) \subset \mathbb{I}(\tilde{V}')$. Since $\mathbb{I}(\tilde{V}')$ is a prime ideal and u is unit then $\tilde{g} \in \mathbb{I}(\tilde{V}')$ and this shows that $g \in \mathbb{I}(V')$, i.e. g is locally true at $(t_1, \ldots, t_m)$. This proof was inspired by [12, Proposition 5.2, Page 183]. Conversely, suppose that g is locally true at $(t_1, \ldots, t_m)$. It follows that $g \in \mathbb{I}(V')$ where V' is an irreducible component of $\mathbb{V}(f_1, \ldots, f_k)$ containing $(t_1, \ldots, t_m)$. Then, applying the map $u_i \mapsto U_i + t_i$, we may assume that $\tilde{V}'$ is an irreducible component of $\mathbb{V}(\tilde{f}_1, \ldots, \tilde{f}_k)$ containing the origin and $\tilde{g} \in \mathbb{I}(\tilde{V}')$. On the other hand, if G is a CSS for $\tilde{\mathcal{I}}$ then there exists a branch (N, W, S) in G so that $(t_1, \ldots, t_m) \in \mathbb{V}(N) \backslash \mathbb{V}(W)$. This yields that S is a standard basis for the ideal $\mathcal{I}$ when we remove the components not containing the origin. From $\tilde{g} \in \mathbb{I}(\tilde{V}')$, we can conclude that the Mora normal form of $\tilde{g}$ by S is zero. $\square$

The next example (from [12, Page 181]) illustrates how the methods developed in this paper can be applied to discuss the validity of a geometric theorem.

Example 4.7. *Let $A = (0, 0), B = (u_1, 0), C = (u_2, u_3), D = (x_1, x_2)$ and $N = (x_3, x_4)$. For the parallelogram below, under which circumstances $AN = DN$ holds?, cf. [12, Pages 181–182].*

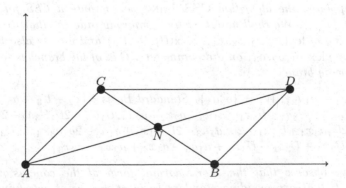

The hypotheses f_1, f_2, f_3, f_4 and the conclusion g are stated by the following equations

$$f_1 = x_2 - u_3 = 0$$
$$f_2 = (x_1 - u_1)u_3 - x_2 u_2 = 0$$
$$f_3 = x_1 x_4 - x_3 x_2 = 0$$
$$f_4 = x_4(u_2 - u_1) - (x_3 - u_1)u_3 = 0$$
$$g = x_1^2 - 2x_1 x_3 - 2x_2 x_4 + x_2^2 = 0.$$

We can check readily that $g \notin \sqrt{\langle f_1, \ldots, f_k \rangle}$ and this shows that g does not generically holds. Now, applying the SINGULAR function cgsdr to compute a CGS of the ideal generated by $\{f_1, f_2, f_3, f_4, 1 - yg\}$ with u_1, u_2, u_3 as parameters and x_1, x_2, x_3, x_4, y as variables gives rise to eight branches among which only one branch has Gröbner basis equals to one. Here, we focus on the following branches with not-equal-to-one Gröbner basis.

(a) $N_1 = \{\}$, $W_1 = \{u_1^2 u_3 + u_1 u_2 u_3\}$, Gröbner Basis $= \{2u_1^3 u_3 y + 4u_1^2 u_2 u_3 y + 2u_1 u_2^2 u_3 y - u_1 u_3, 2u_1 x_4 - u_1 u_3, 2u_3 x_3 - 2u_2 x_4 - u_1 u_3, x_2 - u_3, u_3 x_1 - u_1 u_3 - u_2 u_3\}$,

(b) $N_2 = \{u_3, u_1 + u_2\}$, $W_2 = \{u_2\}$, Gröbner Basis $= \{x_4, x_2, x_1^2 y + 2x_1 x_3 y - 1\}$.

This confirms that even using the concept of CGS's, we are not able to show that when the desired conclusion holds. Applying $u_i \mapsto U_i + a_i$ for $i = 1, 2, 3$ on hypotheses and conclusion polynomials, we get the new polynomials

$$\tilde{f}_1 = x_2 - U_3 - a_3 = 0$$
$$\tilde{f}_2 = (x_1 - U_1 - a_1)(U_3 + a_3) - x_2(U_2 + a_2) = 0$$
$$\tilde{f}_3 = x_1 x_4 - x_3 x_2 = 0$$
$$\tilde{f}_4 = x_4(U_2 + a_2 - U_1 - a_1) - (x_3 - U_1 - a_1)(U_3 + a_3) = 0$$
$$\tilde{g} = x_1^2 - 2x_1 x_3 - 2x_2 x_4 + x_2^2 = 0,$$

where $\tilde{g}, \tilde{f}_i \in \mathbb{K}[a_1, a_2, a_3][U_1, U_2, U_3]_{\langle U_1, U_2, U_3 \rangle}[x_1, x_2, x_3, x_4]$ for $i = 1, 2, 3, 4$. Now, by applying the algorithm CSSMAIN, we compute a CSS for the ideal $\mathcal{I} = \langle \tilde{f}_1, \tilde{f}_2, \tilde{f}_3, \tilde{f}_4 \rangle$. We shall note that the semigroup ordering that we use is the block ordering of $\text{lex}(x_1, x_2, x_3, x_4)$, $\text{alex}(U_1, U_2, U_3)$ and we use also the monomial ordering $\text{lex}(a_1, a_2, a_3)$ on the parameters. One of the branches in this CSS is the following branch

- $N_1 = \{a_1^2 - a_1 a_2\}$, $W_1 = \{a_3^3 a_1\}$, Standard Basis $= \{x_2 - U_3 - a_3, -U_1 U_3 - U_1 a_3 - U_2 x_2 - U_3 a_1 + U_3 x_1 - a_1 a_3 - a_2 x_2 + a_3 x_1, U_1 U_3^2 + 2U_1 U_3 a_3 - 2U_1 U_3 x_4 + U_1 a_3^2 - 2U_1 a_3 x_4 + U_3^2 a_1 + 2U_3 a_1 a_3 - 2U_3 a_1 x_4 + a_1 a_3^2 - 2a_1 a_3 x_4, U_1 U_3 + U_1 a_3 - U_1 x_4 + U_2 x_4 + U_3 a_1 - U_3 x_3 + a_1 a_3 - a_1 x_4 + a_2 x_4 - a_3 x_3\}$.

One can observe that the Mora normal form of the conclusion polynomial $\tilde{g}$ w.r.t. the corresponding standard basis at any point satisfying N_1 and W_1 is zero. For example, the special case $(a_1, a_2, a_3) = (1, 1, 1)$ considered in

[12, Page 181] satisfies the condition sets (N_1, W_1) and this shows that for this case the assertion is true. In addition, for any other values of a_1, a_2, a_3 satisfying the condition sets (N_1, W_1), the conclusion remains true. Finally, if we consider the assertion $CN = NB$, the corresponding conclusion polynomial would be $\tilde{g}' = 2x_3(U_1+a_1) - 2x_3(U_2+a_2) - 2x_4(U_3+a_3) - (U_1+a_1)^2 + (U_2+a_2)^2 + (U_3+a_3)^2 = 0$ and since the Mora normal form of this polynomial w.r.t. the corresponding standard basis at any point satisfying N_1 and W_1 is zero then we can conclude that this assertion is locally true at any point satisfying N_1 and W_1.

Remark 4.8. *Keeping the above notations, assume that $G = \{(N_i, W_i, G_i)\}_{i=1}^{\ell}$ is a CGS for $\mathcal{I} = \langle f_1, \ldots, f_k \rangle$. For a branch $(N_i, W_i, G_i) \in G$ let $A = \mathbb{V}(N_i) \setminus \mathbb{V}(W_i)$. In [52, Theorem 4.3], it is shown that*

"If g is a zero divisor in $\mathbb{K}[u_1, \ldots, u_m, x_1, \ldots, x_n]/\sqrt{\tilde{I}}$, then the geometric statement is true on components under A, where $\tilde{I} = \langle f_1, \ldots, f_k, N \rangle$".

However, the proof of this result is based on the property that "If g is a zero divisor in the quotient ring $\mathbb{K}[u_1, \ldots, u_m, x_1, \ldots, x_n]/\sqrt{\tilde{I}}$ then $\sigma(g)$ is a zero divisor in $\mathbb{K}[x_1, \ldots, x_n]/\langle \sigma(f_1), \ldots, \sigma(f_k) \rangle$ where σ is a specialization satisfying A" which does not hold in general. For example, let $f_1 = x_2 + u_2, f_2 = (u_1 x_1 - u_2)(u_1 x_2 + u_2)$ and $N = \{\}$. Then, it is clear that $g = u_1 x_1 - u_2$ is a zero divisor in $\mathbb{K}[u_1, u_2, x_1, x_2]/\langle f_1, f_2 \rangle$ however it is not a zero divisor in $\mathbb{K}[x_1, x_2]/\langle \sigma(f_1), \sigma(f_2) \rangle$ with $\sigma(u_1) = 1$ and $\sigma(u_2) = 1$.

5 Conclusions

In this paper we introduced the concept of comprehensive standard systems (CSS's) and showed that a straightforward local variant of the Kapur et al. algorithm is applied to compute a CSS. In the second part of the paper, we studied some applications of CSS's in different areas of Mathematics. In particular, we showed how we can use them for detecting all conditions under which a given geometric statement is locally true.

Acknowledgements. The authors would like to thank the anonymous reviewers for their helpful and constructive comments.

References

1. Bahloul, R.: Parametric standard basis, degree bound and local Hilbert-Samuel function. ArXiv:1004.0908, pp. 1–24 (2010)
2. Becker, T.: Standard bases in power series rings: uniqueness and superfluous critical pairs. J. Symb. Comput. **15**(3), 251–265 (1993)
3. Berkesch, C., Schreyer, F.-O.: Syzygies, finite length modules, and random curves. In: Commutative Algebra and Noncommutative Algebraic Geometry. Expository articles, vol. I, pp. 25–52. Cambridge University Press, Cambridge (2015)
4. Botana, F., Montes, A., Recio, T.: An algorithm for automatic discovery of algebraic loci. In: Proceeding of the ADG, pp. 53–59 (2012)

5. Buchberger, B.: A criterion for detecting unnecessary reductions in the construction of Gröbner-bases. In: Ng, E.W. (ed.) Symbolic and Algebraic Computation. LNCS, vol. 72, pp. 3–21. Springer, Heidelberg (1979). https://doi.org/10.1007/3-540-09519-5_52

6. Buchberger, B.: Bruno Buchberger's PhD thesis 1965: An algorithm for finding the basis elements of the residue class ring of a zero dimensional polynomial ideal. J. Symb. Comput. **41**, 3–4 (2006). Translation from the German, 475–511

7. Chen, X., Li, P., Lin, L., Wang, D.: Proving geometric theorems by partitioned-Parametric Gröbner bases. In: Hong, H., Wang, D. (eds.) ADG 2004. LNCS (LNAI), vol. 3763, pp. 34–43. Springer, Heidelberg (2006). https://doi.org/10.1007/11615798_3

8. Chou, S.-C.: Mechanical Geometry Theorem Proving. D. Reidel Publishing Company, Dordrecht (1988)

9. Chou, S.-C., Schelter, W.F.: Proving geometry theorems with rewrite rules. J. Autom. Reasoning **2**, 253–273 (1986)

10. Collins, G.E.: Quantifier elimination for real closed fields by cylindrical algebraic decompostion. In: Caviness, B.F., Johnson, J.R. (eds.) GI-Fachtagung 1975. LNCS, pp. 85–121. Springer, Wien: (1998). https://doi.org/10.1007/978-3-7091-9459-1_4

11. Cox, D., Little, J., O'Shea, D.: Ideals, Varieties, and Algorithms. An Introduction to Computational Algebraic Geometry and Commutative Algebra, 3rd edn. Springer, New York (2007). https://doi.org/10.1007/978-0-387-35651-8

12. Cox, D.A., Little, J., O'Shea, D.: Using Algebraic Geometry, 2nd edn. Springer, New York (2005). https://doi.org/10.1007/978-1-4757-6911-1

13. Dalzotto, G., Recio, T.: On protocols for the automated discovery of theorems in elementary geometry. J. Autom. Reasoning **43**(2), 203–236 (2009)

14. Decker, W., Greuel, G.-M., Pfister, G., Schönemann, H.: Singular 4-1-1 – a computer algebra system for polynomial computations (2018). http://www.singular.uni-kl.de

15. Faugère, J.-C.: A new efficient algorithm for computing Gröbner bases (F_4). J. Pure Appl. Algebra **139**(1–3), 61–88 (1999)

16. Faugère, J.-C.: A new efficient algorithm for computing Gröbner bases without reduction to zero (F_5). In: Proceedings of the 2002 International Symposium on Symbolic and Algebraic Computation, ISSAC 2002, Lille, France, July 07–10, 2002, pp. 75–83. ACM Press, New York (2002)

17. Gazor, M., Kazemi, M.: Singularity: a maple library for local zeros of scalar smooth maps (2016)

18. Gebauer, R., Möller, H.M.: On an installation of Buchberger's algorithm. J. Symb. Comput. **6**(2–3), 275–286 (1988)

19. Gelernter, H., Hanson, J. R., and Loveland, D. W.: Empirical explorations of the geometry-theorem proving machine. In: Proceeding of the West Joint Computer Conference (1960), pp. 143–147

20. Golubitsky, M., Stewart, I., Schaeffer, D.G.: Singularities and groups in bifurcation theory. vols. I, II. Springer-Verlag, New York, 1985, 1988

21. Gordan, P.: Les invariants des formes binaires. J. Math. **6**(5), 141–156 (1900)

22. Grauert, H.: Über die Deformation isolierter Singularitäten analytischer Mengen. Invent. Math. **15**, 171–198 (1972)

23. Guglielmi, A.V.: Foreshocks and aftershocks of strong earthquakes in the light of catastrophe theory. Physics-Uspekhi **58**(4), 384–397 (2015)

24. Hironaka, H.: Resolution of singularities of an algebraic variety over a field of characteristic zero. I, II. Ann. Math. **79**(2), 109–203, 205–326 (1964)

25. Kalkbrener, M.: On the stability of Gröbner bases under specializations. J. Symb. Comput. **24**(1), 51–58 (1997)
26. Kapur, D.: Using Gröbner bases to reason about geometry problems. J. Symb. Comput. **2**, 399–408 (1986)
27. Kapur, D.: An approach for solving systems of parametric polynomial equations. In: Sarawat, V., Van Hentenryck, P. (eds.) Principles and Practice of Constraint Programming, pp. 217–224. MIT Press, Cambridge (1995)
28. Kapur, D., Sun, Y., Wang, D.: A new algorithm for computing comprehensive Gröbner systems. In: Proceedings of the 35th International Symposium on Symbolic and Algebraic Computation, ISSAC 2010, Munich, Germany, July 25–28, 2010, pp. 29–36. Association for Computing Machinery (ACM), New York (2010)
29. Kapur, D., Sun, Y., and Wang, D.: Computing comprehensive Gröbner systems and comprehensive Gröbner bases simultaneously. In: Proceedings of the 36th international symposium on symbolic and algebraic computation, ISSAC 2011, San Jose, CA, USA, June 7–11, 2011, pp. 193–200. Association for Computing Machinery (ACM), New York (2011)
30. Kapur, D., Sun, Y., Wang, D.: An efficient algorithm for computing a comprehensive Gröbner system of a parametric polynomial system. J. Symb. Comput. **49**, 27–44 (2013)
31. Kapur, D., Sun, Y., Wang, D.: An efficient method for computing comprehensive Gröbner bases. J. Symb. Comput. **52**, 124–142 (2013)
32. Kutzler, B., Stifter, S.: On the application of Buchberger's algorithm to automated geometry theorem proving. J. Symb. Comput. **2**, 389–397 (1986)
33. Lazard, D.: Gröbner bases, Gaussian elimination and resolution of systems of algebraic equations. In: van Hulzen, J.A. (ed.) EUROCAL 1983. LNCS, vol. 162, pp. 146–156. Springer, Heidelberg (1983). https://doi.org/10.1007/3-540-12868-9_99
34. Manubens, M., Montes, A.: Improving the DISPGB algorithm using the discriminant ideal. J. Symb. Comput. **41**(11), 1245–1263 (2006)
35. Manubens, M., Montes, A.: Minimal canonical comprehensive Gröbner systems. J. Symb. Comput. **44**(5), 463–478 (2009)
36. Marais, M.S., Steenpaß, A.: The classification of real singularities using Singular. I: Splitting lemma and simple singularities. J. Symb. Comput. **68**, 61–71 (2015)
37. Milnor, J.W.: Singular Points of Complex Hypersurfaces, vol. 61. Princeton University Press, Princeton (1968)
38. Möller, H.M.: On the construction of Gröbner bases using syzygies. J. Symb. Comput. **6**(2–3), 345–359 (1988)
39. Möller, H. M., Mora, T., Traverso, C.: Gröbner bases computation using syzygies, pp. 320–328. In: Proceedings of ISSAC 1992. ACM Press, Baltimore (1992)
40. Montes, A.: A new algorithm for discussing Gröbner bases with parameters. J. Symb. Comput. **33**(2), 183–208 (2002)
41. Montes, A., Recio, T.: Automatic discovery of geometry theorems using minimal canonical comprehensive Gröbner systems. In: Botana, F., Recio, T. (eds.) ADG 2006. LNCS (LNAI), vol. 4869, pp. 113–138. Springer, Heidelberg (2007). https://doi.org/10.1007/978-3-540-77356-6_8
42. Montes, A., Wibmer, M.: Gröbner bases for polynomial systems with parameters. J. Symb. Comput. **45**(12), 1391–1425 (2010)
43. Seidenberg, A.: A new decision method for elementary algebra. Ann. Math. **60**(2), 365–374 (1954)
44. Sit, W.Y.: An algorithm for solving parametric linear systems. J. Symb. Comput. **13**(4), 353–394 (1992)

45. Stewart, I.: Applications of catastrophe theory to the physical sciences. Phys. D **2**(2), 245–305 (1981)
46. Suzuki, A., Sato, Y.: A simple algorithm to compute comprehensive Gröbner bases using Gröbner bases. In: Proceedings of the 2006 International Symposium on Symbolic and Algebraic Computation, ISSAC 2006, Genova, Italy, July 9–12, 2006, pp. 326–331. ACM Press, New York (2006)
47. Tarski, A.: A decision method for elementary algebra and geometry. In: Caviness, B.F., Johnson, J.R. (eds.) Quantifier Elimination and Cylindrical Algebraic Decomposition, pp. 24–84. Springer, Wien (1998). https://doi.org/10. 1007/978-3-7091-9459-1_3
48. Thom, R.: Structural stability and morphogenesis: an outline of a general theory of models. Transl. from the French edition, as updated by the author Fowler, D.H. (ed.) Reprint from the 2nd Engl. ed. Addison-Wesley Publishing Company Inc, Redwood City (1989)
49. Weispfenning, V.: Comprehensive Gröbner bases. J. Symb. Comput. **14**(1), 1–29 (1992)
50. Weispfenning, V.: Canonical comprehensive Gröbner bases. J. Symb. Comput. **36**(3–4), 669–683 (2003)
51. Winkler, F.: Gröbner bases in geometry theorem proving and simplest degeneracy conditions. Math. Pannonica **1**(1), 15–32 (1990)
52. Zhou, J., Wang, D., Sun, Y.: Automated reducible geometric theorem proving and discovery by Gröbner basis method. J. Autom. Reasoning **59**(3), 331–344 (2017)

An Algorithm for Computing Coefficients of Words in Expressions Involving Exponentials and Its Application to the Construction of Exponential Integrators

Harald Hofstätter[1]([✉]) [ID], Winfried Auzinger[2] [ID], and Othmar Koch[3] [ID]

[1] Reitschachersiedlung 4/6, 7100 Neusiedl am See, Austria
hofi@harald-hofstaetter.at

[2] Vienna University of Technology, Institute of Analysis and Scientific Computing,
Wiedner Hauptstraße 8–10, 1040 Wien, Austria
w.auzinger@tuwien.ac.at

[3] University of Vienna, Institute of Mathematics, Oskar-Morgenstern-Platz 1,
1090 Wien, Austria
othmar@othmar-koch.org,
http://harald-hofstaetter.at, http://asc.tuwien.ac.at/~winfried,
http://www.othmar-koch.org

Abstract. This paper discusses an efficient implementation of the generation of order conditions for the construction of exponential integrators like exponential splitting and Magnus-type methods in the computer algebra system Maple. At the core of this implementation is a new algorithm for the computation of coefficients of words in the formal expansion of the local error of the integrator. The underlying theoretical background including an analysis of the structure of the local error is briefly reviewed. As an application the coefficients of all 8th order self-adjoint commutator-free Magnus-type integrators involving the minimum number of 8 exponentials are computed.

Keywords: Splitting methods · Magnus-type integrators ·
Local error · Order conditions · Computer algebra

1 Introduction

In the construction of integration schemes for the numerical solution of evolution equations the coefficients of the schemes are usually determined as solutions of certain systems of polynomial equations. The obvious requirement that all terms up to a certain order $O(\tau^p)$ in the Taylor expansion of the local error with respect to the step-size τ vanish, usually leads, if applied in a naive way, to a far over-determined system of equations. In some cases, however, due to the special structure of the local error, the fact that a small subset of terms vanishes already implies that all terms up to order $O(\tau^p)$ in the Taylor expansion vanish.

© Springer Nature Switzerland AG 2019
M. England et al. (Eds.): CASC 2019, LNCS 11661, pp. 197–214, 2019.
https://doi.org/10.1007/978-3-030-26831-2_14

This leads to a minimal, non-redundant system of equations, the so-called *order conditions*, see for instance [2].

As it turns out, this is in particular the case for (generalized) exponential splitting methods for the numerical solution of evolution equations of the form[1]

$$\partial_t u(t) = Au(t) + Bu(t), \quad t \geq t_0, \quad u(t_0) = u_0, \quad A, B \in \mathbb{C}^{d \times d}$$

(and also for Magnus-type integrators considered below). Here one step $u_{n+1} = \mathcal{S}(\tau)u_n$ with step-size τ is specified by an approximation $\mathcal{S}(\tau) = \mathcal{S}_J(\tau) \cdots \mathcal{S}_1(\tau)$ of the exact solution operator $\mathcal{E}(\tau) = e^{\tau(A+B)}$, where the factors $\mathcal{S}_j(\tau)$ are exponentials whose applications $\mathcal{S}_j(\tau)y$ to vectors $y \in \mathbb{C}^d$ can be effectively computed. Prototypical examples are for instance the classical second order Strang splitting

$$\mathcal{S}(\tau) = e^{\frac{1}{2}\tau B} e^{\tau A} e^{\frac{1}{2}\tau B}, \tag{1}$$

or the 4th order generalized splitting

$$\mathcal{S}(\tau) = e^{\frac{1}{6}\tau B} e^{\frac{1}{2}\tau A} e^{\frac{2}{3}\tau B + \frac{1}{72}\tau^3 [B,[A,B]]} e^{\frac{1}{2}\tau A} e^{\frac{1}{6}\tau B} \tag{2}$$

proposed in [5,13].

The analysis of the structure of the local error

$$\mathcal{L}(\tau) = \mathcal{S}(\tau) - \mathcal{E}(\tau) = \mathcal{S}_J(\tau) \cdots \mathcal{S}_1(\tau) - e^{\tau(A+B)}$$

and the resulting derivation of order conditions is advantageously carried out in a purely formal way by introducing non-commutative symbols A, B representing respectively τA, τB, and considering the formal expression corresponding to $\mathcal{L}(\tau)$ with τA, τB substituted by A, B. Thus, with a slight generalization anticipating an application to Magnus-type integrators, we study expressions of the form

$$X = e^{\Phi_J} \cdots e^{\Phi_1} - e^{\Omega}, \tag{3}$$

where $\Phi_1, \ldots, \Phi_J$, and Ω are linear combinations of non-commutative symbols and commutators thereof. In particular, in Theorem 2 of Sect. 3 the structure of the leading term in the formal series expansion of such expressions is characterized, which leads to the derivation of order conditions in Theorem 3. These theoretical considerations of Sect. 3 are essentially a review of the theory developed in [9, Section 2], which extends and generalizes results from [2].

The main focus of the present paper is on a concrete implementation of the generation of order conditions according to Theorem 3 in the computer algebra system Maple[2]. This can be realized in a very efficient way, utilizing a new algorithm derived in [9] for the computation of coefficients of words (i.e., finite products of non-commutative symbols) in expressions like (3) involving

---

[1] For our considerations, it is sufficient to discuss linear problems. The algebraic structure underlying method construction is the same for nonlinear problems due to the calculus of Lie derivatives [8, Section III.5.1].

[2] We have used Maple 18, Maple is a trademark of Waterloo Maple Inc.

exponentials, whose concise Maple implementation is reproduced in its entirety in Sect. 2.[3] Our approach is thus a relevant contribution compared to previous work on the generation of order conditions, see, e.g., [8, Section III.5] or [3].

Finally in Sect. 4, after a brief review of material from [1,9] on Magnus-type integrators for the numerical solution of non-autonomous evolution equations of the form

$$\partial_t u(t) = A(t)u(t), \quad t \geq t_0, \quad u(t_0) = u_0, \quad A(t) \in \mathbb{C}^{d \times d},$$

we consider a non-trivial application of the theory of Sect. 3: We compute in a systematic way the coefficients of all 8th order commutator-free self-adjoint Magnus-type methods involving the minimum number of 8 exponentials.

2   Coefficients of Words in Expressions Involving Exponentials

Let $\mathcal{A}$ denote a fixed set of non-commutative variables. Given an expression X in these variables involving exponentials like (3) we want to calculate real or complex coefficients

$$c_w = \text{coeff}(w, X), \quad w \in \mathcal{A}^*$$

in the formal expansion

$$X = \sum_{w \in \mathcal{A}^*} c_w w \in \mathbb{C}\langle\langle \mathcal{A} \rangle\rangle.$$

X is thus represented as an element of $\mathbb{C}\langle\langle \mathcal{A} \rangle\rangle$, the algebra of formal power series in the non-commutative variables in $\mathcal{A}$. Here, $\mathcal{A}^*$ denotes the set of all words over the alphabet $\mathcal{A}$, i.e., the set of all finite products (including the empty product Id) of elements of $\mathcal{A}$.

2.1   A Family of Homomorphisms

In [9] an efficient algorithm for the computation of $\text{coeff}(w, X)$ was derived, which is based on a suitably constructed family of maps $\{\varphi_w : w \in \mathcal{A}^*\}$, where for each word $w = w_1 \cdots w_{\ell(w)} \in \mathcal{A}^*$ of length $\ell(w) \geq 1$, $\varphi_w(X)$ is an upper triangular matrix in $\mathbb{C}^{(\ell(w)+1) \times (\ell(w)+1)}$ whose entries are coefficients of subwords of w in X,

$$\varphi_w(X)_{i,j} = \begin{cases} \text{coeff}(w_{i:j-1}, X), & \text{if } i < j, \\ \text{coeff}(\text{Id}, X), & \text{if } i = j, \\ 0, & \text{if } i > j. \end{cases} \tag{4}$$

Here $w_{i:j-1} = w_i w_{i+1} \cdots w_{j-1}$ denotes the subword of w of length $j - i$, starting at position i and ending at position $j - 1$.

---

[3] All Maple code discussed in this paper is also provided by the package Expocon available at [7]. Additionally, this package includes routines for the generation of Lyndon words and Lyndon bases. For simplicity, such words and bases have always been hardcoded whenever needed in the code examples of this paper.

Theorem 1 ([9, Theorem 2.4]). *The map φ_w defined by (4) is an algebra homomorphism*

$$\mathbb{C}\langle\langle\mathcal{A}\rangle\rangle \to \mathbb{C}^{(\ell(w)+1)\times(\ell(w)+1)},$$

i.e.,

(i) φ_w is linear,

$$\varphi_w(\alpha X + \beta Y) = \alpha\varphi_w(X) + \beta\varphi_w(Y), \quad X, Y \in \mathbb{C}\langle\langle\mathcal{A}\rangle\rangle, \ \alpha, \beta \in \mathbb{C};$$

(ii) φ_w preserves the multiplicative structure,

$$\varphi_w(X \cdot Y) = \varphi_w(X) \cdot \varphi_w(Y), \quad X, Y \in \mathbb{C}\langle\langle\mathcal{A}\rangle\rangle.$$

Furthermore, if $\operatorname{coeff}(\mathrm{Id}, X) = 0,$ *then*

$$\varphi_w(\exp X) = \exp \varphi_w(X),$$

where the exponential of the strictly upper triangular and thus nilpotent matrix $\varphi_w(X)$ is exactly computable in a finite number of steps.

2.2   Maple Implementation of the Algorithm

It follows that for a given expression X, a recursive application of φ_w (the recursion terminates with well-defined values $\varphi_w(a)$ for the "atoms" $a \in \mathcal{A}$) yields $\varphi_w(X)$, from which one can read off $\operatorname{coeff}(w, X)$ as the element at the upper right corner,

$$\operatorname{coeff}(w, X) = \varphi_w(X)_{1,\ell(w)+1},$$

cf. (4). By organizing this calculation in a more efficient way, the function phiv defined in the Maple code displayed below computes

$$\texttt{phiv}(w, X, v) = \varphi_w(X) \cdot v$$

for a vector $v \in \mathbb{C}^{\ell(w)+1}$ without explicitly generating the matrix $\varphi_w(X)$. It recursively traverses the expression tree representing the expression X. At each node of the tree the evaluation branches out depending on whether the current node represents

- a non-commutative symbol (the atomic case which terminates the recursion),
- a sum of subexpressions,
- a product of subexpressions,
- a power of a subexpression,
- a commutator of subexpressions, or
- an exponential of a subexpression.

Finally, the function wcoeff[4] computes coeff(w, X) via

$$\text{coeff}(w, X) = \text{first component of } \text{phiv}(w, X, (0, \ldots, 0, 1)^T).$$

The elements of the alphabet $\mathcal{A}$ are represented within Maple as non-commutative symbols, which are provided by the package Physics. Note that except for providing such non-commutative symbols (and the type Commutator) we do not need or use any further feature of the package Physics. Words $w \in \mathcal{A}^*$ are represented as lists of non-commutative symbols.

```
> with(Physics):
> phiv := proc (w, X, v)
     local i, v1, v2, f, zero;
     if type(X, name) and type(X, noncommutative) then
         return [seq('if'(op(i,w)=X, v[i+1], 0), i=1..nops(w)), 0]
     elif type(X, '+') then
         return add(phiv(w, op(i, X), v), i=1..nops(X))
     elif type(X, '*') then
         v1 := v; zero := [0$nops(w)+1];
         for i from nops(X) to 1 by -1 do
             v1 := phiv(w, op(i, X), v1);
             if v1=zero then return zero end if;
         end do;
         return v1
     elif type(X, anything^integer) then
         v1 := v; zero := [0$nops(w)+1];
         for i from 1 to op(2, X) do
             v1 := phiv(w, op(1, X), v1);
             if v1=zero then return zero end if;
         end do;
         return v1
     elif (type(X, function) and
           op(0, X) = Physics[Commutator]) then
         return phiv(w, op(1, X), phiv(w, op(2, X), v))
                - phiv(w, op(2, X), phiv(w, op(1, X), v))
     elif type(X, exp(anything)) then
         v1 := v; v2 := v; zero := [0$nops(w)+1]; f := 1;
         for i from 1 to nops(w) do
             f := f*i; v1 := phiv(w, op(X), v1);
             if v1=zero then return v2 end if;
             v2 := v2 + v1/f;
         end do;
         return v2
     end if;
     return [seq(X*x, x=v)]
```

---

[4] The function was called wcoeff because coeff is already defined in Maple.

```
   end proc:
> wcoeff := proc (w, X)
     return phiv(w, X, [0$nops(w), 1])[1]
   end proc:
```

3   Order Conditions for Exponential Integrators

In this section we review the theory developed in [9, Section 2], which extends and generalizes results from [2]. We consider expressions of the form (3),

$$X = e^{\Phi_J} \cdots e^{\Phi_1} - e^{\Omega}, \tag{5}$$

where the exponents $\Phi_1, \ldots, \Phi_J$, and Ω are linear combinations of non-commutative symbols and commutators thereof, i.e., elements of $[\mathbb{C}\langle\mathcal{A}\rangle]$, the free Lie algebra generated by the non-commutative symbols of a given alphabet $\mathcal{A}$, which in a natural way is embedded in the algebra $\mathbb{C}\langle\langle\mathcal{A}\rangle\rangle$ of formal power series in these symbols.

In the applications we are interested in, $S = e^{\Phi_J} \cdots e^{\Phi_1}$ represents an exponential integrator for the numerical solution of an evolution equation, and $E = e^{\Omega}$ represents the exact local solution operator for this equation. We can interpret (5) as the error of the approximation S of E, i.e., X represents the *local error* of the exponential integrator S.

3.1   Grading of Words and Homogeneous Lie Elements

We consider a grading function on the alphabet $\mathcal{A}$,

$$\mathrm{grade}(a) \in \{1, 2, \ldots\}, \quad a \in \mathcal{A}, \tag{6}$$

and extend it to words $w = w_1 \ldots w_{\ell(w)} \in \mathcal{A}^*$ by

$$\mathrm{grade}(w) = \sum_{j=1}^{\ell(w)} \mathrm{grade}(w_j).$$

We call $\Psi \in [\mathbb{C}\langle\mathcal{A}\rangle]$ a homogeneous Lie element of grade q if it can be expanded in $\mathbb{C}\langle\langle\mathcal{A}\rangle\rangle$ to a linear combination of words all of the same grade q. The decomposition

$$[\mathbb{C}\langle\mathcal{A}\rangle] = \bigoplus_{q=1}^{\infty} \mathfrak{g}_q, \quad \mathfrak{g}_q = \{\text{homogeneous Lie elements of grade } q\} \tag{7}$$

into a direct sum of subspaces makes $[\mathbb{C}\langle\mathcal{A}\rangle]$ a graded Lie algebra, cf. [12].

Remark 1. In the applications we are interested in, the symbols $a \in \mathcal{A}$ represent objects which depend on a (small) parameter $\tau > 0$ (for instance, a time increment). The grading (6) is chosen such that it reflects the order of magnitude of the represented objects,

$$a \simeq O(\tau^{\mathrm{grade}(a)}), \quad a \in \mathcal{A}.$$

For example, in the case of an application to splitting methods with step-size τ,

$$\mathcal{A} = \{\mathrm{A}, \mathrm{B}\}, \quad \mathrm{A} \simeq \tau A = O(\tau), \ \mathrm{B} \simeq \tau B = O(\tau) \ \Rightarrow \ \mathrm{grade}(\mathrm{A}) = \mathrm{grade}(\mathrm{B}) = 1,$$

cf. Sect. 1.

3.2   Leading Error Term

The following theorem states that the leading error term Θ of the approximation $\mathrm{e}^{\Phi_J} \cdots \mathrm{e}^{\Phi_1}$ of e^{Ω} is a homogeneous Lie element of some grade q.

Theorem 2 ([9, Theorem 2.1]). *If for $\Phi_1, \ldots, \Phi_J, \Omega \in [\mathbb{C}\langle \mathcal{A} \rangle]$ the expression $\mathrm{e}^{\Phi_J} \cdots \mathrm{e}^{\Phi_1} - \mathrm{e}^{\Omega}$ is expanded in $\mathbb{C}\langle\langle \mathcal{A} \rangle\rangle$ as*

$$\mathrm{e}^{\Phi_J} \cdots \mathrm{e}^{\Phi_1} - \mathrm{e}^{\Omega} = \sum_{w \in \mathcal{A}^*} c_w w = \Theta + R,$$

where

$$\Theta = \sum_{\mathrm{grade}(w)=q_{\min}} c_w w, \qquad q_{\min} = \min\{\mathrm{grade}(w) : w \in \mathcal{A}^*, c_w \neq 0\} \tag{8}$$

(and the remainder R contains the terms of grade $> q_{\min}$), then Θ can be represented as a linear combination of commutators, i.e., Θ is a homogeneous Lie element of grade $q_{\min}$.

To illustrate Theorem 2 we consider as an example

$$X = \mathrm{e}^{\frac{1}{2}\mathrm{B}} \mathrm{e}^{\mathrm{A}} \mathrm{e}^{\frac{1}{2}\mathrm{B}} - \mathrm{e}^{\mathrm{A}+\mathrm{B}}, \quad \mathcal{A} = \{\mathrm{A}, \mathrm{B}\}, \quad \mathrm{grade}(\mathrm{A}) = \mathrm{grade}(\mathrm{B}) = 1,$$

cf. (1). The following Maple code computes the coefficients of all words of length ≤ 3 (i.e., of all words w with $\mathrm{grade}(w) \leq 3$) in X.

```
> Physics[Setup](noncommutativeprefix = {A, B}):
> X := exp((1/2)*B)*exp(A)*exp((1/2)*B)-exp(A+B):
> W := [[A], [B], [A, A], [A, B], [B, A], [B, B],
        [A, A, A], [A, A, B], [A, B, A], [A, B, B],
        [B, A, A], [B, A, B], [B, B, A], [B, B, B]]:
> seq(wcoeff(w, X), w in W);
```

$$0, 0, 0, 0, 0, 0, 0, \frac{1}{12}, \frac{-1}{6}, \frac{-1}{24}, \frac{1}{12}, \frac{1}{12}, \frac{-1}{24}, 0$$

It follows

$$e^{\frac{1}{2}B} e^A e^{\frac{1}{2}B} - e^{A+B} = \tfrac{1}{12}AAB - \tfrac{1}{6}ABA - \tfrac{1}{24}ABB + \tfrac{1}{12}BAA + \tfrac{1}{12}BAB - \tfrac{1}{24}BBA + \cdots$$

$$= \tfrac{1}{12}[A, [A, B]] - \tfrac{1}{24}[[A, B], B] + \cdots.$$

Here the leading error term is indeed a homogeneous Lie element of grade 3.

3.3 Symmetry

A product of exponentials $S = e^{\Phi_J} \cdots e^{\Phi_1}$, $\Phi_j \in [\mathbb{C}\langle \mathcal{A} \rangle]$ is called self-adjoint or symmetric,[5] if

$$\Phi_{J-j+1} = \sum_k (-1)^{k+1} X_{j,k}, \quad j = 1, \ldots, J$$

holds, where the $X_{j,k} \in \mathfrak{g}_k$ are the components of $\Phi_j = \sum_k X_{j,k}$ with respect to the decomposition (7). It follows that a single exponential e^Φ is self-adjoint, if and only if Φ is a sum of homogeneous Lie elements of odd grade, $\Phi = X_1 + X_3 + \ldots$, $X_k \in \mathfrak{g}_k$. It was proved in [9, Theorem 2.2] that in (8) the grade $q_{\min}$ of the homogeneous Lie element Θ is necessarily odd if $e^{\Phi_J} \cdots e^{\Phi_1}$ and e^Ω are both self-adjoint.

3.4 Lyndon Words and Lyndon Bases

For a homogeneous Lie element Θ of grade q like the one given in (8) let

$$\Theta = \sum_{b \in \mathcal{B}_q} c_b b \tag{9}$$

be its representation in a basis $\mathcal{B}_q$ of the subspace $\mathfrak{g}_q$ of (7). Furthermore, let $\mathcal{W}_q \subset \mathcal{A}^*$ be a set of words of grade q such that the matrix

$$T_q = (\text{coeff}(w, q))_{w \in \mathcal{W}_q, b \in \mathcal{B}_q} \tag{10}$$

is invertible. Then it follows from $c_w = \text{coeff}(w, \Theta) = \sum_{b \in \mathcal{B}_q} c_b \, \text{coeff}(w, b)$ for $w \in \mathcal{W}_q$, and thus $(c_w)_{w \in \mathcal{W}_q} = T_q \cdot (c_b)_{b \in \mathcal{B}_q}$, that the coefficients c_b in (9) can be computed as

$$(c_b)_{b \in \mathcal{B}_q} = T_q^{-1} \cdot (c_w)_{w \in \mathcal{W}_q}. \tag{11}$$

Suitable choices for such a set $\mathcal{W}_q$ and basis $\mathcal{B}_q$ are respectively the set of Lyndon words of grade q and the corresponding Lyndon basis [6,11], see Tables 1 and 2.

---

[5] For an exponential integrator represented by S this definition conforms with the usual definition of a self-adjoint integrator, e.g., $\mathcal{S}(-\tau)\mathcal{S}(\tau) = \text{Id}$ for a generalized splitting method where $\mathcal{S}(\tau)$ is S with $\tau A, \tau B$ substituted for A, B.

Table 1. Lyndon words $\mathcal{W}_q$ of grade q and Lyndon basis $\mathcal{B}_q$ of $\mathfrak{g}_q$ for $\mathcal{A} = \{A,B\}$ and grade(A) = grade(B) = 1.

q	Lyndon words	Lyndon basis
1	A, B	A, B
2	AB	[A, B]
3	AAB, ABB	[A, [A, B]], [[A, B], B]
4	AAAB, AABB, ABBB	[A, [A, [A, B]]], [A, [[A, B], B]], [[[A, B], B], B]
5	AAAAB, AAABB, AABAB, AABBB, ABABB, ABBBB	[A, [A, [A, [A, B]]]], [A, [A, [[A, B], B]]], [[A, [A, B]], [A, B]], [A, [[[A, B], B], B]], [[A, B], [[A, B], B]], [[[[A, B], B], B], B]

3.5   Order Conditions

The following main result of this section is an easy consequence of the previous considerations.

Theorem 3 ([9, Theorem 2.3]). *If for $\Phi_1, \ldots, \Phi_J, \Omega \in [\mathbb{C}\langle \mathcal{A} \rangle]$ the order conditions*

$$c_w = \mathrm{coeff}(w, \mathrm{e}^{\Phi_J} \cdots \mathrm{e}^{\Phi_1} - \mathrm{e}^{\Omega}) = 0, \quad w \in \bigcup_{q=1}^{p} \mathcal{W}_q \qquad (12)$$

are satisfied for all Lyndon words of grade $q \leq p$, then

$$c_w = \mathrm{coeff}(w, \mathrm{e}^{\Phi_J} \cdots \mathrm{e}^{\Phi_1} - \mathrm{e}^{\Omega}) = 0, \quad w \in \mathcal{A}^*, \ \mathrm{grade}(w) \leq p, \qquad (13)$$

and thus $q_{\min} \geq p + 1$ *in* (8).

If $\mathrm{e}^{\Phi_J} \cdots \mathrm{e}^{\Phi_1}$ *and* e^{Ω} *are both self-adjoint, then we may assume that p is even, and* (13) *holds already if the order conditions* (12) *are satisfied only for all Lyndon words of odd grade $q \leq p$.*

In view of Remark 1 in Subsection 3.1 we can interpret (13) as the statement

$$\mathrm{e}^{\Phi_J} \cdots \mathrm{e}^{\Phi_1} - \mathrm{e}^{\Omega} \simeq O(\tau^{p+1}),$$

i.e., $\mathrm{e}^{\Phi_J} \cdots \mathrm{e}^{\Phi_1}$ is an approximation of e^{Ω} of order $p + 1$.

Table 2. Lyndon words $\mathcal{W}_q$ of grade q and Lyndon basis $\mathcal{B}_q$ of $\mathfrak{g}_q$ for $\mathcal{A} = \{A_1, \ldots, A_q\}$ and grade(A_k) = k.

q	Lyndon words	Lyndon basis
1	A_1	A_1
2	A_2	A_2
3	$A_1 A_2$, A_3	$[A_1, A_2]$, A_3
4	$A_1 A_1 A_2$, $A_1 A_3$, A_4	$[A_1, [A_1, A_2]]$, $[A_1, A_3]$, A_4
5	$A_1 A_1 A_1 A_2$, $A_1 A_1 A_3$, $A_1 A_2 A_2$, $A_1 A_4$, $A_2 A_3$, A_5	$[A_1, [A_1, [A_1, A_2]]]$, $[A_1, [A_1, A_3]]$, $[[A_1, A_2], A_2]$, $[A_1, A_4]$, $[A_2, A_3]$, A_5

3.6   Example

For $\mathcal{A} = \{A, B\}$ with $\text{grade}(A) = \text{grade}(B) = 1$ we want to determine the parameters $a, b, c, d \in \mathbb{R}$ such that $S = e^{bB}e^{aA}e^{cB+d[B,[A,B]]}e^{aA}e^{bB}$ is a 5th order approximation of $E = e^{A+B}$. Since the ansatz S and the expression E are both self-adjoint in the sense of Subsection 3.3, we only have to consider Lyndon words of odd grade ≤ 4 (i.e., of odd length ≤ 4 for $\text{grade}(A) = \text{grade}(B) = 1$),

$$\mathcal{W} = \mathcal{W}_1 \cup \mathcal{W}_3 = \{A, B, AAB, ABB\},$$

cf. Table 1. The order conditions (12) lead to 4 polynomial equations in 4 variables a, b, c, d, for which the following Maple code computes a unique solution corresponding to

$$S = e^{\frac{1}{6}B}e^{\frac{1}{2}A}e^{\frac{2}{3}B+\frac{1}{72}[B,[A,B]]}e^{\frac{1}{2}A}e^{\frac{1}{6}B},$$

cf. (2).

```
> Physics[Setup](noncommutativeprefix = {A, B}):
> C := Physics[Commutator]:
> X := exp(b*B)*exp(a*A)*exp(c*B+d*C(B, C(A, B)))*
      exp(a*A)*exp(b*B) - exp(A+B):
> W := [[A], [B], [A, A, B], [A, B, B]]:
> eqs := [seq(simplify(wcoeff(w, X)), w in W)];
```

$$eqs := \left[-1 + 2a, -1 + 2b + c, -\frac{1}{6} + 2a^2b + \frac{1}{2}a^2c, -\frac{1}{6} + \frac{1}{2}ac^2 + acb + ab^2 - d \right]$$

```
> sol := solve(eqs);
```

$$sol := \left\{ a = \frac{1}{2}, b = \frac{1}{6}, c = \frac{2}{3}, d = \frac{1}{72} \right\}$$

Next we compute the leading error term Θ of the approximation S of $E = e^{A+B}$, cf. (8). To this end we take $\mathcal{W}_5$ and $\mathcal{B}_5$ from Table 1 and compute T_5, $(c_w)_{w \in \mathcal{W}_5}$, and $(c_b)_{b \in \mathcal{B}_5}$ according to (10), (11):

```
> W5 := [[A, A, A, A, B], [A, A, A, B, B], [A, A, B, A, B],
         [A, A, B, B, B], [A, B, A, B, B], [A, B, B, B, B]]:
> B5 := [C(A, C(A, C(A, C(A, B)))), C(A, C(A, C(C(A, B), B))),
         C(C(A, C(A, B)), C(A, B)), C(A, C(C(C(A, B), B), B)),
         C(C(A, B), C(C(A, B), B)), C(C(C(C(A, B), B), B), B)]:
> T5 := Matrix([seq([seq(wcoeff(w, b), b in B5)], w in W5)]);
```

$$
T5 := \begin{bmatrix}
1 & 0 & 0 & 0 & 0 & 0 \\
0 & 1 & 0 & 0 & 0 & 0 \\
0 & -2 & 1 & 0 & 0 & 0 \\
0 & 0 & 0 & 1 & 0 & 0 \\
0 & 0 & 0 & -3 & 1 & 0 \\
0 & 0 & 0 & 0 & 0 & 1
\end{bmatrix}
$$

```
> c_w := [seq(wcoeff(w, subs(sol, X)), w in W5)];
```

$$
c\_w := \left[\frac{1}{2880}, \frac{-7}{8640}, \frac{1}{480}, \frac{7}{12960}, \frac{-1}{720}, \frac{-41}{155520} \right]
$$

```
> c_b := evalm(LinearAlgebra[MatrixInverse](T5) &* c_w);
```

$$
c\_b := \left[\frac{1}{2880}, \frac{-7}{8640}, \frac{1}{2160}, \frac{7}{12960}, \frac{1}{4320}, \frac{-41}{155520} \right]
$$

Altogether, we obtain a representation of the leading error term,

$$
e^{\frac{1}{6}B} e^{\frac{1}{2}A} e^{\frac{2}{3}B + \frac{1}{72}[B,[A,B]]} e^{\frac{1}{2}A} e^{\frac{1}{6}B} - e^{A+B} = \Theta + \cdots
$$
$$
= \tfrac{1}{2880}[A,[A,[A,[A,B]]]] - \tfrac{7}{8640}[A,[A,[[A,B],B]]] + \tfrac{1}{2160}[[A,[A,B]],[A,B]]
$$
$$
+ \tfrac{7}{12960}[A,[[[A,B],B],B]] + \tfrac{1}{4320}[[A,B],[[A,B],B]] - \tfrac{41}{155520}[[[[A,B],B],B],B] + \cdots .
$$

Here, the dots represent terms of grade higher than five.

4 Magnus-Type Integrators

In this section we apply the theory of Sect. 3 with the aim of constructing Magnus-type integrators for the numerical solution of non-autonomous evolution equations

$$
\partial_t u(t) = A(t)u(t), \quad t \geq t_0, \quad u(t_0) = u_0, \quad A(t) \in \mathbb{C}^{d \times d}.
$$

One step $(t_n, u_n) \mapsto (t_{n+1}, u_{n+1})$ of step-size τ of such an integrator is given by

$$
t_{n+1} = t_n + \tau, \quad u_{n+1} = \mathcal{S}(\tau, t_n)u_n, \tag{14}
$$

where $\mathcal{S}(\tau, t_n) \approx \mathcal{E}(\tau, t_n)$ approximates the exact local solution operator

$$
\mathcal{E}(\tau, t_n) = e^{\Omega(\tau, t_n)} \tag{15}
$$

given by the *Magnus* series $\Omega = \Omega(\tau, t_n)$, see [8, Section IV.7].

4.1 Legendre Expansions

To construct $\mathcal{S}(\tau, t_n)$ we expand $A(t_n + t)$ on the interval $[t_n, t_n + \tau]$ into a series of shifted Legendre polynomials,

$$A(t_n + t) = A_1 \tilde{P}_0(t) + A_2 \tilde{P}_1(t) + A_3 \tilde{P}_2(t) + \ldots, \quad t \in [0, \tau]$$

with

$$\tilde{P}_k(t) = \frac{1}{\tau} P_k \left(\frac{t}{\tau} \right), \quad P_k(x) = (-1)^k \sum_{j=0}^{k} \binom{k}{j} \binom{k+j}{j} (-1)^j x^j,$$

see [1, Section 3.1]. The matrix-valued coefficients A_k given by

$$A_k = (2k - 1)\tau \int_0^1 P_{k-1}(x) A(t_n + \tau x) \, dx \tag{16}$$

depend on both t_n and τ and satisfy

$$A_k = \mathcal{O}(\tau^k). \tag{17}$$

In terms of these coefficients the Magnus series in (15) is given by

$$\begin{aligned}
\Omega = {}& A_1 - \tfrac{1}{6}[A_1, A_2] + \tfrac{1}{60}[A_1, [A_1, A_3]] - \tfrac{1}{60}[A_2, [A_1, A_2]] \\
& + \tfrac{1}{360}[A_1, [A_1, [A_1, A_2]]] - \tfrac{1}{30}[A_2, A_3] + \ldots.
\end{aligned} \tag{18}$$

see [1, Section 3.2].

4.2 Order Conditions for Magnus-Type Integrators

We consider Magnus-type integrators (14) of the form

$$\mathcal{S}(\tau, t_n) = e^{\tilde{\Phi}_J(\tau, t_n)} \ldots e^{\tilde{\Phi}_1(\tau, t_n)}, \tag{19}$$

where the $\tilde{\Phi}_j$ are linear combinations of (commutators of) approximations $\tilde{A}_k \approx A_k$ obtained by applying a suitable quadrature formula to (16). To apply the theory of Sect. 3 to such integrators, we note that (19) formally corresponds to an expression

$$S = e^{\Phi_J} \ldots e^{\Phi_1}$$

with Lie elements $\Phi_j \in [\mathbb{C}\langle\mathcal{A}\rangle]$ over an alphabet $\mathcal{A} = \{A_1, A_2, \ldots\}$ with symbols A_k representing $\tilde{A}_k \approx A_k$, and with a grading

$$\text{grade}(A_k) = k$$

corresponding to (17), cf. Remark 1 in Subsection 3.1.

To set up order conditions according to Theorem 3 we need the Lyndon words from Table 2, and, furthermore, we have to consider e^Ω with Ω from (18), which is self-adjoint in the sense of Subsection 3.3. For coefficients of words $w \in \mathcal{A}^*$ in e^Ω (which in principle could be calculated with the algorithm of Sect. 2) we use the explicit formula

$$\operatorname{coeff}(A_{d_1} \cdots A_{d_\ell}, e^\Omega) = \sum_{\substack{(k_1,\ldots,k_\ell) \\ 1 \le k_l \le d_l}} \prod_{j=1}^{\ell} \frac{(-1)^{d_j+k_j} \binom{d_j-1}{k_j-1} \binom{d_j+k_j-2}{k_j-1}}{\sum_{i=j}^{\ell} k_i} \qquad (20)$$

proven in [9, Theorem 4.1].

4.3 8th Order Commutator-Free Magnus-Type Integrators

In this section we construct 8th order self-adjoint commutator-free integrators involving a minimum number of exponentials. In an ansatz for such a scheme only the generators A_1, A_2, A_3, A_4 have to be considered, because it can be shown that $\operatorname{coeff}(w, e^\Omega) = 0$ for all words w of grade$(w) \le 8$ containing A_k with $k \ge 5$, see [1, Section 3.3][6]. Corresponding to 22 Lyndon words of odd grade ≤ 8 over the alphabet $\mathcal{A} = \{A_1, A_2, A_3, A_4\}$ there are 22 order-conditions to be considered, see Theorem 3. This implies an ansatz involving 11 exponentials and 22 parameters to be determined; an 8th order scheme of this form was derived in [1, Section 4.4]. Also with this approach we found in [9, Section 4.4] a scheme where some exponentials commute, which can thus be joined together, resulting in an 8th order scheme involving only 8 exponentials. These schemes were found by a rather brute force computation. In contrast, using the following Maple code we are able to compute the coefficients of all 8th order self-adjoint schemes with 8 exponentials in a more systematic and efficient[7] way.

First, we define the self-adjoint ansatz for a scheme involving 8 exponentials.

```
> Physics[Setup](noncommutativeprefix = {A}):
> S := exp(f11*A1-f12*A2+f13*A3-f14*A4)*
        exp(f21*A1-f22*A2+f23*A3-f24*A4)*
        exp(f31*A1-f32*A2+f33*A3-f34*A4)*
        exp(f41*A1-f42*A2+f43*A3-f44*A4)*
        exp(f41*A1+f42*A2+f43*A3+f44*A4)*
        exp(f31*A1+f32*A2+f33*A3+f34*A4)*
        exp(f21*A1+f22*A2+f23*A3+f24*A4)*
        exp(f11*A1+f12*A2+f13*A3+f14*A4):
```

Next we set up the 8 equations corresponding to the 8 Lyndon words of odd grade ≤ 8 involving only the generators A_1, A_2. The right-hand sides of these equations were computed using (20) and are hardcoded here for simplicity.

---

[6] Of course, this follows also from (20) by a direct computation.

[7] Compared with the effort indicated in [1, Section 4.4].

```
> W12 := [[A1], [A1, A2], [A1, A1, A1, A2], [A1, A2, A2],
         [A1, A1, A1, A1, A1, A2], [A1, A1, A1, A2, A2],
         [A1, A1, A2, A1, A2], [A1, A2, A2, A2]]:
> rhs12 := [1, -1/6, -1/40, 1/60, -1/1008, 1/420, 1/2520, -1/840]:
> vars12 := [f11, f21, f31, f41, f12, f22, f32, f42]:
> eqs12 := [seq(expand(wcoeff(w, S)), w in W12)] - rhs12:
```

We now try to solve this system of equations. After a few minutes of computing time on a standard desktop PC, Maple finds a symbolic representation (involving RootOfs, which are Maple representations for roots of polynomial equations) of the general solution of the system, for which we compute all possible values in numerical form. It turns out that this computation (which again takes a few minutes) has to be done with very high precision, otherwise the results do not represent reasonable solutions with small residuals if substituted into the equations.

```
> sols12 := solve(eqs12, vars12):
> Digits := 200:
> FF := seq(evalf(allvalues(sol)), sol in sols12):
```

We obtain 99 solutions altogether, 17 real solutions, and modulo complex conjugation 41 different complex solutions. Each solution determines 8 parameters of the ansatz S. There remain 8 parameters to be determined compared with 14 order conditions corresponding to the $14 = 22 - 8$ remaining Lyndon words over $\mathcal{A} = \{A_1, A_2, A_3, A_4\}$ of odd grade ≤ 8. It is remarkable that the resulting over-determined system of equations always has a solution. To find a theoretical explanation for this fact is the topic of current investigations. Here, however, it is verified by a direct computation.

We select[8] one of the previously obtained 99 sets of 8 parameters, substitute it into the ansatz S, and set up 4 equations corresponding to 4 (out of 9) selected Lyndon words over $\mathcal{A}$ of odd grade ≤ 8 involving A_3 but not A_4. The resulting system of equation is linear and readily solved. That this solution solves also the equations corresponding to the $5 = 9 - 4$ not selected Lyndon words will be verified below.

```
> F12 := FF[78]:
> W3 := [[A3], [A1, A1, A3], [A2, A3], [A1, A1, A1, A1, A3],
         [A1, A1, A2, A3], [A1, A1, A3, A2], [A1, A2, A1, A3],
         [A1, A3, A3], [A2, A2, A3]]:
> rhs3 := [0, 1/60, -1/30, 1/420, -1/168, 1/280, -1/840,
           1/420, -1/210]:
> vars3 := [f13, f23, f33, f43]:
> eqs3 := [seq(expand(wcoeff(w, subs(F12, S))), w in W3[2 .. 5])]
```

---

[8] This is done for the purpose of presentation, the following considerations apply to each of the 99 parameter sets. The solution selected for this presentation leads to a particularly small local error. Note that the selected index 78 may belong to different parameter sets in different runs of the code.

```
            - rhs3[2 .. 5]:
> F3 := op(solve(eqs3, vars3)):
```

Analogously as before, we substitute the 12 already obtained parameters into the ansatz S, set up 4 equations corresponding to 4 (out of 5) selected Lyndon words over $\mathcal{A}$ of odd grade ≤ 8 involving A_4, and solve the resulting linear system of equations. That the obtained solution solves also the equation corresponding to the not selected Lyndon word will again be verified below.

```
> W4 := [[A1, A4], [A1, A1, A1, A4], [A1, A2, A4],
          [A1, A4, A2], [A3, A4]]:
> rhs4 := [0, -1/840, 1/210, -1/140, -1/70]:
> vars4 := [f14, f24, f34, f44]:
> eqs4 := [seq(expand(wcoeff(w, subs(F12, F3, S))),
              w in W4[1 .. 4])] - rhs4[1 .. 4]:
> F4 := op(solve(eqs4, vars4)):
```

Finally we print the calculated solution representing the 16 parameters $f_{j,k}$ of the ansatz S and compute its residual with respect to the order conditions corresponding to all Lyndon words over $\mathcal{A}$ of odd grade ≤ 8 (including those not previously selected). The tiny residual confirms that the obtained scheme indeed satisfies the order conditions for order $p = 8$ of Theorem 3.

```
> for y in [op(F12), op(F3), op(F4)] do
      lprint(evalf(y, 50))
  end do:
  f11 = -1.1210783473381738227756934594506597445892745485109
  f21 = 1.3210319274244662988569102191161576010502669814859
  f31 = -.11488794115695215928140654449977903918312514606917
  f41 = .41493436107065968320018978483428118272213271309425
  f12 = 1.0089705126043564404981241135055937701303470936598
  f22 = -1.1889339712738696420578749909323697235681087890036
  f32 = 0.44866039420480983666929215062389499923245100101695e-1
  f42 = -.13197275582656085011222031954705867101347489961070
  f13 = -.78475484313672167594298542161546182121249218395766
  f23 = .92477328275109744272940525314314765421496759253486
  f33 = 0.24950727790821017623386132247659342458740875944374e-1
  f43 = -.16496916740519678440980596377534517546121628452158
  f14 = .44843133893526952911027738378026389783570981940438
  f24 = -.52881775248948867348601923353730351864984279845615
  f34 = -0.24298790613584639672784191664606712944260031094723e-1
  f44 = .19795913373984127516833047932058800652021234941605
> W := [op(W12), op(W3), op(W4)]:
> RHS := [op(rhs12), op(rhs3), op(rhs4)]:
> printf("%.5e", max(map(abs, [seq(wcoeff(w,
          subs(F12, F3, F4, S)), w in W)]-RHS))):
  8.82689e-143
```

In the scheme

$$S = \prod_{j=8,\ldots,1} \exp\left(\sum_{k=1}^{4} f_{j,k} \mathsf{A}_k\right) \tag{21}$$

with parameters $f_{j,k}$ calculated by the above Maple code, the A_k represent Legendre expansion coefficients defined by integrals (16). To obtain an effective numerical method we have to approximate these integrals using a suitable quadrature formula. Therefore we substitute

$$\mathsf{A}_k \to (2k-1)\tau \sum_{l=1}^{K} w_k P_{k-1}(x_l) A(t_n + \tau x_l)$$

in (21) with Gaussian nodes and weights of order eight,

$$(x_k) = \left(\tfrac{1}{2} - \sqrt{\tfrac{15+2\sqrt{30}}{140}},\ \tfrac{1}{2} - \sqrt{\tfrac{15-2\sqrt{30}}{140}},\ \tfrac{1}{2} + \sqrt{\tfrac{15-2\sqrt{30}}{140}},\ \tfrac{1}{2} + \sqrt{\tfrac{15+2\sqrt{30}}{140}}\right),$$

$$(w_k) = \left(\tfrac{1}{4} - \tfrac{\sqrt{30}}{72},\ \tfrac{1}{4} + \tfrac{\sqrt{30}}{72},\ \tfrac{1}{4} + \tfrac{\sqrt{30}}{72},\ \tfrac{1}{4} - \tfrac{\sqrt{30}}{72}\right),$$

which corresponds to an application of Gaussian quadrature to (16). For the set of parameters $\{f_{j,k}\}$ displayed in the above Maple code we obtain the integrator (cf. (14), (19))

$$\mathcal{S}(t_n,\tau) = \prod_{j=8,\ldots,1} \exp\left(\tau \sum_{k=1}^{4} a_{j,k} A(t_n + \tau x_k)\right) \tag{22}$$

with coefficients $a_{j,k}$ given in Table 3. For these coefficients the positivity condition

Table 3. Coefficients $a_{j,k}$ for an 8th order commutator-free Magnus-type integrator (22).

$k=1$	$k=2$	$k=3$	$k=4$
-1.232611007291861933e+0	1.381999278877963415e-1	-3.352921035850962622e-2	6.861942424401394962e-3
1.452637092757343214e+0	-1.632549976033022450e-1	3.986114827352239259e-2	-8.211316003097062961e-3
-1.783965547974815151e-2	-8.850494961553933912e-2	-1.299159096777419811e-2	4.448254906109529464e-3
-2.982838328015747208e-2	4.530735723950198008e-1	-6.781322579940055086e-3	-1.529505464262590422e-3
-1.529505464262590422e-3	-6.781322579940055086e-3	4.530735723950198008e-1	-2.982838328015747208e-2
4.448254906109529464e-3	-1.299159096777419811e-2	-8.850494961553933912e-2	-1.783965547974815151e-2
-8.211316003097062961e-3	3.986114827352239259e-2	-1.632549976033022450e-1	1.452637092757343214e+0
6.861942424401394962e-3	-3.352921035850962622e-2	1.381999278877963415e-1	-1.232611007291861933e+0

$$\mathrm{Re} f_{j,1} = \mathrm{Re} \sum_{k=1}^{4} a_{j,k} > 0, \quad j = 1,\ldots,8, \tag{23}$$

is not satisfied, in agreement with the fact that this cannot be the case for real coefficients, see [10]. For some applications, however, it is essential for stability reasons that this condition is satisfied. This suggests to consider schemes with complex coefficients, see [4]. As was mentioned above, of the 99 parameter sets $\{f_{j,k}\}$ which can be computed by the above Maple code, $82 = 41 \times 2$ involve complex numbers. One of these parameter sets leads to the scheme (22) with coefficients given in Table 4, for which (23) is satisfied.

Table 4. Real (top) and imaginary (bottom) parts of coefficients $a_{j,k}$ for an 8th order commutator-free Magnus-type integrator (22) satisfying the positivity condition (23).

$k = 1$	$k = 2$	$k = 3$	$k = 4$
5.162172083124911076e-2	-5.787809823308952456e-3	1.404202563971892685e-3	-2.873779919999358082e-4
1.129000600487386325e-1	-1.811008163470541820e-2	8.982553129811831365e-3	-2.544930699554437791e-3
2.631601314221973826e-2	1.983998701294184106e-1	-4.965939955061425298e-2	1.197843408520720342e-2
-1.592059248033346570e-2	1.424220211513735403e-1	4.842122146532602005e-2	-1.013590436679991693e-2
-1.013590436679991693e-2	4.842122146532602005e-2	1.424220211513735403e-1	-1.592059248033346570e-2
1.197843408520720342e-2	-4.965939955061425298e-2	1.983998701294184106e-1	2.631601314221973826e-2
-2.544930699554437791e-3	8.982553129811831365e-3	-1.811008163470541820e-2	1.129000600487386325e-1
-2.873779919999358082e-4	1.404202563971892685e-3	-5.787809823308952456e-3	5.162172083124911076e-2
-1.187198036084005914e-1	1.331082409655082917e-2	-3.229389682031679030e-3	6.609128526175740449e-4
1.359790143178213473e-1	3.226637801235380303e-3	-5.647440118497178834e-3	1.831962429052182520e-3
-1.952925932474600076e-2	4.339859420803126316e-2	4.884840043796339250e-3	-1.849278537972746835e-3
3.513884130112852023e-3	-7.185755041597012718e-2	1.591348406688517315e-2	-1.887432258484616938e-3
-1.887432258484616938e-3	1.591348406688517315e-2	-7.185755041597012718e-2	3.513884130112852023e-3
-1.849278537972746835e-3	4.884840043796339250e-3	4.339859420803126316e-2	-1.952925932474600076e-2
1.831962429052182520e-3	-5.647440118497178834e-3	3.226637801235380303e-3	1.359790143178213473e-1
6.609128526175740449e-4	-3.229389682031679030e-3	1.331082409655082917e-2	-1.187198036084005914e-1

Acknowledgements. This work was supported in part by the Austrian Science Fund (FWF) under grant P30819-N32 and the Vienna Science and Technology Fund (WWTF) under grant MA14–002.

References

1. Alverman, A., Fehske, H.: High-order commutator-free exponential time-propagation of driven quantum systems. J. Comput. Phys. **230**, 5930–5956 (2011)
2. Auzinger, W., Herfort, W.: Local error structures and order conditions in terms of Lie elements for exponential splitting schemes. Opuscula Math. **34**, 243–255 (2014)
3. Auzinger, W., Herfort, W., Hofstätter, H., Koch, O.: Setup of order conditions for splitting methods. In: Gerdt, V.P., Koepf, W., Seiler, W.M., Vorozhtsov, E.V. (eds.) Computer Algebra in Scientific Computing, pp. 30–42. Springer International Publishing, Cham (2016)
4. Blanes, S., Casas, F., Thalhammer, M.: High-order commutator-free quasi-Magnus exponential integrators for nonautonomous linear evolution equations. Comput. Phys. Commun. **220**, 243–262 (2017)
5. Chin, S.: Symplectic integrators from composite operator factorizations. Phys. Lett. A **226**(6), 344–348 (1997)
6. Duval, J.: Géneration d'une section des classes de conjugaison et arbre des mots de Lyndon de longueur bornée. Theoret. Comput. Sci. **60**, 255–283 (1988)

7. Expocon.mpl. https://github.com/HaraldHofstaetter/Expocon.mpl
8. Hairer, E., Lubich, C., Wanner, G.: Geometric Numerical Integration, 2nd edn. Springer-Verlag, Berlin-Heidelberg-New York (2006)
9. Hofstätter, H.: Order conditions for exponential integrators (Feb 2019). submitted (for a preprint see arXiv:1902.11256)
10. Hofstätter, H., Koch, O.: Non-satisfiability of a positivity condition for commutator-free exponential integrators of order higher than four. Numer. Math. **141**(3), 681–691 (2019)
11. Lothaire, M.: Combinatorics on Words, Encyclopedia of Mathematics and its Applications. Cambridge University Press, Cambridge (1997)
12. Munthe-Kaas, H., Owren, B.: Computations in a free Lie algebra. Phil. Trans. R. Soc. Lond. A **357**(1754), 957–981 (1999)
13. Suzuki, M.: New scheme of hybrid exponential product formulas with applications to quantum Monte-Carlo simulations. In: Landau, D.P., Mon, K.K., Schuttler, H.B. (eds.) Computer Simulation Studies in Condensed-Matter Physics VIII, vol. 80, pp. 169–174. Springer, Berlin (1995)

Revisit Sparse Polynomial Interpolation Based on Randomized Kronecker Substitution

Qiao-Long Huang[1,2] and Xiao-Shan Gao[1(✉)]

[1] KLMM, UCAS, Academy of Mathematics and Systems Science Chinese Academy of Sciences, Beijing, China
`xgao@mmrc.iss.ac.cn`
[2] Cheriton School of Computer Science, University of Waterloo, Waterloo, ON, Canada
`q94huang@uwaterloo.ca`

Abstract. In this paper, a new reduction based interpolation algorithm for general black-box multivariate polynomials over finite fields is given. The method is based on two main ingredients. A new Monte Carlo method is given to reduce the black-box multivariate polynomial interpolation problem to the black-box univariate polynomial interpolation problem over any ring. The reduction algorithm leads to multivariate interpolation algorithms with better or the same complexities in most cases when combining with various univariate interpolation algorithms. A modified univariate Ben-Or and Tiwari algorithm over the finite field is proposed, which has better total complexity than the Lagrange interpolation algorithm. Combining our reduction method and the modified univariate Ben-Or and Tiwari algorithm, we give a Monte Carlo multivariate interpolation algorithm, which has better total complexity in most cases for sparse interpolation of black-box polynomial over finite fields.

Keywords: Randomized Kronecker substitution ·
Sparse polynomial interpolation · Black-box · Finite field ·
Monte Carlo algorithm

1 Introduction

The interpolation for a sparse multivariate polynomial

$$f = c_1 m_1 + c_2 m_2 + \cdots + c_t m_t \in \mathcal{R}[x_1, \ldots, x_n] \tag{1}$$

given as a black-box is a basic computational problem, where $\mathcal{R}$ is a ring. Here, the challenge is that both the monomials m_i and the coefficients c_i are unknown and the algorithm needs to take advantage of the sparse structure of f.

Partially supported by an NSFC grant No.11688101.

M. England et al. (Eds.): CASC 2019, LNCS 11661, pp. 215–235, 2019.
https://doi.org/10.1007/978-3-030-26831-2_15

In [1], Zippel gave a probabilistic algorithm which needs an upper bound for the number of terms of f and an upper bound for the degree of f in each variable. In [2], Ben-Or and Tiwari gave a deterministic algorithm over the field of complex numbers, which needs an upper bound for the number of terms in f. After these work, many interesting algorithms were given, such as the computational complexity enhancement [3,4], the interpolation with nonstandard bases [5], the interpolation over finite fields [6–10], the early termination algorithm [11,12], the hybrid interpolation algorithm [13–15], the interpolation for modular black-box polynomials [16,17], and the reduction based methods for black-box and SLP polynomials [8,18–23].

The sparse interpolation algorithms can be roughly divided into two types: (1) the direct methods, such as the Ben-Or and Tiwari algorithm, which find the monomials m_i directly and then find the coefficients; (2) the reduction based methods, such as Zippel's algorithm, which reduce the multivariate interpolation problem into the univariate interpolation problem.

The size of an n-variate polynomial f with a degree bound D, a term bound T, and coefficients c_i is $O(nT \log D + T \log c)$, where $c = \max_{i=1}^{t} |c_i|$. The sparse interpolation algorithms can also be roughly divided into two types according to the complexity in D: (1) the supersparse algorithm whose complexity is polynomial in $\log D$; (2) the exponential algorithm whose complexity is polynomial in D. All algorithms have complexity polynomial in n, T.

Since the value of a polynomial of degree D at any point other than $0, \pm 1$ will have D bits or more, any algorithm whose complexity is proportional to $\log D$ cannot perform such an evaluation over $\mathbb{Q}$ or $\mathbb{Z}$. Even for polynomials over the general finite field $\mathbb{F}_q$, there is no supersparse interpolation algorithms for the standard black-box model. On the other hand, supersparse algorithms do exist for the following special models.

The first model is the straight-line program model [8,18,19,21–24], which uses the arithmetic operations in the $\mathcal{R}[x_1, \ldots, x_n]$ to replace the black-box evaluation.

The second model is the modular black-box model [16,17], which works for the polynomials in $\mathbb{Q}[x_1, \ldots, x_n]$. Given a prime p and an element θ in $\mathbb{F}_p$, the model computes $f(\theta)$ over the field $\mathbb{F}_p$. The cost of the evaluation depends on the size of p.

The third model is the precision accuracy black-box model [13,15,25,26], which allows for evaluations on the unit circle in some representation of a subfield of $\mathbb{C}$ or returns only a limited number of bits of precision for an evaluation.

In this paper, we focus on reduction based methods for general black-box models. Our main contribution is to give a new Monte Carlo reduction method, which leads to multivariate interpolation algorithms with better or the same complexities in most cases comparing to existing reduction based methods. We also propose a modified univariate Ben-Or and Tiwari algorithm over the finite field $\mathbb{F}_q$ costing $O^\sim(D \log q + TB)$ bit operations, where B is the cost of one query of the black-box and is a function of T, D, q. Note that the Lagrange interpolation algorithm costs $O^\sim(D \log q + DB)$ bit operations, which is larger since $T \leq D$. Let f be an n-variate polynomial with a degree bound D and a term bound T. Combining our reduction method and the modified univariate Ben-Or and Tiwari algorithm, we give a multivariate interpolation algorithm whose bit complexity is $O^\sim(nTD \log q + nTB_f)$, where B_f is the cost of evaluating the black-box that gives f and is a function of n, T, D, q.

1.1   Comparing with Other Reduction Based Methods

Our reduction depends on the following Kronecker type substitutions:

$$f(x^{\mathbf{s}}) = f(x^{s_1}, x^{s_2}, \ldots, x^{s_n}) \tag{2}$$
$$f(x^{\mathbf{s}+p\mathbf{I}_k}) = f(x^{s_1}, x^{s_2}, \ldots, x^{s_k+p}, \ldots, x^{s_n}) \tag{3}$$

where p is a prime and $\mathbf{s} = (s_1, \ldots, s_n) \in \mathbb{N}^n$ is a vector of random integers. The substitution (2) introduced in [20] is called *randomized Kronecker substitution*. (3) was introduced in [27]. Our method builds on the work [20,27]. To compare with [20,27], we first explain how these algorithms work. The algorithm in [20] has three main steps. 1. Randomly choose $O(n+\log T)$ substitutions s_i. 2: Find a diversifying set of terms of f such that a diversifying term has the same coefficient after all substitutions. 3: For each term, solve a linear system to obtain its exponents. The algorithm in [27] also has three main steps. 1: Randomly choose $\log T$ substitutions s_i. 2: Find the $f(x^{\mathbf{s}_u})$ with the maximal number of terms. 3: Find a prime p such that $\#f(x^{\mathbf{s}_u}) \bmod (x^p - 1) = \#f(x^{\mathbf{s}_u})$ and half of the terms of f can be recovered from $f(x^{\mathbf{s}_u})$ and $f(x^{\mathbf{s}_u+p\mathbf{I}_k})$, $k = 1, 2, \ldots, n$.

Our algorithm works as follows. 1: Randomly choose $\log T$ primes p_i of size $O^\sim(T \log D)$ and substitutions $s_i \in \mathbb{Z}_{p_i}^n$. 2: Find a u such that $\#f(x^{\mathbf{s}_u}) \bmod (x^{p_u} - 1)$ has the maximal number of terms. 3: Prove that half of the terms of f can be recovered from $f(x^{\mathbf{s}_u})$ and $f(x^{\mathbf{s}_u+p_u\mathbf{I}_k})$, $k = 1, 2, \ldots, n$.

Our method is different from that in [20,27] in the following aspects. Comparing to [20], we do not need to solve linear systems, so our algorithm is linear in n while theirs are linear in n^ω. Also, our algorithm does not need to find the diversifying set, so it works for more general rings. Comparing to [27], our algorithm chooses a prime p_i first and then chooses the substitutions $s_i \in \mathbb{Z}_{p_i}^n$, while in [27], the prime is fixed. As a consequence, the univariate polynomials in our algorithm have degrees $O^\sim(TD)$, while the degrees of the univariate polynomials in [27] contain either T^2 or D^2.

In Table 1, we list some existing reduction methods, where "#Reductions(N)" is the number of univariate interpolations, "Degree($\widetilde{D}$)" is the degree bound of the univariate polynomials, "Extra bit complexity(η)" is the additional complexities needed besides the univariate interpolations. "Type" means whether the algorithm is deterministic (Det), Monte Carlo (MC), or Las Vegas (LV).

We now compare the complexities of multivariate polynomial interpolation algorithms using the reductions given in Table 1. The complexity of the randomized Kronecker substitution algorithm in [20] already can be improved by removing the $O(n^\omega T)$ term in the complexity, by using structured linear algebra. This is an idea attributed to Pernet in [27], see Lemma 5.7.2. We do not list it

Table 1. Reduction of multivariate polynomial interpolations to univariate ones

	#Reductions(N)	Degree ($\widetilde{D}$)	Extra bit cost(η)	Type
Kronecker	1	D^n	$nT \log D$	Det
Zippel [1]	nT	D	$\geq nT \log D$	MC
Klivans-Spielman [6]	n	nT^2D	$nTD^{O(1)}$	MC
Arnold [27]	$n \log T$ $+$ $\log^2 T$	TD $+$ $TD + \overline{D}T \min(D, T\log(TD))$	$nT \log D$	MC
Arnold-Roche [20]	$n + \log T$	TD	$n^\omega T + nT \log D$	MC
Huang-Gao [22]	$n \log T$	nTD	$nT \log D$	MC
This paper (Cor. 1)	$n \log T + \log^2 T$	TD	$nT \log D$	MC

in Table 1 because the result is spelled out for extended-black-box polynomial. According to our analysis, if using this method to general black-box polynomials, N will be increased to at least $O(n \log T(\log D + \log T))$ and $\widetilde{D}$ will be increased to $O(nTD)$. So its cost is more than the algorithm in [22].

Two cases are considered according to the complexity of the univariate interpolation algorithm to be used.

First, assume a univariate interpolation algorithm is supersparse with complexity $\mathbf{SLin}(T^\alpha, \log^\beta D)$, where $\mathbf{SLin}(a, b, \dots)$ means the complexity is softlinear in $a, b, \dots$. For simplicity of analysis, we assume $\mathbf{SLin}(a, b, \dots)$ is the product of $a, b, \dots$. Since $\log D$ is the factor of the size of f, $\beta \geq 1$. Then the complexity of the multivariate interpolation algorithm is $\mathbf{SLin}(NT^\alpha, N \log^\beta \widetilde{D}) + \eta$, where N, $\widetilde{D}$, and η are from Table 1. We list these complexities in Table 2. From the table, we can see that, for the supersparse algorithms, our reduction method is not worse than the methods in [22,27] and the Kronecker substitution, and is better than others.

Table 2. Complexity for supersparse multivariate interpolation algorithms based on reductions

	Complexity	type
Kronecker	$\mathbf{SLin}(n^\beta, T^\alpha, \log^\beta D)$	Det
Zippel [1]	$\mathbf{SLin}(n, T^{\alpha+1}, \log^\beta D)$	MC
Klivans-Spielman [6]	$\mathbf{SLin}(n, T^\alpha, \log^\beta D) + nTD^{O(1)}$	MC
Arnold [27]	$\mathbf{SLin}(n, T^\alpha, \log^\beta D)$	MC
Arnold-Roche [20]	$\mathbf{SLin}(n, T^\alpha, \log^\beta D, n^\omega T)$	MC
Huang-Gao [22]	$\mathbf{SLin}(n, T^\alpha, \log^\beta D)$	MC
This paper (Cor. 1)	$\mathbf{SLin}(n, T^\alpha, \log^\beta D)$	MC

Second, assume a univariate algorithm is exponential and its complexity is $\mathbf{SLin}(T^{\alpha}, D^{\beta})$. Then the complexities of the multivariate interpolation algorithms are $\mathbf{SLin}(NT^{\alpha}, N(\widetilde{D})^{\beta}) + \eta$, which are listed in Table 3. From the table, we can see that, for the exponential algorithms, the complexity of our algorithm is better than all the existed Kronecker-type substitutions [6,20,22,27]. Comparing to Zippel's reduction [1], our method has better, equal, or worse complexities according to $0 < \beta < 1$, $\beta = 1$, or $\beta > 1$, respectively.

Table 3. Complexity for exponential multivariate interpolation algorithms based on reductions

	Complexity	type
Kronecker	$\mathbf{SLin}(T^{\alpha}, D^{n\beta}) + nT\log D$	Det
Zippel [1]	$\mathbf{SLin}(n, T^{\alpha+1}, D^{\beta})$	MC
Klivans-Spielman [6]	$\mathbf{SLin}(n^{\beta+1}, T^{\alpha+2\beta}, D^{\beta}) + nTD^{O(1)}$	MC
Arnold [27]	$\mathbf{SLin}(n, T^{\alpha+\beta}, D^{2\beta})$ or $\mathbf{SLin}(n, T^{\alpha+2\beta}, D^{\beta})$	MC
Arnold-Roche [20]	$\mathbf{SLin}(n, T^{\alpha+\beta}, D^{\beta}) + n^{\omega}T$	MC
Huang-Gao [22]	$\mathbf{SLin}(n^{\beta+1}, T^{\alpha+\beta}, D^{\beta})$	MC
This paper (Cor. 1)	$\mathbf{SLin}(n, T^{\alpha+\beta}, D^{\beta})$	MC

Table 4 is a summary of the comparisons, where "$\sqrt{}$", "$=$", "$\times$" means that our reduction method has better, the same, or worse complexity, respectively. We can see that our reduction method achieves better or the same complexities, except one case: exponential algorithms with $\beta > 1$. Since D is the main factor in the complexity, algorithms with high complexities in D are generally not used.

Table 4. Compare to other reduction based interpolation methods

		Kronecker	Zippel [1]	Klivans-Spielman [6]	Arnold [27]	Arnold-Roche [20]	Huang-Gao(MC) [22]
Supersparse	$\beta = 1$	$=$	$\sqrt{}$	$\sqrt{}$	$=$	$\sqrt{}$	$=$
	$\beta > 1$	$\sqrt{}$	$\sqrt{}$	$\sqrt{}$	$=$	$\sqrt{}$	$=$
Exponential	$0 \leq \beta < 1$	$\sqrt{}$	$\sqrt{}$	$\sqrt{}$	$\sqrt{}$	$\sqrt{}$	$\sqrt{}$
	$\beta = 1$	$\sqrt{}$	$=$	$\sqrt{}$	$\sqrt{}$	$\sqrt{}$	$\sqrt{}$
	$\beta > 1$	$\sqrt{}$	$\times$	$\sqrt{}$	$\sqrt{}$	$\sqrt{}$	$\sqrt{}$

Finally, we remark that for exponential algorithms, the cases $\beta > 1$ and $\beta < 1$ do exist. The original Ben-Or and Tiwari algorithm works for univariate polynomials over the finite field $\mathbb{F}_q$ and costs $O^{\sim}(T^{1.5}\sqrt{D}\log q + T\log^2 q)$ bit operations (Refer to Remark 2), where $\beta = 0.5$. The bit complexity of the Lagrange interpolation algorithm over $\mathbb{Q}$ is $O^{\sim}(D^2)$.

1.2   Comparing with Interpolation Algorithms over Finite Fields

In order to obtain a reduction based multivariate interpolation algorithm, we need univariate interpolation algorithms with best complexities.

Let h be a black-box univariate polynomial in $\mathbb{F}_q[x]$ with a degree bound D and a term bound T. Let B_h be the cost of one query of the black-box. In this paper, we give a modified univariate Ben-Or and Tiwari algorithm which costs $O^\sim(D \log q)$ bit operations and $O(T)$ evaluations of h, so the total cost is $O^\sim(D \log q + T B_h)$. The Lagrange interpolation algorithm costs $O^\sim(D \log q)$ bit operations and $O(D)$ evaluations of h and the total complexity is $O(D \log q + D B_h)$. So, the modified univariate Ben-Or and Tiwari algorithm has lower total complexities than the Lagrange algorithm, since $T \leq D$.

A univariate Ben-Or and Tiwari algorithm over the finite filed was given in [24], whose complexity includes the parameter q. Also, the multivariate Ben-Or and Tiwari algorithm was extended to finite fields [9,10], whose complexities are high (see Table 5).

Combining the modified univariate Ben-Or and Tiwari algorithm and our reduction method, we give a new multivariate interpolation algorithm. Table 5 is a comparison with interpolation algorithms over finite fields. "Probes" is the number of evaluations for the polynomials, "Bit complexity" is the complexity besides the probes, and "Size of $\mathbb{F}_q$" means that the algorithm can work for the finite field whose size satisfies this condition, and in the contrary case, the algorithm need to take values in a proper extension field of $\mathbb{F}_q$.

Table 5. "Soft-Oh" comparison of interpolation algorithms over finite field $\mathbb{F}_q$

	Probes (ρ)	Bit complexity (Θ)	Size of $\mathbb{F}_q$	type
Grigoriev-Karpinski-Singer [7]		$n^2 T^6 \log^2(ntq) + q^{2.5} \log^2 q$		Det
Huang-Rao [9]	$T^2 D$	$(TD)^8((TD)^5 + \log q) \log^2 q + nTD \log q$	$q \geq O(T^2 D^2)$	LV
Javadi and Monagan [10]	nT	$T^2(\log q + nD) \log q$	$\phi(q-1) \geq O(nD^2 T^2)$	MC
Klivans-Spielman [6]	nT	$n^2 T^2 D \log q$	$q \geq O(nT^2 D)$	MC
Arnold-Roche [20]	nT	$nTD \log q + n^\omega T$	$q \geq O(TD)$	MC
Huang-Gao [22]	nT	$n^2 TD \log q$	$q \geq O^\sim(nDT)$	MC
Zippel [1,10]	nTD	$nTD \log q$	$q \geq O(nD^2 T^2)$	MC
This paper (Them. 9)	nT	$nTD \log q$	$q \geq O^\sim(TD)$	MC
This paper (Rem. 2)	nT	$nT^{1.5}\sqrt{D} \log q + nT \log^2 q$	$q \geq O^\sim(TD)$	MC

The total complexity of an algorithm is $O^\sim(\Theta + \rho B)$, where Θ and ρ are from Table 5 and B is the cost of probing the black-box. The bit complexities of the algorithms given in [7,9] are much higher than other algorithms, so we will not compare with them below.

We can see that our algorithm (Theorem 9) has better total complexity than all other methods in [1,6,10,20,22]. Comparing to Zippel's algorithm, our algorithm has the same bit complexity but needs less evaluations and works for a smaller field. Actually, our algorithm is the only one which achieves the best current bounds in all three parameters in Table 5.

The algorithm given in Remark 2 uses the original Ben-Or and Tiwari algorithm for univariate polynomials over the finite field $\mathbb{F}_q$, which costs $O^{\sim}(nT^{1.5}\sqrt{D}\log q + nT\log^2 q)$ bit operations. By Table 4, if using this univariate interpolation algorithm, our reduction method gives a multivariate interpolation algorithm with lowest complexity in D, which can be seen from Table 5.

2   Reduction Based on Randomized Kronecker Substitution

In this section, we give a new Monte Carlo algorithm which reduces multivariate polynomial interpolation to univariate polynomial interpolation based on randomized Kronecker substitutions over any commutative ring with identity.

2.1   Find an "ok" Random Kronecker Substitution

Let $f \in \mathcal{R}[\mathbb{X}]$, where $\mathcal{R}$ is commutative ring with identity and $\mathbb{X} = \{x_1, x_2, \ldots, x_n\}$ is a set of n indeterminates. Denote $\#f$ and $\deg f$ to be the number of terms in f and the total degree of f, respectively. For $\mathbf{s} = (s_1, s_2, \ldots, s_n) \in \mathbb{N}^n$ and a new indeterminate x, let

$$f(x^{\mathbf{s}}) = f(x^{s_1}, x^{s_2}, \ldots, x^{s_n}) \tag{4}$$

$$f_{(p)}^{\mathrm{mod}}(x^{\mathbf{s}}) = f(x^{s_1}, x^{s_2}, \ldots, x^{s_n}) \,\mathrm{mod}\, (x^p - 1). \tag{5}$$

For $\mathbf{s} = (s_1, s_2, \ldots, s_n) \in \mathbb{N}^n$, a term cm_1 of f is said to *collide* in $f(x^{\mathbf{s}})$ (or $f_{(p)}^{\mathrm{mod}}(x^{\mathbf{s}})$) if f has another term em_2 such that $m_1 \neq m_2$ and $m_1(x^{\mathbf{s}}) = m_2(x^{\mathbf{s}})$(or $m_{1(p)}^{\mathrm{mod}}(x^{\mathbf{s}}) = m_{2(p)}^{\mathrm{mod}}(x^{\mathbf{s}})$).

When $\mathbf{s} = (s_1, s_2, \ldots, s_n)$ is chosen randomly, the substitution $x_i = x^{s_i}, i = 1, 2, \ldots, n$ is called a *randomized Kronecker substitution*. For a prime p, a substitution $\mathbf{s}$ is called *"ok" with respect to* p, if a majority, say $\frac{5}{8}$, of the terms of f do not collide in $f_{(p)}^{\mathrm{mod}}(x^{\mathbf{s}})$.

We need the following Hoeffding's inequility for Bernoulli random variables.

Lemma 1. *[28] Let $X = \sum_{i=1}^{n} X_i$, where $X_i, i = 1, 2, \ldots, n$, are independently distributed in $[0, 1]$. Then for all $\varepsilon > 0$, $\Pr[X > \mathbf{E}[X] + \varepsilon] \leq e^{-2\varepsilon^2/n}$ and $\Pr[X < \mathbf{E}[X] - \varepsilon] \leq e^{-2\varepsilon^2/n}$, where $\mathbf{E}[X]$ is the expected value of X.*

We have the following key lemma.

Lemma 2. *Let $f = \sum_{i=1}^{t} c_i m_i \in \mathcal{R}[\mathbb{X}]$, $T \geq \#f$, $D \geq \deg f$, and $N = \max\{31\lfloor(T-1)\log_2 D\rfloor, 1\}$. Let $p_1, p_2, \ldots, p_N$ be N different primes which satisfy $p_i \geq 32(T-1)$. If we randomly choose a prime p in $\{p_1, p_2, \ldots, p_N\}$ and choose $\mathbf{s} \in \mathbb{Z}_p^n$ uniformly at random, where $\mathbb{Z}_p = \{0, 1, \ldots, p-1\}$. Then any fixed term of f collides in $f_{(p)}^{\mathrm{mod}}(x^{\mathbf{s}})$ with probability $\leq \frac{1}{16}$.*

Proof. If $t = 1$ or $D = 1$, then the proof is obvious. So assume $T \geq t \geq 2$ and $D \geq 2$. In this case, $N = 31\lceil (T-1)\log_2 D \rceil$. Assume $m_i = x_1^{e_{i,1}} \cdots x_n^{e_{i,n}}, i = 1, \ldots, t$. Without loss of generality, we consider the first term $c_1 m_1$. Let $h(s_1, \ldots, s_n) = \prod_{i=2}^{t}[(e_{i,1} - e_{1,1})s_1 + \cdots + (e_{i,n} - e_{1,n})s_n]$ which is a polynomial in $\mathbb{Z}[s_1, \ldots, s_n]$ with degree no more than $T - 1$. Assume the variables are ordered as $s_1 \prec \cdots \prec s_n$ and k_i is the largest number such that $e_{i,k_i} - e_{1,k_i} \neq 0$. Then $\prod_{i=2}^{t}[e_{i,k_i} - e_{1,k_i}]s_{k_i}$ is the leading term. Let $C = \prod_{i=2}^{t}[e_{i,k_i} - e_{1,k_i}]$ and let ℓ be the number of different prime factors of C. Since $|e_{i,k_i} - e_{1,k_i}| \leq D$, we have $2^\ell \leq C$ and hence C has at most $\lfloor (T-1)\log_2 D \rfloor$ different prime factors. So if we randomly choose a prime p in $\{p_1, \ldots, p_N\}$, with probability at least $1 - \frac{\lfloor (T-1)\log_2 D \rfloor}{N} = \frac{30}{31}$, $\prod_{i=2}^{t}[e_{i,k_i} - e_{1,k_i}] \bmod p \neq 0$. In this case, $h(s_1, \ldots, s_n) \bmod p$ is a non-zero polynomial in $\mathbb{F}_p[s_1, \ldots, s_n]$. If $h(s_1, \ldots, s_n) \bmod p \neq 0$, then by Zippel's lemma [1], if we choose $\mathbf{s} \in \mathbb{Z}_p^n$ uniformly at random, then $h(\mathbf{s}) \bmod p \neq 0$ with probability at least $1 - \frac{T-1}{p} \geq 1 - \frac{T-1}{32(T-1)} = \frac{31}{32}$. So if we randomly choose a prime p in $\{p_1, \ldots, p_N\}$ and then choose $\mathbf{s} \in \mathbb{Z}_p^n$ uniformly at random, with probability at least $\frac{30}{31} \cdot \frac{31}{32} = \frac{15}{16}$, $h(\mathbf{s}) \bmod p \neq 0$. Now it suffices to show that when $h(\mathbf{s}) \bmod p \neq 0$, $c_1 m_1$ does not collide in $f_{(p)}^{\mathbf{mod}}(x^\mathbf{s})$. Since $h(\mathbf{s}) \bmod p \neq 0$, $(e_{i,1} - e_{1,1})s_1 + \cdots + (e_{i,n} - e_{1,n})s_n \neq 0 \bmod p$. So $(e_{i,1}s_1 + \cdots + e_{i,n}s_n) \bmod p \neq (e_{1,1}s_1 + \cdots + e_{1,n}s_n) \bmod p$, which means that $c_i m_i$ does not collide with $c_1 m_1$ in $f_{(p)}^{\mathbf{mod}}(x^\mathbf{s})$. $\square$

Lemma 3. *Let $B_j, j = 1, 2, \ldots, s$ be nonempty sets of integers and $a_i, i = 1, 2, \ldots, t$ all the different elements in $\cup_{j=1}^{s} B_j$. Let c be the number of a_i satisfying $a_i \in B_j$ and $\#B_j \geq 2$ for some j. Then $t - c \leq s$ and for $s_1 \in [t - c, s] \cap \mathbb{N}$, we have $(t - s_1) \leq c \leq 2(t - s_1)$.*

Proof. B_j is called a single point set if $\#B_j = 1$, and a collision set if $\#B_j \geq 2$. Since $t - c$ is the number of a_i contained in all single point sets, there exist $t - c$ single point sets. So $t - c \leq s$. Since $t - c \leq s_1 \leq s$, we have $(t - s_1) \leq c$. Let k_1 be the number of collision sets. We have $k_1 + t - c = s$. So $c = k_1 + t - s$. Since every collision set contains at least two elements, $k_1 \leq \frac{1}{2}c$. So $c \leq \frac{1}{2}c + t - s$, which is $\frac{1}{2}c \leq t - s \leq t - s_1$. So $c \leq 2(t - s_1)$. $\square$

For $p \in \mathbb{Z}_{>0}$ and $\mathbf{u} \in \mathbb{N}^n$, let $\mathcal{C}_f(p, \mathbf{u})$ be the number of terms of f which collide in $f_{(p)}^{\mathbf{mod}}(x^\mathbf{u})$. We have

Lemma 4. *Let $p_\mathbf{u}, p_\mathbf{v} \in \mathbb{Z}_{>0}$ and $\mathbf{u}, \mathbf{v} \in \mathbb{Z}_{>0}^n$ such that*

$$\#[f_{(p_\mathbf{u})}^{\mathbf{mod}}(x^\mathbf{u})] \geq \#[f_{(p_\mathbf{v})}^{\mathbf{mod}}(x^\mathbf{v})].$$

Then $\mathcal{C}_f(p_\mathbf{u}, \mathbf{u}) \leq 2\mathcal{C}_f(p_\mathbf{v}, \mathbf{v})$.

Proof. Assume $\#[f_{(p_\mathbf{u})}^{\mathbf{mod}}(x^\mathbf{u})] = k_0, \#[f_{(p_\mathbf{v})}^{\mathbf{mod}}(x^\mathbf{v})] = k$ and $f_{(p_\mathbf{u})}^{\mathbf{mod}}(x^\mathbf{u}) = a_1 x^{d_1} + \cdots + a_{k_0} x^{d_{k_0}}, d_i \neq d_j$, when $i \neq j$. Let $f = f_1 + \cdots + f_{k_0} + g$, where $(f_i)_{(p_\mathbf{u})}^{\mathbf{mod}}(x^\mathbf{u}) = a_i x^{d_i}, i = 1, \ldots, k_0$ and $g_{(p_\mathbf{u})}^{\mathbf{mod}}(x^\mathbf{u}) = 0$. Let $B_i, i = 1, \ldots, k_0$ be the set of terms in f_i and B_0 be the set of terms in g. By Lemma 3, we have $(t - k_0) < \mathcal{C}_f(p_\mathbf{u}, \mathbf{u}) \leq 2(t - k_0)$. By the same reason, we have $(t - k) \leq \mathcal{C}_f(p_\mathbf{v}, \mathbf{v}) \leq 2(t - k)$. Now $\mathcal{C}_f(p_\mathbf{u}, \mathbf{u}) \leq 2(t - k_0) \leq 2(t - k) \leq 2\mathcal{C}_f(p_\mathbf{v}, \mathbf{v})$. $\square$

The following theorem gives a method to compute an "ok" substitution, which is similar to [27, Prop.5.4.2] and has two differences. (1). For each substitution, we choose a random prime, while in [27], the prime is fixed. (2). We choose the substitution **s** such that $\#f_{(p)}^{\mathrm{mod}}(x^{\mathbf{s}})$ has the maximal number of terms, while in [27], they choose the one such that $\#f(x^{\mathbf{s}})$ has the maximal number of terms.

Theorem 1. *Let $f(\mathbb{X}) \in \mathcal{R}[\mathbb{X}]$, $T \geq \#f$, $D \geq \deg f$, $N = \max\{31\lceil(T-1)\log_2 D\rceil, 1\}$ and $p_1, \ldots, p_N$ be N different primes which satisfy $p_i \geq 32(T-1)$. Let $\mu \in (0,1)$ and $\ell \geq \lceil 32\ln(T\mu^{-1})\rceil$. For $i = 1, \ldots, \ell$, we randomly choose a prime p_{α_i} in $\{p_1, \ldots, p_N\}$ and then choose $\mathbf{s}_i \in \mathbb{Z}_{p_{\alpha_i}}^n$ uniformly at random. Let $(p, \mathbf{s})$ be the vector in $\{(p_{\alpha_1}, \mathbf{s}_1), \ldots, (p_{\alpha_\ell}, \mathbf{s}_\ell)\}$ such that $\#[f_{(p)}^{\mathrm{mod}}(x^{\mathbf{s}})] = \max_{i=1}^{\ell} \#[f_{(p_{\alpha_i})}^{\mathrm{mod}}(x^{\mathbf{s}_i})]$. Then at least $\frac{5}{8}\#f$ terms of f do not collide in $f_{(p)}^{\mathrm{mod}}(x^{\mathbf{s}})$ with probability at least $1 - \mu$.*

Proof. First we consider a fixed term $c_i m_i$ and let $f_j(x) = f_{(p_{\alpha_j})}^{\mathrm{mod}}(x^{\mathbf{s}_j})$. By Lemma 2, the probability of $c_i m_i$ colliding in $f_j(x)$ is no more than $\frac{1}{16}$. Define $X_j = 1$ to be the event that $c_i m_i$ collides in $f_j(x)$ and $X_j = 0$ to be the event that $c_i m_i$ does not collide in $f_j(x)$ for some j. Define $X = \sum_{j=1}^{\ell} X_j$, then $\mathbf{E}[X] \leq \frac{1}{16}\ell$. By Hoeffding's inequality, we have $\Pr(X > \frac{1}{16}\ell + \varepsilon) \leq \Pr(X > \mathbf{E}[X] + \varepsilon) \leq e^{-2\varepsilon^2/\ell}$. Let $\varepsilon = \frac{1}{8}\ell$, then $\Pr(X > \frac{3}{16}\ell) \leq e^{-\ell/32} \leq e^{\ln(\mu/T)} = \frac{\mu}{T}$. So at probability $\leq 1 - \mu$, for all term $c_i m_i$ of f, $c_i m_i$ collides in at most $\frac{3}{16}\ell$ of $f_j(x), j = 1, \ldots, \ell$. In other words, with probability $\geq 1 - \mu$, at leat $\frac{13}{16}\ell\#f$ terms in $f_{(p_{\alpha_j})}^{\mathrm{mod}}(x^{\mathbf{s}_j}), j = 1, \ldots, \ell$ do not collide. We claim that at least one of $f_j(x)$ has at least $\frac{13}{16}\#f$ non-colliding terms. We prove the claim by contradiction. Assume that each $f_j(x)$ has $< \frac{13}{16}\#f$ non-colliding terms. Then there exist $< \frac{13}{16}\#f\ell$ non-colliding terms in $f_j(x), j = 1, \ldots, \ell$, which contradicts to the fact that these $f_j(x)$ have $\geq \frac{13}{16}\ell\#f$ non-colliding terms. So there must exist one $(p_{\alpha_j}, \mathbf{s}_j)$ for which at most $\frac{3}{16}$ of the terms of f collide. By Lemma 4, the polynomial with maximum $\#f_j(x)$ has at least $\frac{5}{8}\#f$ non-colliding terms. $\square$

2.2 Recover Non-colliding Terms

For $\mathbf{s} = (s_1, s_2, \ldots, s_n) \in \mathbb{N}^n$ and $p \in \mathbb{N}_{>0}$, let

$$f(x^{\mathbf{s}+p\mathbf{I}_k}) = f(x^{s_1}, \ldots, x^{s_k+p}, \ldots, x^{s_n}) \tag{6}$$

to be the univariate polynomial obtained with the substitution: $x_i = x^{s_i}, i = 1, 2, \ldots, n, i \neq k, x_k = x^{s_k+p}$, where $\mathbf{I}_k \in \mathbb{Z}_{\geq 0}^n$ is the k-th unit vector.

In this section, we show how to recover the non-colliding terms of $f \in \mathcal{R}[\mathbb{X}]$ from $f_{(p)}^{\mathrm{mod}}(x^{\mathbf{s}})$, $f(x^{\mathbf{s}})$, and $f(x^{\mathbf{s}+p\mathbf{I}_k})$. Let

$$f_{(p)}^{\mathrm{mod}}(x^{\mathbf{s}}) = a_1 x^{d_1} + \cdots + a_r x^{d_r} \tag{7}$$

Since $f_{(p)}^{\mathbf{mod}}(x^{\mathbf{s}}) = f_{(p)}^{\mathbf{mod}}(x^{\mathbf{s}+p\mathbf{I}_k})$, for $k = 1, 2, \ldots, n$, we can write

$$f(x^{\mathbf{s}}) = f_1 + f_2 + \cdots + f_r + g \tag{8}$$
$$f(x^{\mathbf{s}+p\mathbf{I}_k}) = f_{k,1} + f_{k,2} + \cdots + f_{k,r} + g_k$$

where $f_i \bmod (x^p - 1) = f_{k,i} \bmod (x^p - 1) = a_i x^{d_i}$, $g \bmod (x^p - 1) = g_k$ $\bmod (x^p - 1) = 0$. We define the following key notation

$$TS_{(f,p,\mathbf{s},D)} = \{a_i x_1^{e_{i,1}} \cdots x_n^{e_{i,n}} | a_i \text{ is from (7), and}$$

$$\text{T1}: f_i = a_i x^{u_i}, f_{k,i} = a_i x^{b_{k,i}}, k = 1, 2, \ldots, n. \tag{9}$$

$$\text{T2}: e_{i,k} = \frac{b_{k,i} - u_i}{p} \in \mathbb{N}, k = 1, 2, \ldots, n.$$

$$\text{T3}: u_i = e_{i,1}s_1 + e_{i,2}s_2 + \cdots + e_{i,n}s_n.$$

$$\text{T4}: \sum_{j=1}^{n} e_{i,j} \leq D.\}$$

Lemma 5. *Let* $f = \sum_{i=1}^{t} c_i m_i \in \mathcal{R}[\mathbb{X}]$ *and* $D \geq \deg f$. *If* cm *does not collide in* $f_{(p)}^{\mathbf{mod}}(x^{\mathbf{s}})$, *then* $cm \in TS_{(f,p,\mathbf{s},D)}$.

Proof. It suffices to show that cm satisfies the conditions of the definition of $TS_{(f,p,\mathbf{s},D)}$. Assume $m = x_1^{e_1} \cdots x_n^{e_n}$. Since cm is not a collision in $f_{(p)}^{\mathbf{mod}}(x^{\mathbf{s}})$, without loss of generality, assume $cm(x^{\mathbf{s}}) \bmod (x^p - 1) = a_1 x^{d_1}$, where $a_1 x^{d_1}$ is defined in (7). It is easy to show that cm is also not a collision in $f(x^{\mathbf{s}})$ and in $f(x^{\mathbf{s}+p\mathbf{I}_k})$. Hence, $f_1 = a_1 x^{u_1}$ for $u_1 = \sum_{i=1}^{n} e_i s_i$; $b_{k,1} = u_1 + pe_k$. Clearly, T1, T2 and T3 are correct. Since $\deg m = \sum_{j=1}^{n} e_j \leq D$, T4 is correct. □

Now we give the algorithm to compute $TS_{(f,p,\mathbf{s},D)}$.

Algorithm 2 (TSTerms).

Input:
- Univariate polynomials $f_{(p)}^{\mathbf{mod}}(x^{\mathbf{s}}), f(x^{\mathbf{s}}), f(x^{\mathbf{s}+p\mathbf{I}_k})$, where $k = 1, 2, \ldots, n$.
- A prime p.
- A vector $\mathbf{s} = (s_1, s_2, \ldots, s_n) \in \mathbb{Z}_{\geq 0}^n$.
- Degree bound $D \geq \deg f$.

Output: $TS_{(f,p,\mathbf{s},D)}$.

Step 1: Write $f_{(p)}^{\mathbf{mod}}(x^{\mathbf{s}}), f(x^{\mathbf{s}})$, and $f(x^{\mathbf{s}+p\mathbf{I}_k})$ in the following form

$$f_{(p)}^{\mathbf{mod}}(x^{\mathbf{s}}) = a_1 x^{d_1} + a_2 x^{d_2} + \cdots + a_r x^{d_r}$$
$$f(x^{\mathbf{s}}) = a_1 x^{u_1} + \cdots + a_\gamma x^{u_\gamma} + f_1$$
$$f(x^{\mathbf{s}+p\mathbf{I}_k}) = a_1 x^{b_{k,1}} + \cdots + a_\gamma x^{b_{k,\gamma}} + f_{k,2}$$

where $i = 1, 2, \ldots, \gamma$, $k = 1, \ldots, n$, $a_i x^{u_i}, a_i x^{b_{k,i}}$ are all the terms satisfying: $x^{b_{k,i}}$ is the unique term in $f(x^{\mathbf{s}+p\mathbf{I}_k})$ such that $\mathbf{mod}(b_{k,i}, p) = d_i$ and x^{u_i} is the unique term in $f(x^{\mathbf{s}})$ such that $\mathbf{mod}(u_i, p) = d_i$.

Step 2: Let $S = \{\}$.

Step 3: For $i = 1, 2, \ldots, \gamma$

    a: For $k = 1, 2, \ldots, n$, Let $e_{i,k} = \frac{b_{k,i} - u_i}{p}$. If $e_{i,k} \notin \mathbb{N}$, then break.

    b: If $u_i \neq e_{i,1}s_1 + e_{i,2}s_2 + \cdots + e_{i,n}s_n$, then break;

    c: If $\sum_{j=1}^{n} e_{i,j} > D$, then break;

    d: Let $S = S \bigcup \{a_i x_1^{e_{i,1}} \cdots x_n^{e_{i,n}}\}$.

Step 4: Return S.

Lemma 6. *Algorithm 2 needs $O(nT)$ ring operations in $\mathcal{R}$ and $O^{\sim}(nT \cdot \log(s_{\max}D + pD))$ bit operations, where $s_{\max} = \max\{s_1, s_2, \ldots, s_n\}$.*

Proof. In Step 1, in order to match the terms of $f_{(p)}^{\mathbf{mod}}(x^{\mathbf{s}})$, $f(x^{\mathbf{s}})$, and $f(x^{\mathbf{s}+p\mathbf{I}_k})$, it needs $O^{\sim}(nT \log(s_{\max}D + pD))$ bit operations and $O(nT)$ ring operations in $\mathcal{R}$. In Step 3, **a**, **b** and **c** need $O(nT)$ arithmetic operations in $\mathbb{Z}$. Since the data is $O(s_{\max}D + pD)$, the complexity is $O(nT \log(s_{\max}D + pD))$ bit operations. $\square$

2.3 Algorithms

We will give the reduction algorithm for $f \in \mathcal{R}[\mathbb{X}]$, which works as follows. We first find an "ok" random Kronecker substitution $\mathbf{s}$ based on Theorem 1, then obtain half of the terms of f by applying Algorithm 2, and finally repeat the procedure for at most $\log(\#f)$ times to find f. We assume an interpolation algorithm for univariate polynomials is given in advance.

    We first give an algorithm to obtain the polynomials $g(x^{\mathbf{s}+p\mathbf{I}_k})$, $k = 1, \ldots, n$ from $g(\mathbb{X})$.

Algorithm 3 (PolySubs).

Input: • A polynomial $g \in \mathcal{R}[\mathbb{X}]$.

    • A vector $\mathbf{s} = (s_1, s_2, \ldots, s_n) \in \mathbb{Z}_{\geq 0}^n$.

    • A prime p.

Output: $g(x^{\mathbf{s}+p\mathbf{I}_k})$, $k = 1, 2, \ldots, n$.

Step 1: Assume $g = c_1 m_1 + \cdots + c_t m_t$, where $m_i = x_1^{e_{i,1}} \cdots x_n^{e_{i,n}}$, $i = 1, \ldots, t$.

Step 2: For $i = 1, 2, \ldots, n$, let $h_i = 0$;

Step 3: For $i = 1, 2, \ldots, t$ do

    a: Let $d = 0$.

    b: For $k = 1, 2, \ldots, n$, let $d = d + e_{i,k}s_k$.

    c: For $k = 1, 2, \ldots, n$, let $h_k := h_k + c_i x^{d + e_{i,k}p}$.

Step 4: Return h_i, $i = 1, 2, \ldots, n$;

Lemma 7. *The complexity of Algorithm 3 is $O^{\sim}(nt \log(p + s_{\max}) + nt \log \deg f)$ bit operations and $O(nt)$ ring operations in $\mathcal{R}$, where $s_{\max} = \max\{s_1, \ldots, s_n\}$.*

Proof. In **b** of Step 3, d is the degree of $m_i(x^s)$. In **c**, since $\deg(m_i(x^{s+pI_k})) = \deg(m_i(x^s)) + pe_{i,k}$, h_k is $f(x^{s+pI_k})$ after finishing Step 3. So the correctness is proved. Now we analyse the complexity. In **b** of Step 3, it needs $O(nt)$ arithmetic operations in $\mathbb{Z}$. Since $\deg(m_i(x^s))$ is $O(s_{\max} \deg f)$, the bit operation is $O(nt \log(s_{\max} \deg f))$. In **c**, it needs $O^\sim(nt \log(p \deg f + s_{\max} \deg f))$ bit operations and at most $O(nt)$ arithmetic operations in $\mathcal{R}$. $\qquad\square$

Now we give an algorithm which interpolates at least half of the terms.

Algorithm 4 (HalfPoly).

Input: • A black-box procedure $\mathcal{B}_f$ that computes $f \in \mathcal{R}[x_1, \ldots, x_n]$.
 • An approximation polynomial $f^* \in \mathcal{R}[x_1, \ldots, x_n]$ to f.
 • Term bounds $T \geq \max(\#f, \#f_1), T_1 \geq \#(f - f^*)$ and $T \geq T_1$.
 • Degree bound $D \geq \max(\deg f, \deg f^*)$.
 • A tolerance ν such that $0 < \nu < 1$.
Output: With probability $\geq 1 - \nu$, return a polynomial h such that $\#(f - f^* - h) \leq \lfloor \frac{T_1}{2} \rfloor$.

Step 1: Let $\ell = \lceil 32 \ln(T_1 \nu^{-1}) \rceil$, $N = \max\{31 \lfloor (T_1 - 1) \log_2 D \rfloor, 1\}$. Find the first N primes $\{p_1, p_2, \ldots, p_N\}$ such that $p_i \geq 32(T_1 - 1)$.
Step 2: For $i = 1, \ldots, \ell$, randomly choose p_{α_i} in $\{p_1, \ldots, p_N\}$, then choose $s_i \in \mathbb{Z}_{p_{\alpha_i}}^n$ uniformly at random. Deleting the repeated numbers, we still denote these vectors as $(p_{\alpha_1}, s_1), (p_{\alpha_2}, s_2), \ldots, (p_{\alpha_\ell}, s_\ell)$.
Step 3: For $i = 1, 2, \ldots, \ell$, compute $f(x^{s_i})$ from $\mathcal{B}_f$ by a given univariate interpolation algorithm with degree bound $\|s_i\|_\infty D$ and term bound T. Let $f_i = f(x^{s_i}) - f^*(x^{s_i})$ and $f_i^{\mathrm{mod}} = f_i \bmod (x^{p_{\alpha_i}} - 1)$.
Step 4: Find j_0 such that $\#f_{j_0}^{\mathrm{mod}} = \max\{\#f_i^{\mathrm{mod}} | i = 1, 2, \ldots, \ell\}$.
 If $\#f_{j_0}^{\mathrm{mod}} \geq T_1$, return failure.
Step 5: For $k = 1, 2, \ldots, n$, find $f(x^{s_{j_0} + p_{\alpha_{j_0}} I_k})$ from $\mathcal{B}_f$ by the given univariate interpolation algorithm with degree bound $\|s_{j_0} + p_{\alpha_{j_0}} I_k\|_\infty D$ and term bound T. Let $\{f_1^*, f_2^*, \ldots, f_n^*\} = \mathbf{PolySubs}(f^*, s_{j_0}, p_{\alpha_{j_0}})$.
 Let $g_k = f(x^{s_{j_0} + p_{\alpha_{j_0}} I_k}) - f_k^*$.
Step 6: Let TS $= \mathbf{TSTerms}(f_{j_0}^{\mathrm{mod}}, f_{j_0}, g_1, g_2, \ldots, g_n, p_{\alpha_{j_0}}, s_{j_0}, D)$.
Step 7: Return $h = \sum_{b \in \mathrm{TS}} b$.

Lemma 8. *Algorithm 4 computes h such that $\#(f - f^* - h) \leq \lfloor \frac{T_1}{2} \rfloor$ with probability $\geq 1 - \nu$. The algorithm needs*

 • $O(n + \log T + \log \frac{1}{\nu})$ *interpolations of univariate polynomials of degree $O^\sim(TD)$ and sparseness $\leq T$.*
 • $O^\sim(nT \log \frac{1}{\nu})$ *additional ring operations and $O^\sim(nT \log D \log \frac{1}{\nu})$ additional bit operations.*

Proof. We first show that Algorithm 4 returns the polynomial h such that $\#(f - f^* - h) \leq \lfloor \frac{T_1}{2} \rfloor$ with probability $1 - \nu$. In Step 1, Step 2 and Step 4, by Theorem 1, with probability $1 - \nu$, $\mathcal{C}_{f-f^*}(p_{\alpha_{j_0}}, s_{j_0}) \leq \lfloor \frac{3}{8} T_1 \rfloor$. If j_0 satisfies $\mathcal{C}_{f-f^*}(p_{\alpha_{j_0}}, s_{j_0}) \leq$

$\lfloor \frac{3}{8}T_1 \rfloor$, then by Lemma 5, there are at most $\lfloor \frac{3}{8}T_1 \rfloor$ terms in $f - f^*$ but not in h. Since the terms of h which are not in $f - f^*$ come from at least three terms in $f - f^*$, then there are at most $\frac{1}{3}\lfloor \frac{3}{8}T_1 \rfloor$ terms of h not in $f - f^*$. So $\#(f - f^* - h) \leq \lfloor \frac{3}{8}T_1 \rfloor + \frac{1}{3}\lfloor \frac{3}{8}T_1 \rfloor \leq \frac{1}{2}T_1$. So we have $\#(f - f^* - h) \leq \lfloor \frac{1}{2}T_1 \rfloor$. The first part is proved.

Now we analyse the complexity. In Step 1, use the sieve of Eratosthenes [29, Them.18.10], the cost of finding the N primes bigger than $32(T_1 - 1)$ is $O^\sim(T_1 \log D)$ bit operations.

In Step 2, since probabilistic machines flip coins to decide binary digits, each of these random choices can be simulated with a machine with complexity $O(n \log(T_1 \log D))$. So the complexity of Step 2 is $O(n \log^2 T_1 + n \log T_1 \log \frac{1}{\nu} + n \log T_1 \log \log D + n \log \log D \log \frac{1}{\nu})$ bit operations.

In Step 3, since p_{α_i} is $O^\sim(T_1 \log D)$, the degree of $f(x^{\mathbf{s}_i})$ is $O^\sim(\|\mathbf{s}_i\|_\infty D) = O^\sim(T_1 D)$. So in Step 3, we query $O(\log T_1 + \log \frac{1}{\nu})$ polynomials of degree $O^\sim(T_1 D)$. In order to obtain $f^*(x^{\mathbf{s}_i})$, it needs $O(\ell n T)$ ring operations and $O^\sim(\ell n T \log D)$ bit operations. In order to obtain f_i, it needs $O(\ell T)$ ring operations in $\mathcal{R}$ and $O^\sim(\ell T \log D)$ bit operations. In order to obtain the $f_i^{\mathbf{mod}}$, it needs $O^\sim(\ell T_1 \log D)$ bit operations and $O(\ell T_1)$ ring operations.

So it still needs $O^\sim(n T \log \frac{1}{\nu} + T_1 \log \frac{1}{\nu})$ ring operations and $O^\sim(n T \log D \log \frac{1}{\nu} + T_1 \log D \log \frac{1}{\nu})$ bit operations.

In Step 4, we find the integer j_0. Since $\#f_i^{\mathbf{mod}} \leq T_1$, it needs at most $O^\sim(T_1 \log \frac{T_1}{\nu})$ bit operations to compute all $\#f_i^{\mathbf{mod}}, i = 1, 2, \ldots, \ell$. Find j_0 needs $O^\sim(\ell \log T_1)$ bit operations. So the bit complexity of Step 4 is $O^\sim(T_1 \log \frac{1}{\nu})$.

In Step 5, since the degree of $f(x^{\mathbf{s}_{j_0} + p_{\alpha_{j_0}} \mathbf{I}_k})$ is $O^\sim(T_1 D)$, it queries $O(n)$ polynomials of degrees $O^\sim(T_1 D)$. By Lemma 7, it needs $O^\sim(n T \log D)$ bit operations and $O(n T)$ arithmetic operations in $\mathcal{R}$ to obtain $\{f_1^*, f_2^*, \ldots, f_n^*\}$.

In Step 6, by Lemma 6, the complexity is $O(n T_1)$ ring operations in $\mathcal{R}$ and $O^\sim(n T_1 \log D)$ bit operations. Since $T \geq T_1$, the lemma is proved. $\square$

We now give the complete interpolation algorithm.

Algorithm 5 (MulPolySI).

Input: A Black-box procedure $\mathcal{B}_f$ that computes $f \in \mathcal{R}[\mathbb{X}], T \geq \#f, D \geq \deg f$, and $\mu \in (0, 1)$.
Output: Return f with probability $\geq 1 - \mu$, or failure.

Step 1: Let $h = 0, T_1 = T, \nu = \frac{\mu}{\lceil \log_2 T \rceil + 1}$.
Step 2: While $T_1 > 0$ do
   a: Let $g = \mathbf{HalfPoly}(\mathcal{B}_f, h, T, T_1, D, \nu)$.
   b: Let $h = h + g, T_1 = \lfloor \frac{T_1}{2} \rfloor$.
Step 3: Return h.

Theorem 6. *Algorithm 5 computes f with probability $\geq 1 - \mu$. The algorithm needs*

- $O(n \log T + \log^2 T + \log T \log \frac{1}{\mu})$ interpolations of univariate polynomials with degree $O^\sim(TD)$ and sparseness $\leq T$.
- $O^\sim(nT \log \frac{1}{\mu})$ additional ring operations and $O^\sim(nT \log D \log \frac{1}{\mu})$ additional bit operations.

Proof. In **a** of Step 2, since $\#(f - h) \leq T_1$, by Lemma 8, $\#(f - h - g) \leq \lfloor \frac{T_1}{2} \rfloor$ with probability $\geq 1 - \nu$. Then, Step 2 will run at most $k = \lceil \log_2 T \rceil + 1$ times and return the correct f with probability $\geq (1 - \nu)^k \geq 1 - \mu$. The first part is proved.

Now we analyse the complexity. It is easy to see that the complexity is dominated by Step 2. In Step 2, we call at most $O(\log T)$ times Algorithm 4. Since the terms and degrees of $f - h$ are respectively bounded by T and D, by Lemma 8, it needs $O(n \log T + \log^2 T + \log T \log \frac{1}{\nu})$ queries of degree $O^\sim(TD)$, $O^\sim(nT \log \frac{1}{\nu})$ additional ring operations and $O^\sim(nT \log D \log \frac{1}{\nu})$ additional bit operations.

Since $\nu = \frac{\mu}{\lceil \log_2 T \rceil + 1}$, we have proved the theorem. $\qquad \square$

Corollary 1. *Set $\mu = 1/4$. Then Algorithm 5 computes f with probability at lest $\frac{3}{4}$. The algorithm needs*

- $O(n \log T + \log^2 T)$ *queries of univariate polynomials with degree $O^\sim(TD)$ and sparseness $\leq T$.*
- $O^\sim(nT)$ *additional ring operations and $O^\sim(nT \log D)$ additional bit operations.*

3   Sparse Interpolation over Finite Fields

In this section, we give a sparse interpolation algorithm for black-box multivariate polynomials over general finite fields. We first give a univariate Ben-Or and Tiwari algorithm over finite fields and then combine with Algorithm 5 to give a multivariate interpolation algorithm.

3.1   The Ben-Or and Tiwari Sparse Interpolation Algorithm

Following [3], we give a brief introduction to the multivariate Ben-Or and Tiwari sparse interpolation algorithm over $\mathbb{C}$.

Let $f(x_1, \ldots, x_n) = c_1 m_1 + \cdots + c_t m_t \in \mathbb{C}[\mathbb{X}]$ be the polynomial to be interpolated, where $m_i = x_1^{e_{i,1}} \ldots x_n^{e_{i,n}}$ are distinct monomials, c_i are non-zero coefficients, and $t = \#f$ is the number of terms in f. We assume that f is a *black-box*, which means, for $\forall (q_1, \ldots, q_n) \in \mathbb{C}^n$, we can obtain the value $f(q_1, \ldots, q_n)$. Note that c_i, m_i, t are not known. In order to determine f uniquely, the algorithm needs as input an upper bound $\tau + 1 \geq t$ on the number of terms in f.

The algorithm proceeds in two stages. The monomials m_i are determined first using an auxiliary polynomial $\zeta(z)$. Once the m_i are known, the coefficients c_i can be obtained easily.

We first determine m_i. Let $v_i = p_1^{e_{i,1}} \dots p_n^{e_{i,n}}$ denote the value of the monomial m_i at $(p_1, \dots, p_n)$, where p_i is the i-th prime number. Clearly, different monomials evaluate to different values under this evaluation. Let $a_0, a_1, \dots, a_{2\tau+1}$ be the values of f at the $2(\tau + 1)$ points $\mathbf{p}_i = (p_1^i, \dots, p_n^i)$, $i = 0, 1, \dots, 2\tau + 1$, that is, $a_i = \sum_{j=1}^{t} c_j v_j^i$.

The auxiliary polynomial $\zeta(z)$ is defined as follows.

$$\zeta(z) = \prod_{i=1}^{t}(z - v_i) = z^t + \zeta_{t-1}z^{t-1} + \dots + \zeta_1 z + \zeta_0. \tag{10}$$

Consider the sum $\sum_{i=1}^{t} c_i v_i^j \zeta(v_i) = \sum_{k=0}^{t-1} \zeta_k(c_1 v_1^{k+j} + c_2 v_2^{k+j} + \dots + c_t v_t^{k+j}) + (c_1 v_1^{t+j} + c_2 v_2^{t+j} + \dots + c_t v_t^{t+j}) = a_j \zeta_0 + a_{j+1} \zeta_1 + \dots + a_{j+t-1}\zeta_{t-1} + a_{j+t}$ for $j = 0, \dots, t-1$. Since $\zeta(v_i) = 0$, for $0 \le j \le t - 1$, we have

$$a_j \zeta_0 + a_{j+1}\zeta_1 + \dots + a_{j+t-1}\zeta_{t-1} + a_{j+t} = 0. \tag{11}$$

This is a Toeplitz system $T_{t-1,t-1}\hat{\zeta}_{t-1} = \hat{t}_{2t-1,t-1}$ where

$$T_{u,v} = \begin{pmatrix} a_u & a_{u+1} & \cdots & a_{u+v} \\ a_{u-1} & a_u & \cdots & a_{u+v-1} \\ \vdots & \vdots & \ddots & \vdots \\ a_{u-v} & a_{u-v+1} & \cdots & a_u \end{pmatrix}$$

$\hat{\zeta}_v = (\zeta_0, \zeta_1, \dots, \zeta_v)^\tau$, $\hat{t}_{u,v} = -(a_u, a_{u-1}, \dots, a_{u-v})^\tau$. This system is non-singular as can be seen from the factorization.

$$T_{t-1,t-1} = \begin{pmatrix} 1 & 1 & \cdots & 1 \\ v_1 & v_2 & \cdots & v_t \\ \vdots & \vdots & \ddots & \vdots \\ v_1^{t-1} & v_2^{t-1} & \cdots & v_t^{t-1} \end{pmatrix} \begin{pmatrix} c_1 & 0 & \cdots & 0 \\ 0 & c_2 & \cdots & 0 \\ \vdots & \vdots & \ddots & \vdots \\ 0 & 0 & \cdots & c_t \end{pmatrix} \begin{pmatrix} 1 & v_1 & \cdots & v_1^{t-1} \\ 1 & v_2 & \cdots & v_2^{t-1} \\ \vdots & \vdots & \ddots & \vdots \\ 1 & v_t & \cdots & v_t^{t-1} \end{pmatrix} \tag{12}$$

Since the v_i are distinct, the two Vandermonde matrices are nonsingular and as no c_i is zero, the diagonal matrix is nonsingular, too. If the input value of the upper bound $\tau + 1$ is greater than t, then the coefficients c_k, for $k > t$, can be regarded as zero and the resulting system $T_{\tau,\tau}$ would be singular.

Lemma 9. *[3] If t is the exact number of terms in f, then*

a) $T_{i,t-1}$ *is non-singular for all $i \ge t - 1$.*
b) $T_{i,t+j}$ *is singular for all $i \ge t - 1, j \ge 0$.*

By Lemma 9, when considering $2\tau + 2$ values $a_0, \dots, a_{2\tau+1}$ of f, the coefficients of $\zeta(z)$ can be uniquely recovered from the system $T_{\tau,\tau}\hat{\zeta}_\tau = \hat{t}_{2\tau+1,\tau}$. By finding the roots $v_i = p_1^{e_{i,1}} \dots p_n^{e_{i,n}}$ of $\zeta(z)$, the monomials m_i can be recovered.

By choosing the first t evaluations $a_0, \dots, a_{t-1}$ of f, we obtain the following transposed Vandermonde system $A\hat{c} = \hat{a}$ for the coefficients of f, where

$$A = \begin{pmatrix} 1 & 1 & \cdots & 1 \\ v_1 & v_2 & \cdots & v_t \\ \vdots & \vdots & \ddots & \vdots \\ v_1^{t-1} & v_2^{t-1} & \cdots & v_t^{t-1} \end{pmatrix}, \hat{c} = \begin{pmatrix} c_1 \\ c_2 \\ \vdots \\ c_t \end{pmatrix}, \hat{a} = \begin{pmatrix} a_0 \\ a_1 \\ \vdots \\ a_{t-1} \end{pmatrix} \tag{13}$$

The deterministic Ben-Or and Tiwari's algorithm over $\mathbb{Z}$ needs $O(T)$ evaluations of f plus $O(nT^2d)$ $\mathbb{Z}$-operations and the height of the data is $O(Td)$ [3], where $d = \deg f$.

If the coefficients of the polynomials are from a finite field, then it is difficult to find the exponents from $v_i = p_1^{e_{i,1}} \ldots p_n^{e_{i,n}}$, which is a multi-variate discrete logarithm problem.

3.2 Univariate Ben-Or and Tiwari Algorithm over Finite Field

In this section, we give a modified univariate Ben-Or and Tiwari algorithm over the finite field $\mathbb{F}_q$. Assume $f(x) = \sum_{i=1}^t c_i m_i \in \mathbb{F}_q[x], D \geq \deg f$. Since $f(x)$ is univariate, $\#f \leq D$. We consider two cases: $q > D$ or $q \leq D$.

First, consider the case $q > D$. Let ω be a primitive element of $\mathbb{F}_q$. Assume $m_i = x^{d_i}$ and denote $v_i = \omega^{d_i}$. Let $a_i = \sum_{j=1}^t c_j v_j^i, i = 0, 1, \ldots, 2\tau + 1$. $T_{t-1,t-1}$ still can be factored as (12). Since ω is a primitive element of $\mathbb{F}_q$ and $q > D$, $v_i \neq v_j$ when $i \neq j$. So the two Vandermonde matrices in (12) are nonsingular and Lemma 9 is still correct. Now we can give the algorithm.

Algorithm 7 (UniBoTFq).

Input: A black-box procedure $\mathcal{B}_f$ to compute $f(x) \in \mathbb{F}_q[x]$, $\tau + 1 \geq \#f$, and $D \geq \deg f$.
Output: The polynomial $f = \sum_{i=1}^t c_i m_i$.

Step 1: Let ω be a primitive element of $\mathbb{F}_q$. Evaluate f at the $2(\tau + 1)$ points $\omega^i, i = 0, \ldots, 2\tau + 1$. Let $a_i, i = 0, \ldots, 2\tau + 1$ be the corresponding values.
Step 2: Solve the Toeplitz system $T_{\tau,\tau}\hat{\zeta}_\tau = \hat{t}_{2\tau+1,\tau}$ (or the largest non-singular subsystem $T_{j,2\tau-j}\hat{\zeta}_{2\tau-j} = \hat{t}_{2\tau+1,2\tau-j}$ of $T_{\tau,\tau}$, where j is the smallest positive integer that makes $T_{j,2\tau-j}$ non-singular) to obtain the polynomial $\zeta(z) = \sum_{i=0}^t \zeta_i z^i$.
Step 3: Find the monomial set M of f. $M = \emptyset$. For $i = 0, 1, \ldots, D$, compute ω^i and if $\zeta(\omega^i) = 0$ then let $M = \{x^i\} \cup M$.
Step 4: Find the coefficients c_i by solving the transposed Vandermonde system $A\hat{c} = \hat{a}$ in (13).

Lemma 10. *If $q > D$, Algorithm 7 is correct and it needs $2(\tau + 1)$ evaluations of f plus $O^{\sim}(D \log q)$ bit operations.*

Proof. The correctness comes from Lemma 9. Now we analyse the complexity. Due to the fast integer and polynomial multiplication algorithms [29], one can perform an arithmetic operation in $\mathbb{F}_q$ in $O^{\sim}(\log q)$ bit operations. In Step 1, it needs $O(\tau \log q)$ bit operations to obtain ω^i, $i = 1, 2, \ldots, 2\tau + 1$. In Step 2, it needs $O(M(\tau) \log \tau \log q)$ bit operations, where $M(\tau) = \tau \log(\tau) \log \log(\tau)$ [3].

In Step 3, computing $\omega^i, i = 0, 1, \ldots, D$ needs $O(D \log q)$ bit operations. Then we evaluate $\zeta(\omega^i), i = 0, 1, \ldots, D$, by fast multi-point evaluation method [29, Them.10.6], which needs $O(\frac{D}{T} M(T) \log T \log q) = O^\sim(D \log T \log q)$ bit operations, where $T = \tau + 1$.

In Step 4, it needs $O(M(t) \log t \log q)$ bit operations [3]. So the complexity of the total algorithm is $O^\sim(D \log T \log q + T \log q) = O^\sim(D \log q)$ bit operations, since $\#f \leq D$. $\qquad\square$

Second, consider the case $q < D$. We need evaluate the polynomial in an extended field of $\mathbb{F}_q$. We extends $\mathbb{F}_q$ into $\mathbb{F}_{q^m}$ such that $q^m \geq D + 1$, where $m = \lceil \frac{\log(D+1)}{\log q} \rceil$. Due to the fast integer and polynomial multiplication algorithms [29], one can perform an arithmetic operation in $\mathbb{F}_{q^m}$ in $O^\sim(m \log q) = O^\sim(\log D)$ bit operations, since $m = \lceil \frac{\log(D+1)}{\log q} \rceil$.

Now we can extend Algorithm 7 into the case $q \leq D$. The only change is to replace the primitive element of $\mathbb{F}_q$ by a primitive element of $\mathbb{F}_{q^m}$ in Step 1. Similar to the proof of Lemma 10, the complexity of the algorithm is $O^\sim(D \log Tm \log q + Tm \log q)$, which is $O^\sim(D \log T + T \log D) = O^\sim(D \log D) = O^\sim(D)$ bit operations. We thus have

Lemma 11. *If $q \leq D$, Algorithm 7 needs $2(\tau + 1)$ evaluations of f plus $O^\sim(D)$ bit operations.*

Following Lemmas 10 and 11, we have

Theorem 8. *Let f be a black-box univariate polynomial in $\mathbb{F}_q[x]$ with $T \geq \#f$ and $D \geq \deg f$. We can compute f with $O(T)$ evaluations of f plus $O^\sim(D \log q)$ bit operations.*

Remark 1. In Step 3 of Algorithm 7, we may follow the original Ben-Or and Tiwari algorithm to find the exponents. First, find the roots v_i of $\zeta(z) = 0$, which costs $O^\sim(t \log^2 q)$ bit operations [29] for $t = \#f$. Second, solve the discrete logarithm problem $v_i = \omega^{e_i}$ to find the exponents e_i, which costs $O^\sim(\sqrt{D} \log q)$ bit operations [30]. Therefore, the total complexity of the algorithm is $O^\sim(T \log^2 q + T\sqrt{D} \log q)$ bit operations plus $O(T)$ evaluations.

3.3   Multivariate Polynomial Interpolation over Finite Fields

Combing the reduction algorithm given in Sect. 2 and the univariate interpolation given in Sect. 3.2, we give a multivariate interpolation algorithm over finite fields.

Theorem 9. *Let $f \in \mathbb{F}_q[\mathbb{X}]$ be a black-box polynomial. Given $T \geq \#f$ and $D \geq \deg f$, with probability greater than $\frac{3}{4}$, one can find f using $O^\sim(nTD \log q)$ bit operations plus $O^\sim(nT)$ evaluations of f.*

Proof. We use the Algorithm 5 to compute f and use Algorithm 7 for univariate polynomial interpolation in Step 3 and Step 5 of Algorithm 4.

The complexity consists of two parts. By Corollary 1, we needs $O(n \log T + \log^2 T)$ queries of univariate polynomials with degree $O^{\sim}(TD)$ and sparseness $\leq T$. Then by Theorem 8, we need $O^{\sim}((n \log T + \log^2 T)T) = O^{\sim}(nT)$ evaluations of f and $O^{\sim}((n \log T + \log^2 T)(TD \log q)) = O^{\sim}(nTD \log q)$ bit operations to query these univariate polynomials.

By Corollary 1, we needs additional $O^{\sim}(nT)$ operations in $\mathbb{F}_q$ if $q > D$ (or in $\mathbb{F}_{q^m}$ if $q < D$ for $m = \lceil \frac{\log(D+1)}{\log q} \rceil$) and $O^{\sim}(nT \log D)$ bit operations. $O^{\sim}(nT)$ operations in $\mathbb{F}_q$ costs $O^{\sim}(nT \log q)$ bit operations. $O^{\sim}(nT)$ operations in $\mathbb{F}_{q^m}$ costs $O^{\sim}(nT \log D)$ bit operations. Therefore, the query of f is the dominating step and the bit complexity of the algorithm is $O^{\sim}(nTD \log q)$. □

Remark 2. If using the original Ben-Or and Tiwari algorithm mentioned in Remark 1 to interpolation the univariate polynomials, the total complexity of our algorithm is $O^{\sim}(nT^{1.5}\sqrt{D} \log q + nT \log^2 q)$ bit operations.

Remark 3. Let $f \in \mathbb{F}_q[\mathbb{X}]$ be a black-box polynomial and q a prime. If quantum algorithms can be used, the quantum complexity of finding f is $O^{\sim}(nT \log^2 q)$ plus $O^{\sim}(nT)$ evaluations of f and $O^{\sim}(nT)$ black-box evaluations for solving the discrete logarithm problem.

We need to change Step 3 of Algorithm 7 as follows:

(1) Find the roots v_i of $\zeta(z)$, which costs an expected $O^{\sim}(T \log^2 q)$ bit operations [29, p.368].
(2) Solve the discrete logarithm problem $v_i = \omega^{e_i} \mathbf{mod}\ q$ to find e_i using Shor's quantum algorithm, which costs $O^{\sim}(T \max\{\log^2 D, \log^2 q\})$ plus T black-box evaluations [31, p.238].

By Corollary 1, the total complexity is $O^{\sim}(nT \max\{\log^2 D, \log^2 q\})$.

4   Experimental Results

In this section, practical performances of the interpolation algorithm over finite fields given in Remark 2 will be reported. The algorithm uses Algorithm 5 to reduce multivariate interpolation to univariate interpolation and uses Algorithm 7 for univariate polynomial interpolation. In Algorithm 7, we use the Berlekamp-Massey algorithm to solve the Toeplitz systems, use the command *Roots* in Maple to find the roots, and use the command *mlog* in Maple to solve the discrete logarithm problem.

The data are collected on a desktop with Windows system, 3.60 GHz Core i7-4790 CPU, and 8GB RAM memory. The implementations in Maple can be found in

http://www.mmrc.iss.ac.cn/~xgao/software/rkron.zip

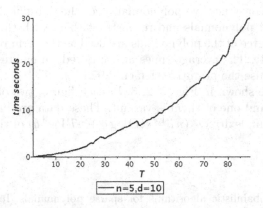

Fig. 1. Average times with varying T

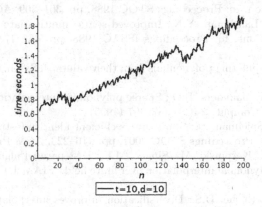

Fig. 2. Average times with varying n

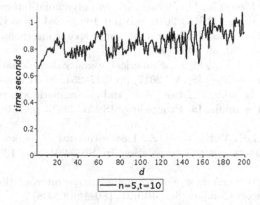

Fig. 3. Average times with varying d

We randomly construct five polynomials over the finite field $\mathbb{F}_q$, then regard them as black-box polynomials and reconstruct them with the algorithm. The actual size and degree of the polynomials are used as the term bound and degree bound, respectively. The average times are collected. In our testing, we fix $q = 30000000001$ and use the primitive element 29 of $\mathbb{F}_q$.

The results are shown in Figs. 1, 2, 3. In each figure, two of the parameters n, T, D are fixed and one of them is variant. These data are basically in accordance with the complexity $O^{\sim}(nT^{1.5}\sqrt{D}\log q + nT\log^2 q)$ of the algorithm.

References

1. Zippel, R.: Probabilistic algorithms for sparse polynomials. In: Ng, E.W. (ed.) Symbolic and Algebraic Computation. LNCS, vol. 72, pp. 216–226. Springer, Heidelberg (1979). https://doi.org/10.1007/3-540-09519-5_73
2. Ben-Or, M., Tiwari, P.: A deterministic algorithm for sparse multivariate polynomial interpolation. In: Proceedings STOC 1988, pp. 301–309. ACM Press (1988)
3. Kaltofen, E.L., Lakshman, Y.N.: Improved sparse multivariate polynomial interpolation algorithms. In: Proceedings ISSAC 1988, pp. 467–474. Springer-Verlag (1988)
4. Zippel, R.: Interpolating polynomials from their values. J. Symb. Comp. **9**, 375–403 (1990)
5. Lakshman, Y.N., Saunders, B.D.: Sparse polynomial interpolation in nonstandard bases. SIAM J. Comput. **24**(2), 387–397 (1995)
6. Klivans, A.R., Spielman, D.: Randomness efficient identity testing of multivariate polynomials. In: Proceedings STOC 2001, pp. 216–223. ACM Press (2001)
7. Grigoriev, D.Y., Karpinski, M., Singer, M.F.: Fast parallel algorithms for sparse multivariate polynomial interpolation over finite fields. SIAM J. Comput. **19**, 1059–1063 (1990)
8. Giesbrecht, M., Roche, D.S.: Diversification improves interpolation. In: Proceedings ISSAC 2011, pp. 123–130, ACM Press (2011)
9. Huang, M.D.A., Rao, A.J.: Interpolation of sparse multivariate polynomials over large finite fields with applications. J. Algorithms **33**, 204–228 (1999)
10. Javadi, S.M.M., Monagan, M.: Parallel sparse polynomial interpolation over finite fields. In: Proceedings PASCO 2010, pp. 160–168. ACM Press (2010)
11. Kaltofen, E.L., Lee, W.S.: Early termination in sparse interpolation algorithms. J. Symbolic Comput. **36**, 365–400 (2003)
12. Hao, Z., Kaltofen, E.L., Zhi, L.: Numerical sparsity determination and early termination. In: Proceedings ISSAC 2017, pp. 247–254. ACM Press (2016)
13. Giesbrecht, M., Labahn, G., Lee, W.: Symbolic-numeric sparse interpolation of multivariate polynomials. In: Proceedings ISSAC 2006, pp. 116–123. ACM Press (2006)
14. Kaltofen, E.L., Lee, W.S., Yang, Z.: Fast estimates of hankel matrix condition numbers and numeric sparse interpolation. In: SNC 2011, pp. 130–136. ACM Press (2011)
15. Cuyt, A., Lee, W.S.: A new algorithm for sparse interpolation of multivariate polynomials. Theor. Comput. Sci. **409**(2), 180–185 (2008)
16. Giesbrecht, M., Roche, D.S.: Interpolation of shifted-lacunary polynomials. Comput. Complex. **19**(3), 333–354 (2010)

17. Bläser, M., Jindal, G.: A new deterministic algorithm for sparse multivariate polynomial interpolation. In: Proceedings ISSAC 2014, pp. 51–58. ACM Press (2014)
18. Arnold, A., Giesbrecht, M., Roche, D.S.: Faster sparse multivariate polynomial interpolation of straight-line programs. J. Symbolic Comput. **75**, 4–24 (2016)
19. Arnold, A., Giesbrecht, M., Roche, D.S.: Faster sparse interpolation of straight-line programs. In: Gerdt, V.P., Koepf, W., Mayr, E.W., Vorozhtsov, E.V. (eds.) CASC 2013. LNCS, vol. 8136, pp. 61–74. Springer, Cham (2013). https://doi.org/10.1007/978-3-319-02297-0_5
20. Arnold, A., Roche, D.S.: Multivariate sparse interpolation using randomized Kronecker substitutions. In: ISSAC 2014, pp. 35–42. ACM Press (2014)
21. Garg, S., Schost, E.: Interpolation of polynomials given by straight-line programs. Theor. Comput. Sci. **410**, 2659–2662 (2009)
22. Huang, Q.L., Gao, X.S.: Faster interpolation algorithms for sparse multivariate polynomials given by straight-Line programs. arXiv 1709.08979v4 (2017)
23. Huang, Q.L.: Sparse polynomial interpolation over fields with large or zero characteristic. In: Proceedings ISSAC 2019, ACM Press, to appear (2019)
24. Avendaño, M., Krick, T., Pacetti, A.: Newton-Hensel interpolation lifting. Found. Comput. Math. **6**(1), 82–120 (2006)
25. Alon, N., Mansour, Y.: Epsilon-discrepancy sets and their application for interpolation of sparse polynomials. Inform. Process. Lett. **54**(6), 337–342 (1995)
26. Mansour, Y.: Randomized interpolation and approximation of sparse polynomials. SIAM J. Comput. **24**(2), 357–368 (1995)
27. Arnold, A.: Sparse polynomial interpolation and testing. PhD Thesis, Waterloo Unversity (2016)
28. Dubhashi, D.P., Panconesi, A.: Concentration of Measure for the Analysis of Randomized Algorithms. Cambridge University Press, Cambridge (2009)
29. Gathen, J., von zur Gerhard, J.: Modern Computer Algebra. Cambridge University Press, Cambridge (1999)
30. Pollard, J.M.: Monte Carlo Methods for Index Computation (mod p). Math. Comput. **32**(143), 918–924 (1978)
31. Nielsen, M.A., Chuang, I.L.: Quantum Computation and Quantum Information. Cambridge University Press, Cambridge (2015)

Root-Finding with Implicit Deflation

Rémi Imbach[1], Victor Y. Pan[2,3]([✉]), Chee Yap[1], Ilias S. Kotsireas[4],
and Vitaly Zaderman[3]

[1] Courant Institute of Mathematical Sciences, New York University, New York, USA
remi.imbach@nyu.edu,yap@cs.nyu.edu
[2] Department of Computer Science,
Lehman College of the City University of New York, Bronx, NY 10468, USA
victor.pan@lehman.cuny.edu
[3] Ph.D. Programs in Mathematics and Computer Science,
The Graduate Center of the City University of New York, New York, NY 10036, USA
vza52@aol.com
[4] Department of Physics and Computer Science, Wilfrid Laurier University,
75 University Avenue West, Waterloo, ON N2L 3C5, Canada
ikotsire@wlu.ca,
http://www.cs.nyu.edu/yap/,
http://comet.lehman.cuny.edu/vpan/,
http://web.wlu.ca/science/physcomp/ikotsireas/

Abstract. Functional iterations such as Newton's are a popular tool for polynomial root-finding. We consider realistic situation where some (e.g., better-conditioned) roots have already been approximated and where further computations is directed to the approximation of the remaining roots. Such a situation is also realistic for root by means of subdivision iterations. A natural approach of applying explicit deflation has been much studied and recently advanced by one of the authors of this paper, but presently we consider the alternative of implicit deflation combined with the mapping of the variable and reversion of an input polynomial. We also show another unexplored direction for substantial further progress in this long and extensively studied area. Namely we dramatically increase the local efficiency of root-finding by means of the incorporation of fast algorithms for multipoint polynomial evaluation and Fast Multipole Method.

Keywords: Polynomial roots · Functional iterations ·
Newton's iterations · Weierstrass's iterations · Ehrlich's iterations ·
Efficiency · Taming wild roots · Deflation · Maps of the variable

2000 Math. Subject Classification: 26C10 · 30C15 · 65H05

© Springer Nature Switzerland AG 2019
M. England et al. (Eds.): CASC 2019, LNCS 11661, pp. 236–245, 2019.
https://doi.org/10.1007/978-3-030-26831-2_16

1   Introduction

Univariate polynomial root-finding, that is, approximation of the roots $x_1, \ldots,$ x_d of a polynomial equation

$$p(x) = 0 \text{ for } p(x) = \sum_{j=0}^{d} p_j x^j = p_d \prod_{i=1}^{d} (x - x_i), \quad p_d \neq 0, \tag{1}$$

has been the central problem of Mathematics for four millennia, since the Sumerian times, is still involved in various areas of modern computation, and is the subject of intensive research worldwide. The user's choice since 2000 has been the package MPSolve (cf. [3,4]), which implements Ehrlich's functional iterations, but other functional iterations such as Newton's and Weierstrass's are also highly popular. Ehrlich's and Weierstrass's iterations converge simultaneously to all complex roots of a polynomial. Newton's iterations converge to a single root, but can be extended to approximation of all roots or the roots in a fixed domain.

Usually root-finding iterations approximate some roots sooner than the other ones; e.g., MPSolve tends to approximate the well-conditioned roots much faster than the ill-conditioned ones (see [3,4]), and then one can deflate an input polynomial and keep updating approximations to the remaining roots.

Efficient methods for explicit deflation can be found in [19,33,35], and references therein. Here we study alternative techniques of implicit deflation, which enable us to exploit the sparseness of an input and to avoid numerical stability problems caused by the coefficient growth in factorization of a polynomial. We enhance the power of implicit deflation by combining it with mapping the variable and reversion of an input polynomial.

In Sect. 8 we point out another promising direction to enhancing the power of root-finding iterations, namely by means of incorporation of superfast multipoint polynomial evaluation and Fast Multipole Method. We demonstrate high promise of this approach by showing that it yields dramatic increase of local efficiency of root-finding iterations.

Otherwise we organize our paper as follows. In the next section we recall some popular functional iterations for polynomial root-finding. In Sect. 3 we comment on partitioning polynomial roots into tame ones (already approximated) and wild ones. In Sect. 4 we compare explicit and implicit deflation and specify implicit deflation for Newton's iterations. We combine implicit deflation with linear maps of the variable and reversion of a polynomial in Sect. 5 and with squaring the variable in Sect. 6, followed by some brief comments on the implementation and potential research impact of implicit deflation in Sect. 7.

2   Some Efficient Functional Iterations for Root-Finding

Among hundreds if not thousands known polynomial root-finders (see up to date coverage in [22,25,33], and the bibliography therein) consider the class of functional iterations. For a fixed set of functions

$$f_1(z), \ldots, f_m(x), \; 1 \leq m \leq d,$$

these iterations recursively refine current approximations $z_1^{(k)}, \ldots, z_m^{(k)}$ to m roots $x_1, \ldots x_m$ of $p(x)$ according to the expressions

$$z_i \leftarrow f_i(z_i), \quad i = 1, \ldots, m. \tag{2}$$

In the case where $m = 1$ write $f(z) = f_1(z)$ and

$$z \leftarrow f(z). \tag{3}$$

These iterations include various interpolation methods, which use no derivatives of $p(x)$ and are recalled in [25, Section 7], for example, Muller's method (see [25, Section 7.4]); methods involving derivative such as Newton's iterations [22, Section 5], and methods involving higher order derivatives [25, Section 7]. We exemplify our study with Newton's iterations (where $m = 1$):

$$z \leftarrow z - N_p(z), \tag{4}$$

$$N_p(x) = p(x)/p'(x), \tag{5}$$

which have efficient extensions to the solution of polynomial systems of equations [2] and to root-finding for various smooth functional equations and systems of equations [10]; Weierstrass's iterations of [38] (rediscovered by Durand in [7] and Kerner in [18]), in which case $m = d$:

$$z_i \leftarrow z_i - W_{p,l}(z_i), \quad i = 1, \ldots, d, \tag{6}$$

$$W_{p,l}(x) = \frac{p(x)}{p_n l'(x)}, \tag{7}$$

$$l(x) = \prod_{i=1}^{d} (x - z_i), \tag{8}$$

and Ehrlich's iterations of [9] (rediscovered by Aberth in [1]), where again $m = d$:

$$z_i \leftarrow z_i - E_{p,i}(z_i), \tag{9}$$

$$E_{p,i}(x) = 0 \text{ if } p(x) = 0; \quad \frac{1}{E_{p,i}(x)} = \frac{1}{N_p(x)} - \sum_{j=1, j \neq i}^{d} \frac{1}{x - z_j} \text{ otherwise}, \tag{10}$$

$i = 1, \ldots, d$, and $N_p(x)$ is defined by (5).

Remark 1. The above root-finders are readily extended to any function $s(x)$ sharing its root set with the polynomial $p(x)$. For example, deduce from the Lagrange interpolation formula that

$$p(x) = l(x)s(x),$$

$$s(x) = p_n + \sum_{i=1}^{d} \frac{W_{p,l}(z_i)}{x - z_i}$$

for any set of d distinct nodes $z_1, \ldots, z_d$. Apply selected iterations to the above *secular rational function* $s(x)$ or the polynomial $l(x)s(x)$. Bini and Robol in [4] show substantial benefits of that application of Ehrlich's iterations to $l(x)s(x)$ rather than $p(x)$, both for convergence acceleration and error estimation.

3   The Problem of Taming Wild Roots

Now suppose that we have applied a fixed functional iteration (2) and have approximated m roots of a polynomial $p(x)$ for $m < d$ (we call them *tame*); next we discuss efficient approximation of the remaining roots; we call them *wild* and call their approximation *taming*.

For example, we face taming problem where functional iterations have approximated a single root of a polynomial $p(x)$ and we seek the other roots.

For another example, Weierstrass's, Ehrlich's, and various other iterations tend to approximate at first the better conditioned roots (that is, the roots stronger isolated from the other roots of $p(x)$); then one can fix these approximations and keep updating the approximations to the remaining wild roots by applying the same iterations (see [3] and [4]).

Likewise Newton's and many other iterations seeking a single root can be applied at a number of initial points in order to approximate all roots, and then some roots can escape from this process. In particular in the paper [36] Newton's iterations initialized at a universal set of $O(d)$ points[1] approximate $t = d - w$ roots of $p(x)$ but leave out a narrow set of w wild roots where $w < 0.001\, d$ for $d < 2^{17}$ and $w < 0.01\, d$ for $d < 2^{20}$. (The paper [36] continued long study traced back to [20] and [16].)

Finally the subdivision root-finding iterations of [5] extend the earlier study in [14, 15, 28, 34, 39], where such iterations are called Quad-tree algorithm. Recently been implemented in [17], it first approximates some sets of tame roots of $p(x)$ in certain domains on the complex plane well-isolated from the other roots and then approximates the remaining wild roots, in particular by combining the subdivision process with complex extension of Abbott's real QIR iterations.

4   Taming Wild Roots by Means of Deflation

An obvious recipe is to tame the wild roots by means of deflation, that is, by applying a selected root-finder to the polynomial

$$q(x) = \sum_{i=0}^{w} q_i x^i = p_d \prod_{j=1}^{w} (x - x_j), \quad p_d \neq 0. \tag{11}$$

In *explicit deflation* we first compute the coefficients of $q(x)$. If the roots of the quotient $q(x)$ are well isolated from the other roots of $p(x)$, we can apply the efficient method of Delves and Lyness [8]. The root-finders of [35] and [19] incorporate its advanced versions; [33] presents them in a concise form.

Bini and Fiorentino argue in [3] that explicit deflation of a polynomial $p(x)$ does not preserve its sparseness and in some cases can be numerically unstable, for instance, in the case of a polynomial $p(x) = x^d \pm 1$ of a large degree d.

---

[1] This set is *universal* for all polynomials $p(x)$ that have all roots lying in the unit disc $D(0,1) = \{z : |z| = 1\}$. Given any polynomial $p(x)$ one can move all its roots into this disc by means of first readily computing a reasonably close upper bound on the absolute values of all roots and then properly shifting and scaling the variable x.

These potential problems somewhat limit the value of explicit deflation, particularly where a polynomial $q(x)$ has large degree w. We can completely avoid these problems by applying *implicit deflation*, that is, applying functional iterations that evaluate $q(x)$ at a point x as the ratio $p(x)/t(x)$ for $t(x) = p_d \prod_{j=1+w}^{d}(x - x_j)$.

We can readily implement this recipe in the case of functional interpolation iterations of [25, Section 7]. Moreover $W_p(x) = W_q(x)$, as we can readily verify by combining Eqs. (1) and (6), and so for Weierstrass's and Ehrlich's iterations (6) implicit deflation amounts to their usual recursive application restricted just to w approximations of the w wild roots.

Let us specify implicit deflation when we apply Newton's iterations and rely on the following well-known identity (cf. [21]),

$$1/N_p(x) = \sum_{j=1}^{n} \frac{1}{x - x_j}. \tag{12}$$

Algorithm 1. Implicit Deflation with Newton's iterations

INPUT: *A polynomial $p(x)$ of (1), a set of its tame roots $x_{w+1}, \ldots, x_d$, an initial approximation z to a wild root of $p(x)$, a Stopping Criterion (see, e.g., [3,4]), and a black-box program $EVAL_p$ that evaluates the ratio $\frac{1}{N_p(z)} = \frac{p'(z)}{p(z)}$ for a polynomial $p(x)$ of (1) and a complex point z.*

OUTPUT: *The updated approximation $z - N_p(z)$ to a root of $p(x)$ (see (4)).*

COMPUTATIONS: *Apply Newton's iteration (4) to the polynomial $q(x)$ defined implicitly, that is, successively compute the values:*

1. *$r = p'(z)/p(z) \leftarrow 1/N_p(z)$,*
2. *$s \leftarrow \sum_{j=w+1}^{d} \frac{1}{z - x_j}$,*
3. *$N_q(z) = \frac{q(z)}{q'(z)} \leftarrow \frac{1}{r - s}$.*
4. *Compute $z_k - N_p(z_k)$. If the fixed Stopping Criterion is met, output z and stop. Otherwise go to stage 1.*

Dario A. Bini (private communication) proposed to improve numerical stability of this algorithm by means of scaling as follows:

$$N_q(z_k) = \frac{1/r_k}{1 - s_k/r_k}.$$

Complexity of Algorithm 1

Stage 1 amounts to m_w invocations of the program $EVAL_p$.

At Stage 2 we perform $(d - m_w)m_w$ divisions and $(2d - 2m_w - 1)m_w$ additions and subtractions.

At Stages 3 and 4 together we perform $2m_w$ subtractions and m_w divisions.

We can readily extend implicit deflation to various other root-finders involving Newton's ratio $N_p(x)$, for example, to Ehrlich's iterations of (9) because (12) implies that $E_{p,j}(x) = E_{q,j}(x)$ for $q(x)$ of (11) and $E_{p,j}(x)$ of (10).

5 Taming Wild Roots by Means of Mapping the Variable with Linear Maps and Reversion

Generally the set of tame roots output by functional iterations varies when an input polynomial $p(x)$ varies. This suggests that we can approximate many or all wild roots if we reapply the same iterations to the polynomials

$$v(z) = v_{a,b,c}(z) = (z + c)^d p\left(a + \frac{b}{z + c}\right) \tag{13}$$

for various triples of complex scalars a, $b \neq 0$, and c. We must limit the overall number of the triples in order to control the overall computational cost.

The following equations map the roots x_j of $p(x)$ to the roots z_j of $v(x)$ and vice versa,

$$x_j = a + \frac{b}{z_j + c}, \; z_j = \frac{b}{x_j - a} - c. \tag{14}$$

Let us specify this recipe for the algorithm of [36], cited in Sect. 3.

Algorithm 2

INITIALIZATION: *Define a polynomial $v(z) = v_{a,b,c}(z)$ by choosing the parameters a, b, and c such that all roots of the polynomial $v(z)$ lie in the unit disc $D(0,1) = \{z : |z| = 1\}$, but do not actually compute the coefficients of this polynomial.*
COMPUTATIONS: *1. Apply Newton's iteration (4) to the polynomial $v(z)$ by using initialization at the universal set of [36] and by expressing the Newton's ratios $N_v(z) = v(z)/v'(z)$ (cf. (4)) via the following equations:*

$$\frac{1}{N_v(z)} = \frac{d}{z + c} - \frac{b}{(z + c)^2 N(x)} \; \text{for } v(z) \text{ of (13) and } x \text{ of (14).} \tag{15}$$

2. Having approximated a root z_j of $v(z)$ for any j, readily recover the root x_j of $p(x)$ from Eq. (14).

In the particular case where $a = c = 0$ and $b = 1$, the above expressions are simplified: $z = 1/x$; $v(z)$ turns into the reverse polynomial of $p(x)$,

$$v(z) = p_{\text{rev}}(z) = \sum_{i=0}^{d} p_{d-i} z^i = z^d p(1/z),$$

$$\frac{1}{N_v(z)} = \frac{v'(z)}{v(z)} = \frac{d}{z} - \frac{1}{z^2 N_p(1/z)},$$

and $p_{\text{rev}}(x) = p_0 \prod_{j=1}^{d} (x - 1/x_j)$ if $p_0 \neq 0$.

6   Taming Wild Roots by Means of Squaring the Variable

One can hope to obtain all roots of $p(x)$ by applying Newton's iterations to the polynomials $v(z) = v_{a,b,c}(z)$ for a reasonable number of triples of a, b, and c, but one can also extend this approach by using more general rational maps $y = r(x)$ (cf., e.g., [24]).

For a simple example, consider the Dandelin's root-squaring map of 1826, rediscovered by Lobachevsky in 1834 and then by Gräffe in 1837 (see [13]):

$$u(y) = (-1)^d p(\sqrt{y})p(-\sqrt{y}) = \prod_{j=1}^{d} (y - x_j^2). \tag{16}$$

In this case one should make a polynomial $p(x)$ of (1) monic by scaling the variable x and then express the Newton's ratio $N_u(y) = u(y)/u'(y)$ as follows:

$$\frac{1}{N_u(y)} = 0.5 \left(\frac{1}{N_p(\sqrt{y})} - \frac{1}{N_p(\sqrt{-y})} \right) y^{-1/2}.$$

Notice that under map (16) the roots lying in the unit disc $D(0,1)$ stay in it.

Having approximated the n roots $y_1, \ldots, y_n$ of the polynomial $u(y)$, we readily recover the n roots $x_1, \ldots, x_n$ of the polynomial $p(x)$ by selecting them from the $2n$ values $\pm\sqrt{y_j}$, $j = 1, \ldots, n$.

We can combine the above maps recursively (a limited number of times, in order to control the overall computational cost); then we can recover the roots from their images in these rational maps by extending the lifting/descending techniques of [27, 30].

7   Two Remarks

Remark 2. For various selected polynomials $p(x)$, $u(y)$, and $v(z)$, one can implement the functional iterations of the previous two sections concurrently, with minimal need for processor communication and synchronization.

Remark 3. The Weierstrass's, Ehrlich's, and some other functional iterations, e.g., the Gauss-Seidel's and Werner's accelerated variations of the Ehrlich's and Weierstrass's iterations (cf. [4] and [40]), converge very fast empirically, but formal support of this empirical observation is a well-known challenge. Can we facilitate obtaining such a support if we allow random maps of the variable x, e.g., if we apply these iterations to the polynomials $v_{a,b,c}(z)$ of (13) for random choice of the parameters a, b, and c? For example, initialization of Newton's iterations at a set of points $\{c + r \exp(\phi_j \mathbf{i}), j = 1, \ldots, s$, of a circle $\{x : |x - c| = r\}$ on the complex plane can be equivalently interpreted as the application of these iterations at a single point $y = c$ to a set of polynomials $p_j(y)$ obtained from $p(x)$ via the linear maps $y \leftarrow x - r \exp(\phi_j \mathbf{i})$, $j = 1, \ldots, s$.

8   Efficiency of Root-Finding Iterations

Since Ostrowski's paper [26], it is customary to measure local efficiency of functional root-finding iterations by the quantity eff $= q^{1/\alpha}$ or sometimes $\log_{10}(\text{eff}) = (1/\alpha)\log_{10} q$ where q denotes the convergence order (rate) and α is the number of function evaluations per iteration and per root. In particular $q = 2$, $\alpha = 2$, and eff $=\sqrt{2} \approx 1.414$ for Newton's and Weierstrass's iterations while $q = 3$, $\alpha = 3$, and eff $=3^{1/3} \approx 1.442$ for Ehrlich's iterations where we assign the same cost to the evaluation of the functions $\sum_{j=1,j\neq i}^{d} \frac{1}{x-z_j}$, $p(x)$, $p'(x)$, and $l'(x)$ at $x = z_i$, noting that $l'(z_i) = \prod_{j=1,j\neq i}^{d}(z_i - z_j)$.

Actually the cost of function evaluation requires further elaboration. Exact evaluation of the values $\sum_{i=1,i\neq j}^{d} \frac{1}{z_j^{(k)}-z_i^{(k)}}$ for $j = 1,\ldots,d$ is Trummer's celebrated problem, whose solution, like exact evaluation of a polynomial $p(x)$ of (1) at d points, involves $O(d\log^2(d))$ arithmetic operations [29, Section 3.1], [11, 23].

Both of these superfast algorithms – for polynomial evaluation and the Trummer's problem – are numerically unstable for $d > 50$, but one can use numerically stable superfast alternatives based on the Fast Multipole Method [6]. Its application to Trummer's problem is well-known [12], but in the case of multipoint polynomial evaluation is more recent and more involved [31] and [32].

Using superfast algorithms for both problems decreases α to the order of $O(\log^2(d)/d)$. Hence local efficiency of Weierstrass's and Ehrlich's iterations grows to the infinity as $d \to \infty$, and similarly for Newton's iterations initialized and applied simultaneously at the order of d points.

The above formal analysis applies locally, where the convergence to the roots becomes superlinear, while the overall computational cost is usually dominant at the previous initial stage, for which only limited formal results are available (see also Remark 3). These limited results favor Ehrlich's iterations, which empirically have milder sufficient conditions for superlinear convergence than both Newton's and Weierstrass's iterations [37].

Acknowledgements. The research of R. Inbach, V. Y. Pan, C. Yap, and V. Zaderman was supported by NSF Grant CCF–1563942. The research of V. Y. Pan and V. Zaderman was also supported by NSF Grants CCF 1116736 and PSC CUNY Award 69813 00 48. The research of Ilias Kotsireas was supported by an NSERC grant.

References

1. Aberth, O.: Iteration methods for finding all zeros of a polynomial simultaneously. Math. Comput. **27**(122), 339–344 (1973)
2. Blum, L., Cucker, F., Shub, M., Smale, S.: Complexity and Real Computation. Springer, New York (1998). https://doi.org/10.1007/978-1-4612-0701-6
3. Bini, D.A., Fiorentino, G.: Design, analysis, and implementation of a multiprecision polynomial rootfinder. Numer. Algorithms **23**, 127–173 (2000)
4. Bini, D.A., Robol, L.: Solving secular and polynomial equations: a multiprecision algorithm. J. Comput. Appl. Math. **272**, 276–292 (2014)

5. Becker, R., Sagraloff, M., Sharma, V., Yap, C.: A near-optimal subdivision algorithm for complex root isolation based on the Pellet test and Newton iteration. J. Symb. Comput. **86**, 51–96 (2018)
6. Barba, L.A., Yokota, R.: How will the fast multipole method fare in Exascale Era? SIAM News **46**(6), 1–3 (2013)
7. Durand, E.: Solutions numériques des équations algébriques, Tome 1: Equations du type F(X)=0. Racines d'un polynôme. Masson, Paris (1960)
8. Delves, L.M., Lyness, J.N.: A numerical method for locating the zeros of an analytic function. Math. Comput. **21**, 543–560 (1967). https://doi.org/10.1090/S0025-5718-1967-0228165-4
9. Ehrlich, L.W.: A modified Newton method for polynomials. Commun. ACM **10**, 107–108 (1967)
10. Hazewinkel, M. (ed.) Encyclopedia of Mathematics. Newton method. Springer Science+Business Media B.V. Kluwer Academic Publishers (1994, first edition), (2000, second edition). ISBN 978-1-55608-010-4
11. Gerasoulis, A., Grigoriadis, M.D., Sun, L.: A fast algorithm for Trummer's problem. SIAM J. Sci. Stat. Comput. **8**(1), 135–138 (1987)
12. Greengard, L., Rokhlin, V.: A fast algorithm for particle simulation. J. Comput. Phys. **73**, 325–348 (1987)
13. Householder, A.S.: Dandelin, Lobachevskii, or Graeffe? Am. Math. Monthly **66**, 464–466 (1959)
14. Henrici, P.: Applied and Computational Complex Analysis. Vol. 1: Power Series, Integration, Conformal Mapping, Location of Zeros. Wiley, New York (1974)
15. Henrici, P., Gargantini, I.: Uniformly convergent algorithms for the simultaneous approximation of all zeros of a polynomial. In: Dejon, B., Henrici, P. (eds.) Constructive Aspects of the Fundamental Theorem of Algebra. Wiley, New York (1969)
16. Habbard, J., Schleicher, D., Sutherland, S.: How to find all roots of complex polynomials by Newton's method. Invent. Math. **146**, 1–33 (2001)
17. Imbach, R., Pan, V.Y., Yap, C.: Implementation of a near-optimal complex root clustering algorithm. In: Davenport, J.H., Kauers, M., Labahn, G., Urban, J. (eds.) ICMS 2018. LNCS, vol. 10931, pp. 235–244. Springer, Cham (2018). https://doi.org/10.1007/978-3-319-96418-8_28
18. Kerner, I.O.: Ein Gesamtschrittverfahren zur Berechung der Nullstellen von Polynomen. Numerische Math. **8**, 290–294 (1966)
19. Kirrinnis, P.: Polynomial factorization and partial fraction decomposition by simultaneous Newton's iteration. J. Complex. **14**, 378–444 (1998)
20. Kim, M.-H., Sutherland, S.: Polynomial root-finding algorithms and branched covers. SIAM J. Comput. **23**(2), 415–436 (1994). https://doi.org/10.1137/S0097539791201587
21. Mahley, H.: Zur Auflösung Algebraisher Gleichngen. Z. Andew. Math. Physik. **5**, 260–263 (1954)
22. McNamee, J.M.: Numerical Methods for Roots of Polynomials, Part I, XIX+354 pages. Elsevier (2007)
23. Moenck, R., Borodin, A.: Fast modular transforms via division. In: Proceedings of 13th Annual Symposium on Switching and Automata Theory (SWAT 1972), pp. 90–96. IEEE Computer Society Press (1972)
24. Mourrain, B., Pan, V.Y.: Lifting/descending processes for polynomial zeros and applications. J. Complex. **16**(1), 265–273 (2000)
25. McNamee, J.M., Pan, V.Y.: Numerical Methods for Roots of Polynomials, Part II, XXI+728 pages. Elsevier (2013)

26. Ostrowski, A.M.: Solution of Equations and Systems of Equations. Academic Press, New York (1966)
27. Pan, V.Y.: Optimal (up to polylog factors) sequential and parallel algorithms for approximating complex polynomial zeros. In: Proceedings of the 27th Annual ACM Symposium on Theory of Computing (STOC 1995), pp. 741–750. ACM Press, New York (1995)
28. Pan, V.Y.: Approximation of complex polynomial zeros: modified quadtree (Weyl's) construction and improved Newton's iteration. J. Complex. **16**(1), 213–264 (2000)
29. Pan, V.Y.: Structured Matrices and Polynomials: Unified Superfast Algorithms. Springer, New York (2001). https://doi.org/10.1007/978-1-4612-0129-8
30. Pan, V.Y.: Univariate polynomials: nearly optimal algorithms for factorization and rootfinding. J. Symb. Comput. **33**(5), 701–733 (2002)
31. Pan, V.Y.: Transformations of matrix structures work again. Linear Algebra Appl. **465**, 1–32 (2015)
32. Pan, V.Y.: Simple and nearly optimal polynomial root-finding by means of root radii approximation. In: Kotsireas, I.S., Martinez-Moro, E. (eds.) ACA 2015. Springer Proceedings in Mathematics and Statistics, Ch. 23: Applications of Computer Algebra, vol. 198, pp. 329–340. Springer, Heidelberg (2017). https://doi.org/10.1007/978-3-319-56932-1_23
33. Pan, V.Y.: Old and new nearly optimal polynomial root-finders. In: CASC (2019). arxiv:1805.12042 [cs.NA], May 2019
34. Renegar, J.: On the worst-case arithmetic complexity of approximating zeros of polynomials. J. Complex. **3**(2), 90 113 (1987)
35. Schönhage, A.: The fundamental theorem of algebra in terms of computational complexity. Technical report, Math. Dept., University of Tübingen, Tübingen, Germany (1982)
36. Schleicher, D., Stoll, R.: Newton's method in practice: finding all roots of polynomials of degree one million efficiently. Theor. Comput. Sci. **681**, 146–166 (2017)
37. Tilli, P.: Convergence conditions of some methods for the simultaneous computation of polynomial zeros. Calcolo **35**, 3–15 (1998)
38. Weierstrass, K.: Neuer Beweis des Fundamentalsatzes der Algebra. Mathematische Werker, Tome III, pp. 251–269. Mayer und Mueller, Berlin (1903)
39. Weyl, H.: Randbemerkungen zu Hauptproblemen der Mathematik. II. Fundamentalsatz der Algebra and Grundlagen der Mathematik. Mathematische Zeitschrift **20**, 131–151 (1924)
40. Werner, W.: Some improvements of classical iterative methods for the solution of nonlinear equations. In: Allgower, E.L., Glashoff, K., Peitgen, H.-O. (eds.) Numerical Solution of Nonlinear Equations. LNM, vol. 878, pp. 426–440. Springer, Heidelberg (1981). https://doi.org/10.1007/BFb0090691

On Linear Invariant Manifolds in the Generalized Problem of Motion of a Top in a Magnetic Field

Valentin Irtegov and Tatiana Titorenko[✉]

Institute for System Dynamics and Control Theory SB RAS,
134, Lermontov street, Irkutsk 664033, Russia
{irteg,titor}@icc.ru

Abstract. Differential equations describing the rotation of a rigid body with a fixed point under the influence of forces generated by the Barnett–London effect are analyzed. They are a multiparametric system of equations. A technique for finding their linear invariant manifolds is proposed. With this technique, we find the linear invariant manifolds of codimension 1 and use them in the qualitative analysis of the equations. Computer algebra tools are applied to obtain the invariant manifolds and to analyze the equations. These tools proved to be essential.

1 Introduction

The problem of rotation of a rigid body with a fixed point in a magnetic field is considered, taking into account the Barnett–London effect [1,2] and the moment of potential forces. As is well known, a rotating "neutral" ferromagnetic is magnetized along its axis of rotation (the Barnett effect). A similar phenomenon occurs when a superconducting body rotates (the London effect). The magnetic moment **B** is related to the angular velocity $\boldsymbol{\omega}$ as follows: $\mathbf{B} = B\boldsymbol{\omega}$ (B is a linear symmetric operator).

There are a lot of works studying the influence of the Barnett–London effect on the motion of the body in various aspects. Similar problems arise in many applications, e.g., in space dynamics [3], in designing instruments having a contactless suspension system [4]. Our interest is in the works related to the qualitative analysis of the equations of motion of the top, e.g., [5]–[7]. In [5] and [6], the integrable cases of the equations have been presented, and an analysis of the problem in these cases has been done. The linear invariant manifold (IM) like the Hess manifold [8] has been found in [7].

In the present work, we study the equations of motion of the top resting on methods and tools of computer algebra. Using the method of undetermined coefficients in combination with computer algebra methods, we find both the existence conditions of linear IMs of codimension 1 and the IMs themselves, and conduct the qualitative analysis of the equations having such solutions. The cases when the system under study is dissipative and when it is conservative

M. England et al. (Eds.): CASC 2019, LNCS 11661, pp. 246–261, 2019.
https://doi.org/10.1007/978-3-030-26831-2_17

are considered. In the latter case, the IMs in question are the first integrals of the problem and they are used to obtain stationary solutions and IMs in the sense of [9] by the Routh–Lyapunov method [10]. All the principal computations concerned with this work have been done with computer algebra system "Mathematica" and the software package developed on its base [11]. The package is intended for the qualitative analysis of dynamical systems possessing first integrals. It is applied as an auxiliary tool when stationary solutions and IMs are investigated on the base of the Routh–Lyapunov method and the Lyapunov stability theorems for linear approximation [12]. Some description of the possibilities of the package is given in [9].

The paper is organized as follows. In Sect. 2, we study the case when the system under consideration is dissipative. We find linear IMs for the equations of motion of the body and conduct the qualitative analysis of the equations having such solutions. In Sect. 3, the same problem is solved when the system is conservative. Finally, we discuss the obtained results and give a conclusion in Sect. 4.

2   On Linear IMs of the Dissipative System

2.1   Formulation of the Problem

The equations of motion of the top in a magnetic field, taking into account the Barnett–London effect and the moment of potential forces, can be written as [7]

$$A\dot{\omega} = A\omega \times \omega + B\omega \times \gamma + \gamma \times (C\gamma - s), \quad \dot{\gamma} = \gamma \times \omega. \tag{1}$$

Equation (1) admit the following first integrals:

$$V_1 = A\omega \cdot \gamma = \kappa, \quad V_2 = \gamma \cdot \gamma = 1. \tag{2}$$

Here $\omega = (\omega_1, \omega_2, \omega_3)$ is the angular velocity of the body, $\gamma = (\gamma_1, \gamma_2, \gamma_3)$ is the direction vector of the magnetic field, $s = (s_1, s_2, s_3)$ is the vector of the center of mass, $A = diag(A_1, A_2, A_3)$ is the inertia tensor, $B = \mathrm{diag}(B_1, B_2, B_3)$ is the matrix characterizing the magnetic moment of the body, $C = \mathrm{diag}(C_1, C_2, C_3)$ is the matrix characterizing the influence of potential forces on the body, κ is the constant of the integral V_1.

When $C_i = \mu A_i$ $(i = 1, 2, 3)$, where μ is some constant, differential Eq. (1) describe the motion of the top under the influence of magnetic and Newtonian fields.

Let us consider the following problem for Eq. (1). Find the IM of the type:

$$F(\omega_1, \omega_2, \omega_3, \gamma_1, \gamma_2, \gamma_3) = f_1\omega_1 + f_2\omega_2 + f_3\omega_3 + f_4\gamma_1 + f_5\gamma_2 + f_6\gamma_3 + f_0 = 0, \tag{3}$$

where f_i $(i = 0, \ldots, 6)$ are constant parameters to be determined.

2.2   Finding the Linear IMs

Let Eq. (3) determine the desired IM for differential Eq. (1). Then, according to IM definition, the derivative of expression (3) calculated by virtue of Eq. (1) must vanish on the given expression.

The derivative of expression (3) calculated by virtue of Eq. (1) is some polynomial $P = P(\omega_j, \gamma_j)$ $(j = 1, 2, 3)$. Using the built-in function *Polynomial-Reduce* as below

$$PolynomialReduce[P, \{F\}, \{\omega_3\}],$$

where ω_3 is the main variable, we can represent the polynomial P as follows:

$$P = QF + R. \tag{4}$$

Here $Q = Q(\omega_1, \omega_2, \gamma_j)$, $R = R(\omega_1, \omega_2, \gamma_j)$ are some polynomials.

Equating the coefficients of similar terms of the polynomial R to zero, we obtain the system of polynomial equations with respect to f_i $(i = 0, \ldots, 6)$:

$$h_1 = h_2 = \ldots = h_{19} = 0, \tag{5}$$

where

$$h_1 = A_2 A_3 (A_3 - A_2) f_0 f_1, \quad h_2 = A_1 A_3 (A_1 - A_3) f_0 f_2,$$
$$h_3 = A_1 A_3 (A_1 - A_3) f_1 f_2, \quad h_4 = A_2 A_3 (A_3 - A_2) f_1 f_2,$$
$$h_5 = A_2 A_3 (A_3 - A_2) f_1^2 + A_1 A_3 (A_1 - A_3) f_2^2 + A_1 A_2 (A_1 - A_2),$$
$$h_6 = A_1 A_3 [(A_1 - A_3) f_2 f_4 + (A_2 f_5 - B_3 f_2) f_1],$$
$$h_7 = A_2 A_3 [(B_3 f_1 - A_1 f_4) f_2 + (A_3 - A_2) f_1 f_5],$$
$$h_8 = A_2 A_3 (B_3 f_1 - A_1 f_4) f_5, \quad h_9 = A_1 A_3 (A_2 f_5 - B_3 f_2) f_4,$$
$$h_{10} = A_2 A_3 (B_3 f_1 - A_1 f_4) f_4 + A_1 A_3 (A_2 f_5 - B_3 f_2) f_5 + A_1 A_2 (C_2 - C_1),$$
$$h_{11} = A_2 A_3 (B_3 f_1 - A_1 f_4) f_1 + A_1 A_3 [(A_1 - A_3) f_2 f_5 - A_2 f_6] + A_1 A_2 B_1,$$
$$h_{12} = A_1 A_3 (A_2 f_5 - B_3 f_2) f_2 + A_2 A_3 [(A_3 - A_2) f_1 f_4 + A_1 f_6] - A_1 A_2 B_2,$$
$$h_{13} = A_2 A_3 [B_2 f_1 - A_1 f_4 + (A_3 - A_2) f_1 f_6],$$
$$h_{14} = A_1 A_3 [A_2 f_5 - B_1 f_2 + (A_1 - A_3) f_2 f_6],$$
$$h_{15} = A_2 A_3 [(C_3 - C_2) f_1 + (B_3 f_1 - A_1 f_4) f_6],$$
$$h_{16} = A_1 A_3 [(C_1 - C_3) f_2 + (A_2 f_5 - B_3 f_2) f_6], \quad h_{17} = A_3 (A_2 s_2 f_1 - A_1 s_1 f_2),$$
$$h_{18} = A_2 [A_3 (B_3 f_1 - A_1 f_4) f_0 - A_2 (A_3 s_3 f_1 - A_1 s_1)],$$
$$h_{19} = A_1 [A_3 (A_2 f_5 - B_3 f_2) f_0 + A_1 (A_3 s_3 f_2 - A_2 s_2)].$$

Without loss of generality, we take $f_3 = 1$ in (5).

So, the problem of finding the linear IMs of differential Eq. (1) is reduced to solving the above system of quadratic algebraic equations.

System (5) is compatible under some constraints on the parameters of the problem. The latter can be found by constructing a Gröbner basis for the polynomials of the system. In particular, it has a non-empty set of solutions with respect to $B_1, B_2, B_3, C_1, s_1, s_2, s_3, f_0, f_1, f_2, f_4, f_5, f_6$. These solutions are given below:

$$I.\ B_1 = \frac{(A_2 - A_1)(C_2 - C_3)}{(A_2 - A_3)f_6} + A_3 f_6,\quad B_2 = \frac{A_2(A_1 - A_2 + A_3)f_6}{2A_1 - A_2},$$

$$B_3 = \frac{C_2 - C_3}{f_6} - \frac{A_1(A_2 - 2A_3)f_6}{2A_1 - A_2},$$

$$C_1 = \frac{A_2(A_1 - A_3)\,C_2 + (A_1 - A_2)(A_2 - 2A_3)C_3}{(2A_1 - A_2)(A_2 - A_3)},$$

$$s_1 = \pm\frac{\sqrt{(A_2 - A_1)\,A_3}\,s_3}{\sqrt{A_1\,(A_3 - A_2)}},\quad s_2 = 0, \tag{6}$$

$$f_0 = 0,\quad f_1 = \pm\frac{\sqrt{A_1\,(A_2 - A_1)}}{\sqrt{(A_3 - A_2)\,A_3}},\quad f_2 = 0,$$

$$f_4 = \pm\frac{\sqrt{A_1\,(A_2 - A_1)}(A_2 - 2A_3)f_6}{(A_2 - 2A_1)\sqrt{(A_3 - A_2)\,A_3}},\quad f_5 = 0. \tag{7}$$

$$II.\ B_1 = \frac{A_1(A_1 - A_2 - A_3)f_6}{A_1 - 2A_2},$$

$$B_2 = -\frac{(A_1 - 2A_2)(A_1 - A_2)(C_2 - C_3)}{A_1(A_2 - A_3)f_6} + A_3 f_6,$$

$$B_3 = -\frac{(A_1 - 2A_2)(A_1 - A_3)(C_2 - C_3)}{A_1(A_2 - A_3)f_6} + \frac{A_2(A_1 - 2A_3)f_6}{A_1 - 2A_2},$$

$$C_1 = \frac{(A_1 - A_2)(A_1 - 2A_3)\,C_3 - (A_1 - 2A_2)(A_1 - A_3)\,C_2}{A_1(A_2 - A_3)},\quad s_1 = 0,$$

$$s_2 = \pm\frac{\sqrt{(A_2 - A_1)\,A_3}\,s_3}{\sqrt{A_2\,(A_1 - A_3)}}, \tag{8}$$

$$f_0 = 0,\quad f_1 = 0,\quad f_2 = \pm\frac{\sqrt{A_2\,(A_2 - A_1)}}{\sqrt{(A_1 - A_3)\,A_3}},\quad f_4 = 0,$$

$$f_5 = \pm\frac{\sqrt{A_2\,(A_2 - A_1)}\,(A_1 - 2A_3)f_6}{(A_1 - 2A_2)\sqrt{(A_1 - A_3)\,A_3}}. \tag{9}$$

Substituting (7) into (3), we obtain:

$$\pm\frac{\sqrt{A_1\,(A_2 - A_1)}}{\sqrt{A_3\,(A_3 - A_2)}}\,\omega_1 + \omega_3 \pm \frac{\sqrt{A_1\,(A_2 - A_1)}\,(A_2 - 2A_3)f_6}{\sqrt{A_3\,(A_3 - A_2)}\,(A_2 - 2A_1)}\,\gamma_1 + f_6\gamma_3 = 0. \tag{10}$$

Relations (10) define two families of IMs of codimension 1 for differential Eq. (1) under constraints (6). Here f_6 is the parameter of the families.

It should be noted that after substitution (6) into (1), we obtain a family of the systems under study which has solutions (10).

When (9) is substituted into (3), we have the equations of two other families of IMs:

$$\pm\frac{\sqrt{A_2\,(A_2 - A_1)}}{\sqrt{A_3\,(A_1 - A_3)}}\,\omega_2 + \omega_3 \pm \frac{\sqrt{A_2\,(A_2 - A_1)}\,(A_1 - 2A_3)f_6}{\sqrt{A_3\,(A_1 - A_3)}\,(A_1 - 2A_2)}\,\gamma_2 + f_6\gamma_3 = 0. \tag{11}$$

Expressions (8) are the existence conditions of these families.

Let us repeat the above calculations. To represent the polynomial P in form (4), take ω_1 as the main variable:

$$PolynomialReduce[P, \{F\}, \{\omega_1\}].$$

As a result, we have yet the two families of IMs of codimension 1 for differential Eq. (1), which differ from families (10) and (11):

$$\omega_1 \pm \frac{\sqrt{A_2\,(A_3 - A_2)}}{\sqrt{A_1\,(A_1 - A_3)}}\,\omega_2 \pm \frac{\sqrt{A_1\,(A_1 - A_3)}\,(A_3 - 2A_2)f_5}{\sqrt{A_2\,(A_3 - A_2)}\,(A_3 - 2A_1)}\,\gamma_1 + f_5\gamma_2 = 0, \quad (12)$$

f_5 is the parameter of the families.

The use of the rest of the variables as main ones to obtain the representation of the polynomial P in form (4) did not give new IMs.

2.3   On Linear IMs of 2nd-Level and Higher

Taking the IMs of Eq. (1) as 1st-level IMs, let us consider the problem of finding the linear IMs of codimension 1 for differential equations on the IMs obtained in Sect. 2.2. Such IMs we call 2nd-level IMs.

The differential equations on the elements of the family of IMs, e.g.,

$$\frac{\sqrt{A_1\,(A_2 - A_1)}}{\sqrt{A_3\,(A_3 - A_2)}}\,\omega_1 + \omega_3 + \frac{\sqrt{A_1\,(A_2 - A_1)}\,(A_2 - 2A_3)f_6}{\sqrt{A_3\,(A_3 - A_2)}\,(A_2 - 2A_1)}\,\gamma_1 + f_6\gamma_3 = 0, \quad (13)$$

can be written as:

$$A_1\dot{\omega}_1 = (A_2 - A_3)\omega_2\bar{\omega}_3 + B_2\omega_2\gamma_3 - B_3\bar{\omega}_3\gamma_2 + (C_3 - C_2)\gamma_2\gamma_3 - s_3\gamma_2,$$
$$A_2\dot{\omega}_2 = (A_3 - A_1)\omega_1\bar{\omega}_3 + B_3\bar{\omega}_3\gamma_1 - B_1\omega_1\gamma_3 + (C_1 - C_3)\gamma_1\gamma_3 - s_1\gamma_3 + s_3\gamma_1,$$
$$\dot{\gamma}_1 = \gamma_2\bar{\omega}_3 - \gamma_3\omega_2, \quad \dot{\gamma}_2 = \gamma_3\omega_1 - \gamma_1\bar{\omega}_3, \quad \dot{\gamma}_3 = \gamma_1\omega_2 - \gamma_2\omega_1. \quad (14)$$

The equations have been derived from the initial ones by eliminating the variable ω_3 from them with (13). Here the following denotations are used:

$$\bar{\omega}_3 = -\left(\frac{\sqrt{A_1\,(A_2 - A_1)}}{\sqrt{A_3\,(A_3 - A_2)}}\,\omega_1 + \frac{\sqrt{A_1\,(A_2 - A_1)}(A_2 - 2A_3)f_6}{(A_2 - 2A_1)\sqrt{(A_3 - A_2)\,A_3}}\,\gamma_1 + f_6\gamma_3\right),$$

$B_1, B_2, B_3, C_1, s_1, s_2$ correspond to expressions (6).

The first integrals of Eq. (14):

$$\bar{V}_1 = A_1\omega_1 + A_2\omega_2 + A_3\bar{\omega}_3 = \bar{\kappa}, \quad V_2 = \gamma_1^2 + \gamma_2^2 + \gamma_3^2 = 1. \quad (15)$$

They have been obtained from integrals (2) by eliminating the variable ω_3 from them with (13).

For differential Eq. (14) by the technique applied in Sect. 2.2, under the condition $s_3 = 0$, we have found the following family of linear IMs:

$$\omega_1 - \frac{(A_2 - 2A_3)f_6}{2A_1 - A_2}\gamma_1 = 0, \tag{16}$$

f_6 is the parameter of the family. It is the family of 2nd-level IMs on 1st-level IMs (13).

The differential equations on the elements of family (16) have the form

$$\dot{\omega}_2 = \frac{2(A_2 - 2A_3)(A_3 - A_1)f_6^2}{A_2(A_2 - 2A_1)}\gamma_1\gamma_3, \quad \dot{\gamma}_1 = -(\omega_2 + f_6\gamma_2)\gamma_3,$$

$$\dot{\gamma}_2 = \frac{2(A_1 - A_3)f_6}{2A_1 - A_2}\gamma_1\gamma_3, \quad \dot{\gamma}_3 = \gamma_1\left(\omega_2 - \frac{(A_2 - 2A_3)f_6}{(2A_1 - A_2)}\gamma_2\right) \tag{17}$$

and admit the first integral

$$V = \omega_2 + f_6\left(\frac{2A_3}{A_2} - 1\right)\gamma_2 = c \tag{18}$$

which is directly derived from the equations themselves.

Expression (18) can also be considered as the equation of IMs family, where c is the parameter of the family. It is the family of 3rd-level IMs on 2nd-level IMs (16).

Besides integral (18), Eq. (17) possess the following two integrals:

$$\tilde{V}_1 = \frac{A_1(A_2 - 2A_3)f_6}{2A_1 - A_2}\gamma_1^2 + A_2\gamma_2\omega_2 - A_3 f_6\gamma_3^2 = \tilde{\kappa}, \quad V_2 = \gamma_1^2 + \gamma_2^2 + \gamma_3^2 = 1. \tag{19}$$

They are derived from relations (15) by eliminating the variable ω_1 from them with (16).

So, system (17) is completely integrable. Let us find its stationary solutions and IMs with the Routh–Lyapunov method.

2.4   Finding Stationary Solutions and IMs

According to the above-mentioned method, the stationary solutions and IMs of differential Eq. (17) can be found from necessary extremum conditions for the first integrals of these equations. Take, e.g., the linear combination of integrals (18) and (19):

$$2K = 2\lambda_0\tilde{V}_1 - \lambda_1 V_2 - \lambda_2 V, \tag{20}$$

where λ_0, λ_1, and λ_2 are some constants, and write the stationary conditions for the integral K with respect to the phase variables:

$$\frac{\partial K}{\partial \omega_2} = 0, \quad \frac{\partial K}{\partial \gamma_i} = 0 \quad (i = 1, 2, 3). \tag{21}$$

These are a system of linear algebraic equations with respect to $\omega_2, \gamma_1, \gamma_2, \gamma_3$ with the parameters $A_1, A_2, A_3, f_6, \lambda_0, \lambda_1, \lambda_2$. Add relation $V_2 = 1$ (19) to them and construct a lexicographical Gröbner basis with respect to $\gamma_1 > \gamma_2 > \gamma_3 > \omega_2 > \lambda_1$ for the polynomials of a resulting system, using the built-in function *GroebnerBasis*. As a result, the system is transformed to a form which enables us to decompose it into the three subsystems:

$$(\text{I}) \quad 2A_3 f_6 \lambda_0 + \lambda_1 = 0,$$
$$\omega_2 = 0, \ \gamma_3 \pm 1 = 0, \ \gamma_1 = 0, \ \gamma_2 = 0; \tag{22}$$

$$(\text{II}) \quad 2A_1(A_2 - 2A_3)f_6\lambda_0 + (A_2 - 2A_1)\lambda_1 = 0,$$
$$\omega_2 = 0, \ \gamma_3 = 0, \ \gamma_2 = 0, \ \gamma_1 \pm 1 = 0; \tag{23}$$

$$(\text{III}) \quad [A_2^2 \lambda_0 + 2(A_2 - 2A_3)f_6\lambda_2]\lambda_0 - \lambda_1\lambda_2 = 0,$$
$$A_2^2 \lambda_0 + (A_2 - 2A_3)f_6\lambda_2 \pm A_2\lambda_2\omega_2 = 0, \ \gamma_3 = 0, \tag{24}$$
$$[A_2^2 \lambda_0 + (A_2 - 2A_3)f_6\lambda_2]\gamma_2 - A_2\lambda_2\omega_2 = 0, \ \gamma_1 = 0.$$

Consider system (22). The latter four equations of this system define the following solutions of differential Eq. 17):

$$\gamma_1 = \gamma_2 = \omega_2 = 0, \ \gamma_3 = \pm 1. \tag{25}$$

Substitute the expression

$$\lambda_1 = -2A_3 f_6 \lambda_0 \tag{26}$$

found from the 1st equation of (22) into K (20). We thereby obtain the family of integrals: $2K_1 = \lambda_0\tilde{V}_1 + 2A_3 f_6\lambda_0 V_2 - \lambda_2 V$. Its elements assume a stationary value on solutions (25). It is easy to verify by direct substitution of solutions (25) into the stationary equations for the integral K_1.

Similarly, the latter four equations of each of systems (23) and (24) define, respectively, the solutions

$$\gamma_2 = \gamma_3 = \omega_2 = 0, \ \gamma_1 = \pm 1 \tag{27}$$

and the families of solutions

$$\gamma_1 = \gamma_3 = 0, \ \gamma_2 = \pm 1, \ \omega_2 = \pm\left[\left(1 - \frac{2A_3}{A_2}\right)f_6 + \frac{A_2\lambda_0}{\lambda_2}\right] \tag{28}$$

for differential Eq. (17). Here λ_0, λ_2, and f_6 are the parameters of the families.

The elements of the families of integrals

$$K_2 = \lambda_0\tilde{V}_1 - \frac{2A_1(A_2 - 2A_3)f_6\lambda_0}{2A_1 - A_2}V_2 - \lambda_2 V$$

and

$$K_3 = \lambda_0 \tilde{V}_1 - \lambda_0 \left(2(A_2 - 2A_3) f_6 + \frac{A_2^2 \lambda_0}{\lambda_2} \right) V_2 - \lambda_2 V$$

take a stationary value on solutions (27) and (28), respectively.

Now, let us find the stationary IMs which solutions (25), (27), and (28) belong to.

First, we solve this problem for solutions (25). For this purpose, substitute λ_1 (26) into Eq. (21) and construct a lexicographical Gröbner basis with respect to $\gamma_1, \omega_2, \lambda_2$ for the polynomials of a resulting system. As a result, we have:

$$-\frac{(A_2 - 2A_3)}{A_2^2} [A_2^2 \lambda_0 + 2(A_2 - A_3) f_6 \lambda_2] = 0, \tag{29}$$

$$\omega_2 + f_6 \gamma_2 = 0, \quad \gamma_1 = 0. \tag{30}$$

Equation (30) together with relation $V_2 = 1$ (19) determine the family of one-dimensional IMs of differential Eq. (17). It is easy to verify by IM definition. Here f_6 is the parameter of the family.

Substituting

$$\lambda_2 = -\frac{A_2^2 \lambda_0}{2(A_2 - A_3) f_6} \tag{31}$$

found from (29) into K_1, we obtain the family of integrals: $2\bar{K}_1 = \tilde{V}_1 + 2A_3 f_6 V_2 + A_2^2/(2(A_2 - A_3) f_6) V$. Its elements take a stationary value on the elements of IMs family (30).

When solutions (25) are substituted into Eq. (30), these turn into identity. Whence it follows that solutions (25) belong to the given family of IMs.

Substituting λ_2 (31) into (28) gives the two subfamilies of these families:

$$\gamma_1 = \gamma_3 = 0, \quad \gamma_2 = \pm 1, \quad \omega_2 = \pm f_6. \tag{32}$$

It is not difficult to show that they also belong to the family of IMs (30).

In the same way, we have found the family of one-dimensional IMs which solutions (27) belong to. Its equations can be written as

$$(2A_1 - A_2) \omega_2 - (A_2 - 2A_3) f_6 \gamma_2 = 0, \quad \gamma_3 = 0, \tag{33}$$

f_6 is the parameter of the family.

When $\lambda_2 = A_2^2 (A_2 - 2A_1) \lambda_0 / (2(A_1 - A_2)(A_2 - 2A_3) f_6)$, the elements of the family of integrals K_2 take a stationary value on the elements of the given family of IMs.

All the solutions found in this Section can be "lifted up" into the original phase space. For this purpose, in the case of IMs, it is sufficient to add Eqs. (13) and (16) to the above-obtained equations of IMs. As to stationary solutions, the values of the variables ω_1, ω_2, which were derived from Eqs. (13) and (16) under the corresponding values of the rest of the variables, should be added to the expressions for the stationary solutions. So, the following families of solutions

$$\omega_1 = \omega_2 = 0, \quad \omega_3 = \pm f_6, \quad \gamma_1 = \gamma_2 = 0, \quad \gamma_3 = \mp 1; \tag{34}$$

$$\omega_1 = \pm\frac{(A_2 - 2A_3)\, f_6}{2A_1 - A_2}, \quad \omega_2 = \omega_3 = 0, \quad \gamma_1 = \pm 1, \quad \gamma_2 = \gamma_3 = 0; \tag{35}$$

$$\omega_1 = 0, \quad \omega_2 = \pm f_6, \quad \omega_3 = 0, \quad \gamma_1 = 0, \quad \gamma_2 = \mp 1, \quad \gamma_3 = 0 \tag{36}$$

correspond to stationary solutions (25), (27), and (32) in the original phase space, f_6 is the parameter of the families. These are the solutions of differential Eq. (1) in which B_1, B_2, B_3, C_1 correspond to expressions (6) and $s_1 = s_2 = s_3 = 0$.

>From a mechanical viewpoint, the elements of the families of solutions (34)–(36) correspond to permanent rotations of the top around one of its principal axes.

2.5   On Stability of the Stationary Solutions and IMs

The integrals and their families which take a stationary value on the above-found solutions are used to investigate the stability of these solutions by the Routh–Lyapunov method. In this case, the problem is to analyze the sign-definiteness conditions for the 2nd variation of the corresponding family of integrals derived in the vicinity of the solution under study on a linear manifold.

Let us investigate the stability of solutions (25), using the integral K_1.

The 2nd variation of the integral K_1 in the vicinity of the solution under study is

$$2\Delta K_1 = \frac{2A_2\,(A_1 - A_3)\, f_6\lambda_0}{2A_1 - A_2}\, y_1^2 + \left(A_3 f_6\lambda_0 - \frac{(A_2 - 2A_3)^2 f_6^2\lambda_2}{A_2^2}\right) y_2^2$$
$$+ 2\left(A_2\lambda_0 + f_6\lambda_2 - \frac{2A_3 f_6\lambda_2}{A_2}\right) y_2 y_4 - \lambda_2\, y_4^2.$$

Here $y_1 = \gamma_1$, $y_2 = \gamma_2$, $y_3 = \gamma_3 \mp 1$, $y_4 = \omega_2$ are the deviations from the unperturbed solution.

We consider the restriction of ΔK_1 to the set defined by the first variations of "conditional" integrals:

$$\delta V_2 = \pm 2 y_3 = 0, \quad \delta V = \left(\frac{2A_3}{A_2} - 1\right) f_6 y_2 + y_4 = 0.$$

On this set, ΔK_1 takes the form:

$$\Delta\tilde{K}_1 = f_6\lambda_0\left[\frac{A_2(A_1 - A_3)}{2A_1 - A_2}\, y_1^2 + (A_2 - A_3)\, y_2^2\right].$$

The conditions

$$f_6 \neq 0, \quad \lambda_0 \neq 0, \quad \frac{A_2(A_1 - A_3)}{2A_1 - A_2} > 0, \quad A_2 - A_3 > 0 \tag{37}$$

for the quadratic form $\Delta\tilde{K}_1$ to be positive definite are sufficient for the stability of the solutions under study.

¿From now on, we apply the built-in function *Reduce* to solve the systems of inequalities and to test their solutions found by hand.

Inequalities (37) are compatible when the following conditions

$$f_6 \neq 0, \lambda_0 \neq 0 \text{ and } \left[\left(A_3 < A_2 < 2A_3 \text{ and } (2A_1 < A_2 \text{ or } A_1 > A_3)\right)\right.$$

$$\text{or } \left(A_2 = 2A_3 \text{ and } (A_1 < A_3 \text{ or } A_1 > A_3)\right)$$

$$\left.\text{or } \left(A_2 > 2A_3 \text{ and } (A_1 < A_3 \text{ or } 2A_1 > A_2)\right)\right]$$

hold.

Next, we investigate the stability of the family of IMs (30), which solutions (25) belong to. The integral $\bar{K}_1$ is used here.

For the equations of perturbed motion, the integral $\bar{K}_1$ in the vicinity of the elements of the family of IMs (30) on the linear manifold $\delta V = y_2 = 0$ is

$$\Delta \bar{K}_1 = \frac{A_2(A_1 - A_3)f_6}{2A_1 - A_2} y_1^2,$$

where $y_1 = \gamma_1$, $y_2 = \omega_2 + f_6\gamma_2$ are the deviations from the elements of the family under study.

Since the quadratic form $\Delta\bar{K}_1$ is positive definite when the following conditions

$$\left(A_1 < A_3 \text{ and } ((A_2 < 2A_1 \text{ and } f_6 < 0) \text{ or } (A_2 > 2A_1 \text{ and } f_6 > 0))\right)$$

$$\text{or } \left(A_1 > A_3 \text{ and } ((A_2 < 2A_1 \text{ and } f_6 > 0) \text{ or } (A_2 > 2A_1 \text{ and } f_6 < 0))\right)$$

hold, then these are sufficient for the stability of the elements of the family of IMs (30).

The same technique was applied to investigate the stability of solutions (27), the families of solutions (28) and (32) as well as the IMs which they belong to. The sufficient conditions of stability for these solutions have been obtained, including stability with respect to part of the variables.

Now, we analyze the stability of the families of solutions (34) which correspond to solutions (25) in the original phase space. Their stability conditions can be derived on the base of the Lyapunov theorems for linear approximation.

The equations of 1st approximation in the case under consideration can be written as

$$\dot{y}_1 = \pm(f_6 y_2 + y_5), \ \dot{y}_2 = \mp(f_6 y_1 + y_4), \ \dot{y}_3 = 0,$$

$$\dot{y}_4 = \pm \frac{(A_2 - 2A_3)f_6}{2A_1 - A_2}[f_6 y_2 + y_5],$$

$$\dot{y}_5 = \mp \frac{(A_2 - 2A_3)\left[(A_2 - A_1)(C_2 - C_3) + A_1(A_2 - A_3)f_6^2\right]}{A_2(2A_1 - A_2)(A_2 - A_3)} y_1$$

$$+ \frac{1}{A_2 f_6}\left[\frac{(A_2 - A_1)(C_2 - C_3)}{A_2 - A_3} \mp (A_1 - 2A_3)f_6^2\right]y_4, \ \dot{y}_6 = 0. \quad (38)$$

Here $y_1 = \gamma_1, y_2 = \gamma_2, y_3 = \gamma_3 \pm 1, y_4 = \omega_1, y_5 = \omega_2, y_6 = \omega_3 \mp f_6$ are the deviations from the elements of the family under study.

The characteristic equation of system (38)

$$\lambda^4 \left[\lambda^2 + \frac{4(A_1 - A_3)(A_2 - A_3) f_6^2}{A_2 (2A_1 - A_2)} \right] = 0$$

has 4 zero roots.

The matrix of system (38) can be transformed to the Jordan form

$$J = \begin{pmatrix} 0 & 0 & 0 & 0 & 0 & 0 \\ 0 & 0 & 1 & 0 & 0 & 0 \\ 0 & 0 & 0 & 0 & 0 & 0 \\ 0 & 0 & 0 & 0 & 0 & 0 \\ 0 & 0 & 0 & 0 & z & 0 \\ 0 & 0 & 0 & 0 & 0 & -z \end{pmatrix}$$

with the built-in function *JordanDecomposition*.

Here $z = 2i\sqrt{(A_1 - A_3)(A_2 - A_3)} f_6 / \sqrt{A_2 (2A_1 - A_2)}$.

As one can see from the structure of the Jordan matrix J, the following Jordan block

$$\begin{pmatrix} 0 & 1 \\ 0 & 0 \end{pmatrix}$$

corresponds to one pair of the zero roots. The latter means that the general solution of system (38) has linear terms with respect to t. So, the elements of the families under consideration are unstable. A similar result has been derived for the families of solutions (35) and (36).

3   On Linear IMs of the Conservative System

Let us consider system (1) when $B_1 = B_2 = B_3 = \alpha = \text{const}$ [6]. Under these conditions, the system becomes conservative and possesses the energy integral

$$H = A\boldsymbol{\omega} \cdot \boldsymbol{\omega} - 2(\mathbf{s} \cdot \boldsymbol{\gamma}) + C\boldsymbol{\gamma} \cdot \boldsymbol{\gamma} = h \tag{39}$$

along with integrals V_1, V_2 (2).

We were solving the same problem for this conservative system as for the dissipative one: finding the IMs of type (3). By the technique applied in Sect. 2.2, under the corresponding constraints on the parameters of the problem, the following relations have been obtained:

(I) $\omega_1 + \frac{\alpha}{A_1} \gamma_1 + f_0 = 0$ when $A_3 = A_2$, $C_3 = C_2$, $s_2 = s_3 = 0$;   (40)

(II) $\omega_2 + \frac{\alpha}{A_2} \gamma_2 + f_0 = 0$ when $A_3 = A_1$, $C_1 = C_3$, $s_1 = s_3 = 0$;   (41)

(III) $\omega_3 + \frac{\alpha}{A_3} \gamma_3 + f_0 = 0$ when $A_1 = A_2$, $C_1 = C_2$, $s_1 = s_2 = 0$.   (42)

These are the first integrals of the system in question under the above constraints. It is easy to verify by the definition of first integral.

Integral (42) was previously found [5] in the absence of the moment of potential forces.

Further, one of the systems having integrals (40)–(42) is studied. We find its stationary solutions and IMs and investigate their stability.

3.1   Finding Stationary Solutions and IMs

Let us analyze, e.g., the differential equations possessing integral (42):

$$A_2\dot{\omega}_1 = -[(A_3 - A_2)\omega_2\omega_3 + (C_2 - C_3)\gamma_2\gamma_3 + \alpha\,(\gamma_2\omega_3 - \gamma_3\omega_2) + s_3\gamma_2],$$
$$A_2\dot{\omega}_2 = (A_3 - A_2)\omega_1\omega_3 + (C_2 - C_3)\gamma_1\gamma_3 + \alpha\,(\gamma_1\omega_3 - \gamma_3\omega_1) + s_3\gamma_1,$$
$$A_3\dot{\omega}_3 = \alpha(\gamma_2\omega_1 - \gamma_1\omega_2),$$
$$\dot{\gamma}_1 = \gamma_2\omega_3 - \gamma_3\omega_2, \ \ \dot{\gamma}_2 = \gamma_3\omega_1 - \gamma_1\omega_3, \ \ \dot{\gamma}_3 = \gamma_1\omega_2 - \gamma_2\omega_1. \tag{43}$$

These have been derived from Eq. (1), taking into account relations (42) and $B_i = \alpha$ $(i = 1, 2, 3)$.

The first integrals of Eq. (43):

$$H = A_2(\omega_1^2 + \omega_2^2) + A_3\omega_3^2 + C_2(\gamma_1^2 + \gamma_2^2) + C_3\gamma_3^2 - 2s_3\gamma_3 = h,$$
$$V_1 = A_2(\gamma_1\omega_1 + \gamma_2\omega_2) + A_3\gamma_3\omega_3 = \kappa,$$
$$V_2 = \gamma_1^2 + \gamma_2^2 + \gamma_3^2 = 1, \ V_3 = \omega_3 + \frac{\alpha}{A_3}\gamma_3 = -f_0. \tag{44}$$

The system in question is completely integrable. In order to find its stationary solutions and IMs by the Routh–Lyapunov method, the following combination of the first integrals

$$K - \lambda_0 H - \lambda_1 V_1 - \lambda_2 V_2 - \lambda_3 V_3^2. \tag{45}$$

is used, λ_i $(i = 0, \ldots, 3)$ are some constants.

As in Sect. 2.4, we write the stationary conditions of the integral K with respect to the phase variables:

$$\frac{\partial K}{\partial \omega_i} = 0, \ \ \frac{\partial K}{\partial \gamma_i} = 0 \ \ (i = 1, 2, 3). \tag{46}$$

These are a system of linear algebraic equations with respect to ω_i, γ_i $(i = 1, 2, 3)$ with the parameters $A_2, A_3, C_2, C_3, \alpha, \lambda_0, \lambda_1, \lambda_2, \lambda_3$.

Next, we add relation $V_2 = 1$ (44) to Eq. (46) and compute a lexicographical basis with respect to $\omega_3 > \omega_2 > \omega_1 > \gamma_3 > \gamma_2 > \gamma_1 > \lambda_3$ for the polynomials of a resulting system, using the built-in function *GroebnerBasis*. The result is a system which can be split up into two subsystems. From the equations of the subsystems, we have found both the families of solutions for differential Eq. (43) and the values of λ_3 under which integral K (45) takes a stationary value on the elements of these families.

The two found families of solutions for differential Eq. (43) are written as

$$\omega_1 = \omega_2 = 0, \ \omega_3 = \pm\frac{\alpha\lambda_1 - 2\lambda_2 + 2\lambda_0(C_3 \mp s_3)}{2\alpha\lambda_0 + A_3\lambda_1}, \ \gamma_1 = \gamma_2 = 0, \ \gamma_3 = \pm 1, \tag{47}$$

where λ_0, λ_1, and λ_2 are the parameters of the families. From a mechanical view point, the elements of these families correspond to permanent rotations of the body around the Oz axis.

The values of λ_3:

$$\lambda_{3_{1,2}} = \frac{A_3^2 \left[4(C_3 \mp s_3)\lambda_0^2 - A_3\lambda_1^2 - 4\lambda_0\lambda_2\right]}{4[(\alpha^2 + A_3(C_3 \mp s_3))\lambda_0 + \alpha A_3\lambda_1 - A_3\lambda_2]}. \tag{48}$$

Now, let us find the stationary IMs which the elements of families (47) belong to. For this purpose, we substitute the values of λ_{3_1} (λ_{3_2}) into Eq. (46) and construct a lexicographical basis with respect to $\omega_3 > \omega_1 > \gamma_3 > \gamma_2 > \lambda_2$ for the polynomials of a resulting system. The basis has the form:

$$4\lambda_0 \left(C_2\lambda_0 - \lambda_2\right) - A_2\lambda_1^2 = 0, \tag{49}$$

$$\lambda_1\gamma_2 - 2\lambda_0\omega_2 = 0, \ 1 \mp \gamma_3 = 0, \ 2\lambda_0\omega_1 - \lambda_1\gamma_1 = 0,$$

$$\pm 2\lambda_0 \left(2\alpha\lambda_0 + A_3\lambda_1\right)\omega_3 + 2\lambda_0 \left[2\lambda_0(C_2 - C_3 \pm s_3) - \alpha\lambda_1\right] - A_2\lambda_1^2 = 0. \tag{50}$$

Equations (50) together with $V_2 = 1$ (44) determine the two families of one-dimensional IMs for differential Eq. (43). It is easy to verify by IM definition. Here λ_0 and λ_1 are the parameters of the families.

Substituting the values of $\lambda_{3_{1,2}}$ and

$$\lambda_2 = C_2\lambda_0 - \frac{A_2\lambda_1^2}{4\lambda_0} \tag{51}$$

found from (49) into (45), we obtain the two families of integrals:

$$K_{1,2} = \lambda_0 H - \lambda_1 V_1 - \left(C_2\lambda_0 - \frac{A_2\lambda_1^2}{4\lambda_0}\right) V_2 - \bar{\lambda}_{3_{1,2}} V_3^2, \tag{52}$$

where $\bar{\lambda}_{3_{1,2}}$ are the expressions for $\lambda_{3_{1,2}}$ into which λ_2 (51) was substituted. The integrals take a stationary value on the elements of the corresponding families of IMs (50).

When λ_2 (51) is substituted into (47), we have the two subfamilies

$$\omega_1 = \omega_2 = 0, \ \omega_3 = \pm\frac{2\lambda_0 \left[\alpha\lambda_1 + 2\lambda_0(C_3 - C_2 \mp s_3)\right] + A_2\lambda_1^2}{2\lambda_0 \left(2\alpha\lambda_0 + A_3\lambda_1\right)},$$

$$\gamma_1 = \gamma_2 = 0, \ \gamma_3 = \pm 1 \tag{53}$$

of the families of solutions (47). The elements of these subfamilies belong to the corresponding families of IMs (50).

Further, we investigate the stability for the elements of families (50) and (53).

3.2   On Stability of the Stationary Solutions and IMs

We use the families of integrals $K_{1,2}$ (52) to analyze the stability of the elements of the families of solutions (53).

Introduce the deviations from the elements of one of the families:

$$y_1 = \omega_1 - \omega_1^0, \ y_2 = \omega_2 - \omega_2^0, \ y_3 = \omega_3 - \omega_3^0, \ y_4 = \gamma_1 - \gamma_1^0, \ y_5 = \gamma_2 - \gamma_2^0,$$

$$y_6 = \gamma_3 - \gamma_3^0.$$

Here ω_i^0 and γ_i^0 $(i = 1, 2, 3)$ are the values of the variables in unperturbed motion (53).

The variation of the integral K_1 (K_2) in the vicinity of the elements of the family under consideration in the deviations y_j $(j = 1, \ldots, 6)$ on the linear manifold $\delta V_2 = \pm 2y_6 = 0$, $\delta V_3 = A_3 y_3 + \alpha y_6 = 0$ is written as

$$4\lambda_0 \Delta K_{1,2} = A_2 \left[\zeta_1^2 + \zeta_2^2 \right], \tag{54}$$

where $\zeta_1 = \lambda_1 y_4 - 2\lambda_0 y_1$, $\zeta_2 = \lambda_1 y_5 - 2\lambda_0 y_2$.

As one can see from (54), the quadratic form $\Delta K_{1,2}$ is sign-definite with respect to the variables ζ_1 and ζ_2 when $\lambda_0 \neq 0$. Whence it follows the stability of the elements of the families under study with respect to the variables $\zeta_1 = \lambda_1 \gamma_1 - 2\lambda_0 \omega_1$, $\zeta_2 = \lambda_1 \gamma_2 - 2\lambda_0 \omega_2$.

Using the same families of the integrals, we investigated the stability of the elements of the families of IMs (50) which solutions (53) belong to.

For the equations of perturbed motion, the integral K_1 (K_2) in the vicinity of the elements of the family under study on the linear manifold $\delta V_3 = y_2 + \alpha y_4 / A_3 = 0$ has the form:

$$\Delta K_{1,2} = \lambda_0 \left(A_2 y_1^2 + \frac{1}{4} A_2 \bar{\lambda}^2 y_3^2 + \left[C_3 - C_2 + \frac{\alpha^2}{A_3} + \alpha \bar{\lambda} + \frac{1}{4} A_2 \bar{\lambda}^2 \right] y_4^2 \right),$$

where y_1, y_2, y_3, y_4 are the deviations from the elements of the family under consideration, $\bar{\lambda} = \lambda_1 / \lambda_0$.

The full list of the conditions for the quadratic form $\Delta K_{1,2}$ to be positive definite, which was derived with the use of the built-in function *Reduce*, is rather bulky. Some of the conditions are given below:

$\lambda_0 > 0$, $C_2 > 0$, $C_3 > 0$ and

$$\left[\left(C_3 > C_2 \text{ and } \left[\alpha < 0 \text{ and } (0 < \lambda_1 \leq \frac{\lambda_0}{\alpha} (C_2 - C_3) \text{ or } \lambda_1 < 0) \right] \right. \right.$$

$$\left. \text{or } \left[\alpha > 0 \text{ and } (\frac{\lambda_0}{\alpha} (C_2 - C_3) \leq \lambda_1 < 0 \text{ or } \lambda_1 > 0) \right] \right)$$

$$\text{or } \left(C_3 = C_2 \text{ and } \left[(\alpha < 0 \text{ and } \lambda_1 < 0) \text{ or } (\alpha > 0 \text{ and } \lambda_1 > 0) \right] \right)$$

$$\text{or } \left(C_3 < C_2, A_2 > 0, 0 < A_3 < \frac{\alpha^2}{C_2 - C_3} \text{ and} \right.$$

$$\left[(\lambda_1 > 0 \text{ and } \alpha < 0 \text{ and } \lambda_0 \geq D) \text{ or} \right.$$

$$\left. \left. (\lambda_1 < 0 \text{ and } \alpha > 0 \text{ and } \lambda_0 \geq D) \right] \right) \right], \quad D = \frac{\alpha A_3 \lambda_1}{A_3 (C_2 - C_3) - \alpha^2}.$$

These are sufficient for the stability of the elements of the families of IMs (50). Note that the constraints on the parameters λ_0 and λ_1 separate the subfamilies from the families of integrals $K_{1,2}$, which enable us to obtain these sufficient conditions.

4   Conclusion

In this work, differential equations describing the rotation of a rigid body with a fixed point in a magnetic field, taking into account the Barnett–London effect and the moment of potential forces, were analyzed. The equations contain 12 parameters. Two cases, when the system under study is dissipative and when it is conservative, were considered. A technique for finding the linear IMs of the equations was proposed. It is based on a combination of undetermined coefficients method with computer algebra methods. In the dissipative case, using this technique, the linear IMs of codimension 1 have been obtained. Most of them are previously unknown. The qualitative analysis of the equations having such solutions was carried out. For this purpose, the reduction of these equations to corresponding equations on the IMs was done. The stationary solutions and IMs of the reduced system have been found and the sufficient conditions of their stability in the Lyapunov sense have been derived. Solutions corresponding to the above solutions in the original phase space were presented and the stability of some of them was investigated.

In the conservative case, the first integrals of the system under study have been obtained. The qualitative analysis of the equations possessing these integrals was performed.

The research technique applied in the present work as well as the obtained results may be of interest in the qualitative analysis of similar multiparametric systems.

References

1. Barnett, S.J.: Magnetization by rotation. Phys. Rev. **6**(4), 239–270 (1915)
2. Egarmin, I.E.: On the magnetic field of a rotating superconducting body. In: Aerophysics and Geocosmic Research. Izd. MOTI, Moscow, pp. 95–96 (1983). (in Russian)
3. Everitt, C.W.F., et al.: Gravity probe B: final results of a space experiment to test general relativity. Phys. Rev. Lett. **106**, 221101 (2011)
4. Urman, Y.M.: Influence of the Barnett-London effect on the motion of a superconducting rotor in a nonuniform magnetic field. Tech. Phys. **43**(8), 885–889 (1998)
5. Samsonov, V.A.: On the rotation of a rigid body in a magnetic field. Izv. Akad. Nauk SSSR. Mekhanika Tverdogo Tela. **4**, 32–34 (1984). (in Russian)
6. Kozlov, V.V.: To the problem of the rotation of a rigid body in a magnetic field. Izv. Akad. Nauk SSSR. Mekhanika Tverdogo Tela. **6**, 28–33 (1985). (in Russian)
7. Gorr, G.V.: A linear invariant relation in the problem of the motion of a gyrostat in a magnetic field. J. Appl. Math. Meth. **61**(4), 549–552 (1997)
8. Hess, W.: Über die Euler'schen Bewegungsgleichungen und über eine neue partikuläre Lösung des Problems der Bewegung eines starren Körpers um einen festen Punkt. Math. Ann. **37**(2), 153–181 (1890)
9. Irtegov, V., Titorenko, T.: On stationary motions of the generalized kowalewski gyrostat and their stability. In: Gerdt, V.P., Koepf, W., Seiler, W.M., Vorozhtsov, E.V. (eds.) CASC 2017. LNCS, vol. 10490, pp. 210–224. Springer, Cham (2017). https://doi.org/10.1007/978-3-319-66320-3_16

10. Lyapunov, A.M.: On permanent helical motions of a rigid body in fluid. Collected Works, USSR Acad. Sci., Moscow-Leningrad. **1**, 276–319 (1954). (in Russian)
11. Banshchikov, A.V., Burlakova, L.A., Irtegov, V.D., Titorenko, T.N.: Software package for finding and stability analysis of stationary sets. Certificate of State Registration of Software Programs. FGU-FIPS. No. 2011615235 (2011)
12. Lyapunov, A.M.: The General Problem of the Stability of Motion. Taylor & Francis, London (1992)

Robust Schur Stability of a Polynomial Matrix Family

Elizaveta Kalinina$^{(\boxtimes)}$, Yuri Smol'kin, and Alexei Uteshev

Faculty of Applied Mathematics, St. Petersburg State University,
7–9 Universitetskaya nab., St. Petersburg 199034, Russia
{e.kalinina,a.uteshev}@spbu.ru, st040343@student.spbu.ru
http://www.apmath.spbu.ru/ru/

Abstract. The problem of robust Schur stability of a polynomial matrix family is considered as that of discovering the structure of the stability domain in parameter space. The algorithms are proposed for establishing whether or not any given box in the parameter space belongs to this domain, and for finding the distance to instability from any internal point of the domain to its boundary. The treatment is performed in the ideology of analytical algorithm for elimination of variables and localization of zeros of algebraic systems. Some examples are given.

Keywords: Matrix polynomials · Robust schur stability ·
Parameters · Discriminant

1 Introduction

For a polynomial $f(z) \in \mathbb{C}[z]$, its Schur stability (or D-stability) property is defined as that of location of all its zeros inside the unit disc of the complex plane:

$$D = \{z \in \mathbb{C} \mid |z| < 1\}. \tag{1}$$

The same definition relates to the matrix $M \in \mathbb{C}^{n \times n}$ with numerical entries if the whole spectrum lies inside D. The D-stability property is of an importance for estimating the behavior of solutions to difference equation systems [11]. There exist several criteria for establishing the Schur stability of a polynomial in terms of its coefficients, for instance the Schur – Cohn or Jury's criteria [17,20,25].

The counterpart of the problem for matrices with entries depending on parameters varying within some set $\mathfrak{B}$ is sometimes referred to as the **robust Schur stability** problem with the meaning that all the matrices of this family should be D-stable. This property is of vital importance for the parameter synthesis in Control Theory. In digital signal processing applications such as sampling rate conversion, echo cancellation, phased-array antenna systems, time delay estimation, timing adjustment in all-digital receivers, modelling of music

Supported by RFBR, project No. **17-29-04288**.

M. England et al. (Eds.): CASC 2019, LNCS 11661, pp. 262–279, 2019.
https://doi.org/10.1007/978-3-030-26831-2_18

instruments, and speech coding and synthesis, there is a need to design a digital filter with predicted characteristics [4, 28, 34]. Therefore, the tolerances are to be estimated for the permissible parameter variations.

In the present paper, we will tackle the D-stability problem for the matrix family

$$\left\{ M(\mu) = [m_{jk}(\mu)]_{j,k=1}^{n} \mid \mu = (\mu_1, \mu_2, \dots, \mu_k) \in \mathfrak{B} \right\}. \tag{2}$$

Here $\{m_{jk}(\mu)\}_{j,k=1}^{n}$ are real polynomials in μ while $\mathfrak{B}$ is a box:

$$\mathfrak{B} = \left\{ \mu_1^- \leq \mu_1 \leq \mu_1^+, \mu_2^- \leq \mu_2 \leq \mu_2^+, \dots, \mu_k^- \leq \mu_k \leq \mu_k^+ \right\} \subset \mathbb{R}^k. \tag{3}$$

For the case of symmetric matrices, the robust Schur stability problem is treated in recent book [26]. In [7, 15], necessary and sufficient conditions for the zeros of arbitrary polynomial matrix to belong to a given region D of the complex plane as a linear matrix inequality (LMI) feasibility problem are formulated. To solve this problem, interior-point methods are used. In [8], analysis and synthesis techniques for quadratic stability in LMI regions that embrace most practically useful stability regions are discussed.

In [6], robust Schur stability of a polynomial matrix family is reduced to the positivity of multivariate polynomials, and the Bernstein expansion method [19] is applied to test positivity of the obtained polynomials.

In Sect. 3, we first detail the structure of the boundary of the set of Schur stable matrices $M(\mu)$ in the parameter space. Then we propose an algebraical approach to the problem of testing the stability of family (2). The algorithm is based on the Le Verrier method for the computation of the characteristic polynomial of a matrix [33] and algebraic procedures for checking the positivity property of multivariate polynomials in the given domain [18]. Another problem dealt with in Sect. 3 is that of finding **the distance to instability** in the parameter space, i.e., the Euclidean distance $d_*(\mu^{(0)})$ from a given point $\mu^{(0)} \in \mathbb{R}^k$ corresponding to a stable matrix $M(\mu^{(0)})$ to the nearest point $\mu_* \in \mathbb{R}^k$ at the boundary of domain of stable matrices [26]. This notion should be distinguished from the one related to the *distance to instability in the matrix space* or **stability radius**. The latter is defined for a stable matrix $A \in \mathbb{R}^{n \times n}$ as the Frobenius norm of the matrix $E \in \mathbb{R}^{n \times n}$ such that the matrix $A + E$ is the nearest to A unstable matrix [1]. This definition can be treated as a particular case of the first one for the matrix set $\left\{ A + [\mu_{j\ell}]_{j,\ell=1}^{n} \right\}$.

In Sect. 4, some numerical examples are presented illuminating the efficiency of the suggested algorithms.

Hereinafter the word *stability* should be understood in the meaning *Schur stability (D-stability)*.

2   Algebraic Preliminaries

Here we give some auxiliary results regarding the properties of the zero sets of polynomials.

2.1   Newton Sums of a Polynomial

Consider a polynomial

$$f(z) = a_0 z^n + a_1 z^{n-1} + a_2 z^{n-2} + \ldots + a_n, \quad \{a_0 \neq 0, a_1, \ldots, a_n\} \subset \mathbb{R}, \quad (4)$$

and denote $\{\alpha_1, \alpha_2, \ldots, \alpha_n\} \subset \mathbb{C}$ its zeros counted with their multiplicities.
The **Newton sums** of the polynomial $f(z)$ are formally defined as

$$s_0 = n; \quad s_k = \sum_{j=1}^{n} \alpha_j^k \text{ for } k \in \mathbb{N},$$

while the **Newton identities** [14]

$$s_0 = n; s_1 = -a_1/a_0;$$

$$s_k = \begin{cases} -(a_1 s_{k-1} + a_2 s_{k-2} + \ldots + a_{k-1} s_1 + k a_k)/a_0 & \text{if } k \in \{2, \ldots, n\}, \\ -(a_1 s_{k-1} + a_2 s_{k-2} + \ldots + a_n s_{k-n})/a_0 & \text{if } k > n. \end{cases} \quad (5)$$

allow one to compute recursively these sums as rational functions (polynomials if $a_0 = 1$) of the coefficients of $f(z)$.

Conversely, if for some reason, the canonical representation (4) of a normalized polynomial is not granted but we are able to compute somehow its Newton sums, the following inversions of the Newton identities

$$a_1 = -s_1; a_2 = -(s_2 + a_1 s_1)/2;$$
$$a_k = -(s_k + a_1 s_{k-1} + a_2 s_{k-2} + \ldots + a_{k-1} s_1)/k, \text{ if } k \in \{3, \ldots, n\} \quad (6)$$

allow one to restore the coefficients of $f(z)$. This opportunity gives rise to the Le Verrier method [12,14] for computation of the characteristic polynomial of a matrix $A \in \mathbb{C}^{n \times n}$. Indeed, the Newton sums of this polynomial can be evaluated via computation of the traces of powers of the matrix A:

$$s_k = \mathrm{Tr}(A^k) \quad \text{for } k \in \mathbb{N}. \quad (7)$$

It turns out that with the aid of the sequence of Newton sums of a polynomial, one can express some symmetric function of the pairs of zeros of this polynomial. We will utilize further two such functions.

Theorem 1. *Set the Newton sums of the polynomial with zeros $\alpha_\ell \alpha_k$ of degree $n(n-1)/2$*

$$S_j := \sum_{1 \leq \ell < k \leq n} \alpha_\ell^j \alpha_k^j \quad \text{for } j \in \mathbb{N}.$$

Then

$$S_j = (s_j^2 - s_{2j})/2. \quad (8)$$

Proof. One has

$$s_j^2 = \left(\sum_{k=1}^{n} \alpha_k\right)^2 = \sum_{k=1}^{n} \alpha_k^2 + 2 \sum_{1 \leq \ell < k \leq n} \alpha_\ell^j \alpha_k^j = s_{2j} + 2S_j. \qquad \square$$

The second symmetric function of the zeros of the polynomial $f(z)$ is formally defined as

$$\mathcal{D}(f(z)) := a_0^{2n-2} \prod_{1 \le \ell < k \le n} (\alpha_\ell - \alpha_k)^2$$

and is known as the **discriminant** of the polynomial $f(z)$. It vanishes iff the polynomial $f(z)$ possesses a multiple zero (or, equivalently, iff $f(z)$ possesses a common zero with $f'(z)$). For the aim of expressing $\mathcal{D}(f(z))$ via the Newton sums of $f(z)$, we introduce the **Hankel determinant**

$$H_k := \det[s_{j+\ell}]_{j,\ell=0}^{k-1} = \begin{vmatrix} s_0 & s_1 \; s_2 & \cdots & s_{k-1} \\ s_1 & s_2 \; s_3 & \cdots & s_k \\ \vdots & \vdots \;\; \vdots & & \vdots \\ s_{k-1} & s_k \; s_{k+1} & \cdots & s_{2k-2} \end{vmatrix} . \tag{9}$$

Theorem 2. *The following equality is valid:*

$$\mathcal{D}(f(z)) = a_0^{2n-2} H_n . \tag{10}$$

It turns out that the sequence of Hankel determinants $\{H_k\}_{k=1}^n$ introduced by (9) permits one to establish the exact numbers of real zeros for the polynomial $f(z)$. Moreover, a slight modification of these determinants allows one to construct a sequence of polynomials that localize all its real zeros. For this aim, introduce the parameter dependent determinant

$$\mathcal{H}_k(z) := \begin{vmatrix} s_0 & s_1 \; s_2 & \cdots & s_k \\ s_1 & s_2 \; s_3 & \cdots & s_{k+1} \\ \vdots & \vdots \;\; \vdots & & \vdots \\ s_{k-1} & s_k \; s_{k+1} & \cdots & s_{2k-1} \\ 1 & z \;\; z^2 & \cdots & z^k \end{vmatrix} . \tag{11}$$

Its expansion by the last row yields the polynomial in z

$$\mathcal{H}_k(z) \equiv \sum_{j=0}^k h_{kj} z^{k-j},$$

which is sometimes called the kth **Hankel polynomial** generated by the sequence $\{s_j\}$. It is evident that $h_{k0} = H_k$.

Theorem 3 (Jacobi, Joachimsthal [16, 20, 27]).
Let $H_n = 0, \ldots, H_{\mathfrak{r}+1} = 0, H_{\mathfrak{r}} \ne 0, \ldots, H_1 \ne 0$. Then
  (i) *The number of distinct zeros of $f(z)$ equals $\mathfrak{r}$;*
  (ii) *The number of distinct real zeros of $f(z)$ equals*

$$\mathcal{P}(1, H_1, \ldots, H_{\mathfrak{r}}) - \mathcal{V}(1, H_1, \ldots, H_{\mathfrak{r}}) ;$$

  (iii) *The number of distinct real zeros of $f(z)$ lying in the interval $[a, b], a < b$ equals*

$$\mathcal{V}(1, \mathcal{H}_1(a), \ldots, \mathcal{H}_{\mathfrak{r}}(a)) - \mathcal{V}(1, \mathcal{H}_1(b), \ldots, \mathcal{H}_{\mathfrak{r}}(b)) .$$

Here $\mathcal{P}$ (or $\mathcal{V}$) stands for the number of permanences (variations) of sign for the given sequences[1].

Corollary 1. *The following identity is valid:*

$$\mathcal{H}_n(z) \equiv H_n f(z)/a_0 .$$

Remark. One may notice an evident relationship of the part (iii) of Theorem 3 to the algorithm of localization of zeros of the polynomial $f(z)$ based on the Sturm – Habicht sequence construction [2]. The principal distinction in the two procedures is that the Sturm – Habicht algorithm constructs the sequence of polynomials with decreasing degrees starting with the polynomial $f(z)$, whilst the Jacobi – Joachimsthal sequence is composed of polynomials with increasing degrees, with the final one coinciding with $f(z)$. This distinction is of importance for the class of problems where the polynomial $f(z)$ is not a priory represented in canonical form like the above mentioned problem related to characteristic polynomial of a matrix. With Newton sums evaluated via (7), computation of sequence $\mathcal{H}_1(z), \ldots, \mathcal{H}_n(z)$ results in this polynomial and furnishes, *free of an additional charge*, an opportunity to locate its real zeros.

Corollary 2. *[29]. If $H_n = 0, H_{n-1} \neq 0$, then the polynomial $f(z)$ possesses a single multiple zero with its multiplicity equal to 2. This zero can be expressed via the two coefficients of the polynomial $\mathcal{H}_{n-1}(z)$:*

$$\alpha = s_1 + h_{n-1,1}/H_{n-1} . \tag{12}$$

If $H_n = H_{n-1} = \ldots = H_{n-k+1} = 0, H_{n-k} \neq 0, k > 1$, then $\gcd(f(z), f'(z))$ is of the order k and can be expressed with the aid of the kth order minors of H_n.

Computation of the sequence of Hankel polynomials in part (iii) of Theorem 3 can be optimized with the following result.

Theorem 4 (Jacobi, Joachimsthal [16,30]). *Let $k \in \{3, \ldots, n\}$. If $H_{k-1} \neq 0, H_k \neq 0$, then the following identity is valid:*

$$\frac{H_k}{H_{k-1}} \mathcal{H}_{k-2}(z) - \left(z - \frac{h_{k-1,1}}{H_{k-1}} + \frac{h_{k1}}{H_k} \right) \mathcal{H}_{k-1}(z) + \frac{H_{k-1}}{H_k} \mathcal{H}_k(z) \equiv 0. \tag{13}$$

Formula (13) permits one to compute the sequence of polynomials $\{\mathcal{H}_k(z)\}_{k=1}^n$ recursively with every polynomial computed as linear combination of two preceding ones and with the two constants involved also evaluated via the coefficients of these polynomials (v. [30]):

$$\begin{cases} H_k = & s_{k-1}h_{k-1,k-1} + s_k h_{k-1,k-2} + \ldots + s_{2k-2}h_{k-1,0}, \\ h_{k1} = -(s_k h_{k-1,k-1} + s_{k+1} h_{k-1,k-2} + \ldots + s_{2k-1}h_{k-1,0}). \end{cases} \tag{14}$$

---

[1] $\mathcal{P}(A_1, \ldots, A_K) := \sum_{j=1}^{K-1} \mathcal{P}(A_j, A_{j+1})$ where $\mathcal{P}(A_j, A_{j+1}) := 1$ if $A_j A_{j+1} > 0$ and $\mathcal{P}(A_j, A_{j+1}) := 0$ if $A_j A_{j+1} < 0$. $\mathcal{V}(A_1, \ldots, A_K)$ is defined similarly with $\mathcal{V}(A_j, A_{j+1}) := 1 - \mathcal{P}(A_j, A_{j+1})$.

Given polynomial (4) and a polynomial

$$g(z) = b_0 z^m + b_1 z^{m-1} + \cdots + b_{m-1} z + b_m \in \mathbb{C}[z], \ b_0 \neq 0$$

with zeros $\beta_1, \ldots, \beta_m$, the **resultant** of these polynomials is formally defined by the formula

$$\mathcal{R}(f,g) = a_0^m b_0^n \prod_{\ell=1}^{n} \prod_{j=1}^{m} (\alpha_\ell - \beta_j), \tag{15}$$

while practically can be expressed as a polynomial in the coefficients of $f(z)$ and $g(z)$ using several determinantal representations. The mostly known is the Sylvester representation

$$\mathcal{R}(f,g) = \begin{vmatrix} a_0 & a_1 & a_2 & \cdots & & a_n & 0 & 0 & \cdots & 0 \\ 0 & a_0 & a_1 & \cdots & & & a_n & 0 & \cdots & 0 \\ & \cdots & & & & & & & & \\ 0 & 0 & 0 & \cdots & & a_0 & a_1 & a_2 & \cdots & a_n \\ b_0 & b_1 & \cdots & & b_m & 0 & \cdots & & & 0 \\ 0 & b_0 & b_1 & \cdots & & b_m & 0 & 0 & \cdots & 0 \\ & \cdots & & & & & & & & \\ 0 & 0 & 0 & \cdots & & b_0 & b_1 & b_2 & \cdots & b_m \end{vmatrix} \begin{matrix} \left.\vphantom{\begin{matrix}a\\a\\a\\a\end{matrix}}\right\} m \text{ rows} \\[2em] \left.\vphantom{\begin{matrix}a\\a\\a\\a\end{matrix}}\right\} n \text{ rows} \end{matrix}$$

Obviously, polynomials $f(z)$ and $g(z)$ have a common zero iff $\mathcal{R}(f,g) = 0$.

Remark. In some further formulas involving discriminant or resultant, we will occasionally specify the variable of the considered polynomials with subscripts like $\mathcal{D}_z$ or $\mathcal{R}_z$.

2.2 Positivity of a Multivariate Polynomial

We present here a multidimensional counterpart for Theorem 3. We wish to establish the relative position of the manifold given implicitly in $\mathbb{R}^k, k \geq 2$ by an algebraic equation to a specified box

$$\mathfrak{B} = \left\{ x_1^- \leq x_1 \leq x_1^+, x_2^- \leq x_2 \leq x_2^+, \ldots, x_k^- \leq x_k \leq x_k^+ \right\} \tag{16}$$

in this space. For the proof of the following result, we refer to [18].

Theorem 5 *Consider a multivariate polynomial*

$$G(X) \in \mathbb{R}[X], X := (x_1, \ldots, x_k) \in \mathbb{C}^k, k \geq 2, \deg G = N \geq 1$$

and expand it in powers of one of the variables, say x_1:

$$G(X) \equiv a_0(x_2, \ldots, x_k) x_1^N + \ldots + a_N(x_2, \ldots, x_k).$$

Let $a_0(x_2, \ldots, x_k) \neq 0$ in the box

$$\mathfrak{B}_1 = \left\{ x_2^- \leq x_2 \leq x_2^+, \ldots, x_k^- \leq x_k \leq x_k^+ \right\}.$$

Polynomial $G(X)$ does not vanish in the box (16) iff the following conditions hold:

(i) *For any $j \in \{2, 3, \ldots, k\}$, the polynomials*

$$G_j^- := G(x_1, \ldots, x_{j-1}, x_j^-, x_{j+1}, \ldots, x_k)$$

and

$$G_j^+ := G(x_1, \ldots, x_{j-1}, x_j^+, x_{j+1}, \ldots, x_k)$$

do not vanish in the box

$$\mathcal{B}_j = \{\mu_1^- \le \mu_1 \le \mu_1^+, \ldots, \mu_{j-1}^- \le \mu_{j-1} \le \mu_{j-1}^+,$$
$$\mu_{j+1}^- \le \mu_{j+1} \le \mu_{j+1}^+, \ldots, \mu_k^- \le \mu_k \le \mu_k^+\}.$$

(ii) *The system of equations*

$$G(X) = 0, \; \partial G(X)/\partial x_2 = 0, \ldots, \; \partial G(X)/\partial x_k = 0 \qquad (17)$$

does not have real zeros in the box (16).

For the case of a bivariate polynomial $G(x, y)$, the condition of the part (ii) of the above theorem can be immediately verified via application of Theorem 3. As for the condition (ii), system (17) takes then the form

$$G(x, y) = 0, \quad \partial G(x, y)/\partial y = 0. \qquad (18)$$

One can eliminate the variable y from this system using the discriminant technique introduced in Sect. 2.1. Assuming that this discriminant $F_1(x) := \mathcal{D}_y(G(x, y))$ is not identically zero (i.e., the set of zeros of (18) is zero-dimensional), it is a univariate polynomial. Under additional assumption of the absence of multiple zeros for $F_1(x)$, its real zeros yield the x-component of the real zeros of system (18). To establish the (non)existence of the latter in the box

$$\mathcal{B} = \{a^- \le x \le a^+, b^- \le y \le b^+\}$$

one can first verify, via Theorem 3, if the polynomial $F_1(x)$ does not have real zeros in $[a^-, a^+]$. If it does, then any such a zero can be localized with the aid of the mentioned theorem within an arbitrary prescribed accuracy. Application of the result of Corollary 2 permits one to evaluate y-component of the corresponding zero of system (18) and to verify if it does not lie in $[b^-, b^+]$.

Remark. The suggested approach has an evident relationship to a real quantifier elimination (QE) problem [3, 24].

The just outlined scheme for checking the conditions of Theorem 5 for the bivariate case can be extended for the general case of a multivariate polynomial. It is connected with one notion that is introduced at the end of the next subsection.

2.3   Distance from a Point to an Algebraic Manifold

We treat here the problem of Euclidean distance evaluation from a point X_0 to the algebraic manifold defined implicitly by the equation

$$G(X) = 0 \qquad (19)$$

in $\mathbb{R}^k, k \in \{2,3\}, \deg G > 1$. For this aim, we utilize the construction of the so-called **distance equation**, i.e., the algebraic univariate equation whose zero set coincides with that of the critical values of the squared distance function from the point to the manifold [31].

Theorem 6. *Let $G(0,0) \neq 0$, and $G(x,y)$ be an even polynomial in y. Expand G in powers of y^2 and denote $\widetilde{G}(x,y^2) \equiv G(x,y)$. Equation $G(x,y) = 0$ does not define a real curve if*
  (i) *equation $G(x,0) = 0$ does not have real zeros and*
  (ii) *distance equation*

$$\mathcal{F}(z) := \mathcal{D}_x(\widetilde{G}(x, z - x^2)) = 0 \qquad (20)$$

does not possess positive zeros. If any of these conditions fails, then the distance from $X_0 = (0,0)$ to the curve $G(x,y) = 0$ equals the minimal of the two values:
  1. the minimal absolute value of real zeros of the equation $G(x,0) = 0$,
  2. the square root of the minimal positive zero of Eq. (20) provided that this zero is not a multiple one.

Remark. The condition of simplicity of the minimal positive zero of the distance equation appeared in theorems of the present subsection is essential since in some (fortunately exceptional) cases, this zero is generated by a pair of imaginary points in the manifold [32].

The generalization of this result to the case of an arbitrary polynomial $G(x,y)$, not necessarily even in any of its variables, can be performed by reduction to the just treated one via artificial *evenization* of the problem. Unfortunately, this causes the appearance of an extraneous factor in the distance equation.

Theorem 7. *Let $G(0,0) \neq 0$, and $G(x,y)$ be not an even polynomial in y. Split G into the sum of even and odd terms in this variable:*

$$G(x,y) \equiv G_1(x,y^2) + yG_2(x,y^2), \quad \{G_1, G_2\} \subset \mathbb{R}[x,y^2].$$

Let

$$\widetilde{G}(x,y^2) := G(x,y)G(x,-y) \equiv G_1^2(x,y^2) - y^2 G_2^2(x,y^2)$$

and compute the polynomial $\mathcal{F}(z)$ via (20). The latter is reducible over $\mathbb{R}$:

$$\mathcal{F}(z) \equiv \mathcal{F}_1(z)\mathcal{F}_2^2(z) \text{ with } \mathcal{F}_2(z) := \mathcal{R}_x(G_1(x, z - x^2), G_2(x, z - x^2)).$$

Equation $G(x, y) = 0$ does not define a real curve if
  (i) *equation $G(x, 0) = 0$ does not possess real zeros and*
  (ii) *distance equation*

$$\mathcal{F}_1(z) = 0 \tag{21}$$

does not possess positive zeros.

If any of these conditions fails, then the distance from $X_0 = (0, 0)$ to the curve $G(x, y) = 0$ equals the minimal of the two values:

1. the minimal absolute value of real zeros of the equation $G(x, 0) = 0$,

2. the square root of the minimal positive zero of the Eq. (21) provided that this zero is not a multiple one.

Conditions **(i)** and **(ii)** of Theorems 6 and 7 can be verified using symbolic algebraic algorithms from Theorem 3. Equations (20) and (21) are the distance equations for the point $X_0 = (0, 0)$ and the curve $G(x, y) = 0$. For arbitrary point $X_0 = (x_0, y_0)$, this equation can be extracted from the corresponding theorem via shifting the origin: $\widehat{G}(x, y) := G(x + x_0, y + y_0)$. Generically, one gets $\deg \mathcal{F}_1(z) = (\deg G)^2$ for (21).

The treatment of the problem for the case of surfaces in $\mathbb{R}^3$ can be organized in a similar manner. Consider, for instance, a polynomial that is even in one of its variables, say $G(x_1, x_2, x_3^2)$. The distance equation for the point $(0, 0, 0)$ and the surface $G = 0$ can be obtained as a result of elimination of the variables x_1, x_2 from the system of equations

$$\widetilde{G} = 0, \ \partial\widetilde{G}/\partial x_1 = 0, \ \partial\widetilde{G}/\partial x_2 = 0 \quad \text{for } \widetilde{G}(x_1, x_2, z) := G(x_1, x_2, z - x_1^2 - x_2^2).$$

One may notice the similarity of the obtained system with system (17). The common underlying notion is known as the **multivariate discriminant**, i.e., an algebraic function of the coefficients of a multivariate polynomial $G(x_1, \ldots, x_n)$ responsible for the existence of a multiple zero for this polynomial, i.e., zero for the system

$$G = 0, \ \partial G/\partial x_1 = 0, \ldots, \partial G/\partial x_n = 0.$$

There are different approaches for constructive computation of this object with the universal one based on the Gröbner basis construction. For computations in Example 3 considered further, we utilize the procedure of the multivariate polynomial resultant computation based on a certain determinantal representation via its coefficients [5].

3   Stability Domain in the Parameter Space

3.1   Structure of the Boundary

Theorem 8. *For matrix (2), consider the characteristic polynomial and its reciprocal:*

$$f(z; \mu) := \det(zI - M(\mu)) \quad \text{and} \quad f^*(z; \mu) \equiv z^n f(1/z; \mu).$$

Assume that matrix (2) is nonsingular for μ in $\mathfrak{B}$ with the probable exception of manifold of codimension 1. Family (2) is stable iff its arbitrary member is stable and the polynomial

$$\Phi(\mu) := \mathcal{R}_z(f(z;\mu), f^*(z;\mu)) \tag{22}$$

is positive for $\mu \in \mathfrak{B}$.

Proof. If we denote by $\alpha_1(\mu), \ldots, \alpha_n(\mu)$ the zeros of

$$f(z;\mu) = z^n + a_1(\mu)z^{n-1} + \cdots + a_{n-1}(\mu)z + a_n(\mu),$$

then those of $f^*(z;\mu)$ are $1/\alpha_1(\mu), \ldots, 1/\alpha_n(\mu)$. Due to the definition of resultant (15), one obtains

$$\Phi(\mu) = \mathcal{R}_z(f(z;\mu), f^*(z;\mu)) = a_n^n(\mu) \prod_{j,k=1}^n (\alpha_j(\mu) - 1/\alpha_k(\mu))$$

$$= \frac{(-1)^{n^2} a_n^n(\mu)}{\prod_{j=1}^n \alpha_j^n(\mu)} \prod_{j,k=1}^n (1 - \alpha_j(\mu)\alpha_k(\mu)) = \prod_{j,k=1}^n (1 - \alpha_j(\mu)\alpha_k(\mu))$$

$$= \prod_{j=1}^n (1 - \alpha_j(\mu)) \prod_{j=1}^n (1 + \alpha_j(\mu)) \prod_{1 \le j < k \le n} (1 - \alpha_j(\mu)\alpha_k(\mu))^2. \tag{23}$$

If the matrix $M(\mu)$ is stable for some specialization of parameter $\mu = \mu_0 \in \mathfrak{B}$, then $\Phi(\mu_0) > 0$. When the parameter μ varies continuously within the (simply connected domain) $\mathfrak{B}$ starting from this value, the eigenvalues $\{\alpha_j(\mu)\}_{j=1}^n$ of the matrix drift continuously within disk (1). The inequality $\Phi(\mu) > 0$ keeps to be valid until either any real eigenvalue α_j reaches ± 1, or a pair of complex conjugate eigenvalues $\{\alpha_j(\mu), \alpha_k(\mu) = \overline{\alpha_j(\mu)}\}$ reaches the unit circle. Therefore, the condition $\Phi(\mu) > 0$ prevents the spectrum of the matrix to leave disc (1). $\square$

Thus, the boundary of the set of stable matrices $M(\mu)$ in the parameter space $\mathbb{R}^k$ is given by the equation

$$\Phi(\mu) = 0.$$

Next we determine the structure of this manifold.

Theorem 9. *Under conditions and in notation of Theorem 8, one has*

$$\Phi(\mu) \equiv \det f^*(M(\mu); \mu)$$

$$= \det(I + a_1(\mu)M(\mu) + \cdots + a_{n-1}(\mu)M^{n-1}(\mu) + a_n(\mu)M^n(\mu)). \tag{24}$$

Proof. It follows from the more general result [14]. For any polynomial $g(z) = b_0 z^m + \cdots + b_m \in \mathbb{C}[z]$ and any matrix $A \in \mathbb{C}^{n \times n}$, the following equality is valid

$$\det g(A) = \mathcal{R}_z(\det(zI - A), g(z)).$$

$\square$

For low order matrices $M(\mu)$, computation of $\Phi(\mu)$ via formula (24) does not cause troubles. However, for higher orders, any simplification of computations is valuable. One of the opportunities for such a potential simplification is provided by representation (23).

Corollary 3. *Polynomial (22) is reducible over* $\mathbb{R}$:

$$\Phi(\mu) \equiv f(1; \mu) f(-1; \mu) \Phi_1^2(1; \mu). \tag{25}$$

Here

$$\Phi_1(z; \mu) \equiv \prod_{1 \leq j < k \leq n} (z - \alpha_j(\mu)\alpha_k(\mu)). \tag{26}$$

Being symmetric functions of the zeros of the polynomial $f(z; \mu)$, the coefficients of $\Phi_1(z; \mu)$ can be expressed as polynomials over $\mathbb{Z}$ in the coefficients of $f(z; \mu)$.

Corollary 4. *Family (2) is stable iff* $M(\mu^{(0)})$ *is stable for a particular specialization of the parameter* $\mu = \mu^{(0)}$ *in (3) and polynomials*

$$f(1; \mu), \ f(-1; \mu), \ \Phi_1(1; \mu) \tag{27}$$

are positive in (3). Equations

$$f(1; \mu) = 0, \ f(-1; \mu) = 0, \ \Phi_1(1; \mu) = 0 \tag{28}$$

define implicit manifolds in $\mathbb{R}^k$ *that form the boundary for the* **stability domain** *in the parameter space, i.e., of parameter specializations responsible for stability of the matrix* $M(\mu)$.

Example 1. In terms of the coefficients of the characteristic polynomial $f(z) := z^n + \sum_{j=1}^{n} a_j z^{n-j}$, one has

$$\Phi_1(1; a_1, a_2, a_3) \equiv -a_3^2 + a_3 a_1 - a_2 + 1 \quad \text{for} \quad n = 3,$$
$$\Phi_1(1; a_1, a_2, a_3, a_4) \equiv (a_4 - 1)^2(1 - a_2 + a_4) + (a_1 a_4 - a_3)(a_3 - a_1)$$
$$\text{for} \quad n = 4.$$

From (26) and Viète formulas, one can notice that the degree of $\Phi_1(1; a_1, \ldots, a_n)$, treated as a polynomial in all the coefficients of $f(z)$, equals $n-1$, and it contains the term $(-1)^{n(n-1)/2} a_n^{n-1}$. $\qquad\square$

To obtain the general expression for $\Phi_1(1; \mu)$ for any n via factorization of polynomial (24), looks like a nontrivial task even if we know two its factors from identity (25). There exist several approaches to factoring polynomials (see [22] and references therein). All the known algorithms run in polynomial time or are conjectured so (using randomization and for dense polynomials). Their running time bounds, however, seem to have high exponents [23] (Theorem 3.12). An alternative procedure for the construction of the polynomial $\Phi_1(1; \mu)$ can be suggested on the base of Theorem 1. We describe it in the next section.

3.2   The Algorithm

To check the robust stability of family (2), perform the following steps

0. Take arbitrary point $\mu^{(0)}$ in $\mathfrak{B}$. If $M(\mu^{(0)})$ is not stable, then the claim is wrong. Otherwise
1. Calculate the powers of matrix M and their traces: $s_k = \text{Tr}(M^k)$ for $k \in \{1, 2, \ldots, n(n-1)\}$.
2. By (6), calculate the coefficients of $f(z; \mu) := \det(zI - M(\mu))$.
3. By formulae (8), calculate the Newton sums S_k for $k \in \{1, 2, \ldots, n(n-1)/2\}$.
4. By (6), calculate the coefficients of polynomial $\Phi_1(z; \mu)$ defined by (26).
5. By Theorem 5, verify that polynomials $f(1; \mu)$, $f(-1; \mu)$ and $\Phi_1(1; \mu)$ do not have real zeros in the box $\mathfrak{B}$.

First, consider the computational complexity of the first 4 steps of the algorithm. Here, matrix multiplication is the most expensive operation. The square matrix multiplication has an asymptotic complexity of $O(n^3)$, if carried out naively, and the complexity of $O(n^{\log_2 7}) \approx O(n^{2.807})$ if utilized Strassen's algorithm. The exponent appearing in the complexity of matrix multiplication has been improved several times, and a final (up to date) complexity of $O(n^{2.3728639})$ has the Le Gall algorithm that generalizes the Coppersmith – Winograd algorithm [9].

To compute M^k for $k \in \{0, 1, 2, \ldots, n^2 - n\}$, we have to perform $n^2 - n - 1$ matrix multiplications, so totally we have $O(n^5)$ operations. Then we find traces of matrices M^k for $k \in \{0, 1, 2, \ldots, n^2 - n\}$ and coefficients of (27). This yields $\approx O(n^4)$ operations in total. Hence, there arc $O(n^5)$ operations, if we do not take into account operations for testing positiveness of polynomials (27) in the box $\mathfrak{B}$.

For the same computations, the most expensive operation in the algorithm described in [6] is calculation of the determinant of matrix $I - M(\mu) \cdot M(\mu)$, where bialternate product is defined as

$$M \cdot M := [n_{ij,k\ell}] \quad \text{where } n_{ij,k\ell} := \begin{vmatrix} m_{ik} & m_{i\ell} \\ m_{jk} & m_{j\ell} \end{vmatrix}$$

for the indices ordered lexicographically. Hence, the matrix $M \cdot M$ is of the order $(n^2 - n)/2$, and one needs $O(n^6)$ operations to compute its determinant.

Therefore, the algorithm presented here is more efficient in its first part, i.e., for computation of polynomials (27).

Now consider the last step of the algorithm. To test the positivity of the obtained polynomials in [6] numerical procedures based on Bernstein expansion method [10,13] are used. Even applied for the order 3 matrices with only two parameters (with a sample one considered in the next section), they require more than 100 iterations [6].

The algebraic approach that we propose allows one to find all the required values with arbitrary precision.

4   Numerical Examples

Example 2. It is shown in [6] that the family

$$M(\mu_1, \mu_2) = \begin{bmatrix} -0.14 & 0.235 & 0.29 \\ -0.94 & -0.811 & 1.246 \\ -0.22 & -0.35 & 0.95 \end{bmatrix} + \mu_1 \begin{bmatrix} -0.3 & 0.15 & 0.275 \\ -0.275 & -0.3 & 0.55 \\ -0.35 & -0.25 & 0.625 \end{bmatrix}$$

$$+ \mu_2 \begin{bmatrix} 0.4 & -0.1 & -0.4 \\ -0.6 & -0.325 & 0.225 \\ 0.725 & 0.225 & -0.45 \end{bmatrix}$$

is stable for $(\mu_1, \mu_2) \in [-1, 1] \times [-1, 1]$. We will demonstrate that it is stable in the larger box $\mathfrak{B} := [-2.8, 1] \times [-1.03, 1.1]$ and find the distance to instability from $\mu^{(0)} = (0, 0)$.

Solution. It can be verified that the matrix $M(0, 0)$ is stable. To check the other conditions of Corollary 4, we compute the traces of powers of the matrix[2] $M(\mu_1, \mu_2)$

$$s_1 = \frac{1}{40}\mu_1 - \frac{3}{8}\mu_2 - \frac{1}{1000};$$

$$s_2 = \frac{297}{20000}\mu_1 + \frac{9317}{20000}\mu_2 + \frac{33}{1600}\mu_1^2 + \frac{13}{160}\mu_1\mu_2 + \frac{7}{64}\mu_2^2 + \frac{138221}{1000000};$$

$$\cdots$$

$$s_6 = -\frac{487405339}{2048000000}\mu_1^3\mu_2^3 + \frac{519633}{204800000}\mu_1^5\mu_2 + \frac{526764063}{4096000000}\mu_1^2\mu_2^4$$

$$+ \frac{85443417}{2048000000}\mu_1\mu_2^5 + \frac{81273}{4096000000}\mu_1^6 - \frac{1399171}{2048000000}\mu_2^6 + \cdots$$

$$+ \frac{167430804832241561}{1000000000000000000}$$

and then restore by (6) the coefficients of its characteristic polynomial $f(z; \mu)$:

$$a_1 = -s_1;$$

$$a_2 = -\frac{1}{100}\mu_1^2 - \frac{1}{20}\mu_1\mu_2 + \frac{1}{64}\mu_2^2 - \frac{149}{20000}\mu_1 - \frac{4651}{20000}\mu_2 - \frac{6911}{100000};$$

$$a_3 = -\frac{31}{16000}\mu_1^3 - \frac{2249}{12800}\mu_1^2\mu_2 + \frac{13131}{64000}\mu_1\mu_2^2 - \frac{139}{16000}\mu_2^3 + \cdots - \frac{117957}{500000}.$$

Then compute the sums S_1, S_2, S_3 by (8):

$$S_1 = a_2;$$

$$S_2 = \frac{1}{2560000}(124\mu_1^4 - 8340\mu_2^4 + 9385\mu_1^3\mu_2 - 181806\mu_1^2\mu_2^2 + 197521\mu_1\mu_2^3) + \cdots$$

$$S_3 = \frac{54880506251}{256000000000}\mu_1^3\mu_2 + \frac{19234505889}{2560000000000}\mu_1^2\mu_2^2 - \frac{255226105707}{3200000000000}\mu_1^3$$

$$+ \frac{171599531007}{1600000000000}\mu_2^3 + \cdots$$

---

[2] We treat the matrix entries as rational fractions.

and coefficients A_1, A_2, A_3 of polynomial $\Phi_1(z; \mu)$:

$$A_1 = -S_1;$$

$$A_2 = \frac{1}{2560000}(124\mu_1^4 - 8340\mu_2^4 + 9385\mu_1^3\mu_2 - 181806\mu_1^2\mu_2^2 + 197521\mu_1\mu_2^3) + \ldots$$

$$A_3 = -\frac{961}{256000000}\mu_1^6 - \frac{19321}{256000000}\mu_2^6 - \frac{123193537}{4096000000}\mu_1^4\mu_2^2 + \frac{147589151}{2048000000}\mu_1^3\mu_2^3$$

$$- \frac{69719}{102400000}\mu_1^5\mu_2 - \frac{184927601}{4096000000}\mu_1^2\mu_2^4 + \frac{1825209}{512000000}\mu_1\mu_2^5 + \ldots$$

Next we get polynomials (27)

$$f(1; \mu) = 1 + a_1 + a_2 + a_3$$

$$= \frac{1}{2^9 5^6}\big(-15500\mu_1^3 - 1405625\mu_1^2\mu_2 + 1641375\mu_1\mu_2^2 - 69500\mu_2^3$$

$$- 1013085\mu_1^2 - 3658925\mu_1\mu_2 + 3113455\mu_2^2 - 2963542\mu_1 - 52708\mu_2$$

$$+ 5567808\big);$$

$$f(-1; \mu) = 1 - a_1 + a_2 - a_3 = \ldots$$

$$\Phi_1(1; \mu) = -\frac{1}{15500^2}\big[(-15500\mu_1^3 - 1405625\mu_1^2\mu_2 + 1641375\mu_1\mu_2^2 - 69500\mu_2^3$$

$$- 933085\mu_1^2 - 3258925\mu_1\mu_2 + 2988455\mu_2^2 - 2603942\mu_1$$

$$- 2692308\mu_2 - 1891312)^2 - 16000000(40625\mu_1^2 + 181250\mu_1\mu_2$$

$$+ 78125\mu_2^2 + 29750\mu_1 + 930950\mu_2 + 4276441)\big].$$

To verify that these polynomials do not possess real zeros in the box $\mathfrak{B}$, we utilize the algorithm from Theorem 5. For our particular example, system (17) takes the form

$$G(\mu_1, \mu_2) = 0, \quad \partial G(\mu_1, \mu_2)/\partial\mu_2 = 0.$$

Using discriminant (10), the first component of any zero to this system satisfies the univariate equation $\mathcal{D}_{\mu_2}(G) = 0$. For $G(\mu) := f(1; \mu)$, this equation, up to a numerical factor, is as follows

$$19121246654477782265625\mu_1^6 + 2329084130401050056250000\mu_1^5$$

$$+ 112097000075943500892281250\mu_1^4 + 2329629849954313800682500000\mu_1^3$$

$$+ 116857964228863141511533325\mu_1^2 - 2471562942327467134374 8560\mu_1$$

$$- 2700128240066472549538801 6 = 0.$$

From part (iii) of Theorem 3, it follows that in $[-2.8, 1]$, it has a single real zero. We can improve its approximation like, for instance, $\mu_1^{(1)} = -2.121108 \pm 10^{-6}$. For this value, the polynomial $G(\mu_1^{(1)}, \mu_2)$ in μ_2 has a multiple zero. It can be evaluated via (12) as $\mu_2^{(1)} = -4.888746 \pm 10^{-6}$. The point $(\mu_1^{(1)}, \mu_2^{(1)})$ does not belong to the box $\mathfrak{B}$. Therefore, the conditions (ii) of Theorem 5 are satisfied.

The absence of real zeros for boundary univariate polynomials $G(\mu_1, -1.03)$, $G(\mu_1, 1.1)$, $G(-2.8, \mu_2)$, and $G(1, \mu_2)$ in the corresponding intervals composing the box $\mathfrak{B}$ can be established via application of part (iii) of Theorem 3. Hence, the polynomial $f(1; \mu)$ does not possess zeros in $\mathfrak{B}$.

Analogously, it can be verified that polynomials $f(-1; \mu)$ and $\Phi_1(1; \mu)$ do not have real zeros in the box $\mathfrak{B}$.

For the curve $G(\mu_1, \mu_2) := f(1; \mu_1, \mu_2) = 0$ and the point $(\mu_1, \mu_2) = (0,0)$, distance equation (21) is as follows:

$$\mathcal{F}_1(z) := {\scriptstyle 1656437927789246676302557628670718002047330141067504882812} 50\, z^9 + \ldots$$
$$- {\scriptstyle 1418671204757960500059517541617921148525764211238280835065504137216} = 0.$$

According to Theorem 3, it has 3 real (and positive) zeros with the minimal one equal to $z_* = 1.225741 \pm 10^{-6}$. Distance to this curve equals $d_* = \sqrt{z_*} = 1.107132 \pm 10^{-6}$ and is achieved at $(\mu_1^*, \mu_2^*) = (1.055645 \pm 10^{-6}, 0.333698 \pm 10^{-6})$. Minimal positive zero of the distance equation constructed for the curve $f(-1; \mu_1, \mu_2) = 0$ equals 1.524132 ± 10^{-6}. For the curve $\Phi_1(1; \mu_1, \mu_2) = 0$, distance equation is of the order 24, and it possesses 6 real (and positive) zeros with the minimal one equal to 1.509424 ± 10^{-6}. Therefore, the distance to instability from $(\mu_1, \mu_2) = (0,0)$ equals d_* (Fig. 1).

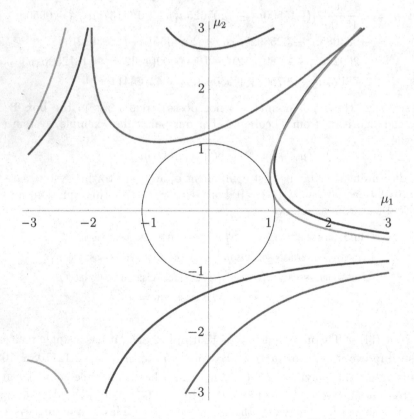

Fig. 1. Example 2. Plots of $f(1; \mu) = 0$ (green); $f(-1; \mu) = 0$ (blue); $\Phi_1(1; \mu) = 0$ (red) (Color figure online)

One can notice that the asymptotes for the three drawn curves look identical. This is indeed the case. It is known [21] that the asymptotes for an algebraic curve $G(x, y) = 0, \deg G := N > 1$ are determined by the coefficients of the two highest order forms in expansion of $G(x, y)$ in the decreasing powers of variables:

$$G(x, y) \equiv G_N(x, y) + G_{N-1}(x, y) + \dots$$

If for $(x_0, y_0) \in \mathbb{R}^2, (x_0, y_0) \neq (0, 0)$, the following conditions hold

$$G_N(x_0, y_0) = 0, \quad (\partial G_N(x_0, y_0)/\partial x)^2 + (\partial G_N(x_0, y_0)/\partial y)^2 \neq 0,$$

then the equation

$$x \partial G_N(x_0, y_0)/\partial x + y \partial G_N(x_0, y_0)/\partial y + G_{N-1}(x_0, y_0) = 0$$

gives an asymptote for the curve. Generically, the degree of the polynomial $a_n(\mu_1, \mu_2) \equiv (-1)^n \det M(\mu_1, \mu_2)$ is much higher than those of the other coefficients of $f(z; \mu_1, \mu_2)$. Its forms of the two highest orders coincide with those of the polynomials $f(\pm 1; \mu_1, \mu_2)$. According to the remark in Example 1, the two highest order forms of $\Phi_1(1; \mu_1, \mu_2)$ coincide up to a sign with those of $a_n^{n-1}(\mu_1, \mu_2)$.   □

The complexity of computations increases drastically with the number of parameters and degrees of the matrix entries.

Example 3. For the matrix family

$$M(\mu_1, \mu_2, \mu_3) = \begin{bmatrix} -0.3 & -0.1 & 0 \\ 0.2 & 0.3 & 0.3 \\ -0.1 & 0 & 0.3 \end{bmatrix} + \mu_3 \begin{bmatrix} 0.1 & -0.2 & -0.3 \\ 0.3 & 0.1 & 0.2 \\ 0.1 & 0.3 & 0.2 \end{bmatrix}$$

$$+ \mu_2 \begin{bmatrix} -0.2 & 0 & 0.1 \\ -0.3 & 0.1 & -0.3 \\ -0.1 & 0.1 & -0.3 \end{bmatrix} + \mu_2 \mu_3 \begin{bmatrix} 0.2 & 0.3 & 0.1 \\ 0 & 0.1 & 0.2 \\ 0.1 & 0 & 0 \end{bmatrix}$$

$$+ \mu_1^2 \begin{bmatrix} 0.1 & -0.2 & 0 \\ 0.3 & 0.3 & 0.1 \\ 0 & 0 & -0.2 \end{bmatrix},$$

find the distance to instability from $\mu^{(0)} = (0, 0, 0)$.

Solution. Here the matrix entries are even polynomials in μ_1, and the distance equations can be constructed using the approach outlined in Sect. 2.3. These equations are polynomials over $\mathbb{Z}$ with the magnitude of some of their coefficients exceeding 10^{500}.

Surface	distance $(\pm 10^{-6})$	achieved at $(\pm 10^{-6})$	distance equation degree
$f(1, \mu) = 0$	3.775852	$(\pm 2.062413, 0.572611, -3.110566)$	51
$f(-1; \mu) = 0$	2.322106	$(\pm 2.322106, 0.6986351, -0.305716)$	51
$\Phi_1(1; \mu) = 0$	1.749203	$(\pm 1.620479, -0.450343, 0.480575)$	162

The distance to instability from $\mu^{(0)} = (0, 0, 0)$ equals 1.749203 ± 10^{-6}.   □

5   Conclusions

We have investigated Schur stability property for the matrices with the entries polynomially depending on parameters. The first task has been stated as that of the description of the domain of stability in the parameter space, i.e., finding its boundaries. The second task has been aimed at the estimation of possible tolerances for the parameter specializations that would not disturb the stability property of a particular matrix. In this paper, the purely algebraic procedures based on symbolic algorithms for the elimination of variables and localization of the real zeros for algebraic equation systems have been suggested for solving the stated problems. This provides one with precise information on the obtained solution, i.e., the results do not depend on the precision of calculations and round-off errors (which are known to make a tight bottleneck for any numerical algorithm relating the problems solved). For further investigation, it remains the optimization of computational efficiency of developed algorithms.

Acknowledgments. The authors are grateful to Prof Evgenii Vorozhtzov and to the anonimous referees for valuable suggestions that helped to improve the quality of the paper.

References

1. Ackermann, J.: Robust Control: The Parameter Space Approach. Springer-Verlag, London (2002)
2. Akritas, A.G.: Elements of Computer Algebra with Applications. Wiley, New York (1989)
3. Basu, S., Pollack, R., Roy, M.-F.: Algorithms in Real Algebraic Geometry. Springer-Verlag, Heidelberg (2010)
4. Bhattacharya, S.P., Chapellat, H., Keel, L.: Robust Control: The Parametric Approach. Prentice-Hall, New Jersey (1995)
5. Bikker, P., Uteshev, A.Y.: On the Bézout construction of the resultant. J. Symb. Comput. **28**, 45–88 (1999)
6. Büyükköroğlu, T., Çelebi, G., Dzhafarov, V.: On the robust stability of polynomial matrix families. Electron. J. Linear Algebra **30**, 905–915 (2015)
7. Chesi, G.: Exact robust stability analysis of uncertain systems with a scalar parameter via LMIs. Automatica **49**, 1083–1086 (2013)
8. Chilali, M., Gahinet, P., Apkarian, P.: Robust pole placement in LMI regions. IEEE Trans. Automat. Control. **44**(12), 2257–2270 (1999)
9. Coppersmith, D., Winograd, S.: Matrix multiplication via arithmetic progressions. J. Symb. Comput. **9**(3), 251–280 (1990)
10. Dzhafarov, V., Büyükköroğlu, T.: On nonsingularity of a polytope of matrices. Linear Algebra Appl. **429**, 1174–1183 (2008)
11. Elaydi, S.: An Introduction to Difference Equations. Springer, New York (2005)
12. Faddeev, D.K., Faddeeva, V.N.: Computational Methods of Linear Algebra. Freeman, San Francisco (1963)
13. Farouki, R.T.: The Bernstein polynomial basis: a centennial retrospective. Comput. Aided Geom. Des. **29**(6), 379–419 (2012)
14. Gantmakher, F.R.: The Theory of Matrices, vol. I, II. Chelsea, New York (1959)

15. Henrion, O., Bachelier, O., Sebek, M.: D-stability of polynomial matrices. Int. J. Control **74**(8), 845–856 (2001)
16. Joachimsthal, F.: Bemerkungen über den Sturmschen Satz. J. Reine Angew. Math. **48**, 386–416 (1854)
17. Jury, E.I.: Innors and Stability of Dynamic Systems. Wiley, New York (1974)
18. Kalinina, E.A.: Stability and D-stability of the family of real polynomials. Linear Algebra Appl. **438**, 2635–2650 (2013)
19. Keel, L.H., Bhattacharya, S.P.: Robust stability via sign-definite decomposition. IEEE Trans. Automat. Control **56**(1), 140–145 (2011)
20. Krein, M.G., Naimark, M.A.: The method of symmetric and Hermitian forms in the theory of the separation of the roots of algebraic equations. Linear Multilinear A. **10**, 265–308 (1981)
21. Kunz, E.: Introduction to Plane Algebraic Curves. Birkhäuser, Boston (2005)
22. Lee, M. M.-D.: Factorization of Multivariate Polynomials. Technische Universität Kaiserslautern (2013)
23. Lenstra, A.K.: Factoring multivariate polynomials over finite fields. J. Comput. Syst. Sci. **30**(2), 235–248 (1985)
24. Mishra, B.: Algorithmic Algebra. Springer-Verlag, New York (1993)
25. Schur, J.: Über Potenzreihen, die im Innern des Einheitskreises beschränkt sind. J. Reine Angew. Math., **147**, 205–232 (1917), **148**, 122–145 (1918)
26. Skalna, I.: Parametric Interval Algebraic Systems. Springer, Cham (2018)
27. Uteshev, A.Y., Shulyak, S.G.: Hermite's method of separation of solutions of systems of algebraic equations and its applications. Linear Algebra Appl. **177**, 49–88 (1992)
28. Smith III, J.O.: Introduction to Digital Filters: with Audio Applications. W3K Publishing, USA (2007)
29. Uteshev, A.Y., Cherkasov, T.M.: The search for the maximum of a polynomial. J. Symb. Comput. **25**(5), 87–618 (1998)
30. Uteshev, A.Y., Borovoi, I.I.: Solution of a rational interpolation problem using Hankel polynomials. Vestn. St.-Peterbg. Univ. Ser. 10 Prikl. Mat. Inform. Protsessy Upr. **4**, 31–43 (2016). (in Russian)
31. Uteshev, A.Yu., Goncharova, M.V.: Metric problems for algebraic manifolds: analytical approach. In: Constructive Nonsmooth Analysis and Related Topics – CNSA 2017 Proceedings, 7974027 (2017)
32. Uteshev, A.Y., Yashina, M.V.: Metric problems for quadrics in multidimensional space. J. Symb. Comput. **68**, 287–315 (2015)
33. Wayland, H.: Expansion of determinantal equations into polynomial form. Quart. Appl. Math. **2**(4), 277–305 (1945)
34. Yli-Kaakinen, J., Saramaki, T.: Efficient recursive digital filters with variable magnitude characteristics. In: Proceedings of 7th Nordic Signal Processing Symposium – NORSIG 2006, pp. 30–33 (2006)

Symbolic and Numerical Methods for Searching Symmetries of Ordinary Differential Equations with a Small Parameter and Reducing Its Order

Alexey A. Kasatkin$^{(\boxtimes)}$ (iD) and Aliya A. Gainetdinova (iD)

Ufa State Aviation Technical University, Ufa, Russia
kasatkin@ugatu.su

Abstract. Two programs for computer algebra systems are described that deal with Lie algebras of generators admitted by systems of ordinary differential equations (ODEs). The first one allows to find the generators of admitted transformations in the specified form. This program is written in Python and based on SciPy library. It does not require solving partial differential equations symbolically and can also analyze equations with Riemann–Liouville fractional derivatives and find approximate symmetries for systems of equations with a small parameter. The second program written as a package for Maple computes the operator of invariant differentiation in special form for given Lie algebra of generators. This operator is used for order reduction of given ODE systems.

Keywords: Lie algebras of generators ·
Point symmetries of differential equation · Fractional derivatives ·
Differential invariants · Operators of invariant differentiation ·
Computer algebra

1 Introduction

Group analysis of differential equations (see, e.g., [1–4]) provides efficient analytical methods for studying and solving differential equations, which arise in different areas of science.

In this paper, we consider the systems of ordinary differential equations (ODEs), focusing on nonlinear ones. The Lie symmetry method for solving nonlinear ODE is one of the most universal methods because it can work for equations not matching any particular form. Each symmetry of equation (or system of equations) describes the one-parametric group of point transformations that leaves the equation (or system) invariant. When the Lie algebra of symmetry generators is found, it can be used to reduce the order of the given equation

Supported by the Ministry of Science and High Education of the Russian Federation (State task No. 1.3103.2017/4.6).

or system. The order reduction algorithm for single equation is well-known and Lie method is implemented and often used by default for solving ODEs in computer algebra software like Maple and SymPy Python library.

However, the modifications of order reduction algorithm for systems of ODEs (see, e.g., [5,6]) are not universal. Recently, a new order reduction algorithm was proposed in [7,8]. It uses the operator of invariant differentiation (OID) with multiplier of special type. In [9], a modification of this algorithm is proposed for application to systems of ODEs with a small parameter that admit approximate Lie algebras of generators (theory of approximate Lie groups is introduced in [4]). To simplify the calculations, a program for constructing an OID with a multiplier of a special type was developed in Maple computer algebra system (Maple was chosen because it has a powerful solver module for partial differential equations). In Sect. 3.2, we describe this program and illustrate its usage with some examples.

Methods based on symmetries have been recently applied to classes of equations with fractional derivatives [10–12]. Systems of ordinary fractional differential equations (FDEs) are often considered in modern control system theory and mathematical models of different processes. Such systems are also considered when finding particular solutions of fractional partial differential equations by invariant subspace method [13–15]. Known symmetries allow one to construct invariant solutions of the considered systems. However, there are only few works on symmetries of ordinary FDEs and systems, for example [16,17], there are many technical difficulties and limitations because the fractional derivatives are non-local integro-differential operators. For FDEs, the current method of constructing symmetries is restricting the form of operators and the resulting Lie algebra is usually formed by combinations of some standard generators (usually corresponding to translation, scaling, rotation and projective transformations).

Constructing symmetries for the considered systems requires a lot of calculations. To find the symmetries of equation or a system, one needs to solve the so-called determining equations, which are the linear first-order partial differential equations (PDEs) [1–4,11]. There are multiple existing packages for finding Lie symmetries of different types [18–20] (some of them support finding approximate symmetries), and one for equations with fractional derivatives [21]. They are usually based on powerful first-order PDE solvers but there is no guarantee that the symmetry generators can be found in closed form for arbitrary form of system, for example, including functions of derivatives.

The program described in Sect. 3.1 of this paper uses a semi-numerical approach for finding symmetries. Symmetry generators are found as linear combinations of given basis operators X_i. This approach is usually implemented for polynomial form of coefficients. For example, it is available inside Maple PDEtools package and in SymPy library for single equation. Recent application for dynamical systems is also described in [22]. However, in this work, we do not restrict coefficients to have polynomial form. The numerical algorithms like SVD decomposition are used to find the space of solutions for automatically constructed determining equations. This approach allows to work with very complicated forms of nonlinear systems. Symbolic calculations are only used to differentiate equation one time, substitute the coefficients of given operators X_i into prolongation formula and collect the terms in linear expressions, the final steps of computation are numerical.

Another advantage of using the fixed-form representations of symmetry generator is the possibility to analyze symmetries of the first-order differential equations that are used in many mathematical models. Finding the symmetries analytically in this case is a complicated task because the space of symmetry generators is usually infinite-dimensional. Restricting the form of infinitesimal generator (for example, by assuming polynomial coefficients) allows one to construct solutions or conservation laws even without finding the complete set of symmetries.

2   Theoretical Section

2.1   Point Symmetries of Differential Equations

Let us consider the system of ordinary differential equations

$$u^{(p)}(t) = f\left(t, u(t), \ldots, u^{(p-1)}(t)\right), \tag{1}$$

where $u = u(t)$ is a vector of m unknown functions $u_1(t), \ldots, u_m(t)$, f is a vector function with components f_μ and $u^{(k)}$ is a vector of kth derivatives of u by t.

To define the symmetry of (1), the local Lie group of point transformations T_a with parameter a is also considered:

$$T_a : (t, u) \to (\widetilde{t}, \widetilde{u}), \qquad \widetilde{t} = \varphi(t, u; a), \quad \widetilde{u} = \psi(t, u; a), \quad T_{a+b} = T_a T_b. \tag{2}$$

The group of transformations can be defined by its generator

$$X = \xi(t, u)\frac{\partial}{\partial t} + \eta(t, u)\frac{\partial}{\partial u}, \tag{3}$$

where functions $\xi(t, u) = \left.\dfrac{\partial \varphi}{\partial a}\right|_{a=0}$ and $\eta(t, u) = \left.\dfrac{\partial \psi}{\partial a}\right|_{a=0}$ are the coordinates of the tangent vector field. To study the symmetry properties of system (1), we find the prolonged group acting in space, where the coordinates of some point are t, u and all derivatives of u up to the pth order. This group is defined by prolonged generator which has the form

$$X^{(p)} = \xi(t, u)\frac{\partial}{\partial t} + \sum_{\mu=1}^{m} \eta^\mu(t, u)\frac{\partial}{\partial u_\mu} + \sum_{\mu=1}^{m}\sum_{k=1}^{p} \zeta_k^\mu(t, u, \ldots, u^{(k)})\frac{\partial}{\partial u_\mu^{(k)}}, \tag{4}$$

where the coefficients ζ_k^μ are given by the prolongation formula

$$\zeta_k^\mu = D_t^k\left(\eta - \xi u_\mu'\right) + \xi u_\mu^{(k+1)}. \tag{5}$$

The coefficients ζ_k^μ describe the infinitesimal transformations of derivatives:

$$\tilde{u}_\mu^{(k)}(\tilde{t}) = u_\mu^{(k)}(t) + a\zeta_k^\mu + o(a).$$

Here D_t is the total derivative operator:

$$D_t = \frac{\partial}{\partial t} + \sum_{k=0}^{p}\sum_{\mu=1}^{m} u_\mu^{(k+1)}(t)\frac{\partial}{\partial u_\mu^{(k)}}.$$

The transformation group (2) is called *point symmetry* of system (1) if the system is transformed to itself. The group generator is said to be *admitted* by system (1). It is known [1–3] that ODE (1) admits generator (3), if and only if

$$X^{(p)}\left(u_\mu^{(p)} - f_\mu\left(t, u, \dots, u^{(p-1)}\right)\right)\Big|_{(1)} = 0, \quad \mu = 1, \dots, m. \qquad (6)$$

From equation (6), after splitting with respect to powers of the derivatives, one can obtain a system of *determining equations*. It is a linear overdetermined system of partial differential equations for ξ, η coefficients. The solution of this system gives a complete set of generators (3) admitted by equation (1).

The set of generators of the form (3) together with operation of commutation

$$[X_1, X_2] = X_1(X_2) - X_2(X_1) \qquad (7)$$

forms a Lie algebra of generators if it is a vector space and the commutator of every two operators belongs to the same space. The conditions of bilinearity antisymmetry and Jacobi identity are satisfied (see e.g. [3]). It is known that generators admitted by an equation or a system always form a Lie algebra.

In this work we consider the ODE systems, which order is equal to dimension of admitting Lie algebra L_r of generators, i.e., $r = mp$.

2.2   Invariant Representation of Differential Equations

Some function $J^{(k)} = J(t, u, \dots, u^{(k)})$, $J^{(k)} \neq$ const, is kth-order differential invariant of r-dimensional Lie algebra of generators L_r, if

$$X_j^{(k)}\left(J^{(k)}\right) = 0, \ j = 1, \dots, r. \qquad (8)$$

We can rewrite system (1) by differential invariants of admitted Lie algebra L_r, if and only if

$$rank\left(\xi_j\ \eta_j\ \dots\ \zeta_j^{(p)}\right)\Big|_{(1)} = r, \ j = 1, \dots, r,$$

i.e., the general rank of the matrix formed by the coordinates of the prolonged generators does not change on the manifold defined by system (1) (see e.g. [2]). The system can then be rewritten so that every equation has the form

$$J^{(p)} = F\left(J^{(q)}\right) \qquad (9)$$

with some function F, here $J^{(p)}$ and $J^{(q)}$ are some pth-order and qth-order differential invariants of L_r, $q < p$.

2.3 Order Reduction for Differential Equations

For order reduction, in [7,8] authors suggested to seek auxiliary function $\Phi \neq$ const and construct the operator of invariant differentiation (OID) (see [2]) in the form

$$\frac{1}{D_t \Phi} D_t. \tag{10}$$

The function Φ is obtained as any particular solution of system

$$D_t \left(X_j^{(p-1)} (\Phi) \right) = 0, \tag{11}$$

or

$$X_j^{(p-1)} (\Phi) = C_j, \ j = 1, \ldots, r, \tag{12}$$

where constants C_j satisfy the relations

$$\sum_{l=1}^{r} a_{ij}^l C_l = 0,$$

and a_{ij}^l are structural constants of Lie algebra L_r (see [1–3]).

Then by using this OID, one obtains

$$\frac{1}{D_t \Phi} D_t \left(J^{(q)} \right) \bigg|_{(9)} \equiv \Psi \left(J^{(q)}, J^{(p)} \right) \bigg|_{(9)} = H \left(J^{(q)} \right).$$

The last expression can be rewritten as a first-order ODE

$$\frac{dJ^{(q)}}{d\Phi} = H \left(J^{(q)} \right).$$

General solution of this equation is also the first integral of ODE (1).

A similar method can also be used to reduce the order and integrate the systems of ODEs [7,8].

2.4 Symmetries, Invariants, and OID for Equations with a Small Parameter

Studying equations with a small parameter, i.e., ODEs of the form

$$u^{(p)} = f_0(t, u, \ldots, u^{(p-1)}) + \varepsilon f_1(t, u, \ldots, u^{(p-1)}), \tag{13}$$

we can use the theory of approximate transformation groups (see e.g. [4]). We consider only the case of linearity on ε.

Such equations admit the generators of two types:

$$X_{(0)} + \varepsilon X_{(1)}, \ \varepsilon Y_{(0)}.$$

They can be obtained by solving Eq. (6) after neglecting ε^2 terms and splitting by ε together with derivatives. The generators form approximate Lie algebra

[4]. We assume that system (13) has r_0 operators of the first type, and $r - r_0$ operators of the second type.

Invariants of the approximate Lie algebra can also have two forms [24]:

$$J_{(0)}^{(k)} + \varepsilon J_{(1)}^{(k)}, \quad \varepsilon J_{(0)}^{(n)},$$

where the term $J_{(0)}^{(s)}$ satisfies the equations

$$X_{i,(0)}^{(s)} J_{(0)}^{(s)} = 0, \quad i = 1, \ldots, r_0,$$

$$Y_{j,(0)}^{(s)} J_{(0)}^{(s)} = 0, \quad j = 1, \ldots, r - r_0,$$

and $J_{(1)}^{(s)}$ satisfies equations

$$X_{i,(0)}^{(s)} J_{(1)}^{(s)} + X_{i,(1)}^{(s)} J_{(0)}^{(s)} \approx 0.$$

In the work [9], the OID for the approximate Lie algebra was introduced. It is shown that it can be obtained in the form

$$\frac{D_t \Phi_0 - \varepsilon D_t \Phi_1}{(D_t \Phi_0)^2} D_t,$$

where the functions Φ_0 and Φ_1 are particular solutions of systems

$$X_{i,(0)}^{(p-1)} \Phi_0 = C_{i,(0)}, \quad Y_{j,(0)}^{(p-1)} \Phi_0 = C_{j,(0)}$$

and

$$X_{i,(0)}^{(p-1)} \Phi_1 + X_{i,(1)}^{(p-1)} \Phi_0 \approx C_{i,(1)},$$

respectively.

2.5   Symmetries of Equations with Fractional Derivatives

In the last decade, Lie group analysis methods were applied to the specific case of integro-differential equations – equations with fractional derivatives. The survey of proposed techniques and results may be found, for example, in [11,12].

The Riemann–Liouville fractional derivative is a nonlocal operator defined as

$$D_t^\alpha u(t, x) = \frac{1}{\Gamma(n - \alpha)} \left(\frac{\partial}{\partial t} \right)^n \int_0^t \frac{u(z, x)}{(t - z)^{\alpha - n + 1}} dz, \quad n - 1 \leq \alpha < n. \tag{14}$$

Here x can be a vector of all other independent variables, n is a natural number.

Let the generator of point transformation group have the form

$$X = \xi^0(t, x, u) \frac{\partial}{\partial t} + \sum_{i=1}^{n} \xi^i(t, x, u) \frac{\partial}{\partial x_i} + \sum_{\mu=1}^{m} \eta^\mu(t, x, u) \frac{\partial}{\partial u_\mu}. \tag{15}$$

The infinitesimal transformation of fractional derivative $D_t^\alpha u$ can be written as

$$D_{\tilde t}^\alpha \tilde u_\mu = D_t^\alpha u_\mu + a\zeta_\mu^\alpha + o(a).$$

The prolongation formula for the fractional derivative has the general form

$$\zeta_\alpha^\mu = D_t^\alpha \left(\eta - \frac{\partial u_\mu}{\partial t}\xi^0 - \sum_{i=1}^n \frac{\partial u_\mu}{\partial x_i}\xi^i \right) + \xi^0 D_t^{\alpha+1} u_\mu + \sum_{i=1}^n \xi^i D_t^\alpha \frac{\partial u_\mu}{\partial x_i}. \tag{16}$$

with the restrictions $\xi^0|_{t=0} = 0$.

The main problem here is that ζ_μ^α does not depend on finite set of differential variables like it was for operator (4). To propose constructive algorithm for finding symmetries, we have to restrict generator coefficients, considering the *linearly autonomous class of symmetries* with

$$\xi^0 = \xi^0(x), \quad \eta^\mu = \sum_{\nu=1}^m p^{\mu,\nu}(t,x)u_\nu + q^\mu(t,x). \tag{17}$$

Then the prolongation formula (16) can be written in the form

$$\zeta_\alpha^\mu = D_t^\alpha q^\mu + \sum_{\nu=1}^m \sum_{k=0}^\infty \binom{\alpha}{k} \left(D_t^k p^{\mu,\nu} + \delta_\nu^\mu \frac{k-\alpha}{k+1} D_t^{k+1}\xi^0 \right) D_t^{\alpha-k} u_\nu +$$

$$+ \sum_{i=1}^n \sum_{k=1}^\infty \binom{\alpha}{k} \left(D_t^k \xi^i \right) D_t^{\alpha-k} \frac{\partial u_\mu}{\partial x_i}. \tag{18}$$

When only one independent variable t is considered, the last term in prolongation formula vanishes.

After writing the invariance criterion like (6), we consider all integer derivatives $u_\mu^{(k)}$ and fractional derivatives or integrals $D_t^{(\alpha-k)} u_\mu$ to be independent variables, m of them are to be substituted from the given system. This is another assumption that allows constructive symmetry finding.

The resulting system of determining equations is now infinite-dimensional. Usually, but not always, it becomes finite after finding that ξ^0 is a polynomial function of t.

3   Algorithms and Computer Algebra Modules

3.1   Determining Lie Symmetries

To simplify the procedure of calculating symmetries, the computer algebra systems are often used.

There are many programs for calculating symmetry generators for equations and systems developed using Maple, Mathematica, Reduce, and Maxima computer algebra systems, see, for example, [18]. Modern examples of such packages implemented in Maple are GeM [19] and DESOLVEII [20]. The package FracSym [21] also supports fractional differential equations.

Here we describe a new package for finding symmetries based on free and open source Python SymPy library [25]. It is not based on any of the existing modules, only SymPy differentiation, substitution, and singular value decomposition from Numpy are used.

The program part that constructs the determining equation is analogous to other packages but includes specific modifications for using prolongation formulas suitable for fractional derivatives (18). Working with approximate symmetries [4] is also supported by automatic collecting powers of ε and eliminating terms with ε^2.

Note that SymPy does not seem to have built-in "jet notation" for derivatives, so it was implemented manually. For example, the symbol `u_tttt` describes the derivative $u^{(4)}(t)$. The notation for fractional differential variable was also introduced. SymPy symbols and expressions can be used in given system.

While constructing the determining equations, the program also detects fractional derivatives in original system and restricts generators to be linearly autonomous. For each fractional derivative, some unknown coefficient η^{μ} is replaced by its form (17) and the number of independent variables in corresponding coefficient ξ^i is reduced.

The determining equations for system $F = 0$ are constructed in general form

$$\left(\tilde{X}[\xi, \eta, t, u]F_\mu\right)\Big|_{F=0} = 0, \quad \mu = 1, \ldots, m, \tag{19}$$

Here $\tilde{X}[\xi, \eta, t, u]$ is the prolonged generator given by (4) including prolongations to derivatives of fractional order. It depends on functions ξ, η linearly but contains the arbitrary combinations of u and its derivatives with respect to independent variables t. SymPy built-in PDE solver is not powerful enough to solve this overdetermined linear system automatically, but this is not a problem for proposed approach. Instead of splitting the determining equations and solving them symbolically, we use semi-numerical method described below.

We search for the generator as a combination of given basis operators:

$$X = \sum_{i=1}^{N} C_i X_i.$$

This idea is commonly used for polynomial coefficients of symmetry generators implemented for single ODE in *sympy.solvers.ode.lie_heuristic_bivariate*, for example. Maple package PDETools also contains an option for polynomial symmetries finding. Similar method is described in the work [22]. However, here we do not require coefficients ξ_i, η_i of X_i to be polynomial. Arbitrary known terms can be included in the form of generator, which is a more universal way.

Introducing the notation $\Psi_\mu[\xi, \eta, t, u] = (\tilde{X}F_\mu)|_{F=0}$, one can rewrite Eq. (19) as a linear equation for $C_1, \ldots, C_N$:

$$\sum_{i=1}^{N} C_i \Psi_\mu[\xi_i, \eta_i, t, u] = 0. \tag{20}$$

For each μ and i, $\Psi_{\mu,i}[t,u] = \Psi_\mu[\xi_i, \eta_i, t, u]$ is a function of independent variables t, dependent variables u and their derivatives (integer or fractional).

The program substitutes each ξ_i, η_i (or ξ_i, p_i, and q_i for linearly autonomous symmetry) into the determining equation symbolically and builds the list of functions $\Psi_{\mu,i}$ for each number of dependent variable μ.

The most general way of reducing (20) to algebraic system is using multivariate Taylor series expansion of $\Psi_{\mu,i}[t,u]$ with respect to all of its independent variables. However, to speed up calculations, the program can also substitute random arguments into (20) enough times to get overdetermined linear system for C_i. This substitution can be done numerically by using the fast NumPy library, all constants and arbitrary functions should be specified exactly at this stage. As a result, rectangular matrix M with R rows and N columns is obtained.

To solve the overdetermined system $MC = 0$ numerically, it is convenient to use singular value decomposition procedure. After using `np.linalg.svd` function to compute the decomposition of the form

$$M = U\Sigma V, \quad \Sigma = \text{diag}\{\sigma_1, ..., \sigma_r, 0, ..., 0\}$$

with orthogonal matrices U, V and r non-zero singular values $\sigma_1, ..., \sigma_r$, the fundamental set of solutions C is given by last rows of matrix V (with numbers starting from $r + 1$), see [23].

Every found vector C describes the symmetry generator, but to present the symmetries in convenient form, the matrix formed by coefficients C_i is transformed to reduced row echelon form. The basis columns are chosen automatically to maximize the number of coefficients close to zero. SymPy function `nsimplify` is then called to guess rational fractions like $1/7$ and roots like $\sqrt{2}$ from the corresponding numerical values.

Hereafter we assume that our module and SymPy are imported as follows:

```
> import symmetries as s
> from sympy import *
```

All commands entered by user are marked by starting > sign, lines after code without the sign shows its output converted to LaTeX. The superscript with dependent variable symbol is used in the program instead of dependent variable number, for example, $\xi^t = \xi^0$, $\eta^u = \eta^1$, $p^{uv} = p^{1,2}$ in (15) and (17).

The process of calculating symmetries contains several steps described below.

1. Procedure `LieSymmSetup(indepsList, depsList)` is used to define the lists of independent and dependent variables. For example, to study the ODE systems with $u(t)$ and $v(t)$, one should call

```
> s.LieSymmSetup(indepsList=['t'], depsList=['u','v'])
```

2. Then the user defines the form of the system. The system can be presented directly as Python dictionary with strings or SymPy objects for left-hand side and right-hand side of equations, for example,

```
> sys={'u_t': 'v_xx-u', 'v_t': 'u_xx-v'}
```

3. The function `detPDES` is used to compose the determining Eq. (19). For example, determining equation for $u''(t) = u'(t)$ is calculated as $(\zeta_2 - \zeta_1)|_{u''=u'}$:

```
> s.LieSymmSetup(indepsList=['t'], depsList=['u'])
> XFS=s.detPDES({'u_tt':'u_t'},UseJetVars2=True)
```

$$\eta_t^u - \eta_{tt}^u - 2\eta_{tu}^u u_t - \eta_{uu}^u u_t^2 + u_t^3 \xi_{uu}^t + 2u_t^2 \xi_{tu}^t + 2u_t^2 \xi_u^t + u_t \xi_t^t + u_t \xi_{tt}^t$$

To find approximate symmetries, one should call the function `detPDE_Approx`. When the equation contains fractional derivatives, the function `prolongUn` that implements prolongation formula (18) is used internally. The named parameter `UseJetVars2=True` is used here to show result in the short form and should not be used for further calculations.

4. The list `BaseOps` of basis operators X_i is constructed to specify the chosen form of symmetry generator. Operators can be manually added to `BaseOps` list or chosen in multivariate polynomial form by `AddPolynomialBasis` function as shown in the examples.

5. The `build_C_Coeffs(XFS)` substitutes all basis operators into constructed determining Eq. (19) and calculates coefficients `XFS_C_coeffs` of system (20) in symbolic form.

6. Using the procedure `solveNumeric`, overdetermined algebraic system for coefficients C_i is constructed from (20) and solved numerically. The best combinations of basis operators are chosen and stored as `BestForm` matrix. To view the symmetries in convenient form looking like differential operators, one calls the function `getAdmittedOperators`.

Example 1. Let us consider the system of fractional differential equations

$$D_t^\alpha u = uv, \quad D_t^\alpha v = v^2.$$

According to [17], any symmetry of (1) is a combination of basis operators

$$X_1 = u\partial_u, \quad X_2 = v\partial_u, \quad X_3 = u\partial_v, \quad X_4 = v\partial_v, \quad X_5 = t\partial_t, \quad X_6 = t^2\partial_t.$$

To construct the determining equations (steps 1–3), we use the code

```
> s.LieSymmSetup(indepsList=['t'], depsList=['u','v'])
> alpha=sympify('1/5');
> dau = s.fracD('u','t',alpha); dav = s.fracD('v','t',alpha)
> XFS=s.detPDES({dau: 'u*v',dav: 'v*v'},sumN=2)
```

The function `s.fracD('u','t',alpha)` gives $Da_t[u](1/5)$, which is the representation of $D_t^{1/5}u$ (differential variable in the code). The determining equations stored in `XFS` contain multiple terms, one of them is calculated internally as fractional prolongation by calling `s.prolongUn('u','t',alpha,sumN=2)`

$$p^{uu} Da_t[u]\left(\frac{1}{5}\right) + \frac{p_t^{uu}}{5} Da_t[u]\left(-\frac{4}{5}\right) - \frac{2p_{tt}^{uu}}{25} Da_t[u]\left(-\frac{9}{5}\right) +$$

$$p^{uv} Da_t[v]\left(\frac{1}{5}\right) + \frac{p_t^{uv}}{5} Da_t[v]\left(-\frac{4}{5}\right) - \frac{2p_{tt}^{uv}}{25} Da_t[v]\left(-\frac{9}{5}\right) - \frac{\xi_t^t}{5} Da_t[u]\left(\frac{1}{5}\right) +$$

$$\frac{2\xi_{tt}^t}{25} Da_t[u]\left(-\frac{4}{5}\right) - \frac{6\xi_{ttt}^t}{125} Da_t[u]\left(-\frac{9}{5}\right) + Da_t[q^u]\left(\frac{1}{5}\right)$$

with substituted $Da_t[u] = uv$, $Da_t[v] = v^2$. The parameter sumN restricts the number of considered terms in prolongation formula (18). Note that no simplifications are done automatically to speed up calculations. During the following steps, t, u, and v and all other remaining fractional derivatives and integrals of u and v are considered as independent variables, see [11].

Using the code with provided basis operators

```
> s.BaseOps=['u*D_u','v*D_u','u*D_v','v*D_v','t*D_t','t**2*D_t']
> s.build_C_Coeffs(XFS)
> display(s.XFS_C_coeffs)
```

one gets the lists of $\Psi_{\mu,i}$ stored as XFS_C_coeffs that define pairs of equations $\sum C_i \Psi_{1,i} = 0, \sum C_i \Psi_{2,i} = 0$ for C_i:

$$\left[\left[0,\ \ 0,\ \ u^2,\ \ uv,\ \ \frac{uv}{5},\ \ \frac{2t}{5}uv - \frac{4}{25}Da_t[u]\left(-\frac{4}{5}\right)\right],\right.$$

$$\left.\left[0,\ \ 0,\ \ uv,\ \ v^2,\ \ \frac{v^2}{5},\ \ \frac{2t}{5}v^2 - \frac{4}{25}Da_t[v]\left(-\frac{4}{5}\right)\right]\right].$$

Zeros in the first and second columns mean that C_1 and C_2 can be arbitrary constants and X_1, X_2 are symmetry generators. All combinations of basis operators that form symmetries are found by constructing and solving the overdetermined algebraic system (step 6):

```
> s.solveNumeric()
> display(s.getAdmittedOperators())
    9.977099577364706
    Shape of matrix of overdetermined system for finding C_k:
      (22, 6), doing SVD decomposition
    Singular values:
    [ 3.36e+00  1.07e+00  3.50e-01  3.41e-17  9.30e-18 -0.00e+00]
    Finding best combinations of 3 basis operators...OK
```

The resulting 3 operators are found correctly, the code prints

$$[uD_u,\ \ vD_u,\ \ -5tD_t + vD_v].$$

Example 2. Let us consider equation from [16] that contains fractional derivatives of different orders:

$$D_x^{\alpha+1} y = -\frac{\alpha+1}{x} D_x^\alpha y + y\left(\frac{D_x^\alpha y}{y}\right)^{\frac{\alpha+1}{\alpha}}.$$

```
> s.LieSymmSetup(indepsList=['x'], depsList=['y'])
> x,y = symbols('x y'); alpha = sympify('1/3')
> s.BaseOps=[]; s.AddPolynomialBasis(3)
> DAy = s.fracD('y','x',alpha); DAy1 = s.fracD('y','x',alpha+1)
> XFS = s.detPDES({ DAy1 : -(alpha+1)*DAy/x +
    y*(DAy/y)**((alpha+1)/alpha)})
> s.build_C_Coeffs(XFS);s.solveNumeric();s.getAdmittedOperators()
```

All 3 admitted operators are found correctly:

$$\left[xD_x, \quad x^2 D_x - \frac{2x}{3} y D_y, \quad y D_y \right].$$

Note that although the right-hand side depends on fractional derivative, it causes no problems for the code because we do not solve partial differential equations analytically.   □

Example 3. The method also works when operators contain non-polynomial terms. For equation

$$D_x^\alpha y = x^{-1-\alpha} e^{yx^{1-\alpha}}$$

the symmetries are found correctly and include the term $x^{\alpha-1}\partial_y$:

```
> s.LieSymmSetup(indepsList=['x'], depsList=['y'])
> x,y = symbols('x y'); alpha = sympify('1/3')
> s.BaseOps=[]; s.AddPolynomialBasis(2)
> s.BaseOps.append('x**(1/3-1)*D_y');
> s.BaseOps.append('x**(1/3)*D_y'); DAy = s.fracD('y','x',alpha)
> XFS = s.detPDES({DAy: x**(-1-alpha)* exp(y*x**(1-alpha))})
> s.build_C_Coeffs(XFS);s.solveNumeric();s.getAdmittedOperators()
```

$$\left[x^2 D_x - \frac{2x}{3} y D_y, \quad 3x D_x + \left(-2y + \frac{1}{x^{\frac{2}{3}}} \right) D_y \right].$$

Note that the program should be able to automatically calculate $D^\alpha q$ for given functions $q(t)$ by using its internal `fractional.D_RL` function.   □

Example 4. In the work [15], the system of fractional differential equations

$$D_t^\alpha a = \mu b^2/2, \quad D_t^\alpha b = 2\mu bc, \quad D_t^\alpha c = 2\mu c^2$$

is considered. It is obtained when searching for exact solutions of the time fractional Korteweg–de Vries equation

$$D_t^\alpha u = \frac{\mu}{2} \left(\frac{\partial u}{\partial x} \right)^2 + \frac{\partial^3 u}{\partial x^3}$$

by invariant subspace method [13], using the form of solution

$$u(t,x) = a(t) + b(t)x + c(t)x^2.$$

The program successfully finds 3 symmetries for arbitrary order α and 4 symmetries for order $\alpha = 1/3$:

```
> t,a,b,c = symbols('t a b c')
> alpha = sympify('1/3'); mu = sympify('7/8');
> s.LieSymmSetup(indepsList=['t'], depsList=['a','b','c'])
> dax = s.fracD('a','t',alpha); day = s.fracD('b','t',alpha);
> daz = s.fracD('c','t',alpha);
> XFS = s.detPDES({dax: mu*b*b/2,day: mu*2*b*c, daz: mu*2*c**2 })
> s.BaseOps = [];  s.AddPolynomialBasis(2); s.build_C_Coeffs(XFS)
> s.solveNumeric()); display(s.getAdmittedOperators())
```

The displayed symmetries are

$$[bD_a + 2cD_b, \quad aD_a - cD_c + 3tD_t,$$
$$atD_a + btD_b + ctD_c - \frac{3t^2}{2}D_t, bD_b + 2cD_c - 6tD_t].$$

It can be shown similarly to [17] that all of the linearly autonomous symmetry generators have polynomial coefficients, so the complete Lie algebra is obtained. These symmetries are found for the first time and allow one to construct more invariant solutions of the considered system.   □

Example 5. To calculate approximate symmetries for ODEs with a small parameter ε, one needs to add the terms with ε into basis operators X_i. The determining equations are automatically split with respect to ε and the number of equations doubles. For example, consider equation from [9]

$$x'' = F_0(x') + \varepsilon \left(F_1(x') - F_0(x')(3xx' - t) + F_0'(x')x'(xx' - t)\right)$$

constructed from approximate differential invariants. It has four approximate symmetries for arbitrary functions $F_0(x')$, $F_1(x')$. For example, let us test the procedure for specific functions $F_0 = \sin x'$, $F_1 = (x')^2$:

```
> s.LieSymmSetup(indepsList=['t'], depsList=['x'])
> dx = Symbol('x_t'); d2x = Symbol('x_tt')
> f0 = sin(dx); f01 = cos(dx); f1 = dx**2
> sys = {d2x: f0+ s.e*(f1-f0*(3*x*dx-t)+f01*dx*(x*dx-t))}
> XFS = s.detPDE_Approx(sys)
> s.BaseOps=[]; s.AddPolynomialBasis(3,approx=True)
> s.build_C_Coeffs_Approx(XFS);  s.solveNumeric()
> display(s.getAdmittedOperators())
```

$$[\varepsilon x D_x + D_t, \quad \varepsilon x D_t + D_x, \quad \varepsilon D_t, \quad \varepsilon D_x]$$

The found operators are correct. The procedure gives the same result for any other forms of F_0, F_1, even very complicated.

Here `BaseOps` includes operators like $\varepsilon x^2 t^3 \partial_t$ and `XFS` contains ε^0 and ε^1 terms (ε^2 are treated like zeros in this theory [4]).   □

Example 6. Let us check the program on a more complex system composed of some differential invariants (the functions F_0, G_0, F_1, and G_1 are arbitrary):

$$\begin{cases} \ddot{x} = \frac{\dot{x}^2}{x} F_0(\dot{y}) + \varepsilon \left(\frac{\dot{x}^2(t\dot{y}-y)}{x} F_0'(\dot{y}) + \frac{\dot{x}^2}{x} F_1(\dot{y}, x) - \dot{x}^3 \right), \\ \ddot{y} = \frac{\dot{x}}{x} G_0(\dot{y}) - \varepsilon \left(\frac{t\dot{x}}{x} G_0(\dot{y}) - \frac{\dot{x}(t\dot{y}-y)}{x} G_0'(\dot{y}) + \frac{\dot{x}}{x} G_1(\dot{y}, x) \right), \end{cases} \quad (21)$$

```
> s.LieSymmSetup(indepsList=['t'], depsList=['x','y'])
> F0=sympify('z'); G0=sympify('z**3')
> F0y=F0.subs({'z':'y_t'}); DF0y=F0.diff('z').subs({'z':'y_t'})
> G0y=G0.subs({'z':'y_t'}); DG0y=G0.diff('z').subs({'z':'y_t'})
> dx = Symbol('x_t'); dy = Symbol('y_t')
> F1=sympify('sin(x*y_t)'); G1=sympify('y_t**3-x')
> RS1=dx**2/x*F0y + s.e*(dx*dx*(t*dy-y)/x*DF0y+ dx*dx/x * F1-2*dx)
> RS2=dx/x*G0y - s.e*(t*dx/x * G0y- dx*(t*dy-y)/x*DG0y+dx/x*G1)
> XFS=s.detPDE_Approx({'x_tt':RS1,'y_tt':RS2})
> s.BaseOps=[]; s.AddPolynomialBasis(2, approx=True)
> s.build_C_Coeffs_Approx(XFS); s.solveNumeric()
> display(s.getAdmittedOperators())
```

The following admitted operators are found:

$$[-\varepsilon y D_y + D_t, \quad (t^2\varepsilon + t) D_t + (t\varepsilon y + y) D_y,$$

$$(t\varepsilon + 1) D_y, \quad \varepsilon D_t, \quad t\varepsilon D_t + \varepsilon y D_y, \quad \varepsilon x D_x, \quad \varepsilon D_y].$$

Combinations of 162 basis operators were considered. The calculations took about 15 s on a laptop with Intel Core i7-4500U.   □

The obvious limitation of the method is in suggesting the fixed form of operator. It is not possible to include all terms like $x^\beta \partial_u$ or $\sin(\omega t)\partial_u$ with unknown β or ω. Therefore, if the equations admit such specific form of generators, this approach can be used only to check the symmetries (helping to avoid mistakes in calculations). However, most of nonlinear, approximate, and fractional differential equations have rather simple form of symmetries. For fractional differential equations, the fixed form of symmetry generator is the most common case [11].

The main advantage of the approach is that it works for a very wide class of equations including ones with nonlinear functions of derivatives. The program can easily be modified to compute different kinds of symmetries. It makes the described package a suitable tool for express-analysis. It also works when the group is infinite-dimensional, for example, if the first-order equations are considered, when constructing full symmetry algebra analytically is a very complicated problem.

3.2 Computing the Operator of Invariant Differentiation in the Specified Form

The algorithm for constructing the differential invariants and the OID for an approximate Lie algebra of operators is realized in Maple system as the program

"PR-OID: construction of differential invariants and an operator of invariant differentiation for an approximate Lie algebra of operators", which is registered by Rospatent ([26]).

The program is a set of procedures. In this work, we present it with some modifications.

The procedure commut(S1,S2,DepVars) takes two vectors with coordinates of the corresponding infinitesimal generators (for example, the vectors $[\tau_1(t,x,y),$ $\xi_1(t,x,y),\eta_1(t,x,y)]$ and $[\tau_2(t,x,y),\xi_2(t,x,y),\eta_2(t,x,y)]$ and a list of dependent variables (for example, $[x(t),y(t)]$). The procedure calculates the commutator of given generators by formula (7) and returns the coordinate vector of the resulting generator.

The procedure is_algebra(S,DepVars) takes a set of coordinate vectors of infinitesimal generators and a list of dependent variables. The procedure checks that every commutator of given generators belongs to the same vector space. After calculations, the procedure returns the string "It is a Lie algebra", if the operators generate an Lie algebra, and "It is not a Lie algebra" otherwise.

The procedure prolong(S,DepVars,n) takes the vector of coordinates of the infinitesimal generator, the list of dependent variables and the order of derivatives, to which the generator must be continued. The procedure calculates the coordinates of the prolonged generator using standard prolongation formulas (4) and returns the vector of coordinates for the prolonged generator.

The procedure acting(S,f,DepVars,n) is auxiliary. It takes the vector of coordinates of the infinitesimal operator, the tested function, the list of dependent variables and the order of the highest derivative in f. The procedure returns the result of the action of the prolonged generator on the given function.

The procedure approx_invariants(S,DepVars,n,eps) takes a set of coordinate vectors of generators, a list of dependent variables, the required order of differential invariants and indicator of case (exact or approximate). The procedure returns a set of all independent invariants up to the n-th order.

The procedure approx_OID(S,DepVars,eps) takes a set of coordinate vectors for infinitesimal generators, a list of dependent variables and indicator of case (exact or approximate). The procedure returns the multiplier for the OID, the function Φ used to construct this OID, as well as the result of the action of given generators on the obtained function Φ.

Example 7. Let us consider the generators

$$X_1 = (1+\varepsilon t)\frac{\partial}{\partial t}, \quad X_2 = \varepsilon x\frac{\partial}{\partial x}, \quad X_3 = (1+\varepsilon t)\frac{\partial}{\partial y},$$
$$X_4 = (t+\varepsilon t^2)\frac{\partial}{\partial t} + (y+\varepsilon ty)\frac{\partial}{\partial y} \tag{22}$$

and write them in a Maple worksheet:

```
> S:=[[1+epsilon*t,0,0],[0,epsilon*x,0],[0,0,1+epsilon*t],
    [t+epsilon*t^2,0,y+epsilon*t*y]];
```

$$S := [[\varepsilon t + 1, 0, 0], \ [0, \varepsilon x, 0], \ [0, 0, \varepsilon t + 1], \ [\varepsilon t + t, 0, \varepsilon t y + y]]$$

If we check them with procedure

```
> is_algebra(S,[x,y](t));
```

we obtain the following structural constants of a certain Lie algebra:

$$["It \, is \, a \, Lie \, algebra"], [\{a_{1,2,1} = 0, a_{1,2,2} = 0, a_{1,2,3} = 0, a_{1,2,4} = 0\},$$
$$\{a_{1,3,1} = 0, a_{1,3,2} = 0, a_{1,3,3} = \varepsilon, a_{1,3,4} = 0\},$$
$$\{a_{1,4,1} = 1, a_{1,4,2} = 0, a_{1,4,3} = 0, a_{1,4,4} = \varepsilon\},$$
$$\{a_{2,3,1} = 0, a_{2,3,2} = 0, a_{2,3,3} = 0, a_{2,3,4} = 0\},$$
$$\{a_{2,4,1} = 0, a_{2,4,2} = 0, a_{2,4,3} = 0, a_{2,4,4} = 0\},$$
$$\{a_{3,4,1} = 0, a_{3,4,2} = 0, a_{3,4,3} = 1, a_{3,4,4} = 0\}]$$

Note that each constant $a_{i,j,k}$ is a sum $a_{i,j,k,(0)} + \varepsilon a_{i,j,k,(1)}$, i.e., if we obtain $a_{1,3,3} = \varepsilon$, it means that $a_{1,3,3,(0)} = 0$, $a_{1,3,3,(1)} = 1$, and if we obtain $a_{3,4,3} = 1$, it means $a_{3,4,3,(0)} = 1$, $a_{3,4,3,(1)} = 0$. So, the basis of Lie algebra is six-dimensional and consists of the following generators:

$$X_1, \ X_2, \ X_3, \ X_4, \ \varepsilon X_3, \ \varepsilon X_4.$$

For obtaining the differential invariants up to the second order, we use

```
>approx_invariants(S,[x,y](t),2,1);
```

Its result is four invariants

$$y_t + \varepsilon\,(t y_t - y), \quad \frac{x_{t,t}x}{x_t^2} + \frac{2\varepsilon x}{x_t}, \quad \frac{y_{t,t}x}{x_t} + \frac{\varepsilon y_{t,t}tx}{x_t}, \quad \varepsilon x.$$

The OID for approximate Lie algebra generated by six operators can be obtained by the expression

```
>approx_OID(S,[x,y](t),1);
```

which returns the multiplier for OID, the function Φ and the result of acting generators on Φ:

$$\left[\frac{x}{x_t}\right], \quad [\ln(x)], \quad [0, \varepsilon, 0, 0].$$

Let us consider the system of ODEs (21), which admits the given generators. Using obtained invariants, we can rewrite this system in the form

$$J_1^{(2)} \approx F_0(J^{(1)}) + \varepsilon F_1(J_{(0)}^{(1)}, J_{(0)}^{(0)}), \quad J_1^{(2)} \approx G_0(J^{(1)}) + \varepsilon G_1(J_{(0)}^{(1)}, J_{(0)}^{(0)}).$$

Applying OID to invariants $J^{(1)}$ and $J^{(0)}$, we get

$$\begin{cases} \dfrac{x}{\dot{x}} D_t(\dot{y}) = H_0(\dot{y}), \\[2mm] \dfrac{x}{\dot{x}} D_t(t\dot{y} - y) = H_0(\dot{y})\dfrac{x}{t\dot{x}} + H_0'(\dot{y}) \cdot (t\dot{y} - y) + H_1(\dot{y}, x), \\[2mm] \dfrac{x}{\dot{x}} D_t(x) = x. \end{cases}$$

General solution of these equations is the approximate first integral of considered system. Thus, we obtain the reducing system

$$J^{(1)} \approx W_0(\Phi) + \varepsilon W_1(\Phi_0), \quad J^{(2)} \approx \hat{F}_0(\Phi) + \varepsilon \hat{F}_1(\Phi_0),$$

which admits generators X_1, X_3, and X_4. Repeating the procedure, we can obtain the approximate solution of given system.　　□

The program can also be used to study the exact Lie algebra and construct its differential invariants and the invariant differentiation operator.

Example 8. Let us consider the generators

$$X_1 = x\frac{\partial}{\partial t}, \ X_2 = x\frac{\partial}{\partial x}, \ X_3 = y\frac{\partial}{\partial t}, \ X_4 = y\frac{\partial}{\partial y} \tag{23}$$

and write them in Maple worksheet:

```
> S:=[[x,0,0],[0,x,0],[y,0,0],[0,0,y]];
```

If we check them with procedure

```
> is_algebra(S,[x,y](t));
```

we obtain the following structural constants of a certain Lie algebra:

$$[\text{"}It\,is\,a\,Lie\,algebra\text{"}], [\{a_{1,2,1} = -1, a_{1,2,2} = 0, a_{1,2,3} = 0, a_{1,2,4} = 0\},$$
$$\{a_{1,3,1} = 0, a_{1,3,2} = 0, a_{1,3,3} = 0, a_{1,3,4} = 0\},$$
$$\{a_{1,4,1} = 0, a_{1,4,2} = 0, a_{1,4,3} = 0, a_{1,4,4} = 0\},$$
$$\{a_{2,3,1} = 0, a_{2,3,2} = 0, a_{2,3,3} = 0, a_{2,3,4} = 0\},$$
$$\{a_{2,4,1} = 0, a_{2,4,2} = 0, a_{2,4,3} = 0, a_{2,4,4} = 0\},$$
$$\{a_{3,4,1} = 0, a_{3,4,2} = 0, a_{3,4,3} = -1, a_{3,4,4} = 0\}]$$

The first- and second-order differential invariants of this Lie algebra are obtained by

```
>approx_invariants(S,[x,y](t),2,0);
```

and they have the form

$$\frac{xy_t}{x_t y}, \ \frac{(x_t y_{t,t} - x_{t,t} y_t) x^2}{x_t^3 y}, \ \frac{x_{t,t}(-ty_t + y) + y_{t,t}(tx_t - x)}{x_t y_{t,t} - x_{t,t} y_t}.$$

The OID for Lie algebra generated by given operators can be found by

```
>approx_OID(S,[x,y](t),0);
```

which returns the multiplier for OID, the function Φ, and the result of acting generators on Φ:

$$\left[\frac{x}{C_2 x_t}\right], \quad [C_2 \ln(x)], \quad [0, C_2, 0, 0]$$

with arbitrary nonzero constant C_2. Let $C_2 = 1$.

Applying OID to the first-order invariant, we obtain

$$\frac{x}{\dot{x}} D_t \left(\frac{x\dot{y}}{\dot{x}y}\right) = \frac{x\dot{y}}{\dot{x}y} - \left(\frac{x\dot{y}}{\dot{x}y}\right)^2 + \frac{(\dot{x}\ddot{y} - \dot{y}\ddot{x})x^2}{\dot{x}^3 y}.$$

So, for the system of ODEs

$$\begin{cases} \ddot{x} = \dfrac{x\dot{y} + t\dot{x}^2\dot{y}^2 - x\dot{x}\dot{y}^2}{x\left(\dot{y}(y - t\dot{y}) + \dot{x}(x - t\dot{x})\right)}, \\[3mm] \ddot{y} = \dfrac{y\dot{x}\dot{y}^2 - t\dot{x}\dot{y}^3 - x\dot{x}}{x\left(\dot{y}(y - t\dot{y}) + \dot{x}(x - t\dot{x})\right)}, \end{cases}$$

which admits these generators, one obtains

$$\frac{dJ^{(1)}}{d\Phi} = 2J^{(1)} - (J^{(1)})^2, \quad J^{(1)} = \frac{x\dot{y}}{\dot{x}y}.$$

And the first integral of given system is

$$\Phi = \frac{1}{2}\ln\left|\frac{J^{(1)}}{J^{(1)} - 2}\right| - \frac{1}{2}\ln C_1.$$

The reduced system is

$$\begin{cases} \dot{x} = \dfrac{C_1 x^2 \dot{y} - y}{2C_1 xy}, \\[3mm] \ddot{y} = \dfrac{y\dot{x}\dot{y}^2 - t\dot{x}\dot{y}^3 - x\dot{x}}{x\left(\dot{y}(y - t\dot{y}) + \dot{x}(x - t\dot{x})\right)}. \end{cases}$$

These equations admit 3 generators: $x\dfrac{\partial}{\partial t}$, $y\dfrac{\partial}{\partial t}$, $y\dfrac{\partial}{\partial y}$. The order reduction procedure can be repeated. $\quad\square$

4 Conclusion

In this work, we have presented two programs for automatic symmetry calculation and constructing OID for ODE system order reduction by Lie group methods.

The procedure of finding symmetries is realized semi-numerically. It allows one to compute symmetries of equations of any complex form. Another feature of the program is that it can be used for finding the symmetries of ODEs with small parameter and fractional ODEs and their systems. The form of symmetry generator is chosen by the user, it can contain non-polynomial terms. No analytical solving of PDE systems is required. The examples show the correct behavior of the program for FDEs and ODEs with a small parameter for equations with known symmetries. In example 4, new symmetries are found for FDE system which were not published before.

The second program implements the recently developed algorithm of OID construction. It is the first realization of this algorithm. The program can be used for constructing OIDs of Lie algebras of generators, which are admitted by standard ODE systems as well as systems of ODEs with a small parameter. In the second case, the OID is constructed for corresponding approximate Lie algebra of generators.

In the future, we plan to combine described programs on one platform (maybe using SAGE as it has interfaces to other systems) and automate the algorithm for order reduction of the ODE system with and without small parameter.

Acknowledgments. We are grateful to Prof. R.K. Gazizov and Prof. S.Yu. Lukashchuk for constructive discussion. Also we thank the referees whose comments helped us a lot to improve the early draft of this paper.

References

1. Olver, P.J.: Applications of Lie Groups to Differential Equations, 1st edn. Springer-Verlag, New York (1986). https://doi.org/10.1007/978-1-4684-0274-2
2. Ovsyannikov, L.V.: Group Analysis of Differential Equations, 1st edn. Academic Press, New York (1982)
3. Ibragimov, N.H.: Elementary Lie Group Analysis and Ordinary Differential Equations. John Wiley and Sons, Chichester (1999)
4. Baikov, V.A., Gazizov, R.K., Ibragimov, N.H.: Approximate groups of transformations. Differ. Uravn. **29**(10), 1712–1732 (1993). (in Russian)
5. Ayub, M., Mahomed, F.M., Khan, M., Qureshi, M.N.: Symmetries of second-order systems of ODEs and integrability. Nonlinear Dyn. **74**, 969–989 (2013). https://doi.org/10.1007/s11071-013-1016-3
6. Wafo Soh, C., Mahomed, F.M.: Reduction of order for systems of ordinary differential equations. J. Nonlinear Math. Phys. **11**(1), 13–20 (2004). https://doi.org/10.2991/jnmp.2004.11.1.3
7. Gainetdinova, A.A., Gazizov, R.K.: Integrability of systems of two second-order ordinary differential equations admitting four-dimensional Lie algebras. Proc. Roy. Soc. A: Math. Phys. Eng. Sci. **473**(2197), 20160461 (2017). https://doi.org/10.1098/rspa.2016.0461. 13 pp
8. Gazizov, R.K., Gainetdinova, A.A.: Operator of invariant differentiation and its application for integrating systems of ordinary differential equations. Ufa Math. J. **9**(4), 12–21 (2017). https://doi.org/10.13108/2017-9-4-12
9. Gainetdinova, A.A.: Integration of systems of ordinary differential equations with a small parameter which admit approximate Lie algebras. Vestnik Udmurtskogo Universiteta: Matematika, Mekhanika, Komp'yuternye Nauki **28**(2), 143–160 (2018). https://doi.org/10.20537/vm180202. (in Russian)

10. Kilbas, A.A., Srivastava, H.M., Trujillo, J.J.: Theory and Applications of Fractional Differential Equations. Elsevier, The Netherlands (2006)
11. Gazizov, R.K., Kasatkin, A.A., Lukashchuk, S.Yu.: Symmetries and group invariant solutions of fractional ordinary differential equations. In: Kochubei, A., Luchko, Yu. (eds.) Fractional Differential Equations, pp. 65–90. De Gruyter, Berlin (2019). https://doi.org/10.1515/9783110571660
12. Gazizov, R.K., Kasatkin, A.A., Lukashchuk, S.Yu.: Symmetries, conservation laws and group invariant solutions of fractional PDEs. In: Kochubei, A., Luchko, Yu. (eds.) Fractional Differential Equations, pp. 353–382. De Gruyter, Berlin (2019). https://doi.org/10.1515/9783110571660
13. Galaktionov, V.A., Svirshchevskii, S.R.: Exact Solutions and Invariant Subspaces of Nonlinear Partial Differential Equations in Mechanics and Physics. Chapman and Hall/CRC (2006). https://doi.org/10.1201/9781420011623
14. Gazizov, R.K., Kasatkin, A.A.: Construction of exact solutions for fractional order differential equations by the invariant subspace method. Comput. Math. Appl. **66**(5), 576–584 (2013). https://doi.org/10.1016/j.camwa.2013.05.006
15. Sahadevan, R., Bakkyaraj, T.: Invariant subspace method and exact solutions of certain nonlinear time fractional partial differential equations. Fractional Calc. Appl. Anal. **18**(1), 146–162 (2015). https://doi.org/10.1515/fca-2015-0010
16. Gazizov, R.K., Kasatkin, A.A., Lukashchuk, S.Yu.: Linearly autonomous symmetries of the ordinary fractional differential equations. In: Proceedings of 2014 International Conference on Fractional Differentiation and Its Applications (ICFDA 2014), pp. 1–6. IEEE (2014). https://doi.org/10.1109/ICFDA.2014.6967419
17. Kasatkin, A.A.: Symmetry propertics for systems of two ordinary fractional differential equations. Ufa Math. J. **4**(1), 71–81 (2012)
18. Hereman, W.: Review of symbolic software for Lie symmetry analysis. Math. Comput. Model. **25**(8–9), 115–132 (1997). https://doi.org/10.1016/S0895-7177(97)00063-0
19. Cheviakov, A.F.: Symbolic computation of local symmetries of nonlinear and linear partial and ordinary differential equations. Math. Comput. Sci. **4**(2–3), 203–222 (2010). https://doi.org/10.1007/s11786-010-0051-4
20. Vu, K.T., Jefferson, G.F., Carminati, J.: Finding higher symmetries of differential equations using the MAPLE package DESOLVII. Comput. Phys. Commun. **183**(4), 1044–1054 (2012). https://doi.org/10.1016/j.cpc.2012.01.005
21. Jefferson, G.F., Carminati, J.: FracSym: automated symbolic computation of Lie symmetries of fractional differential equations. Comput. Phys. Commun. **185**(1), 430–441 (2014). https://doi.org/10.1016/j.cpc.2013.09.019
22. Merkt, B., Timmer, J., Kaschek, D.: Higher-order Lie symmetries in identifiability and predictability analysis of dynamic models. Phys. Rev. E, Stat. Nonlinear Soft Matter Phys. **92**(1), 012920 (2015). https://doi.org/10.1103/PhysRevE.92.012920
23. Golub, G.H., Van Loan, C.F.: Matrix Computations. JHU Press, Baltimore (2012)
24. Bagderina, Y.Y., Gazizov, R.K.: Invariant representation and symmetry reduction for differential equations with a small parameter. Commun. Nonlinear Sci. Num. Simul. **9**(1), 3–11 (2004). https://doi.org/10.1016/S1007-5704(03)00010-8
25. Meurer, A., et al.: SymPy: symbolic computing in Python. PeerJ Comput. Sci. **3**, e103 (2017). https://doi.org/10.7717/peerj-cs.103
26. Gainetdinova, A.A.: Computer program registration certificate 2018618063. Federal Service for Intellectual Property (Rospatent). Registered on 07 September 2018

An Algorithm for Computing Invariant Projectors in Representations of Wreath Products

Vladimir V. Kornyak[(✉)] [iD]

Laboratory of Information Technologies, Joint Institute for Nuclear Research,
141980 Dubna, Russia
vkornyak@gmail.com

Abstract. We describe an algorithm for computing the complete set of primitive orthogonal idempotents in the centralizer ring of the permutation representation of a wreath product. This set of idempotents determines the decomposition of the representation into irreducible components. In the formalism of quantum mechanics, these idempotents are projection operators into irreducible invariant subspaces of the Hilbert space of a multipartite quantum system. The C implementation of the algorithm constructs irreducible decompositions of high-dimensional representations of wreath products. Examples of computations are given.

Keywords: Wreath product · Irreducible decomposition ·
Invariant projectors · Multipartite quantum system

1 Introduction

A typical description of a physical system usually includes a *space* X, on which a group of *spatial symmetries* $G = G(X)$ acts, and a set of *local states* V with a group of *local* (or *internal*) *symmetries* $F = F(V)$. The sets X and V and the group F can be thought of, respectively, as the *base*, the *typical fiber* and the *structure group* of a *fiber bundle*. The *sections* of the bundle are the set of functions from X to V, denoted by V^X. The set V^X describes the whole states of the physical system. A natural symmetry group that acts on V^X and preserves the structure of the fiber bundle is the *wreath product* [1,2] of F and G

$$\widetilde{W} = F \wr G \cong F^X \rtimes G.$$

The *primitive*[1] action of $\widetilde{W}$ on V^X is described by the rule

$$v(x)(f(x), g) = v(xg^{-1}) f(xg^{-1}), \tag{1}$$

---

[1] Another canonical action of the wreath product, the *imprimitive* action, acts on the fibers, i.e., on the set $V \times X$. We will not consider this action here.

© Springer Nature Switzerland AG 2019
M. England et al. (Eds.): CASC 2019, LNCS 11661, pp. 300–314, 2019.
https://doi.org/10.1007/978-3-030-26831-2_20

where $v \in V^X$, $f \in F^X$, $g \in G$; the *right-action* convention is used for all group actions.

The wreath product is an important mathematical construct. In particular, the *universal embedding theorem* (also known as the *Kaloujnine-Krasner theorem*) states that any extension of group A by group B is isomorphic to a subgroup of $A \wr B$, i.e., the wreath product is a universal object containing all extensions. Wreath products play an important role [3] in the influential *O'Nan-Scott theorem*, which classifies *maximal subgroups* of the *symmetric* group, yet another universal object. The wreath product $S_m \wr S_n$ is the automorphism group of the *hypercubic graph* or *Hamming scheme* $H(n, m)$ in coding theory [4].

A quantum description of a physical system is obtained by introducing a Hilbert space $\mathcal{H}$ spanned by the "classical" states of the system. Accordingly, the action of the symmetry group on classical states goes over to the unitary representation of the group in the Hilbert space $\mathcal{H}$. The next important and natural step in the study of the quantum system is the decomposition of the representation into irreducible components.

Among the problems of quantum mechanics, the study of multipartite quantum systems is of particular interest because they manifest such phenomena as entanglement and non-local correlations. In particular, the very possibility of quantum computing is based on these phenomena. The Hilbert space of a multipartite system consisting of N identical constituents has the form

$$\widetilde{\mathcal{H}} = \mathcal{H}^{\otimes N}, \tag{2}$$

where $\mathcal{H}$ is the Hilbert space of a single constituent. Assuming that the local group F acts through a representation in $\mathcal{H}$, and the group $G \leq S_N$ permutes the constituents, we come to the representation of the wreath product $F \wr G$ in the space $\widetilde{\mathcal{H}}$.

In [5], we proposed an algorithm for decomposing representations of finite groups into irreducible subrepresentations. The algorithm computes a complete set of mutually orthogonal irreducible invariant projectors. In fact, this is a special case of a general construction — a *Peirce decomposition* of a ring with respect to an orthogonal system of idempotents (see [6,7] for more details). Invariant projection operators are important in problems of quantum physics, since they define invariant inner products in invariant subspaces of a Hilbert space. Computer implementation of the approach has proved to be very effective in many cases. For example, the program coped with many high dimensional representations of simple groups and their "small" extensions, presented in the ATLAS [8], in the computationally difficult case of characteristic zero. The algorithm in [5] uses polynomial algebra methods, which by their nature are algorithmically hard. The number of polynomial variables is equal to the rank – defined as the dimension of the centralizer ring – of the representation to be split. Representations of simple (or "close" to simple) groups usually have low ranks, and in such cases the algorithm works well. However, wreath products, which contain all possible extensions, are far from simple groups and have high ranks. Therefore, the approach from [5] is not applicable in the case of wreath products.

In this paper, we propose an algorithm for calculating irreducible invariant projectors in the representation of a wreath product in the form of tensor product polynomials with the projectors of a local group representation as variables. We will consider here permutation representations – the most fundamental type of representations.

2 Irreducible Invariant Projectors of Wreath Product

Let $X \cong \{1, \ldots, N\}$ and $V \cong \{1, \ldots, M\}$. This implies that $G(X) \leq S_N$ and $F(V) \leq S_M$. The functions $v \in V^X$ and $f \in F^X$ can be thought as arrays (ordered lists) $[v_1, \ldots, v_N]$ and $[f_1, \ldots, f_N]$, respectively. Accordingly, the wreath product element $\widetilde{w} \in \widetilde{W}$ can be written as the pair $([f_1, \ldots, f_N]; g)$, where $g \in G$. The action (1) takes the form

$$[v_1, \ldots, v_N] \xrightarrow{\left([f_1, \ldots, f_N];\, g\right)} \left[v_{1g^{-1}} f_{1g^{-1}}, \ldots, v_{Ng^{-1}} f_{Ng^{-1}}\right].$$

The permutation representation $\widetilde{P}$ of the wreath product is a representation of $\widetilde{W}$ by $(0, 1)$-matrices of the size $M^N \times M^N$ that have the form

$$\widetilde{P}(\widetilde{w})_{u,v} = \delta_{u\widetilde{w},v}, \text{ where } \widetilde{w} \in \widetilde{W};\ u, v \in V^X;\ \delta \text{ is the } Kronecker\ delta. \tag{3}$$

We assume that the representation space is an M^N-dimensional Hilbert space $\widetilde{\mathcal{H}}$ over some abelian extension of the field $\mathbb{Q}$. This extension $\mathcal{F}$ is a splitting field of the local group F.

2.1 Centralizer Ring of Wreath Product

Let

$$\widetilde{A}_1, \ldots, \widetilde{A}_{\widetilde{R}} \tag{4}$$

be the basis elements of the centralizer ring of the representation (3), $\widetilde{R}$ denotes the rank of the representation. The basis elements (4) are solutions of the system of equations (invariance condition)

$$\widetilde{P}(\widetilde{w}^{-1}) \widetilde{A} \widetilde{P}(\widetilde{w}) = \widetilde{A}, \ \widetilde{w} \in \widetilde{W}. \tag{5}$$

A more detailed analysis of (5) shows that the elements (4) are in one-to-one correspondence with the orbits of $\widetilde{W}$ on the Cartesian square $V^X \times V^X$. Such orbits are called *orbitals*.

Let us present the Cartesian square in the form $(V \times V)^X$, i.e., as the array

$$[(V \times V)_1, \ldots, (V \times V)_N]. \tag{6}$$

To construct orbitals, consider the structure of the group $\widetilde{W} = F \wr G$ in more detail. Its subgroup

$$\widetilde{F^X} = (F^X; 1_G) \cong F^X \tag{7}$$

is called the *base group* of the wreath product. The group $\widetilde{F^X}$ is a *normal* (or *invariant*) subgroup of $\widetilde{W}$. This means that $\widetilde{w}^{-1}\widetilde{F^X}\widetilde{w} = \widetilde{F^X}$ for any $\widetilde{w} \in \widetilde{W}$, and this is denoted by $\widetilde{F^X} \lhd \widetilde{W}$. The subgroup

$$\widetilde{G} = \left(1_F^X; G\right) \cong G \tag{8}$$

is a complement of $\widetilde{F^X}$ in $\widetilde{W}$, i.e.,

$$\widetilde{W} = \widetilde{F^X} \cdot \widetilde{G} \text{ and } \widetilde{F^X} \cap \widetilde{G} = 1_{\widetilde{W}} \equiv \left(1_F^X; 1_G\right).$$

Thus, we can construct the orbits on the set $(V \times V)^X$ acting first by the elements of the base group (7), and then by the elements of the complement (8). Further, we note that F^X being the direct product of N copies of F, $F^X = F_1 \times \cdots \times F_N$, can be applied to the array (6) component wise independently. The action of the local group F splits the set $V \times V$ into R disjoint subsets

$$\Delta_1, \ldots, \Delta_R,$$

which we will call *local orbitals*. Calculating local orbitals is a simple task, since the local group is exponentially smaller than the wreath product. Let $\overline{R} = \{1, \ldots, R\}$ and $\overline{R}^X$ be the set of all mappings from X into $\overline{R}$. We define the action of $g \in G$ on the mapping $r \in \overline{R}^X$ by $rg = [r_{1g}, \ldots, r_{Ng}]$. This action decomposes the set $\overline{R}^X$ into orbits, and we can write the orbital of the wreath product as

$$\widetilde{\Delta}_r = \bigsqcup_{q \in rG} \Delta_{q_1} \times \cdots \times \Delta_{q_N},$$

where rG denotes the orbit of the mapping r. To translate from the language of sets to the language of matrices, we must replace the orbitals with the basis elements of the local centralizer ring, union by summation, and Cartesian products by the tensor products. Thus, we obtain the expression for the basis element of the centralizer ring of the wreath product

$$\widetilde{A}_r = \sum_{q \in rG} A_{q_1} \otimes \cdots \otimes A_{q_N}, \tag{9}$$

where $A_1, \ldots, A_R$ are basis elements of the local centralizer ring. It is easy to show that the basis elements (9) form a complete system, i.e.,

$$\sum_{i=1}^{\widetilde{R}} \widetilde{A}_{r^{(i)}} = \mathbb{J}_{M^N},$$

where $\mathbb{J}_{M^N}$ is the $M^N \times M^N$ *all-ones matrix*, and $r^{(i)}$ denotes some numbering of the orbits of G on $\overline{R}^X$.

2.2   Complete Set of Irreducible Invariant Projectors

The complete set of irreducible invariant projectors is a subset of the centralizer ring, specified by the conditions of idempotency and mutual orthogonality. A similar construction in ring theory is called a *complete set of primitive orthogonal idempotents*. An arbitrary ring with a complete set of orthogonal idempotents can be represented as a direct sum of indecomposable rings. This is called a *Peirce decomposition* [6,7].

Before constructing the complete set of primitive orthogonal idempotents for the centralizer ring of the permutation representation of the wreath product, we recall some properties of the tensor (Kronecker) product [9]:

1. $(A \otimes B) \otimes C = A \otimes (B \otimes C)$,
2. $(A + B) \otimes (C + D) = A \otimes C + A \otimes D + B \otimes C + B \otimes D$,
3. $(\alpha A) \otimes B = A \otimes (\alpha B) = \alpha (A \otimes B)$,
4. $(A \otimes B)(C \otimes D) = (AC) \otimes (BD)$,
5. $(S \otimes T)^{-1} = S^{-1} \otimes T^{-1}$,

where A, B, C and D are matrices, S and T are invertible matrices and α is a scalar. It follows immediately from these properties that

1. if A and B are both invariant, then $A \otimes B$ is invariant;
2. if A and B are both idempotents, then $A \otimes B$ is idempotent;
3. if $A' = S^{-1} A S$ and $B' = T^{-1} B T$, then

$$A' \otimes B' = (S \otimes T)^{-1} (A \otimes B)(S \otimes T) \equiv (A \otimes B)'.$$

This relation means that we can freely change the bases in the factors of the tensor product to more convenient ones.

Using the above relations, their consequences and some additional technical considerations allows us to come to the main result of this section.

Let $B_1, \ldots, B_K$ be the complete set of irreducible invariant projectors of the permutation representation of the local group F. Let $\overline{K} = \{1, \ldots, K\}$ and $\overline{K}^X$ be the set of all mappings from X into $\overline{K}$. The action of $g \in G$ on the mapping $k \in \overline{K}^X$ is defined as $kg = [k_{1g}, \ldots, k_{Ng}]$. Then we have

Proposition 1. *An irreducible invariant projector in the permutation representation of the wreath product takes the form*

$$\widetilde{B}_k = \sum_{\ell \in kG} B_{\ell_1} \otimes \cdots \otimes B_{\ell_N}, \tag{10}$$

where kG denotes the G-orbit of the mapping k on the set $\overline{K}^X$.

The easily verifiable completeness condition $\sum_{i=1}^{\widetilde{K}} \widetilde{B}_{k^{(i)}} = \mathbb{1}_{M^N}$ holds. Here $\widetilde{K}$ is the number of irreducible components of the wreath product representation, $\mathbb{1}_{M^N}$

is the identity matrix in the representation space, $k^{(i)}$ denotes some numbering of the orbits of G on $\overline{K}^X$.

To calculate the basis elements (9) of the centralizer ring and irreducible invariant projectors (10), we wrote a program in C. The input data for the program are the generators of the spatial and local groups, and the complete set of irreducible invariant projectors of the local group (obtained, for example, using the program described in [5]).

3   Calculation Example: S_4 (*octahedron*) $\wr$ A_5 (*icosahedron*)

We give here the calculation of the centralizer ring and invariant projectors for the permutation representation of the wreath product of the rotational symmetry groups of the octahedron and icosahedron. This representation has dimension $M^N = 2\,176\,782\,336$ and rank $\widetilde{R} = 122\,776$.

3.1   Space Group

In our example, the space X is represented by the icosahedron, see Fig. 1. The full symmetry group of the icosahedron is the product $A_5 \times C_2$. As a group of spatial symmetries we take the orientation-preserving factor A_5, which describes the *rotational* (or *chiral*) *symmetries* of the icosahedron. The order of A_5 is equal to 60. The points of the space X are the vertices of the icosahedron: $X \cong \{1, \ldots, 12\}$. The space symmetry group $G(X)$ can be generated by two permutations. For example, for the vertex numbering as in Fig. 1, we may use the following presentation

$$G(X) = \langle (1,7)(2,8)(3,12)(4,11)(5,10)(6,9),\ (2,3,4,5,6)(8,9,10,11,12) \rangle \cong A_5$$

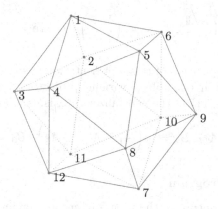

Fig. 1. Icosahedron.

3.2   Local Group

The set of local states is represented by the vertices of the octahedron (Fig. 2): $V \cong \{1, \ldots, 6\}$. The full octahedral symmetry group is isomorphic to $S_4 \times C_2$. Note that the full octahedral group itself is a wreath product, namely, $S_2 \wr S_3$.

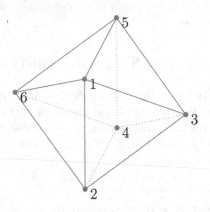

Fig. 2. Octahedron.

The group of rotational symmetries of the octahedron is S_4. The order of S_4 is equal to 24. For the vertex numbering of Fig. 2, the local symmetry group $F(V)$ has the following presentation by two generators

$$F(V) = \langle (1,3,5)(2,4,6), (1,2,4,5) \rangle \cong S_4.$$

The six-dimensional permutation representation $\underline{6}$ of $F(V)$ has rank 3, and the basis of the centralizer ring is

$$A_1 = \mathbb{1}_6, \ A_2 = \begin{pmatrix} 0_3 & \mathbb{1}_3 \\ \mathbb{1}_3 & 0_3 \end{pmatrix}, \ A_3 = \begin{pmatrix} Y & Y \\ Y & Y \end{pmatrix}, \ \text{where } Y = \begin{pmatrix} 0 & 1 & 1 \\ 1 & 0 & 1 \\ 1 & 1 & 0 \end{pmatrix}. \quad (11)$$

The irreducible decomposition of this representation has the form $\underline{6} = \underline{1} \oplus \underline{2} \oplus \underline{3}$. The complete set of irreducible invariant projectors for representation $\underline{6}$, expressed in the basis (11) of the centralizer ring, is as follows

$$B_1 = \frac{1}{6}(A_1 + A_2 + A_3), B_2 = \frac{1}{3}\left(A_1 + A_2 - \frac{1}{2}A_3\right), B_3 = \frac{1}{2}(A_1 - A_2). \quad (12)$$

3.3   Applying C Program

Below we present the output of the program, accompanying parts of the output with comments. An example with a larger dimension of representation is given in Appendix A. Both calculations were performed on PC with 3.30 GHz CPU and 16 GB RAM.

Part 1. The program reads the generators of the spatial and local groups and prints related information: 'Name' is the name of the file containing the generators, 'Number of points' is the dimension of the permutation representation, etc.

```
Space G(X) group:
   Name = "A5_on_icosahedron"
   Number of points = 12
   Comment = "Action of A_5 on 12 vertices of icosahedron"
   Size = "60"
   Number of generators = 2
Local F(V) group:
   Name = "S4_on_octahedron"
   Number of points = 6
   Comment = "Action of S_4 on 6 vertices of octahedron"
   Size = "24"
   Number of generators = 2
```

Part 2. The program constructs the wreath product generators from the generators of the constituent groups and creates a file for the constructed generators. This part of the calculation is optional and can be disabled.

```
Whole F(V).wr.G(X) group:
   Name = "S4_on_octahedron_wr_A5_on_icosahedron"
   Number of points V^X = 2176782336
   Size = "24^12*60"
   Number of generators = 4
```

Part 3. The program prints a common header (in LaTeX):
Wreath product S_4 (*octahedron*) $\wr$ A_5 (*icosahedron*)
Representation dimension: **2 176 782 336**

Part 4. The program computes the basis elements of the centralizer ring of the wreath product. This optional computation can be skipped. If it is enabled, the following items are calculated:

1. *Rank* of the representation (dimension of the centralizer ring).
2. The set of *suborbit lengths*. For a permutation group, suborbits are the orbits of the stabilizer of a single point of the permutation domain [3]. The sizes of orbitals are equal to the suborbit lengths multiplied by the representation dimension. The sum of suborbit length is equal to the representation dimension — we use this fact to verify calculation by displaying 'Checksum'. The multiplicities of the suborbit lengths are shown as superscripts.
3. The basis elements of the centralizer ring for the wreath product representation. We display here only a few elements of the large array. The expressions for the basis elements are tensor product polynomials in the matrices A_1, A_2 and A_3, shown explicitly in (11).

Rank: **122 776**
Number of different suborbit lengths: 46
Wreath suborbit lengths:

$1^{35}, 2^{249}, 3^{11}, 4^{258}, 5^{16}, 6^{403}, 8^{1442}, 16^{1890}, 24^{2418}, 32^{3943}, 48^{43}, 64^{5082}, 80^{51}, 96^{6482},$
$128^{6629}, 256^{8858}, 384^{10735}, 512^{7237}, 768^{67}, 1024^{9901}, 1280^{58}, 1536\underline{^{12006}}, 2048^{5611},$
$4096^{8209}, 6144^{9603}, 8192^{3093}, 12288^{39}, 16384^{4795}, 20480^{46}, 24576^{5558}, 32768^{1225},$
$65536^{2171}, 98304^{2389}, 131072^{310}, 196608^{15}, 262144^{674}, 327680^{14}, 393216^{743},$
$524288^{77}, 1048576^{146}, 1572864^{177}, 2097152^{18}, 4194304^{16}, 5242880^{2}, 6291456^{24},$
16777216^{7}
Checksum = 2176782336 Maximum multiplicity = 12006

Wreath invariant basis forms:

$$\widetilde{A}_1 = A_1^{\otimes 12}$$
$$\widetilde{A}_2 = A_1^{\otimes 5} \otimes A_2 \otimes A_1^{\otimes 2} \otimes A_2 \otimes A_1^{\otimes 3}$$
$$\widetilde{A}_3 = A_1^{\otimes 4} \otimes A_2^{\otimes 2} \otimes A_1^{\otimes 2} \otimes A_2^{\otimes 2} \otimes A_1^{\otimes 2}$$

$$\vdots$$

$$\begin{aligned}
\widetilde{A}_{61387} = &\, A_2^{\otimes 3} \otimes A_1 \otimes A_3 \otimes A_1^{\otimes 2} \otimes A_3^{\otimes 2} \otimes A_2^{\otimes 2} \otimes A_3 \\
&+ A_2 \otimes A_1 \otimes A_2^{\otimes 2} \otimes A_1 \otimes A_3 \otimes A_1 \otimes A_2 \otimes A_3^{\otimes 3} \otimes A_2 \\
&+ A_1 \otimes A_2^{\otimes 2} \otimes A_3^{\otimes 3} \otimes A_2 \otimes A_1 \otimes A_3 \otimes A_1 \otimes A_2^{\otimes 2} \\
&+ A_1 \otimes A_3^{\otimes 2} \otimes A_2^{\otimes 2} \otimes A_3 \otimes A_2^{\otimes 2} \otimes A_1 \otimes A_3 \otimes A_1 \otimes A_2
\end{aligned}$$

$$\begin{aligned}
\widetilde{A}_{61388} = &\, A_2 \otimes A_1 \otimes A_2^{\otimes 2} \otimes A_1 \otimes A_3 \otimes A_1 \otimes A_2^{\otimes 2} \otimes A_3^{\otimes 3} \\
&+ A_2^{\otimes 2} \otimes A_3 \otimes A_1 \otimes A_2 \otimes A_1^{\otimes 2} \otimes A_3^{\otimes 2} \otimes A_2^{\otimes 2} \otimes A_3 \\
&+ A_1 \otimes A_3^{\otimes 2} \otimes A_2^{\otimes 2} \otimes A_3 \otimes A_2^{\otimes 2} \otimes A_1 \otimes A_2 \otimes A_1 \otimes A_3 \\
&+ A_1 \otimes A_2 \otimes A_3^{\otimes 3} \otimes A_2^{\otimes 2} \otimes A_1 \otimes A_3 \otimes A_1 \otimes A_2^{\otimes 2}
\end{aligned}$$

$$\begin{aligned}
\widetilde{A}_{61389} = &\, A_2 \otimes A_1 \otimes A_2^{\otimes 2} \otimes A_1 \otimes A_3 \otimes A_2^{\otimes 3} \otimes A_3^{\otimes 3} \\
&+ A_2^{\otimes 2} \otimes A_3 \otimes A_2^{\otimes 2} \otimes A_1^{\otimes 2} \otimes A_3^{\otimes 2} \otimes A_2^{\otimes 2} \otimes A_3 \\
&+ A_1 \otimes A_3^{\otimes 2} \otimes A_2^{\otimes 2} \otimes A_3 \otimes A_2^{\otimes 2} \otimes A_1 \otimes A_2^{\otimes 2} \otimes A_3 \\
&+ A_2^{\otimes 2} \otimes A_3^{\otimes 3} \otimes A_2^{\otimes 2} \otimes A_1 \otimes A_3 \otimes A_1 \otimes A_2^{\otimes 2}
\end{aligned}$$

$$\vdots$$

$$\widetilde{A}_{122774} = A_3^{\otimes 2} \otimes A_2 \otimes A_3^{\otimes 9} + A_3^{\otimes 3} \otimes A_2 \otimes A_3^{\otimes 8} + A_3^{\otimes 10} \otimes A_2 \otimes A_3 + A_3^{\otimes 11} \otimes A_2$$

$$\begin{aligned}
\widetilde{A}_{122775} = &\, A_3^{\otimes 4} \otimes A_2 \otimes A_3^{\otimes 7} + A_3^{\otimes 5} \otimes A_2 \otimes A_3^{\otimes 6} + A_3^{\otimes 8} \otimes A_2 \otimes A_3^{\otimes 3} \\
&+ A_3^{\otimes 9} \otimes A_2 \otimes A_3^{\otimes 2}
\end{aligned}$$
$$\widetilde{A}_{122776} = A_2 \otimes A_3^{\otimes 11} + A_3 \otimes A_2 \otimes A_3^{\otimes 10} + A_3^{\otimes 6} \otimes A_2 \otimes A_3^{\otimes 5} + A_3^{\otimes 7} \otimes A_2 \otimes A_3^{\otimes 4}$$

Part 5. This main part of the program computes the complete set of irreducible invariant projectors (10). The output below contains

1. Information on whether the irreducible decomposition is multiplicity free or contains subrepresentations with nontrivial multiplicity.
2. The total number of the irreducible invariant projectors.
3. The number of different dimensions.
4. Complete list of irreducible dimensions. The superscripts represent the numbers of irreducible projectors of respective dimensions. The total sum of dimensions must be equal to the dimension of the wreath product representation, and this is verified by the direct calculation of 'Checksum'.
5. A few expressions for irreducible invariant projectors, which are tensor product polynomials in matrices of local projectors (12).

```
Wreath product decomposition is multiplicity free
Number of irreducible components: 122 776
Number of different dimensions: 134
Irreducible dimensions:
```

$1, 4^6, 6^3, 8^6, 9^3, 12^{15}, 16^{32}, 18^7, 20, 24^{70}, 32^{41}, 36^{86}, 45, 48^{191}, 54^{26}, 64^{84}, 72^{298},$
$80^4, 81^7, 96^{412}, 108^{223}, 128^{114}, 144^{913}, 162^{54}, 180^8, 192^{704}, 216^{926}, 243^4, 256^{104},$
$288^{1804}, 320^7, 324^{504}, 384^{772}, 405^4, 432^{2517}, 486^{99}, 512^{76}, 576^{2508}, 648^{1909}, 720^{17},$
$729^9, 768^{705}, 864^{4303}, 972^{818}, 1024^{51}, 1152^{2562}, 1280^3, 1296^{4455}, 1458^{141}, 1536^{479},$
$1620^{16}, 1728^{5322}, 1944^{2712}, 2048^{20}, 2187^4, 2304^{1935}, 2592^{6708}, 2880^{14}, 2916^{961},$
$3072^{223}, 3456^{4575}, 3645^7, 3888^{5495}, 4096^4, 4374^{136}, 4608^{1004}, 5120, 5184^{6924},$
$5832^{2754}, 6144^{59}, 6480^{18}, 6561^9, 6912^{2719}, 7776\underline{^{6966}}, 8192^3, 8748^{822}, 9216^{329},$
$10368^{4760}, 11520^{10}, 11664^{4695}, 12288^{19}, 13122^{98}, 13824^{1011}, 14580^{13}, 15552^{5781},$
$17496^{1999}, 18432^{83}, 19683^3, 20736^{2085}, 23328^{4826}, 25920^{16}, 26244^{511}, 27648^{260},$
$31104^{2964}, 32805^3, 34992^{2775}, 36864^5, 39366^{55}, 41472^{534}, 46080, 46656^{3012},$
$52488^{1023}, 55296^{15}, 58320^{19}, 59049^5, 62208^{877}, 69984^{2173}, 78732^{242}, 82944^{48},$
$93312^{1038}, 103680^4, 104976^{1079}, 118098^{27}, 124416^{102}, 131220^8, 139968^{905},$
$157464^{355}, 186624^{130}, 209952^{568}, 233280^7, 236196^{84}, 279936^{148}, 295245,$
$314928^{254}, 354294^6, 419904^{116}, 472392^{79}, 524880^3, 531441, 629856^{62}, 708588^{15},$
$944784^{26}, 1180980, \underline{\mathbf{1417176}}^9$

```
Checksum = 2176782336 Maximum number of equal dimensions = 6966

Wreath irreducible projectors:
```

$$\widetilde{B}_1 = B_1^{\otimes 12}$$

$$\widetilde{B}_2 = B_1^{\otimes 3} \otimes B_2 \otimes B_1^{\otimes 6} \otimes B_2 \otimes B_1$$

$$\widetilde{B}_3 = B_1^{\otimes 9} \otimes B_2 \otimes B_1^{\otimes 2} + B_1^{\otimes 4} \otimes B_2 \otimes B_1^{\otimes 7}$$

$$\vdots$$

$$\widetilde{B}_{61387} = B_2 \otimes B_3 \otimes B_1 \otimes B_2^{\otimes 2} \otimes B_1^{\otimes 3} \otimes B_3^{\otimes 2} \otimes B_2 \otimes B_3$$
$$+ B_3 \otimes B_2 \otimes B_1 \otimes B_2 \otimes B_1 \otimes B_3 \otimes B_1 \otimes B_2 \otimes B_3 \otimes B_1 \otimes B_3 \otimes B_2$$

$$+ B_1^{\otimes 2} \otimes B_3 \otimes B_2 \otimes B_3^{\otimes 2} \otimes B_2 \otimes B_3 \otimes B_1 \otimes B_2^{\otimes 2} \otimes B_1$$

$$+ B_1 \otimes B_2^{\otimes 2} \otimes B_3 \otimes B_1 \otimes B_3^{\otimes 2} \otimes B_2 \otimes B_3 \otimes B_1 \otimes B_2 \otimes B_1$$

$$\widetilde{B}_{61388} = B_2^{\otimes 2} \otimes B_3 \otimes B_1^{\otimes 2} \otimes B_3 \otimes B_1 \otimes B_2 \otimes B_3 \otimes B_2^{\otimes 2} \otimes B_3$$

$$+ B_1 \otimes B_2 \otimes B_3 \otimes B_2^{\otimes 2} \otimes B_3 \otimes B_2^{\otimes 2} \otimes B_3 \otimes B_1^{\otimes 2} \otimes B_3$$

$$\widetilde{B}_{61389} = B_1^{\otimes 2} \otimes B_2^{\otimes 3} \otimes B_1 \otimes B_2 \otimes B_3 \otimes B_2 \otimes B_1 \otimes B_3^{\otimes 2}$$

$$+ B_2 \otimes B_1 \otimes B_2 \otimes B_3 \otimes B_2 \otimes B_1 \otimes B_3 \otimes B_2 \otimes B_1^{\otimes 2} \otimes B_2 \otimes B_3$$

$$+ B_1 \otimes B_3 \otimes B_2 \otimes B_1^{\otimes 3} \otimes B_3 \otimes B_2^{\otimes 3} \otimes B_3 \otimes B_2$$

$$+ B_3 \otimes B_2 \otimes B_3 \otimes B_2 \otimes B_1^{\otimes 2} \otimes B_2 \otimes B_1^{\otimes 2} \otimes B_2 \otimes B_3 \otimes B_2$$

$$+ B_2 \otimes B_3^{\otimes 3} \otimes B_1 \otimes B_2 \otimes B_1^{\otimes 3} \otimes B_2^{\otimes 3}$$

$$+ B_3 \otimes B_2^{\otimes 2} \otimes B_3 \otimes B_2^{\otimes 2} \otimes B_1 \otimes B_3 \otimes B_1^{\otimes 3} \otimes B_2$$

$$\vdots$$

$$\widetilde{B}_{122774} = B_3^{\otimes 2} \otimes B_2 \otimes B_3^{\otimes 9} + B_3^{\otimes 3} \otimes B_2 \otimes B_3^{\otimes 8} + B_3^{\otimes 10} \otimes B_2 \otimes B_3 + B_3^{\otimes 11} \otimes B_2$$

$$\widetilde{B}_{122775} = B_3^{\otimes 2} \otimes B_2 \otimes B_3 \otimes B_2 \otimes B_3^{\otimes 7} + B_3^{\otimes 2} \otimes B_2 \otimes B_3^{\otimes 2} \otimes B_2 \otimes B_3^{\otimes 6}$$

$$+ B_3^{\otimes 3} \otimes B_2 \otimes B_3^{\otimes 4} \otimes B_2 \otimes B_3^{\otimes 3} + B_3^{\otimes 5} \otimes B_2 \otimes B_3^{\otimes 4} \otimes B_2 \otimes B_3$$

$$+ B_3^{\otimes 8} \otimes B_2 \otimes B_3^{\otimes 2} \otimes B_2 + B_3^{\otimes 9} \otimes B_2 \otimes B_3 \otimes B_2$$

$$\widetilde{B}_{122776} = B_3^{\otimes 3} \otimes B_2^{\otimes 2} \otimes B_3^{\otimes 7} + B_2 \otimes B_3^{\otimes 4} \otimes B_2 \otimes B_3^{\otimes 6}$$

$$+ B_3 \otimes B_2 \otimes B_3^{\otimes 3} \otimes B_2 \otimes B_3^{\otimes 6} + B_3^{\otimes 6} \otimes B_2 \otimes B_3 \otimes B_2 \otimes B_3^{\otimes 3}$$

$$+ B_3^{\otimes 7} \otimes B_2^{\otimes 2} \otimes B_3^{\otimes 3} + B_3^{\otimes 9} \otimes B_2^{\otimes 2} \otimes B_3$$

```
Time: 0.58 s
Maximum number of tensor monomials: 531441
```

4   Application Remarks

One of the main goals of the work was to develop a tool for the study of models of multipartite quantum systems. The projection operators obtained by the program are supposed to be used to calculate quantum correlations in such models. These operators are matrices of huge dimension (for example, about ten trillion for projectors in Appendix A). Obviously, the explicit calculation of such matrices is impossible. However, the expression of projectors for wreath products in the form of tensor polynomials makes it possible to reduce the computation of quantum correlations to a sequence of computations with small matrices of local projectors. To demonstrate this, recall that the computation of quantum correlations is ultimately reduced to the computation of scalar products. Let

$$\widetilde{\Phi} = \sum_{m \in V^X} \varphi_m \, |e_{m_1}\rangle \otimes \cdots \otimes |e_{m_N}\rangle \in \widetilde{\mathcal{H}} \text{ and } \widetilde{\Psi} = \sum_{m \in V^X} \psi_n \, |e_{n_1}\rangle \otimes \cdots \otimes |e_{n_N}\rangle \in \widetilde{\mathcal{H}}$$

be vectors of the Hilbert space (2), where $|e_1\rangle, \ldots, |e_M\rangle$ is a basis in the local

Hilbert space $\mathcal{H}$, φ_m and ψ_n are arbitrary scalars from the base field $\mathcal{F}$. Then we can calculate the scalar product of these vectors in the invariant subspace defined by the projector (10) as follows

$$
\left\langle \widetilde{\Phi} \middle| \bar{B}_k \middle| \widetilde{\Psi} \right\rangle
$$

$$
= \sum_{\substack{m \in V^X \\ n \in V^X \\ \ell \in kG}} \overline{\varphi}_m \psi_n \, \langle e_{m_1} | \otimes \cdots \otimes \langle e_{m_N} | \, B_{\ell_1} \otimes \cdots \otimes B_{\ell_N} \, |e_{n_1}\rangle \otimes \cdots \otimes |e_{n_N}\rangle
$$

$$
= \sum_{\substack{m \in V^X \\ n \in V^X \\ \ell \in kG}} \overline{\varphi}_m \psi_n \, \langle e_{m_1} | B_{\ell_1} |e_{n_1}\rangle \cdots \langle e_{m_N} | B_{\ell_N} |e_{n_N}\rangle .
$$

Acknowledgments. I am grateful to Yu.A. Blinkov for help in preparing the article and V.P. Gerdt for fruitful discussions and valuable advice.

A   Computing Invariant Projectors for $\mathbf{A_5}$ (*icosahedron*) $\wr$ $\mathbf{A_5}$ (*icosahedron*)

The representation $\underline{\mathbf{12}}$ of A_5 on the icosahedron has rank 4, and the basis of the centralizer ring is

$$
A_1 = \mathbb{1}_{12}, \quad A_2 = \begin{pmatrix} \mathbb{0}_6 & \mathbb{1}_6 \\ \mathbb{1}_6 & \mathbb{0}_6 \end{pmatrix}, \quad A_3 = \begin{pmatrix} Y & Z \\ Z & Y \end{pmatrix}, \quad A_4 = \begin{pmatrix} Z & Y \\ Y & Z \end{pmatrix}, \tag{13}
$$

where $Y = \begin{pmatrix} 0&1&1&1&1&1 \\ 1&0&1&0&0&1 \\ 1&1&0&1&0&0 \\ 1&0&1&0&1&0 \\ 1&0&0&1&0&1 \\ 1&1&0&0&1&0 \end{pmatrix}$, $Z = \begin{pmatrix} 0&0&0&0&0&0 \\ 0&0&0&1&1&0 \\ 0&0&0&0&1&1 \\ 0&1&0&0&0&1 \\ 0&1&1&0&0&0 \\ 0&0&1&1&0&0 \end{pmatrix}$. The irreducible decompo-

sition of this representation has the form $\underline{\mathbf{12}} = \underline{\mathbf{1}} \oplus \underline{\mathbf{3}} \oplus \underline{\mathbf{3}'} \oplus \underline{\mathbf{5}}$. The complete set of irreducible invariant projectors for representation $\underline{\mathbf{12}}$, expressed in the basis (13) of the centralizer ring, is as follows

$$
B_1 = \frac{1}{12} \left(A_1 + A_2 + A_3 + A_4 \right),
$$

$$
B_3 = \frac{1}{4} \left(A_1 - A_2 + \frac{1}{\sqrt{5}} A_3 - \frac{1}{\sqrt{5}} A_4 \right),
$$

$$
B_{3'} = \frac{1}{4} \left(A_1 - A_2 - \frac{1}{\sqrt{5}} A_3 + \frac{1}{\sqrt{5}} A_4 \right),
$$

$$
B_5 = \frac{1}{12} \left(5A_1 + 5A_2 - A_3 - A_4 \right).
$$

```
Local F(V) group:
  Name = "A5_on_icosahedron"
  Number of points = 12
  Comment = "Action of A_5 on 12 vertices of icosahedron"
  Number of generators = 2

Space G(X) group:
  Name = "A5_on_icosahedron"
  Number of points = 12
  Comment = "Action of A_5 on 12 vertices of icosahedron"
  Number of generators = 2

Whole group F(V).wr.G(X)
  Number of points V^X = 8916100448256
```

Wreath product A_5 (*icosahedron*) $\wr A_5$ (*icosahedron*)
Representation dimension: **8 916 100 448 256**
Rank: **3 875 157**
Wreath product decomposition is multiplicity free
Number of irreducible components: 3 875 157
Number of different dimensions: 261
Irreducible dimensions:

$1, 6^6, 9^6, 10^3, 18^{28}, 25^3, 30^{29}, 36^{16}, 45^2, 50^6, 54^{176}, 60^{16}, 81^{29}, 90^{248}, 100^4, 108^{81},$
$125, 125, 150^{115}, 162^{715}, 180^{122}, 225^{30}, 243^{15}, 250^{21}, 270^{1361}, 300^{62}, 324^{475}, 405^{16},$
$450^{941}, 486^{2401}, 500^{10}, 540^{1005}, 625^7, 675^{14}, 729^{71}, 750^{293}, 810^{5584}, 900^{743}, 972^{1791},$
$1125^{16}, 1250^{31}, 1350^{5254}, 1458^{6178}, 1500^{246}, 1620^{4615}, 1875^4, 2025^{111}, 2187^{42},$
$2250^{2441}, 2430^{17059}, 2500^{25}, 2700^{4607}, 2916^{4071}, 3125^4, 3645^{47}, 3750^{547}, 4050^{19890},$
$4374^{11761}, 4500^{2254}, 4860^{12862}, 5625^{54}, 6075^{63}, 6250^{50}, 6561^{128}, 6750^{12386},$
$7290^{37925}, 7500^{546}, 8100^{16086}, 8748^{7733}, 10125^{66}, 11250^{4247}, 12150^{52986}, 12500^{53},$
$13122^{16751}, 13500^{10685}, 14580^{27576}, 15625^9, 16875^{34}, 18225^{259}, 18750^{786}, 19683^{61},$
$20250^{40875}, 21870^{62340}, 22500^{3861}, 24300^{41325}, 26244^{10200}, 28125^{32}, 31250^{61},$
$32805^{51}, 33750^{18678}, 36450^{101494}, 37500^{742}, 39366^{17904}, 40500^{34194}, 43740^{41406},$
$46875^4, 50625^{187}, 54675^{118}, 56250^{5024}, 59049^{91}, 60750^{93760}, 62500^{50},$
$65610^{74687}, 67500^{16874}, 72900^{71901}, 78125^7, 78732^{9419}, 91125^{103}, 93750^{731},$
$101250^{53180}, 109350^{138609}, 112500^{4992}, 118098^{13935}, 121500^{71716}, 131220^{42937},$
$140625^{64}, 151875^{101}, 156250^{45}, 164025^{227}, 168750^{18946}, 177147^{46}, 182250^{148591},$
$187500^{814}, 196830^{64140}, 202500^{44570}, 218700^{85846}, 236196^{5558}, 253125^{72},$
$281250^{4058}, 295245^{55}, 303750^{100768}, 312500^{57}, 328050^{132909}, 337500^{17918},$
$354294^{7569}, 364500^{100099}, 390625^9, 393660^{28803}, 421875^{30}, 455625^{223}, 468750^{487},$
$492075^{117}, 506250^{44563}, 531441^{35}, 546750^{\underline{162353}}, 562500^{4467}, 590490^{38051},$
$607500^{75302}, 656100^{64773}, 703125^{26}, 708588^{2079}, 781250^{19}, 820125^{136},$
$843750^{12709}, 911250^{127883}, 937500^{642}, 984150^{87505}, 1012500^{37970}, 1062882^{2545},$
$1093500^{86513}, 1171875^3, 1180980^{11554}, 1265625^{103}, 1366875^{136}, 1406250^{2224},$
$1476225^{94}, 1518750^{67801}, 1562500^{37}, 1594323^{11}, 1640250^{120077}, 1687500^{12838},$
$1771470^{14025}, 1822500^{76014}, 1953125^3, 1968300^{28555}, 2125764^{258}, 2278125^{121},$
$2343750^{212}, 2531250^{24163}, 2657205^{16}, 2733750^{108068}, 2812500^{2782}, 2952450^{35413},$

$303750^{46259}, 3188646^{403}, 3280500^{42505}, 3515625^{26}, 3542940^{1796}, 3796875^{71},$
$3906250^9, 4100625^{115}, 4218750^{5748}, 4428675^{43}, 4556250^{66794}, 4687500^{346},$
$4920750^{53856}, 5062500^{19542}, 5314410^{2418}, 5467500^{42675}, 5904900^{5082}, 6328125^{66},$
$7031250^{839}, 7381125^{51}, 7593750^{28831}, 781250^{16}, 8201250^{54394}, 8437500^{5668},$
$8857350^{6482}, 9112500^{30331}, 9765625^5, 9841500^{8743}, 10546875^{18}, 11390625^{75},$
$11718750^{83}, 12301875^{67}, 12656250^{8656}, 13668750^{38539}, 14062500^{1040},$
$14762250^{10735}, 15187500^{15435}, 16402500^{9901}, 17578125^{16}, 19531250^4, 20503125^{58},$
$21093750^{1717}, 22781250^{19527}, 23437500^{102}, 24603750^{12006}, 25312500^{5605},$
$27337500^{8134}, 31640625^{31}, 34171875^{39}, 35156250^{245}, 37968750^{7111}, 39062500^3,$
$41006250^{9603}, 42187500^{1419}, 45562500^{4795}, 48828125, 56953125^{46}, 58593750^{26},$
$63281250^{1800}, 68343750^{5558}, 70312500^{215}, 75937500^{2140}, 87890625^{10}, 94921875^{15},$
$105468750^{336}, 113906250^{2389}, 117187500^{20}, 126562500^{674}, 158203125^{14},$
$175781250^{42}, 189843750^{743}, 195312500^3, 210937500^{136}, 244140625, 316406250^{177},$
$351562500^{16}, 439453125^2, 527343750^{24}, \mathbf{585937500}^6$

Checksum = 8916100448256 Maximum number of equal dimensions = 162353

Wreath irreducible projectors:

$$\widetilde{B}_1 = B_1^{\otimes 12}$$
$$\widetilde{B}_2 = B_1^{\otimes 11} \otimes B_{3'} + B_1^{\otimes 2} \otimes B_{3'} \otimes B_1^{\otimes 9}$$
$$\widetilde{B}_3 = B_1^{\otimes 9} \otimes B_{3'} \otimes B_1^{\otimes 2} + B_1^{\otimes 4} \otimes B_{3'} \otimes B_1^{\otimes 7}$$

$\vdots$

$$\widetilde{B}_{1937578} = B_5 \otimes B_{3'}^{\otimes 2} \otimes B_3^{\otimes 2} \otimes B_1^{\otimes 3} \otimes B_5 \otimes B_{3'} \otimes B_3 \otimes B_5$$
$$+ B_{3'} \otimes B_1 \otimes B_3 \otimes B_{3'} \otimes B_1 \otimes B_5 \otimes B_3 \otimes B_5 \otimes B_3$$
$$\otimes B_1 \otimes B_5 \otimes B_{3'}$$
$$+ B_1^{\otimes 2} \otimes B_5 \otimes B_3 \otimes B_{3'} \otimes B_5^{\otimes 2} \otimes B_{3'} \otimes B_1 \otimes B_3^{\otimes 2} \otimes B_{3'}$$
$$+ B_3 \otimes B_5 \otimes B_{3'} \otimes B_5 \otimes B_1 \otimes B_3 \otimes B_{3'} \otimes B_1 \otimes B_5$$
$$\otimes B_1 \otimes B_{3'} \otimes B_3$$

$$\widetilde{B}_{1937579} = B_5 \otimes B_3 \otimes B_1 \otimes B_{3'} \otimes B_5^{\otimes 2} \otimes B_3 \otimes B_1 \otimes B_3^{\otimes 2} \otimes B_1 \otimes B_{3'}$$
$$+ B_3 \otimes B_1 \otimes B_{3'} \otimes B_1 \otimes B_3^{\otimes 2} \otimes B_5 \otimes B_3 \otimes B_5^{\otimes 2} \otimes B_{3'} \otimes B_1$$
$$+ B_3^{\otimes 2} \otimes B_1 \otimes B_3 \otimes B_5 \otimes B_3 \otimes B_{3'} \otimes B_5^{\otimes 2} \otimes B_1 \otimes B_{3'} \otimes B_1$$
$$+ B_{3'} \otimes B_5 \otimes B_1 \otimes B_{3'} \otimes B_1 \otimes B_5 \otimes B_3^{\otimes 3} \otimes B_5 \otimes B_3 \otimes B_1$$

$$\widetilde{B}_{1937580} = B_3 \otimes B_{3'}^{\otimes 2} \otimes B_3 \otimes B_5^{\otimes 2} \otimes B_1^{\otimes 3} \otimes B_{3'}^{\otimes 2} \otimes B_5$$
$$+ B_{3'} \otimes B_1 \otimes B_{3'}^{\otimes 2} \otimes B_5 \otimes B_1 \otimes B_3^{\otimes 2} \otimes B_5 \otimes B_1 \otimes B_5 \otimes B_{3'}$$
$$+ B_3^{\otimes 2} \otimes B_{3'} \otimes B_5 \otimes B_1 \otimes B_5 \otimes B_{3'} \otimes B_1^{\otimes 2} \otimes B_5 \otimes B_{3'}^{\otimes 2}$$
$$+ B_1^{\otimes 2} \otimes B_5 \otimes B_{3'}^{\otimes 2} \otimes B_1 \otimes B_3 \otimes B_{3'} \otimes B_5^{\otimes 2} \otimes B_3 \otimes B_{3'}$$

$\vdots$

$$\widetilde{B}_{3875155} = B_3 \otimes B_5^{\otimes 11} + B_5 \otimes B_3 \otimes B_5^{\otimes 10} + B_5^{\otimes 6} \otimes B_3 \otimes B_5^{\otimes 5} + B_5^{\otimes 7}$$
$$\otimes B_3 \otimes B_5^{\otimes 4}$$

$$\widetilde{B}_{3875156} = B_5^{\otimes 4} \otimes B_{3'} \otimes B_5^{\otimes 7} + B_5^{\otimes 5} \otimes B_{3'} \otimes B_5^{\otimes 6} + B_5^{\otimes 8} \otimes B_{3'} \otimes B_5^{\otimes 3}$$
$$+ B_5^{\otimes 9} \otimes B_{3'} \otimes B_5^{\otimes 2}$$

$$\widetilde{B}_{3875157} = B_5^{\otimes 4} \otimes B_3 \otimes B_5^{\otimes 7} + B_5^{\otimes 5} \otimes B_3 \otimes B_5^{\otimes 6} + B_5^{\otimes 8} \otimes B_3 \otimes B_5^{\otimes 3}$$
$$+ B_5^{\otimes 9} \otimes B_3 \otimes B_5^{\otimes 2}$$

Time: 7.35 s
Maximum number of tensor monomials: 16777216

References

1. Meldrum, J.D.P.: Wreath Products of Groups and Semigroups. Longman/Wiley, New York (1995)
2. James, G.D., Kerber, A.: The Representation Theory of the Symmetric Group. Encyclopedia of Mathematics and its Applications, vol. 16. Addison-Wesley, Reading (1981)
3. Cameron, P.J.: Permutation Groups. Cambridge University Press, Cambridge (1999)
4. Eiichi, B., Tatsuro, I.: Algebraic Combinatorics I: Association Schemes. The Benjamin/Cummings Publishing Co., Inc., Menlo Park (1984)
5. Kornyak, V.V.: Splitting permutation representations of finite groups by polynomial algebra methods. In: Gerdt, V.P., Koepf, W., Seiler, W.M., Vorozhtsov, E.V. (eds.) CASC 2018. LNCS, vol. 11077, pp. 304–318. Springer, Cham (2018). https://doi.org/10.1007/978-3-319-99639-4_21
6. Jacobson, N.: Structure of Rings, vol. 37. American Mathematical Society, Providence, R.I. (1956)
7. Rowen, L.H.: Ring Theory. Academic Press, Inc., Boston (1991)
8. Wilson, R., et al.: Atlas of finite group representations. http://brauer.maths.qmul.ac.uk/Atlas/v3
9. Steeb, W.-H.: Matrix Calculus and the Kronecker Product with Applications and C++ Programs. World Scientific Publishing Co., Inc., River Edge (1997)

PAF Reconstruction with the Orbits Method

Ilias S. Kotsireas[1], Youtong Liu[2], and Jing Yang[2(✉)]

[1] Wilfrid Laurier University, Waterloo, ON N2L 3C5, Canada
ikotsire@wlu.ca
[2] SMS-HCIC, Guangxi University for Nationalities, Nanning 530006,
People's Republic of China
youtongliu@163.com, yangjing0930@gmail.com

Abstract. The maximal determinant problem for $-1/+1$ matrices has been studied extensively and upper bounds for the determinant are known for various classes of matrices. These upper bounds are attained by specific kinds of combinatorial matrices and D-optimal matrices are one such case. One of the key issues in the search for D-optimal matrices is to reconstruct a $-1/+1$ sequence of length n from a given sequence of periodic autocorrelation function (PAF) values. In turn, this is reduced to solving a quadratic system with $\lfloor n/2 \rfloor$ equations over $\{-1, +1\}^n$. In this paper, a method for reconstructing a special class of $-1/+1$ sequences is proposed by making use of some combinatorial properties of PAF values and the orbits method based on group actions. Furthermore, we apply additional filtering criteria to enhance the effectiveness of the method. Experiments show that the new approach can solve relatively large-scale problems and can help to generate solutions for many D-optimal problems.

Keywords: Periodic autocorrelation function · D-optimal design · Orbits method

1 Introduction

Hadamard's maximal determinant problem, which was first posed by J. Hadamard in his 1893 paper [8], asks for the largest possible determinant of any $N \times N$ matrix with entries equal to -1 or $+1$. Despite being extensively studied in the past century, Hadamard's maximal determinant problem remains an open problem till today.

There are various known bounds on maximal determinants (see [1,3,4,17]). Another aspect of the maximal determinant problem is how to construct a matrix that attains the upper bound. D-optimal matrices are one such case where $N \equiv 2 \mod 4$, i.e., there exist an odd n such that $N = 2n$. For the remainder of the paper, we assume n is an odd number. The construction of D-optimal matrices of

© Springer Nature Switzerland AG 2019
M. England et al. (Eds.): CASC 2019, LNCS 11661, pp. 315–329, 2019.
https://doi.org/10.1007/978-3-030-26831-2_21

this type was first given by Ehlich in [3]. He proved that if A and B are circulant $-1/+1$ matrices of order n such that

$$AA^T + BB^T = 2(n-1)I_n + 2J_n \tag{1}$$

where I_n is the identity matrix of order n and J_n is a matrix of order n with entries equal to 1, then the following matrix

$$H = \begin{pmatrix} A & B \\ -B^T & A^T \end{pmatrix} \tag{2}$$

has maximal determinant. We say the matrix H constructed in this way is of circulant type. It is clear that in order to construct a circulant-type D-optimal matrix, one has to find two sequences a and b such that the circulant matrices A and B whose first rows are given by these two sequences, satisfy (1). Then one can generate the other rows via making a right cyclic shift by one of the previous row. Let a and b be the first rows of A and B, respectively. One may easily derive from (1) that $|a|^2 + |b|^2 = 4n - 2$ where $|a|$ and $|b|$ are the sums of elements in a and b, respectively. By solving the Diophantine equation, one may obtain a necessary condition under which a circulant-type matrix H of the form (2) exists. A comprehensive table of all odd $n < 200$ for which D-optimal matrices of order $2n$ are known can be found in [15]. The lowest outstanding order is currently $N = 198$ (i.e., $n = 99$). In addition, there are two infinite classes of D-optimal matrices of orders $n = q^2 + q + 1$ and $n = 2q^2 + 2q + 1$ (see [15] for more details). When $n > 100$, only a few circulant-type D-optimal matrices are known and the corresponding values of n are 103, 113, 121, 131, 145, 157, 181 and 241 (see [2,5–7,9,14–16]).

   The main difficulty for constructing D-optimal matrices lies in the exponential size of the search space in terms of N where N is the order of the desired D-optimal matrices. In priori, the search space consists of 2^{N^2} possibilities. Some strategies for constructing D-optimal matrices use several combinatorial properties of the aforementioned sequences a and b, in order to significantly reduce the search space. One such property is related with the periodic autocorrelation function (PAF) that we explain below.

   Let $a = [a_0, \dots, a_{n-1}]$ and $b = [b_0, \dots, b_{n-1}]$ be the first rows of A and B. Then (1) is equivalent to

$$\begin{cases} |a|^2 + |b|^2 = 4n - 2, \\ \mathrm{PAF}(a,k) + \mathrm{PAF}(b,k) = 2, \quad k = 1, \dots, \lfloor n/2 \rfloor, \end{cases} \tag{3}$$

where $|a| = \sum_{i=0}^{n-1} a_i$, $|b| = \sum_{i=0}^{n-1} b_i$, and $\mathrm{PAF}(a,k)$ (or $\mathrm{PAF}(b,k)$) is called the periodic autocorrelation function of a (or b) and is defined as

$$\mathrm{PAF}(a,k) = \sum_{i=0}^{n-1} a_i a_{i+k}, \text{ for } k = 0, 1, \dots, \lfloor n/2 \rfloor$$

(where $i + k$ is taken modulo n when $i + k \geq n$). The resulting algorithm to construct circulant-type D-optimal matrices H is to search for $-1/+1$ sequences

a and b such that (3) holds, where one of the key steps is to reconstruct a $-1/+1$ sequence of length n from a sequence of PAF values. In turn, this is reduced to solving a quadratic system with $\lfloor n/2 \rfloor$ equations over $\{-1, +1\}^n$. More explicitly, we will consider the following problem:

PAF Profile Problem (PPP). For a specific length n and element sum $|a|$, given a specific sequence of numbers, $P = [p_0, p_1, \ldots, p_{\lfloor n/2 \rfloor}]$,[1] is there a binary $\{-1, +1\}$ sequence of length n, with a constant sum of elements, whose PAF profile is equal to P?

For example, let $n = 9$, $|a| = -1$ and $P = [-3, 1, -3, 1]$. We are looking for $a = [a_0, \ldots, a_{n-1}] \in \{-1, +1\}^n$ which satisfies

$$\begin{cases} |a| = a_0 + a_1 + \cdots + a_7 + a_8 = -1 \\ \text{PAF}(a, 1) = a_0 a_1 + a_1 a_2 + \cdots + a_8 a_0 = -3, \\ \text{PAF}(a, 2) = a_0 a_2 + a_1 a_3 + \cdots + a_8 a_1 = 1, \\ \text{PAF}(a, 3) = a_0 a_3 + a_1 a_4 + \cdots + a_8 a_2 = -3, \\ \text{PAF}(a, 4) = a_0 a_4 + a_1 a_5 + \cdots + a_8 a_3 = 1. \end{cases}$$

The PAF profile problem was first investigated in [10] where the authors encoded a $-1/+1$ sequence into a partition of n via the concept of runs and identified some combinatorial features the partition possesses from the given PAF values. However, this method can only solve relatively small-scale systems of equations, say $n < 50$. In order to solve large-scale systems, we will adopt the orbits method based on group actions to reconstruct $-1/+1$ sequences which are generated by using the orbits method from a set of PAF values. Furthermore, we apply the power spectral density (PSD) test and ranking/unranking technique to enhance the effectiveness of the method. Experiments show that the new approach can solve relatively large-scale problems and can help to generate solutions for many D-optimal problems.

This paper is structured as follows. In Sect. 2, we recall some combinatorial concepts and prove a new combinatorial property of the PAF sequences which is used to shrink the search space of solutions to the PAF profile problem. In Sect. 3, the orbits method and several practical techniques related to it are reviewed. In Sect. 4, an algorithm is presented to solve a special class of PPP and an example is given to illustrate how the algorithm works. It is followed by benchmark experiments in Sect. 5 to show the efficiency of the proposed algorithm. The paper is concluded in Sect. 6.

2 Combinatorial Concepts and Properties

In this section, we first recall some combinatorial concepts and some nice properties related to PAF values and $-1/+1$ sequences, which may help reduce the size of search space when solving a PPP.

Suppose $a = [a_0, \ldots, a_{n-1}]$. Clearly, $\text{PAF}(a, k) = \text{PAF}(a', k)$ where a' is a rotation or reverse rotation of a, i.e., the PAF values remain unchanged under

---

[1] Obviously, $p_0 = n$. It is omitted unless it is used.

rotations and reverse rotations of the sequence. In this case, we say a and a' are equivalent sequences. Therefore, without loss of generality, we assume a starts with $+1$ and ends with -1.

In coding theory, $-1/+1$ sequences are described using the concept of a run, see [12]. A *run* of a $-1/+1$ sequence is defined as a fragment of the sequence such that: (a) its elements are all $+1$'s or all -1's; (b) the last element before the fragment and the first element after the fragment are both with the opposite sign to the elements in the fragment. If the length of a run is k, we call it a *k-run*. For example, $a = [1, -1, -1, 1, 1, -1, -1]$ has four runs, i.e., $[1], [-1, -1], [1, 1], [-1, -1]$, among which $[1]$ is the only 1-run in a. In general, every $-1/+1$ sequence of length n can be encoded into an ordered partition of n by describing each run with the number of elements therein; and vice versa. For example, $a = [1, -1, -1, 1, 1, -1, -1]$ can be encoded into $[1, 2, 2, 2]$ which is a partition of 7; given a partition $[1, 2, 1, 3]$ of 7, one can decode it into a $-1/+1$ sequence $[1, -1, -1, 1, -1, -1, -1]$. Therefore, finding a $-1/+1$ sequence of length n for a PAF profile problem is equivalent to finding an ordered partition of n such that its decoded sequence has the given PAF profile. Due to the assumption that a starts with $+1$ and ends with -1, the partition obtained from encoding a always has even length. The following theorem [10] gives more features of the partition which generates a solution of the given PPP.

Theorem 1. *Let a be a $-1/+1$ sequence of length n and $p_k = \mathrm{PAF}(a, k)$. Then*

- $\mathrm{PAF}(a, k) \equiv n \mod 4$, *for $k = 1, \ldots, \lfloor n/2 \rfloor$;*
- *the number of runs in a is $(p_0 - p_1)/2$;*
- *the number of 1-runs in a is $(p_0 + p_2 - 2p_1)/4$.*

Example 1. Let $n = 9$ and $a = (1, 1, -1, 1, -1, 1, -1, -1, -1)$. By calculation, $P = (-3, 1, -3, 1)$. Moreover, it can be verified that

- $\mathrm{PAF}(a, 1) = \mathrm{PAF}(a, 3) = -3 \equiv 9 \mod 4$;
  $\mathrm{PAF}(a, 2) = \mathrm{PAF}(a, 4) = 1 \equiv 9 \mod 4$;
- the number of runs in a is $(p_0 - p_1)/2 = 6$;
- the number of 1-runs in a is $(p_0 + p_2 - 2p_1)/4 = 4$.

Given an ordered partition $\delta = (n_1, \ldots, n_t)$ of even length, we may find fragments of the form (2) or $(1, k)$ $(k \geq 2)$ in δ. The following theorem reveals the relationship between the number of such fragments and the PAF sequences of the decoded $-1/+1$ sequence from δ.

Theorem 2. *Let $\delta = (n_1, \ldots, n_t)$ be an ordered partition of n with even number of elements and #2, #1k denote the numbers of fragments of the form (2) and $(1, k)$ $(k \geq 2)$ in δ, respectively.[2] Let a be the decoded $-1/+1$ sequence from δ and $p_i = \mathrm{PAF}(a, i)$. Then $\#2 + \#1k = (p_0 - p_1 - p_2 + p_3)/4$.*

---

[2] If $n_1 > 1$ and $n_t = 1$, (n_t, n_1) is regarded as a fragment of the form $(1, k)$.

For proving this property, we introduce two new sequences, i.e., $b = [b_0, \ldots, b_{n-1}]$ associated to a with $b_i = (a_i + 1)/2$ and $c = [c_0, \ldots, c_{n-1}]$ associated to b with $c_i = 1 - b_i$. Then

$$b_i = \begin{cases} 0, \text{ if } a_i = -1; \\ 1, \text{ if } a_i = 1; \end{cases} \quad c_i = \begin{cases} 1, \text{ if } b_i = 0; \\ 0, \text{ if } b_i = 1. \end{cases}$$

Let $p_k = \text{PAF}(a, k)$. We have

$$q_k := \sum_{i=0}^{n-1} b_i b_{i+k} = (p_k - 2|a| + n)/4, \tag{4}$$

$$q'_k := \sum_{i=0}^{n-1} c_i c_{i+k} = (p_k - 2|a| + n)/4. \tag{5}$$

where $k = 0, \ldots, \lfloor n/2 \rfloor$. Now we are ready to prove Theorem 2 with the help of b and c.

Proof. Let b and c be defined as above. Obviously, the encoded partitions of b and c have the same numbers of fragments of the form (2) and $(1, k)$ with the encoded partition of a. Noting (4) and (5), we only need to show that $s := \#2 + \#1k = q_0 - q_1 - q_2 + q_3$ or $\#2 + \#1k = q'_0 - q'_1 - q'_2 + q'_3$, which can be proved by induction on s.

Case 1. $s = 0$. Obviously, $n_i \geq 3$. By calculation, we have

$$q_1 = q_0 - t, \quad q_2 = q_0 - 2t, \quad q_3 = q_0 - 3t.$$

Thus $q_0 - q_1 - q_2 + q_3 = 0$, which is equal to s.

Case 2. Suppose the conclusion holds for $s' < s$. We prove it is also true for s.

If δ only contains the fragments of the form (2), then at least one of the following fragments appears in b.

$$
\begin{array}{ll}
(2.1) \ 0,0,0,\underline{1},\underline{1},0,0,0 & (2.2) \ 0,0,0,\underline{1},\underline{1},0,0,1 \\
(2.3) \ 1,0,0,\underline{1},\underline{1},0,0,0 & (2.4) \ 1,0,0,\underline{1},\underline{1},0,0,1 \\
(2.5) \ 0,1,1,\underline{0},\underline{0},1,1,0 & (2.6) \ 0,1,1,\underline{0},\underline{0},1,1,1 \\
(2.7) \ 1,1,1,\underline{0},\underline{0},1,1,0 & (2.8) \ 1,1,1,\underline{0},\underline{0},1,1,1
\end{array}
$$

In what follows, we only show that the conclusion holds for Cases (2.1)–(2.4). Similar strategies can be used to prove Cases (2.5)–(2.8) with the help of c. First we choose one such fragment in b and replace $\underline{1}$ with 0. Denote the new sequence with $\bar{b}$. In the encoded partition of $\bar{b}$, assume $s' = \#2 + \#1k$ ($k \geq 2$). Let $\bar{q}_k = \sum_{i=0}^{n-1} \bar{b}_i \bar{b}_{i+k}$. Further calculation leads to
Taking the hypothesis $\bar{q}_0 - \bar{q}_1 - \bar{q}_2 + \bar{q}_3 = s' (< s)$ into account, one can immediately verify that $q_0 - q_1 - q_2 + q_3 = s$ for Cases (2.1)–(2.4).

Case s		q_0	q_1	q_2	q_3
(2.1)	$s' + 1$	$\bar{q}_0 + 2$	$\bar{q}_1 + 1$	$\bar{q}_2$	$\bar{q}_3$
(2.2)	$s' + 2$	$\bar{q}_0 + 2$	$\bar{q}_1 + 1$	$\bar{q}_2$	$\bar{q}_3 + 1$
(2.3)	$s' + 2$	$\bar{q}_0 + 2$	$\bar{q}_1 + 1$	$\bar{q}_2$	$\bar{q}_3 + 1$
(2.4)	$s' + 3$	$\bar{q}_0 + 2$	$\bar{q}_1 + 1$	$\bar{q}_2$	$\bar{q}_3 + 2$

If δ has one or more fragments of the form $(1, k)$, then at least one of the following fragments appears in b. In what follows, we only show that the conclusion holds for Cases (2.9)–(2.12). Similar strategies can be used to prove Cases (2.13)–(2.16) with the help of c.

$$(2.9)\ 0, 0, 0, \underline{1}, 0, 0, 1 \quad (2.10)\ 0, 1, 0, \underline{1}, 0, 0, 1$$

$$(2.11)\ 1, 0, 0, \underline{1}, 0, 0, 1 \quad (2.12)\ 1, 1, 0, \underline{1}, 0, 0, 1$$

$$(2.13)\ 0, 0, 1, \underline{0}, 1, 1, 0 \quad (2.14)\ 0, 1, 1, \underline{0}, 1, 1, 0$$

$$(2.15)\ 1, 0, 1, \underline{0}, 1, 1, 0 \quad (2.16)\ 1, 1, 1, \underline{0}, 1, 1, 0$$

Again we choose one such fragment in b and replace $\underline{1}$ with 0. Denote the new sequence with $\bar{b}$. In the encoded partition of $\bar{b}$, assume $s' = \#2 + \#1k$ $(k \geq 2)$. Let $\bar{q}_k = \sum_{i=0}^{n-1} \bar{b}_i \bar{b}_{i+k}$. Further calculation results in

Case	s	q_0	q_1	q_2	q_3
(2.9)	$s' + 2$	$\bar{q}_0 + 1$	$\bar{q}_1$	$\bar{q}_2$	$\bar{q}_3 + 1$
(2.10)	$s' + 1$	$\bar{q}_0 + 1$	$\bar{q}_1$	$\bar{q}_2 + 1$	$\bar{q}_3 + 1$
(2.11)	$s' + 3$	$\bar{q}_0 + 1$	$\bar{q}_1$	$\bar{q}_2$	$\bar{q}_3 + 2$
(2.12)	$s' + 2$	$\bar{q}_0 + 1$	$\bar{q}_1$	$\bar{q}_2 + 1$	$\bar{q}_3 + 2$

Combining the induction $\bar{q}_0 - \bar{q}_1 - \bar{q}_2 + \bar{q}_3 = s' (< s)$, one can immediately verify that $q_0 - q_1 - q_2 + q_3 = s$ for Cases (2.9)-(2.12).

Example 2. Consider the sequence a in Example 1. Recall $P = (-3, 1, -3, 1)$. Thus $(p_0 - p_1 - p_2 + p_3)/4 = 2$, which is equal to the total number of 2-runs and 1-runs followed by a k-run $(k \geq 2)$ in a.

3   Technical Details

In this section, we first review the orbits method which is used to formulate a new PPP solving algorithm. The algorithm is specially designed for reconstructing a $-1/+1$ sequence which is obtained by the orbits method. We will detail the $-1/+1$ generation and then reconstruct it from its PAF values using its special properties. The PSD test and the ranking/unranking technique are also introduced to further enhance the effectiveness of the algorithm.

3.1   The Orbits Method

Let $\mathbb{Z}_n$ be the ring of integers mod n, i.e., $\mathbb{Z}_n = \{\overline{0}, \ldots, \overline{n-1}\}$. Let $\mathbb{Z}_n^*$ bet the group of all invertible elements in $\mathbb{Z}_n$, i.e., $\mathbb{Z}_n^* = \{\overline{k} : 1 \leq k \leq n, \gcd(k,n) = 1\}$. For the sake of simplicity, we use k instead of $\overline{k}$ for elements of Z_n. It is well-known that the number of elements in $\mathbb{Z}_n^*$ is equal to $\phi(n)$ where ϕ denotes the Euler function. Let H be a subgroup of $\mathbb{Z}_n^*$. Then H acting on $\mathbb{Z}_n$ produces a series of orbits denoted by

$$\mathcal{O}_1 = \{0\}, \mathcal{O}_2, \ldots, \mathcal{O}_m$$

and the disjoint union of $\mathcal{O}_i$'s ($1 \leq i \leq m$) is equal to $\mathbb{Z}_n$. The orbits method is frequently used for constructing D-optimal matrices [5–7,13,14,16]. This is because sometimes, the two sequences a and b can be expressed as the unions of some orbits associated to a proper subgroup of $\mathbb{Z}_n^*$. More explicitly, the indices appearing in such a union indicate the position of -1's and the remaining positions are for $+1$'s.

Example 3. Consider the case when $n = 13$, $|a| = 1$, $|b| = 7$ and $H = \{1, 3, 9\} \trianglelefteq \mathbb{Z}_{13}^*$. We compute the orbits of H acting on $\mathbb{Z}_{13}$ and obtain

$$\mathcal{O}_1 = \{0\}, \quad \mathcal{O}_2 = \{1, 3, 9\}, \quad \mathcal{O}_3 = \{2, 5, 6\}, \quad \mathcal{O}_4 = \{4, 10, 12\}, \quad \mathcal{O}_5 = \{7, 8, 11\}.$$

Note that the number of -1's in a is $(n - |a|)/2 = 6 = 3 \cdot 2$. Thus we choose two 3-orbits (e.g., $\mathcal{O}_2$ and $\mathcal{O}_3$) and form a set

$$I_a = \mathcal{O}_2 \cup \mathcal{O}_3 = \{1, 2, 3, 5, 6, 9\},$$

which indicates that

$$a_1 = a_2 = a_3 = a_5 = a_6 = a_9 = -1.$$

Therefore, $a = [1, -1, -1, -1, 1, -1, -1, 1, 1, -1, 1, 1, 1]$. Similarly, b can be constructed in the same manner.

When n is given, $|a|$ and $|b|$ are obtained by solving the Diophantine equation $|a|^2 + |b|^2 = 4n - 2$. Let k be a divisor of $\phi(n)$. One can construct a subgroup of $\mathbb{Z}_n^*$ which has k elements, and a number of a's and b's from the orbits obtained by the subgroup acting on $\mathbb{Z}_n$. To find a pair of a and b which can be used to form a D-optimal matrix, one needs to check the condition $\mathrm{PAF}(a, k) + \mathrm{PAF}(b, k) = 2$ ($1 \leq k \leq \lfloor n/2 \rfloor$) for all possible pairs of a and b.

In general, the above approach for generating a's and b's via the orbits methods produces an enormous amount of data. For $n > 50$, it is computationally infeasible to detect sequences that satisfy the PAF conditions using only the computer memory as the storage device. Therefore, for $n > 50$, it is preferable to write the candidate PAF sequences for a and b in two distinct result files and subsequently execute a matching algorithm which will find any pairs of sequences whose PAF values sum to 2, if they exist. It should be pointed out that the PAF sequences stored in the file of b's are subtracted by 2. Then one may sort the files and find the matching PAF sequences in linear time. For each sequence, we may recover the sequences a and b by using the algorithm PPPsolving in Sect. 4 and construct a circulant-type D-optimal matrix from a and b.

3.2   PSD Test

The special construction of a and b implies some nice properties of the power spectral densities of the two sequences. Now we first recall the definition of the power spectral density (PSD) of a complex sequence $x = [x_0, \ldots, x_{n-1}]$. Let $\omega = e^{2\pi i/n}$ denote the primitive n-th root of unity. Then the discrete Fourier transformation of x is defined as

$$\mathrm{DFT}(x, k) = \sum_{i=0}^{n-1} x_i \omega^{ik}, \quad k = 0, \ldots, n-1$$

and the power spectral density value $\mathrm{PSD}(x, k)$ $(0 \le k \le n)$ is defined as

$$\mathrm{PSD}(x, k) = |\mathrm{DFT}(x, k)|^2 = \mathrm{DFT}(x, k) \cdot \overline{\mathrm{DFT}(x, k)},$$

which indicates its nonnegativity. Furthermore, we have the following theorem [16].

Theorem 3. *Let k and k' be chosen from the same orbit $\mathcal{O}_r \in \mathbb{Z}_n$ and sequences a and b are constructed by the orbits method. Then*

(i) $\mathrm{PSD}(a, k) + \mathrm{PSD}(b, k) = 2n - 2$ $(1 \le k \le n - 1)$ *if a and b are solutions for constructing the D-optimal matrix;*
(ii) $\mathrm{PSD}(a, k) = \mathrm{PSD}(a, k')$.

One may derive $\mathrm{PSD}(a, k) \le 2n - 2$ from (i) in Theorem 3 because $\mathrm{PSD}(b, k) \ge 0$. Similarly, $\mathrm{PSD}(b, k) \le 2n - 2$. Experiments show that this filtering criterion helps to discard about 95% sequences generated by the orbits method. However, the cost for computing the PSD values is very expensive and (ii) plays an important role in reducing the time cost. In particular, one only needs to compute one PSD value for all indices in each orbit.

3.3   Ranking/Unranking Technique

During the process of solving a PPP, the ranking/unranking technique is adopted when the numbers of a's and b's are too big (e.g., hundreds of millions). In such circumstance, it is impossible to store the combinatorial sets which indicate how to choose a group of orbits in the computer memory. Therefore, we resort to the ranking/unranking technique in [11, Chapter 2.3] to generate the index set for choosing the orbits one by one.

A ranking algorithm determines the position (or rank) of a combinatorial object among all the objects (with respect to a given order) while an unranking algorithm finds the object having a specified rank. Ranking and unranking can be considered as inverse operations. Suppose that S is a finite set and $N = \#S$. A ranking function is a bijection $rank : S \rightarrow \{0, \ldots, N-1\}$. In what follows, we use a simple example to illustrate how the ranking/unranking technique works.

Example 4. Consider Example 3. For generating a, we need to choose two orbits from four 3-orbits. It is easy to calculate $\binom{4}{2} = 6$ and the ranking algorithm will give the six choices different ranks, i.e.,

$$rank(1,2) = 1, rank(1,3) = 2, rank(2,3) = 3,$$
$$rank(1,4) = 4, rank(2,4) = 5, rank(3,4) = 6.$$

When one needs to restore a choice from a given rank (e.g., $rank = 4$), we can call the unranking algorithm and get the index set $\{1, 4\}$.

The ranking/unranking technique helps to resolve the problem of memory overflow and can be used for any m where m is the number of orbits no matter how big it is.

4   Algorithm and Example

In this section, we adopt the techniques described above to design an orbit-based algorithm for solving a special class of PPP. In the input of the algorithm, the PAF sequence should be associated to a $-1/+1$ sequence a which is obtained by the orbits method. Otherwise, the algorithm fails and returns $\{\}$.

Algorithm 1. (PPPsolving).

Input: n, a natural number;

$\quad\quad$ r, an integer which indicates the number of -1's in a;

$\quad\quad$ $p = [p_1, \ldots, p_{\lfloor n/2 \rfloor}]$, a sequence of integers;

$\quad\quad$ H, a subset of $\mathbb{Z}_n^*$ used to generate the orbits.

Output: a, a solution to the PPP determined by n, r, p and which can be generated by the orbits of H acting on $\mathbb{Z}_n$, if it exists; $\{\}$, otherwise.

---

1. $\quad t \leftarrow (n - p_1)/2,$ $\quad\quad$ /* Check the number of runs

$\quad\quad$ $N_1 \leftarrow (n - 2p_1 + p_2)/4,$ $\quad\quad$ /* Check the number of 1-runs

$\quad\quad$ $s \leftarrow (n - p_1 - p_2 + p_3)/4.$ $\quad\quad$ /* Check the number #2 + #1k

2. $\quad$ If one of the following occurs, return $\{\}$.

$\quad\quad$ (a) t is fractional/odd/negative;

$\quad\quad$ (b) N_1 is fractional/negative;

$\quad\quad$ (c) s is fractional/negative.

3. $\quad$ Compute the orbits of H acting on $\mathbb{Z}_n$, denoted by $\mathcal{O}_1 = \{0\}, \mathcal{O}_2, \ldots, \mathcal{O}_m$.

4. $\quad$ For every combination C of $\{1, \ldots, m\}$ such that $\#(\cup_{c \in C} \mathcal{O}_c) = r$, execute the following procedure.

$\quad\quad$ 4.1 Restore a candidate sequence a from C.

$\quad\quad$ 4.2 If a doesn't have the combinatorial features t, N_1 and s, go to next loop.

   4.3 Carry out the PSD test for a on each orbit; if the test fails, go to next
        loop.
   4.4 Check whether a satisfies $\text{PAF}(a, k) = p_k$ for $k = 4, \ldots, \lfloor n/2 \rfloor$; if the
        test is passed, return a.
5.   Return $\{\}$.

In the above algorithm, Step 4.3 can be neglected if the comparison is carried
out in memory instead of writing the candidate sequences as well as their PAF
values into a file. In this case, we may also use the ranking/unranking technique
to generate the combination C in Step 4.

Example 5. Reconsider Example 3. Given $n = 13$, $r = 6$, $p = [1, -3, 1, 1, -3, -3]$
and a is obtained by using the orbits method where $H = \{1, 3, 9\}$. We will
reconstruct a with the above algorithm.

1. By calculation, $t = 6$, $N_1 = 2$, $s = 4$.
2. Since t, N_1 and s are all positive integers, we go to Step 3.
3. Compute the orbits of H acting on $\mathbb{Z}_{13}$ and get

$$\mathcal{O}_1 = \{0\}, \quad \mathcal{O}_2 = \{1, 3, 9\}, \quad \mathcal{O}_3 = \{2, 5, 6\}, \quad \mathcal{O}_4 = \{4, 10, 12\}, \quad \mathcal{O}_5 = \{7, 8, 11\}.$$

4. Note that the number of -1's in a is $6 = 3 \cdot 2$. Thus we will consider all the
   possibilities of two 3-orbits, i.e., C is chosen from

$$\{(2, 3), (2, 4), (2, 5), (3, 4), (3, 5), (4, 5)\}.$$

   – For $C = (2, 3)$,

$$a = [1, -1, -1, -1, 1, -1, -1, 1, 1, -1, 1, 1, 1]$$

   and a has 6 runs, which passes the combinatorial feature test. We calculate the PSD values of a and get $\text{PSD}(a, 1) \doteq 21.20$, $\text{PSD}(a, 2) \doteq 6.79$,
   $\text{PSD}(a, 4) \doteq 21.2$, which also succeeds in the PSD test. It can be further
   verified that a also passes the PAF test.
   – When $C = (2, 4)$,

$$a = [1, -1, 1, -1, -1, 1, 1, 1, 1, -1, -1, 1, -1]$$

   and it fails the combinatorial feature test.
   – For $C = (2, 5)$,

$$a = [1, -1, 1, -1, 1, 1, 1, -1, -1, -1, 1, -1, 1]$$

   and it fails the combinatorial feature test.
   – For $C = (3, 4)$,

$$a = [1, 1, -1, 1, -1, -1, -1, 1, 1, 1, -1, 1, -1]$$

   and it fails the combinatorial feature test.

– For $C = (3, 5)$,

$$a = [1, 1, -1, 1, 1, -1, -1, -1, -1, 1, 1, -1, 1].$$

a passes all the three test.
– For $C = (4, 5)$,

$$a = [1, 1, 1, 1, -1, 1, 1, -1, -1, 1, -1, -1, -1].$$

a passes all the three tests. It turns out that a constructed from $C = (4, 5)$ is the reverse rotation of a constructed from $C = (2, 3)$, which implies their equivalence.

To sum up, we find two non-equivalent solutions for the given PPP. One is

$$[1, -1, -1, -1, 1, -1, -1, 1, 1, -1, 1, 1, 1]$$

and the other is

$$[1, 1, -1, 1, 1, -1, -1, -1, -1, 1, 1, -1, 1].$$

5   Experiments: Đoković's D-optimal Solutions

In this section, we invoke the PPPsolving algorithm for some large-scale benchmark problems chosen from [14, 16] to test the effectiveness and efficiency of the algorithm. The algorithm is implemented in C and the experiments are performed on a Linux Box with an Intel(R) Core(TM) i7-6700U CPU @3.40 GHz and 8 GB RAM. The examples presented here cannot be solved within acceptable time by using the algorithm in [10].

5.1 $n = 73$

We are given $(|a|, |b|) = (1, 17)$ which is a solution of the Diophantine equation $|a|^2 + |b|^2 = 290 = 4 \cdot 73 - 2$ and whose PAF sequences are

$$\begin{aligned} \mathrm{PAF}_a = &[1, 1, 1, 1, 1, 1, 1, 1, 1, 1, -7, 1, -7, 1, -7, 1, 1, 1, 1, 1, \\ &-7, -7, 1, 1, 1, -7, 1, 1, -7, -7, -7, 1, 1, 1, 1, 1] \end{aligned}$$

and

$$\begin{aligned} \mathrm{PAF}_b = &[1, 1, 1, 1, 1, 1, 1, 1, 1, 1, 9, 1, 9, 1, 9, 1, 1, 1, 1, 1, 9, 9, \\ &1, 1, 1, 9, 1, 1, 9, 9, 9, 1, 1, 1, 1, 1]. \end{aligned}$$

Each of PAF_a and PAF_b contains $(73 - 1)/2 = 36$ PAF values. From $|a|$ and $|b|$, we immediately know that the numbers of -1's in a and b are $r_a = 36$ and $r_b = 28$. Furthermore, it is also known that a and b are obtained from the orbits method with $H = \{1, 8, 64\}$. By calling the algorithm, we first get all the cosets of H acting on $\mathbb{Z}_{73}$ below.

$$
\begin{array}{lll}
H \cdot 0 = \{0\}, & H \cdot 1 = \{1, 8, 64\}, & H \cdot 2 = \{2, 16, 55\}, \\
H \cdot 3 = \{3, 24, 46\}, & H \cdot 4 = \{4, 32, 37\}, & H \cdot 5 = \{5, 28, 40\}, \\
H \cdot 6 = \{6, 19, 48\}, & H \cdot 7 = \{7, 10, 56\}, & H \cdot 9 = \{9, 65, 72\}, \\
H \cdot 11 = \{11, 15, 47\}, & H \cdot 12 = \{12, 23, 38\}, & H \cdot 13 = \{13, 29, 31\}, \\
H \cdot 14 = \{14, 20, 39\}, & H \cdot 17 = \{17, 63, 66\}, & H \cdot 18 = \{18, 57, 71\}, \\
H \cdot 21 = \{21, 22, 30\}, & H \cdot 25 = \{25, 54, 67\}, & H \cdot 26 = \{26, 58, 62\}, \\
H \cdot 27 = \{27, 49, 70\}, & H \cdot 33 = \{33, 45, 68\}, & H \cdot 34 = \{34, 53, 59\}, \\
H \cdot 35 = \{35, 50, 61\}, & H \cdot 36 = \{36, 41, 69\}, & H \cdot 42 = \{42, 44, 60\}, \\
H \cdot 43 = \{43, 51, 52\}. & &
\end{array}
$$

The output of the algorithm for a gives two nonequivalent solutions with the positions of -1's determined by the union of some sets chosen from the above cosets whose indices are given by:

$$
J_1 = \{1, 2, 4, 5, 7, 9, 11, 14, 18, 21, 36, 42\},
$$
$$
J_2 = \{3, 5, 6, 7, 11, 12, 14, 17, 21, 33, 34, 42\}
$$

while the output for b can be reduced to only one non-equivalent solution where positions of -1's are given by the union of some cosets of H chosen with the following indices

$$
K = \{0, 3, 5, 6, 7, 12, 14, 25, 27, 35\}.
$$

Since a and b satisfy (3), they can be used to construct circulant-type D-optimal matrices of order 146.

The total numbers of the tested sequences for computing a and b are about 2.70 millon and 1.31 million, and the time costs for computing a and b are 3.603 s and 1.591 s, respectively.

5.2 $n = 79$

We are given $(|a|, |b|) = (5, 17)$ which is a solution of the Diophantine equation $|a|^2 + |b|^2 = 314 = 4 \cdot 79 - 2$ and whose PAF sequences are

$$
\begin{aligned}
\mathrm{PAF}_a = [&11, 11, 7, 3, -1, -5, 7, 7, -1, 7, 7, -5, 3, -5, 7, 7, 3, -1, -1, -5, -1, \\
&- 5, 11, 11, -5, 7, 7, -5, 7, -1, 11, -5, 11, 7, 7, -1, -1, -1, -5]
\end{aligned}
$$

and

$$
\begin{aligned}
\mathrm{PAF}_b = [&-9, -9, -5, -1, 3, 7, -5, -5, 3, -5, -5, 7, -1, 7, -5, -5, -1, 3, 3, 7, \\
&3, 7, -9, -9, 7, -5, -5, 7, -5, 3, -9, 7, -9, -5, -5, 3, 3, 3, 7]
\end{aligned}
$$

Each of PAF_a and PAF_b contains $(79 - 1)/2 = 39$ PAF values. From $|a|$ and $|b|$, we can calculate the numbers of -1's in a and b which are $r_a = 48$ and $r_b = 42$. Furthermore, a and b are obtained from the orbits method with $H = \{1, 23, 55\}$.

By calling the PPPsolving algorithm, we first get all the cosets of H acting on $\mathbb{Z}_{79}$ below.

$$
\begin{array}{lll}
H \cdot 0 = \{0\}, & H \cdot 1 = \{1, 23, 55\}, & H \cdot 2 = \{2, 31, 46\}, \\
H \cdot 3 = \{3, 7, 69\}, & H \cdot 4 = \{4, 13, 62\}, & H \cdot 5 = \{5, 36, 38\}, \\
H \cdot 6 = \{6, 14, 59\}, & H \cdot 8 = \{8, 26, 45\}, & H \cdot 9 = \{9, 21, 49\}, \\
H \cdot 10 = \{10, 72, 76\}, & H \cdot 11 = \{11, 16, 52\}, & H \cdot 12 = \{12, 28, 39\}, \\
H \cdot 15 = \{15, 29, 35\}, & H \cdot 17 = \{17, 66, 75\}, & H \cdot 18 = \{18, 19, 42\}, \\
H \cdot 20 = \{20, 65, 73\}, & H \cdot 22 = \{22, 25, 32\}, & H \cdot 24 = \{24, 56, 78\}, \\
H \cdot 27 = \{27, 63, 68\}, & H \cdot 30 = \{30, 58, 70\}, & H \cdot 33 = \{33, 48, 77\}, \\
H \cdot 34 = \{34, 53, 71\}, & H \cdot 37 = \{37, 60, 61\}, & H \cdot 40 = \{40, 51, 67\}, \\
H \cdot 41 = \{41, 43, 74\}, & H \cdot 44 = \{44, 50, 64\}, & H \cdot 47 = \{47, 54, 57\}.
\end{array}
$$

The output of the algorithm can be reduced to a non-equivalent solution of a which is constructed from the following index set of the cosets for indicating the positions of -1's:

$$
J = \{2, 4, 5, 6, 9, 11, 12, 15, 18, 20, 30, 33, 37, 41, 44, 47\}.
$$

Other solutions output by the algorithm are equivalent to it. The output of the algorithm for b can also be reduced to one single solution where positions of -1's are given by the union of the above cosets chosen with the following indices

$$
K = \{1, 2, 4, 5, 8, 10, 11, 15, 22, 27, 30, 40, 41, 44\}.
$$

The total numbers of the tested sequences for computing a and b are about 5.31 million and 9.66 million, and the time costs for computing a and b are 9.266 sec and 13.509 sec.

5.3 $n = 93$

We are given $(|a|, |b|) = (3, 19)$ which is a solution of the Diophantine equation $|a|^2 + |b|^2 = 370 = 4 \cdot 93 - 2$ and whose PAF sequences are

$$
\begin{aligned}
\text{PAF}_a = [&13, 1, -3, -11, -7, 1, -11, -3, 1, 5, -11, -3, -3, -3, -3, -3, \\
& -3, -3, 5, 1, -3, -3, -3, -7, 13, 13, -7, -3, 5, 1, 21, -7, -3, \\
& -3, 1, 1, -7, 1, 1, -3, 1, -7, 1, -3, 1, -3]
\end{aligned}
$$

and

$$
\begin{aligned}
\text{PAF}_b = [&-11, 1, 5, 13, 9, 1, 13, 5, 1, -3, 13, 5, 5, 5, 5, 5, 5, 5, -3, \\
& 1, 5, 5, 5, 9, -11, -11, 9, 5, -3, 1, -19, 9, 5, 5, 1, 1, 9, 1, \\
& 1, 5, 1, 9, 1, 5, 1, 5].
\end{aligned}
$$

Each of PAF_a and PAF_b contains $(93 - 1)/2 = 46$ PAF values. We start with calculating the numbers of -1's in a and b from $|a|$ and $|b|$, which results in $r_a = 45$ and $r_b = 37$. Furthermore, it is known that a and b are obtained from the orbits method with $H = \{1, 25, 67\}$. By calling the PPPsolving algorithm, we get all the cosets of H acting on $\mathbb{Z}_{93}$ below.

$$
\begin{array}{lll}
H \cdot 0 = \{0\}, & H \cdot 1 = \{1, 25, 67\}, & H \cdot 2 = \{2, 41, 50\}, \\
H \cdot 3 = \{3, 15, 75\}, & H \cdot 4 = \{4, 7, 82\}, & H \cdot 5 = \{5, 32, 56\}, \\
H \cdot 6 = \{6, 30, 57\}, & H \cdot 8 = \{8, 14, 71\}, & H \cdot 9 = \{9, 39, 45\}, \\
H \cdot 10 = \{10, 19, 64\}, & H \cdot 11 = \{11, 86, 89\}, & H \cdot 12 = \{12, 21, 60\}, \\
H \cdot 13 = \{13, 34, 46\}, & H \cdot 16 = \{16, 28, 49\}, & H \cdot 17 = \{17, 23, 53\}, \\
H \cdot 18 = \{18, 78, 90\}, & H \cdot 20 = \{20, 35, 38\}, & H \cdot 22 = \{22, 79, 85\}, \\
H \cdot 24 = \{24, 27, 42\}, & H \cdot 26 = \{26, 68, 92\}, & H \cdot 29 = \{29, 74, 83\}, \\
H \cdot 31 = \{31\}, & H \cdot 33 = \{33, 72, 81\}, & H \cdot 36 = \{36, 63, 87\}, \\
H \cdot 37 = \{37, 61, 88\}, & H \cdot 40 = \{40, 70, 76\}, & H \cdot 43 = \{43, 52, 91\}, \\
H \cdot 44 = \{44, 65, 77\}, & H \cdot 47 = \{47, 59, 80\}, & H \cdot 48 = \{48, 54, 84\}, \\
H \cdot 51 = \{51, 66, 69\}, & H \cdot 55 = \{55, 58, 73\}, & H \cdot 62 = \{62\}.
\end{array}
$$

There is only one non-equivalent solution for a and the indices of the cosets for indicating the positions of -1's in a are

$$J = \{0, 1, 3, 9, 13, 16, 17, 22, 24, 26, 29, 31, 44, 47, 48, 55, 62\}.$$

The output of the algorithm for b is also reduced to one single solution where positions of -1's are given by the union of some cosets of H chosen with the following index set

$$K = \{0, 2, 3, 6, 8, 20, 24, 37, 43, 47, 48, 51, 55\}.$$

The total numbers of the tested sequences for computing a and b are about 300.54 million and 259.48 million, and the time costs for computing a and b are 488.686 sec and 356.803 sec, respectively.

6   Conclusion

In this paper, we continue to explore combinatorial properties of PAF sequences. By making use of such properties and the orbits method based on group actions, we propose an algorithm for solving PAF profile problems, which is also stated as reconstructing $-1/+1$ sequences from a given PAF sequence. In addition, the PSD test and the ranking/unranking technique are employed to further enhance the effectiveness and efficiency of the algorithm, especially for nontrivial large problems. It is shown that the PPPsolving algorithm based on the orbits method is able to solve highly non-trivial problems and can be used to construct D-optimal matrices of high order.

Acknowledgements. This work was made possible by the facilities of the CARGO Lab at Wilfrid Laurier University, and the SMS International as well as the Key Laboratory of Software Engineering (2018-18XJSY-03) at Guangxi University for Nationalities. ISK's work is supported by NSERC grants. JY and YL's work is supported by NSFC (No. 10801101), the Special Fund for Guangxi Bagui Scholars, Guangxi Science and Technology Program (AD18126010), and the Startup Foundation for Advanced Talents in Guangxi University for Nationalities (2015MDQD018).

References

1. Barba, G.: Intorno al teorema di hadamard sui determinanti a valore massimo. Giorn. Mat. Battaglini **71**, 70–86 (1933)
2. Craigen, R., Kharaghani, H.: Hadamard matrices and hadamard designs. In: Colbourn, C.J., Dinitz, J.H. (eds.) Handbook of Combinatorial Designs, Discrete Mathematics and its Applications (Boca Raton), 2nd edn. pp. 273–280. Chapman & Hall/CRC, Boca Raton (2007)
3. Ehlich, H.: Determinantenabsschätzungen für binäre matrizen. Math. Z. **83**, 123–132 (1964)
4. Ehlich, H.: Determinantenabsschätzungen für binäre matrizen mit n3 mod 4. Math. Z. **84**, 438–447 (1964)
5. Gysin, M.: Combinatorial Designs, Sequences and Cryptography. Ph.D. thesis, University of Wollongong (1997)
6. Gysin, M.: New D-optimal designs via cyclotomy and generalised cyclotomy. Australas. J. Comb. **15**, 247–256 (1997)
7. Gysin, M., Seberry, J.: An experimental search and new combinatorial designs via a generalisation of cyclotomy. J. Comb. Math. Comb. Comput. **27**, 143–160 (1997)
8. Hadamard, J.: Résolution dùne question relative aux déterminants. Bulletin des Sciences Mathématiques **17**, 240–246 (1893)
9. Kotsireas, I., Pardalos, P.: D-optimal matrices via quadratic integer optimization. J. Heuristics **19**, 617–627 (2013)
10. Kotsireas, I.S., Yang, J.: Autocorrelation via runs. In: Fleuriot, J., Wang, D., Calmet, J. (eds.) AISC 2018. LNCS (LNAI), vol. 11110, pp. 195–205. Springer, Cham (2018). https://doi.org/10.1007/978-3-319-99957-9_13
11. Kreher, D.L., Stinson, D.R.: Combinatorial Algorithms, Generation, Enumeration and Search (CAGES), 1st edn. CRC Press, Boca Raton (1999)
12. Mertens, S.: Exhaustive search for low-autocorrelation binary sequences. J. Phys. A: Math. Gen. **29**(18), 473–481 (1996)
13. Đoković, D.: On maximal $(1, -1)$-matrices of order $2n$, n odd. Rad. Mat. **7**(2), 371–378 (1991)
14. Đoković, D.: Some new D-optimal designs. Australas. J. Comb. **15**, 221–231 (1997)
15. Đoković, D., Kotsireas, I.S.: D-optimal matrices of orders 118, 138, 150, 154 and 174. In: Colbourn, C.J. (ed.) Algebraic Design Theory and Hadamard Matrices, vol. 133, pp. 71–82. Springer, Cham (2015). https://doi.org/10.1007/978-3-319-17729-8_6
16. Đoković, D., Kotsireas, I.: New results on D-optimal matrices. J. Comb. Des. **20**, 278–289 (2012)
17. Wojtas, M.: On Hadamard's inequality for the determinants of order non-divisible by 4. Colloq. Math. **712**, 73–83 (1964)

Analytic Complexity of Hypergeometric Functions Satisfying Systems with Holonomic Rank Two

V. A. Krasikov[✉]

Laboratory of Artificial Intelligence, Plekhanov Russian University of Economics,
Stremyanny 36, Moscow 115054, Russia
Krasikov.VA@rea.ru

Abstract. We investigate the analytic complexity of solutions to holonomic bivariate hypergeometric systems of the Horn type by means of a Mathematica package. We classify hypergeometric systems with holonomic rank two by the polygons of the Ore–Sato coefficients up to transformations of the defining matrices which do not affect the analytic complexity of solutions. We establish an upper bound for the analytic complexity of solutions to bivariate hypergeometric systems with holonomic rank two.

Keywords: Hypergeometric systems of partial differential equations ·
Holonomic rank · Analytic complexity · Differential polynomial ·
Hypergeometry package

1 Introduction

The notion of the analytic complexity of a bivariate holomorphic function has been introduced by V.K. Beloshapka in [2]. It stems from the 13th Hilbert problem on the representation of continuous (originally algebraic) functions as superpositions of finitely many continuous functions of two variables [16]. Hilbert's 13th problem was solved in 1957 by V.I. Arnold and A.N. Kolmogorov in a positive way – they proved that every continuous function can be represented as a composition of finitely many univariate functions and a single bivariate function which can be chosen to be the addition [1]. It is well known that for analytic functions, this statement is false [16]. The reason for this is that the space of analytic functions in k variables is not vast enough for their finite superpositions to comprise the space of analytic functions in $k + 1$ variables [16, Section 4]. A concrete example is given by the function $\xi(x, y) = \sum\limits_{n=1}^{\infty} \frac{x^n}{n^y}$ (see [9]). For the analytic functions that admit such a representation, the analytic complexity reflects the number of summations and univariate functions in this representation.

A computation of the analytic complexity for an arbitrary analytic function is, in general, a very difficult problem [15]. For every function that belongs to the

© Springer Nature Switzerland AG 2019
M. England et al. (Eds.): CASC 2019, LNCS 11661, pp. 330–342, 2019.
https://doi.org/10.1007/978-3-030-26831-2_22

classes of finite analytic complexity, there exists a differential polynomial which annihilates it. However, a computation of such polynomials can already be a challenge for functions with very few summations [3]. Symmetries in the functions of finite analytic complexity admit the existence of corresponding Lie symmetry groups for the differential equations which define these functions, though the computation of such groups is another non-trivial problem.

In this article, we calculate the analytic complexity of the solutions to hypergeometric systems with a small holonomic rank. Systems under consideration in this article are hypergeometric systems in the sense of Horn [7,11]. This choice of the class of functions under consideration is due to the importance of hypergeometric functions in mathematical physics and its applications. Furthermore, both the functions of the finite analytic complexity and the hypergeometric functions are differentially algebraic, although the systems of differential polynomials annihilating them may differ a lot.

Numerous theoretical results [5,6] and extensive computer experiments [8] suggest that the analytic complexity of solutions to any bivariate hypergeometric system of partial differential equations is finite.

It has been conjectured in [14] that the holonomic rank of a hypergeometric system is closely related to the analytic complexity of the holomorphic solutions of this system. In this article, we describe all bivariate hypergemetric systems with the holonomic rank two. We prove that for these systems, the analytic complexity of their solutions (possibly, after a certain uniformization) cannot exceed two.

2 Notation and Definitions

We choose a matrix $A \in \mathbb{Z}^{m \times n} = (A_{ij} | i = 1, \ldots, m, j = 1, \ldots, n)$ and a vector of parameters $c = (c_1, c_2, \ldots, c_m) \in \mathbb{C}^m$. Let us denote the rows of this matrix by $\mathbf{A}_i$, $i = 1, \ldots, m$.

Definition 1. The *Horn system* Horn(A, c) is the following system of partial differential equations:

$$x_j P_j(\theta) f(x) = Q_j(\theta) f(x), \ j = 1, \ldots, n, \tag{1}$$

where

$$P_j(s) = \prod_{i:A_{ij}>0} \prod_{l_j^{(i)}=0}^{A_{ij}-1} \left(\langle \mathbf{A}_i, s \rangle + c_i + l_j^{(i)} \right),$$

$$Q_j(s) = \prod_{i:A_{ij}<0} \prod_{l_j^{(i)}=0}^{|A_{ij}|-1} \left(\langle \mathbf{A}_i, s \rangle + c_i + l_j^{(i)} \right),$$

and $\theta = (\theta_1, \ldots, \theta_n)$, $\theta_j = x_j \frac{\partial}{\partial x_j}$.

This system is a far-going generalization of the classical systems of partial differential equations investigated in [7].

Definition 2. The system of equations $\mathrm{Horn}(A,c)$ is called *nonconfluent* if $\sum_{i=1}^{m} \mathbf{A}_i = 0$.

Let D be the Weyl algebra $\mathbb{C}\langle x_1, \ldots, x_n, \partial_{x_1}, \ldots, \partial_{x_n} \rangle$ of linear partial differential operators with polynomial coefficients in the variables $(x_1, \ldots, x_n) \in \mathbb{C}^n$.

Definition 3. For a left ideal $J \subset D$, *the holonomic rank of J* is the dimension of the $K = \mathbb{C}$ vector space $K\langle \partial_{x_1}, \ldots, \partial_{x_n} \rangle / K \langle \partial_{x_1}, \ldots, \partial_{x_n} \rangle J$. The holonomic rank of the ideal J is denoted by $\mathrm{rank}(J)$. It coincides with the dimension of the space of holomorphic solutions to the associated system of linear partial differential equations at a generic point.

Definition 4. Let l_i denote a generator of the sublattice $\{s \in \mathbb{Z}^n : \langle \mathbf{A}_i, s \rangle = 0\}$ and let k_i be the number of elements in the set $\{\mathbf{A}_1, \ldots, \mathbf{A}_m\}$, which coincide with $\mathbf{A}_i$. *The polygon of the Ore–Sato coefficient* (see [13]) is defined to be the integer convex polygon whose sides are translations of the vectors $k_i l_i$, the vectors $\mathbf{A}_1, \ldots, \mathbf{A}_m$ being the outer normals to its sides. The nonconfluency condition implies that there exists the unique polygon (up to a translation) with these properties.

In this article, we only consider the case of bivariate systems that is, $n = 2$.

The main tool for the calculation of the holonomic rank of bivariate hypergeometric systems is Theorem 2.5 in [4] which states that

$$\mathrm{rank}(\mathrm{Horn}(A,c)) = d_1 d_2 - \sum_{\mathbf{A}_i, \mathbf{A}_j \text{ lin. dependent}} \nu_{ij}.$$

Here $d_j = \sum_{i=1}^{m} \max(A_{ij}, 0)$, $j = 1, 2$ and

$$\nu_{ij} = \begin{cases} \min(|A_{i1} A_{j2}|, |A_{j1} A_{i2}|), & \text{if } \mathbf{A}_i, \mathbf{A}_j \text{ are in opposite open quadrants of } \mathbb{Z}^2, \\ 0, & \text{otherwise.} \end{cases}$$

Recall that a *zonotope* is the Minkowski sum of segments. In the case when the polygon of the Ore–Sato coefficient for the system is a zonotope, we can compute its rank by a generalization of the formula in [11, Section 6].

Proposition 1. Let A be the matrix for the Horn system $\mathrm{Horn}(A,c)$ and the polygon of the Ore–Sato coefficient for $\mathrm{Horn}(A,c)$ is a zonotope. We divide the matrix A into blocks of the form $A^i = \begin{pmatrix} \alpha_1^i a_i & \alpha_1^i b_i \\ \cdots & \cdots \\ \alpha_{p_i}^i a_i & \alpha_{p_i}^i b_i \end{pmatrix}$, $\quad a_i > 0, \quad \alpha_j^i \in \mathbb{Z}$,

$\sum_{j=1}^{p_i} \alpha_j^i = 0$, each block having rank 1. Let l be the number of different blocks, then

$$\text{rank(Horn}(A, c)) = \sum_{\substack{i, j = 1 \\ i \neq j}}^{l} a_i |b_j| \sum_{\substack{q = 1 \\ \alpha_q^i > 0}}^{p_i} \sum_{\substack{r = 1 \\ \alpha_r^j > 0}}^{p_j} \alpha_q^i \alpha_r^j. \tag{2}$$

Proof. Each of the matrices A^i contributes to the holonomic rank formula in the both parts: the product $d_1 d_2$ includes the sum of $\alpha_q^i a_i > 0$ multiplied by $\alpha_r^j b_j > 0$ from all $A_j, j = 1, \dots, l$, and linearly dependent vectors from the block contribute to the sum $\sum \nu_{ij}$.

Denote by

$$C_+(A^i, A^j) = \sum_{\alpha_q^i > 0} \alpha_q^i a_i \sum_{\alpha_r^j b_j > 0} \alpha_r^j b_j = \sum_{\substack{\alpha_q^i > 0 \\ \alpha_r^j b_j > 0}} \alpha_q^i \alpha_r^j a_i b_j$$

the function of the positive contribution, by $C_-(A^i) = \sum_{\substack{\alpha_q^i > 0 \\ \alpha_r^i b_i < 0}} \alpha_q^i a_i |\alpha_r^i b_i|$ the

function of the negative contribution. Note that $C_+(A^i, A^i) = C_-(A^i)$ for all $i = 1, \dots, l$. Indeed, due to the nonconfluency condition, $\sum_{\alpha_q^i > 0} \alpha_q^i = \sum_{\alpha_r^i < 0} |\alpha_r^i|$,

that is, $\sum_{\alpha_q^i > 0} \alpha_q^i |b_i| = \sum_{\alpha_r^i < 0} |\alpha_r^i b_i|$.

The nonconfluency condition furthermore implies that

$$C_+(A^i, A^j) = \sum_{\substack{\alpha_q^i > 0 \\ \alpha_r^j b_j > 0}} \alpha_q^i \alpha_r^j a_i b_j = \sum_{\substack{\alpha_q^i > 0 \\ \alpha_r^j > 0}} \alpha_q^i \alpha_r^j a_i |b_j|,$$

since for $b_j < 0$, the sum by $\alpha_j^r < 0$ coincides with the sum by $\alpha_j^r > 0$.

Hence $\text{rank(Horn}(A, c)) = \sum_{i,j=1}^{l} C_+(A^i, A^j) - \sum_{i=1}^{l} C_-(A^i) = \sum_{\substack{i, j = 1 \\ i \neq j}}^{l} C_+(A^i, A^j).$

Substituting the formula for $C_+(A^i, A^j)$ into the right-hand side of the above equation, we arrive at (2). □

In Sect. 4, we compute the analytic complexity of holomorphic solutions to bivariate hypergeometric systems with the holonomic rank two. Let us recall the main definitions of the analytic complexity theory.

Definition 5 (See [2]). The class Cl_0 of functions of the analytic complexity zero is defined to comprise the functions that depend on at most one of the variables. A function $f(x, y)$ is said to belong to *the class Cl_n of functions with the analytic complexity $n > 0$* if there exists a point $(x_0, y_0) \in \mathbb{C}^2$ and a germ $f(x, y) \in \mathcal{O}(U(x_0, y_0))$ of this function holomorphic at (x_0, y_0) such that $f(x, y) = c(a(x, y) + b(x, y))$ for some germs of holomorphic functions $a, b \in Cl_{n-1}$ and $c \in Cl_0$. If there is no such representation for any finite n, then the function f is said to be *of the infinite analytic complexity*.

Example 1. A generic element of the first complexity class Cl_1 is a function of the form $f_3(f_1(x) + f_2(y))$. A function in Cl_2 can be represented in the form $f_7(f_5(f_1(x) + f_2(y)) + f_6(f_3(x) + f_4(y)))$, where $f_i(\cdot)$ are univariate holomorphic functions, $i = 1, \ldots, 7$.

Definition 6. We will call a holomorphic function *the function of analytic complexity* n, if it belongs to Cl_n, but does not belong to Cl_{n-1}.

3   Hypergeometric Systems of Holonomic Rank Two

Let us describe the set of matrices which define hypergeometric systems with the holonomic rank two and analyze its structure.

Consider some elementary transformations of matrices of hypergeometric systems that do not make a significant impact on solutions, in particular, do not change their analytic complexity. Further, we do not distinguish the matrices obtained from each other by means of any composition of such transformations.

I. A transposition of columns. In terms of the hypergeometric Eq. (1), a transposition of columns of the matrix corresponds to a transposition of variables, which does not affect the analytic complexity of solutions. The polygon of the Ore–Sato coefficient under this transformation is reflected with respect to the bisector of the first quadrant of the integer lattice.

II. A transposition of rows. It corresponds to a transposition of elements in the vector of parameters c, which leads to a transposition of these parameters in the formula of the solution at a generic point. The polygon of the Ore–Sato coefficient does not change under this transformation.

III. A multiplication of ith column by -1. It corresponds to the monomial change of variables $x_i \mapsto \frac{1}{x_i}$ in the solution. The polygon of the Ore–Sato coefficient under this transformation is reflected with respect to the vertical axis if $i = 1$ and with respect to the horizontal axis if $i = 2$.

IV. An addition of proportional rows with the coefficient of proportionality greater than zero. This transformation decreases the size of the matrix but does not affect the polygon of the Ore–Sato coefficient. Furthermore, the difference between the maximal and the minimal value of the analytic complexity for the set of solutions to the systems obtained one from another with this transformation does not exceed 1. In the case, when the analytic complexity changes this way, we consider its minimal value.

Analyzing the polygons of the Ore–Sato coefficients of the matrices we conclude that there are two different kinds of polygons of the Ore–Sato coefficients for the systems of rank two. First, we consider the polygons defined in the sense

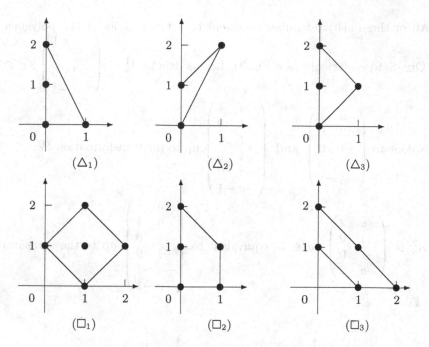

Fig. 1. The polygons of the Ore–Sato coefficients for the systems defined by the matrices $A_{\triangle_1}, A_{\triangle_2}, A_{\triangle_3}, A_{\square_1}, A_{\square_2}, A_{\square_3}$

of Definition 4 by the matrices whose elements have absolute values at most 2 (see Fig. 1). The matrices that correspond to these configurations are as follows:

$$
A_{\triangle_1} = \begin{pmatrix} -2 & 0 \\ 0 & -1 \\ 2 & 1 \end{pmatrix}, \quad
A_{\triangle_2} = \begin{pmatrix} -1 & 0 \\ -1 & 1 \\ 2 & -1 \end{pmatrix}, \quad
A_{\triangle_3} = \begin{pmatrix} -2 & 0 \\ 1 & -1 \\ 1 & 1 \end{pmatrix}.
$$

$$
A_{\square_1} = \begin{pmatrix} -1 & -1 \\ -1 & 1 \\ 1 & -1 \\ 1 & 1 \end{pmatrix}, \quad
A_{\square_2} = \begin{pmatrix} -2 & 0 \\ 0 & -1 \\ 1 & 0 \\ 1 & 1 \end{pmatrix}, \quad
A_{\square_3} = \begin{pmatrix} -1 & 0 \\ -1 & -1 \\ 0 & -1 \\ 2 & 2 \end{pmatrix}.
$$

Second, there are families of the matrices with a parameter:

$$
\begin{pmatrix} a & 1 \\ 2 & 0 \\ -2 & 0 \\ -a & -1 \end{pmatrix},
\begin{pmatrix} a & 2 \\ 1 & 0 \\ -1 & 0 \\ -a & -2 \end{pmatrix},
\begin{pmatrix} a & 1 \\ 2 & 0 \\ -1 & 0 \\ -1 & 0 \\ -a & -1 \end{pmatrix},
\begin{pmatrix} a & 1 \\ 1 & 0 \\ 1 & 0 \\ -1 & 0 \\ -1 & 0 \\ -a & -1 \end{pmatrix},
\begin{pmatrix} a & 1 \\ a & 1 \\ 1 & 0 \\ -1 & 0 \\ -a & -1 \\ -a & -1 \end{pmatrix}, \quad a \in \mathbb{Z}.
$$

All of these matrix families represent only two families of the polygons of the Ore–Sato coefficients (see Fig. 2). Let us denote $A_a^1 = \begin{pmatrix} a & 1 \\ 2 & 0 \\ -2 & 0 \\ -a & -1 \end{pmatrix}$, $a \in \mathbb{Z}$,

equivalent to $\begin{pmatrix} a & 1 \\ 2 & 0 \\ -1 & 0 \\ -1 & 0 \\ -a & -1 \end{pmatrix}$ and $\begin{pmatrix} a & 1 \\ 1 & 0 \\ 1 & 0 \\ -1 & 0 \\ -1 & 0 \\ -a & -1 \end{pmatrix}$ up to the transformation IV.

$A_a^2 = \begin{pmatrix} a & 2 \\ 1 & 0 \\ -1 & 0 \\ -a & -2 \end{pmatrix}$, $a \in \mathbb{Z}$, equivalent to $\begin{pmatrix} a & 1 \\ a & 1 \\ 1 & 0 \\ -1 & 0 \\ -a & -1 \\ -a & -1 \end{pmatrix}$ up to the transforma-

tion IV.

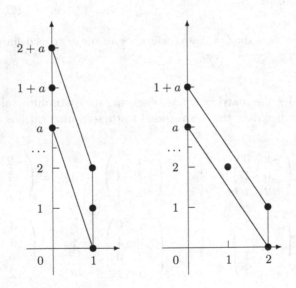

Fig. 2. The polygons of the Ore–Sato coefficients for the hypergeometric systems defined by the matrices A_a^1 and A_a^2

Theorem 1. There are no other matrices for the systems of the holonomic rank two, except for $A_{\triangle_i}, A_{\square_i}, i = 1, 2, 3$ and $A_a^1, A_a^2, a \in \mathbb{Z}$.

Proof. Consider the matrices without pairs of linearly dependent rows. The formula of the holonomic rank for such matrices does not include ν_{ij}, then from $\text{rank}(\text{Horn}(A, c)) = 2$ follows that $d_1 d_2 = 2$, and this is possible only if $d_1 \leq 2$,

$d_2 \leq 1$ (a transposition of rows has no impact on the solutions). This condition generates a finite number of matrices, and we distinguish only three of them keeping in mind equivalent transformations: $A_{\triangle_1}, A_{\triangle_2}, A_{\triangle_3}$. The only matrix (up to the transformations I-IV) which defines a system of the holonomic rank one without pairs of linearly dependent rows is $\begin{pmatrix} 1 & 0 \\ 0 & 1 \\ -1 & -1 \end{pmatrix}$.

Note that adding a nonconfluent block of the form A^i from Proposition 1 can only result in increasing the holonomic rank of the system. It follows from the same idea, that was used in the proof of Proposition 1, that every block contributes to $d_1 d_2$ more than to the sum of ν_{ij}. This statement is also correct for the nonconfluent blocks containing rows, which are pairwise linearly dependent with one of the rows of the initial matrix. Hence we cannot build matrices with the holonomic rank two including pairs of linearly dependent rows by adding such blocks.

Then the only way of adding pairs of linearly dependent rows is to replace a row of the initial matrix with two rows from its sum representation. By doing so and excluding matrices equivalent to each other up to the transformation IV, we obtain the matrices $A_{\square_2}, A_{\square_3}$.

In the case when the polygon of the Ore–Sato coefficient is a zonotope, we use the representation $\text{rank}(\text{Horn}(A,c)) = \sum_{\substack{i,j=1 \\ i \neq j}}^{l} C_+(A^i, A^j)$ from Proposition 1. It follows directly from it that matrices for the systems of the holonomic rank two cannot consist of more than two blocks, since the holonomic rank formula for the three blocks matrices includes six summands: $\sum_{\substack{i,j=1 \\ j>i}}^{3} C_+(A^i, A^j) + C_+(A^j, A^i)$, only three of them can be simultaneously equal to zero (the case when the first column in one of the blocks and the second one in another block contains only zeros), which means that the minimal rank in this case equals three.

When the matrix consists of two blocks (the polygon of the Ore–Sato coefficient is a parallelogram), the holonomic rank formula contains only two summands: $\text{Horn}(A,c)) = C_+(A^1, A^2) + C_+(A^2, A^1)$. Note that there are only three decompositions of 2 into the sum of two natural numbers: $2 = 2+0 = 1+1 = 0+2$.

The decomposition $2 = 1 + 1$ gives the matrix $A_{\square_1}$ with $C_+(A^1, A^2) = 1 \cdot 1 = 1$, $C_+(A^2, A^1) = 1 \cdot 1 = 1$ and the decompositions $2 = 0+2 = 2+0$ lead us to the matrices A_a^1 and A_a^2 with $C_+(A^1, A^2) = a \cdot 0 = 0$, $C_+(A^2, A^1) = 2 \cdot 1 = 2$. $\square$

The dependence of the holonomic rank on the absolute values of the elements of the matrix A can be expressed in the following way. If we consider a matrix defining the holonomic rank of a system as a function of two variables by choosing two elements as parameters, counting other elements of the matrix constant, this function is convex and under certain circumstances even piecewise quadratic.

For example, consider the matrix

$$M(a,b) = \begin{pmatrix} 10 & a \\ b & 4 \\ -2 & 3 \\ -b-8 & -a-7 \end{pmatrix}. \tag{3}$$

The last row contains parameters a, b to satisfy the nonconfluency condition. The holonomic rank function for the hypergeometric system defined by the matrix $M(a,b)$ is shown in Fig. 3.

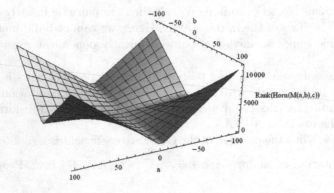

Fig. 3. The holonomic rank function of the hypergeometric system defined by matrix (3)

4   Analytic Complexity of Solutions to Systems of Holonomic Rank Two

To solve the systems above, we use the Wolfram Mathematica package HyperGeometry for solving hypergeometric systems. The package is available for free public use at https://www.researchgate.net/publication/318986894_HyperGeometry, the description of available functions is given in [12]. A computation of the analytic complexity of a function is, in general, a very difficult task [10, Chapter 3]. However, for the first class of the analytic complexity Cl_1, there exists the membership criterion defined by the differential polynomial

$$\Delta_1 = f'_{x_1}(f'_{x_2})^2 f'''_{x_1 x_1 x_2} - (f'_{x_1})^2 f'_{x_2} f'''_{x_1 x_2 x_2} + f''_{x_1 x_2}(f'_{x_1})^2 f''_{x_2 x_2} - f''_{x_1 x_2}(f'_{x_2})^2 f''_{x_1 x_1}.$$

That is the differential polynomial Δ_1 vanishes on a bivariate function $f(x_1, x_2)$ if and only if $f(x_1, x_2) \in Cl_1$ (see [2] and the references therein).

The systems $(\triangle_1), (\triangle_2), (\triangle_3)$ represent the simplicial configurations (see Proposition 4.4 in [11]), so the solutions to these systems can be found by means of formula (4.2) in [11]:

$$x^{-\tilde{A}^{-1}c}\left(1 + x^{-\tilde{A}^{-1}e_1} + x^{-\tilde{A}^{-1}e_2}\right)^{-|c|},$$

where $e_1 = (1,0), e_2 = (0,1)$, $\tilde{A}$ is the 2×2 matrix consisting of first two rows of the matrix which defines the hypergeometric system.

The solutions to the systems $(\triangle_1), (\triangle_2), (\triangle_3)$ are given by $x_1^{\frac{c_1}{2}} x_2^{c_2}(1 + \sqrt{x_1} + x_2)^{-c_1-c_2-c_3}$, $x_1^{c_1} x_2^{c_1-c_2}(1 + \frac{1}{x_2} + x_2)^{-c_1-c_2-c_3}$, and $x_1^{\frac{c_1}{2}} x_2^{\frac{c_1}{2}+c_2}(1 + \sqrt{x_1 x_2} + x_2)^{-c_1-c_2-c_3}$, respectively. All of these functions belong to Cl_2. For example, the $(\triangle_1)$ solution is the product of Cl_1 functions $x_1^{\frac{c_1}{2}} x_2^{c_2}$ and $(1 + \sqrt{x_1} + x_2)^{-c_1-c_2-c_3}$, and the product function in the terms of the analytic complexity classes is equivalent to the addition (see [2]). Similar statements are true for other solutions. The correct choice of parameters c_i can put these functions into Cl_1.

The basis in the space of holomorphic solutions to the system $(\square_1)$ (see [12]) is given by the Puiseux polynomials $-4 + \frac{1}{x_1} + \frac{1}{x_2} + x_1 + x_2$, $\frac{(x_1-1)(x_2-1)}{\sqrt{x_1 x_2}}$. Both of these functions have analytic complexity 1. For an arbitrary parameter vector $c = (c_1, c_2, c_3, c_4)$, the solution to the system $(\square_1)$ is given by

$$x_1^{\frac{c_1}{2}+\frac{c_2}{2}} x_2^{\frac{c_1}{2}-\frac{c_2}{2}}\left(1 + \frac{\sqrt{x_1}}{\sqrt{x_2}}\right)^{-c_2-c_4}(1 + \sqrt{x_1}\sqrt{x_2})^{-c_1-c_3}.$$

This function belongs to Cl_2 for any parameter values as the product of the Cl_1 function $(x_1 x_2)^{\frac{c_1}{2}}(1 + \sqrt{x_1 x_2})^{-c_1-c_3}$ of the argument $x_1 x_2$ and the Cl_1 function $\left(\frac{x_1}{x_2}\right)^{\frac{c_2}{2}}\left(1 + \sqrt{\frac{x_1}{x_2}}\right)^{-c_2-c_4}$ of the argument $\frac{x_1}{x_2}$.

Example 2. Let us compute the analytic complexity of the solutions to the system $(\square_2)$. Using the HyperGeometry package, we conclude that the solution to $(\square_2)$ is given by $\frac{1}{2} + \frac{x_1}{2} + x_2 + \frac{x_1 x_2}{2} + \frac{x_2^2}{2}$. According to the definition of the holonomic rank, a basis in the solution space to this system consists of two elements. To find the second solution we write the system $(\square_2)$ explicitly:

$$\begin{cases} (x_1^3 - 4x_1^2)f''_{x_1 x_1} + x_1^2 x_2 f''_{x_1 x_2} - (2x_1^2 + 2x_1)f'_{x_1} - x_1 x_2 f'_{x_2} + 2x_1 f = 0, \\ x_1 x_2 f'_{x_1} + (x_2^2 + x_2)f'_{x_2} - 2x_2 f = 0. \end{cases}$$

The solution to the second equation is $f(x_1, x_2) = x_1^2 \phi(\frac{1+x_2}{x_1})$, where $\phi(\cdot)$ is any differentiable function. Substituting it into the first equation in $(\square_2)$, we obtain the ordinary differential equation for the function $\phi(t)$

$$(t - 4t^2)\phi'' + 10t\phi' - 12\phi = 0.$$

The basis in the linear space of its solutions is given by

$$\phi_1(t) = t(1 + t), \quad \phi_2 = 24t(t + 1)\text{arctg}\sqrt{4t - 1} - (26t + 1)\sqrt{4t - 1}.$$

After the substitution $t = \frac{1+x_2}{x_1}$ we see that $x_1^2 \phi_1(\frac{1+x_2}{x_1})$ is the solution we have already found by means of the HyperGeometry package, and hence the second element of the basis is $x_1^2 \phi_2(\frac{1+x_2}{x_1})$. It follows directly from the definition of the analytic complexity classes that the analytic complexity of the both functions does not exceed 2.

Example 3. Consider the matrix $\widetilde{A}_{\square_3} = \begin{pmatrix} -1 & 0 \\ 0 & -1 \\ -1 & -1 \\ 1 & 1 \\ 1 & 1 \end{pmatrix}$ which is equivalent to $A_{\square_3}$

up to transformation IV. Using the HyperGeometry package we conclude that the first element of the basis of the solution space to the system defined by the matrix $\widetilde{A}_{\square_3}$ and the parameter vector $(0, 0, 1, -2, -2)$ is given by $x_1 + x_2 + \frac{x_1^2}{2} + x_1 x_2 + \frac{x_2^2}{2}$. Note that it belongs to the first complexity class Cl_1 since it admits the representation $(x_1 + x_2) + \frac{(x_1 + x_2)^2}{2}$. Varying the parameter vector, for its value $\widetilde{c} = (0, 0, 3, -1, -4)$ we obtain the system with the polynomial basis in its solution space (see Fig. 4). The big circles in Fig. 4 represent the support of the first solution $x_1 + x_2 - \frac{1}{2}$, while the small circles correspond to the support of the second solution $\frac{x_1^3}{3} + x_1^2 x_2 + x_1 x_2^2 + \frac{x_2^3}{3} - \frac{x_1^4}{6} - \frac{2x_1^3 x_2}{3} - \frac{2x_1 x_2^3}{3} - \frac{x_2^4}{6}$. Both of these functions depend on the sum of variables $x_1 + x_2$, and hence the whole solution space of the hypergeometric system $\mathrm{Horn}(\widetilde{A}_{\square_3}, \widetilde{c})$ consists of functions whose analytic complexity is at most 1. The sum of the corresponding series with the parameter vector $(0, 0, c_3, c_4, c_5)$ is given by $\frac{\Gamma(c_4)\Gamma(c_5)}{\Gamma(c_3)} {}_2F_1(c_5, c_4, c_3, x_1 + x_2)$.

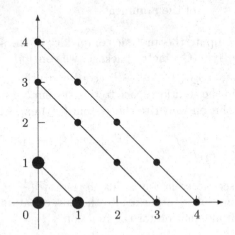

Fig. 4. The polygon of the Ore–Sato coefficients for the system defined by the matrix $\widetilde{A}_{\square_3}$

The sets of systems defined by the matrices A_a^1 and A_a^2 are the parallelepipedal configurations, so we use formula (4.7) in [11] to solve them:

$$x^{-\tilde{A}^{-1}(c_1,c_2)}\left(1+x^{-\tilde{A}^{-1}e_1}\right)^{-c_1-c_3}\left(1+x^{-\tilde{A}^{-1}e_2}\right)^{-c_2-c_4}.$$

The solutions to the systems defined by the matrices A_a^1 are given by

$$x_1^{\frac{c_4}{2}}x_2^{-c_1+\frac{ac_2}{2}}\left(1+\frac{1}{x_2}\right)^{-c_1-c_3}\left(\sqrt{x_1}+x_2^{a/2}\right)^{-c_2-c_4},$$

and the solutions to the systems defined by the matrices A_a^2 are given by

$$x_1^{c_4}x_2^{-\frac{c_1}{2}+\frac{ac_2}{2}}\left(1+\frac{1}{\sqrt{x_2}}\right)^{-c_1-c_3}\left(x_1+x_2^{a/2}\right)^{-c_2-c_4}.$$

The both sets of solutions contain only functions from Cl_2 since the solutions to A_a^1 are the products of the Cl_1 functions $x_1^{\frac{c_4}{2}}x_2^{-c_1+\frac{ac_2}{2}}\left(1+\frac{1}{x_2}\right)^{-c_1-c_3}$ and $\left(\sqrt{x_1}+x_2^{a/2}\right)^{-c_2-c_4}$, and the solutions to A_a^2 are the products of the Cl_1 functions $x_1^{c_4}x_2^{-\frac{c_1}{2}+\frac{ac_2}{2}}\left(1+\frac{1}{\sqrt{x_2}}\right)^{-c_1-c_3}$ and $\left(x_1+x_2^{a/2}\right)^{-c_2-c_4}$. The differential polynomial $\mathbf{\Delta}_1$ vanishes on these solutions only with the particular set of the parameter values (for example, $a=0$ or $c_2=c_4=0$), and only for these values, solutions to A_a^1 and A_a^2 are the Cl_1 functions. Summarizing all the computations, we conclude that for any system in the set of the hypergeometric systems $A_{\Delta_i}, A_{\square_i}, A_a^1, A_a^2, i=1,2,3, a\in\mathbb{Z}$, there exists a basis in the space of holomorphic solutions to this system consisting only of functions which belong to the analytic complexity class Cl_2.

Proposition 2. The analytic complexity of any analytic solution to a hypergeometric system $\mathrm{Horn}(\hat{A},\hat{c})$ with the holonomic rank two does not exceed four.

Proof. By Theorem 1, the matrix $\hat{A}$ is equivalent up to the transformations I–IV to one of the matrices from the set $\{A_{\Delta_i}, A_{\square_i}, i=1,2,3\}\cup\{A_a^1, A_a^2, a\in\mathbb{Z}\}$. Let $\{f_1(x,y), f_2(x,y)\}$ be the basis in the space of holomorphic solutions to $\mathrm{Horn}(\hat{A},\hat{c})$, where $f_i(x,y)\in Cl_2$. Then any analytic solution to $\mathrm{Horn}(\hat{A},\hat{c})$ at a generic point (x,y) can be represented as a linear combination $\alpha f_1(x,y)+\beta f_2(x,y)$, $\alpha,\beta\in\mathbb{C}$ up to multiplication of $f_i(x,y)$ by a monomial $x^{l_i}y^{m_i}$, $l_i, m_i\in\mathbb{Z}$. From the equivalence of the addition and the product in the definition of the analytic complexity classes, it follows that the analytic complexity of a function $x^{l_i}y^{m_i}f_i(x,y), i=1,2$ does not exceed three. By the definition of the analytic complexity classes, the analytic complexity of the function $\alpha x^{l_1}y^{m_1}f_1(x,y)+\beta x^{l_2}y^{m_2}f_2(x,y)$ does not exceed four. $\square$

Acknowledgements. This research was performed in the framework of the basic part of the scientific research state task in the field of scientific activity of the Ministry of Education and Science of the Russian Federation, project "Intellectual analysis of big textual data in finance, business, and education by means of adaptive semantic models", grant No. 2.9577.2017/8.9.

References

1. Arnold, V.I.: On the representation of continuous functions of three variables by superpositions of continuous functions of two variables. Mat. Sb. **48**(1), 3–74 (1959)
2. Beloshapka, V.K.: Analytic complexity of functions of two variables. Russ. J. Math. Phys. **14**(3), 243–249 (2007)
3. Beloshapka, V.K.: On the complexity of differential algebraic definition for classes of analytic complexity. Math. Notes **105**(3), 323–331 (2019)
4. Dickenstein, A., Matusevich, L.F., Sadykov, T.M.: Bivariate hypergeometric D-modules. Adv. Math. **196**, 78–123 (2005)
5. Dickenstein, A., Sadykov, T.M.: Algebraicity of solutions to the Mellin system and its monodromy. Dokl. Math. **75**(1), 80–82 (2007)
6. Dickenstein, A., Sadykov, T.M.: Bases in the solution space of the Mellin system. Sb. Math. **198**(9), 59–80 (2007)
7. Horn, J.: Über die Konvergenz der hypergeometrischen Reihen zweier und dreier Veränderlichen. Math. Ann. **34**, 544–600 (1889)
8. Krasikov, V.A., Sadykov, T.M.: On the analytic complexity of discriminants. Proc. Steklov Inst. Math. **279**, 78–92 (2012)
9. Ostrowski, A.: Über Dirichletsche Reihen und algebraische Differentialgleichungen. Math. Z. **8**, 241–298 (1920)
10. Robertz, D.: Formal Algorithmic Elimination for PDEs. Lecture Notes in Mathematics. Springer, Cham (2014). https://doi.org/10.1007/978-3-319-11445-3
11. Sadykov, T.M., Tanabe, S.: Maximally reducible monodromy of bivariate hypergeometric systems. Izv. Math. **80**(1), 221–262 (2016)
12. Sadykov, T.M.: Computational problems of multivariate hypergeometric theory. Program. Comput. Soft. **44**(2), 131–137 (2018)
13. Sadykov, T.M.: The Hadamard product of hypergeometric series. Bulletin des Sciences Mathematiques **126**(1), 31 (2002)
14. Sadykov, T.M.: On the analytic complexity of hypergeometric functions. Proc. Steklov Inst. Math. **298**(1), 248–255 (2017)
15. Stepanova, M.A.: Jacobian conjecture for mappings of a special type in $\mathbb{C}^2$. J. Siberian Fed. Univ. Math. Phys. **11**(6), 776–780 (2018)
16. Vitushkin, A.G.: On Hilbert's thirteenth problem and related questions. Russ. Math. Surv. **59**(1), 11–25 (2004)

On Explicit Difference Schemes
for Autonomous Systems of Differential
Equations on Manifolds

E. A. Ayryan[1], M. D. Malykh[2(✉)] [iD], L. A. Sevastianov[1,2] [iD], and Yu Ying[2,3]

[1] Joint Institute for Nuclear Research (Dubna), Joliot-Curie, 6, Dubna,
Moscow Region 141980, Russia
ayrjan@jinr.ru
[2] Department of Applied Probability and Informatics, Peoples' Friendship University
of Russia (RUDN University), 6 Miklukho-Maklaya St, Moscow 117198, Russia
{malykh_md,sevastianov_la}@rudn.university
[3] Department of Algebra and Geometry, Kaili University,
3 Kaiyuan Road, Kaili 556011, China
yingy6165@gmail.com

Abstract. The problem of the existence of explicit and at the same
time conservative finite difference schemes that approximate a system
of ordinary differential equations is investigated. An autonomous system
of nonlinear ordinary differential equations on an algebraic manifold V
is considered. A difference scheme for solving this system is called conservative, if the calculations of this scheme do not go beyond V, i.e.,
preserve it exactly. An explicit scheme is understood as such a difference
scheme in which a system of linear equations is required to proceed to
the next layer. We formulate the problem of constructing an explicit conservative scheme approximating a given autonomous system on a given
manifold. For the case of 1-manifold, a solution to this problem is given
and geometric obstacles to the existence of such difference schemes are
indicated. Namely, it is proved that the scheme exists only if the genus
of the integral curve is 1 or 0.

Keywords: Finite difference method · Elliptic and Abelian functions ·
Algebraic correspondence

1 Introduction

One of the most common and relevant mathematical models is the Cauchy problem for an autonomous system of ordinary differential equations. Analytical
methods allow one to find algebraic integrals of motion [1] for such systems,

The publication has been prepared with the support of the "RUDN University Program
5-100" and funded by RFBR according to the research projects Nos. 18-07-00567,
18-51-18005, and 19-01-00645 A.

© Springer Nature Switzerland AG 2019
M. England et al. (Eds.): CASC 2019, LNCS 11661, pp. 343–361, 2019.
https://doi.org/10.1007/978-3-030-26831-2_23

and numerical methods provide approximate plotting of particular solutions [2]. Analytical and numerical methods are not always consistent with each other. It often turns out that the algebraic integral of motion is known, but a difference scheme is used which does not preserve this integral. It is particularly annoying when such a quantity is the total mechanical energy, the variation of which is a property not inherent in the original physical model.

In this paper, we investigate the question of what hinders the construction of difference schemes that preserve precisely defined integral manifolds. In order to avoid excessive abstraction, we will illustrate the presentation with examples from the theory of motion of a rigid body with one fixed point.

2   Autonomous Systems on Manifolds

Consider an autonomous system of differential equations in an affine space of dimension r

$$\frac{dx}{dt} = f(x). \tag{1}$$

Here $x = (x_1, \ldots, x_r)$ is a point of an affine space, $f = (f_1, \ldots, f_r)$ is a set of rational functions from $\mathbb{Q}(x)$. Such systems often possess algebraic integrals of motion, or at least Darboux polynomials [1].

Definition 1. *An algebraic manifold V will be called integral for system (1) if every integral curve of this system that has at least one common point with the manifold V belongs to this manifold entirely.*

Example 1. By definition, the Darboux polynomial for system (1) is a polynomial g, for which it is possible to specify a polynomial h, such that

$$\sum_{i=1}^{r} f_i \frac{\partial g}{\partial x_i} = hg.$$

If $x(t)$ is a particular solution of system (1), then

$$\frac{dg(x(t))}{dt} = \sum_{i=1}^{r} \frac{\partial g}{\partial x_i} f_i = h(x(t)) \cdot g(x(t))$$

and, therefore, turning of $g(x(t))$ into zero at a certain value of t leads to turning of this expression into zero at all admissible values of t. Therefore, the hypersurface $g(x) = 0$ in the affine space is an integral manifold for system (1).

The search for the solutions of system (1) belonging to a known integral manifold will be referred to as a problem of autonomous system integration on the manifold. A perfect example of such a problem is the problem of top rotation.

Example 2. The motion of a top is described by six variables: three components of the angular velocity vector p, q, and r with respect to the principal axes with the origin at the fixed point of the top, and three direction cosines of one of the principal axes $\gamma, \gamma', \gamma''$. These variables satisfy a system of six autonomous equations with quadratic right-hand side, which has three classical integrals of motions. By analogy with the Bruns theorem, in the beginning of the last century it was proved that the fourth algebraic integral exists only in three particular cases [3]. Therefore, generally this system of differential equations should be considered not in the entire six-dimensional space, but in a three-dimensional manifold embedded in it.

3   Numerical Integration and Conservation Laws

A standard numerical method of solving systems of ordinary differential equations is the finite difference method. Within this method, a system of differential equations is replaced with a system of equations that describe a transition from the value x of the solution of Eq. (1) at a certain moment of time t to the approximate value $\hat{x}$ of the solution at the moment of time $t + \Delta t$. Below we consider x and $\hat{x}$ as points of two adjacent layers, and the difference scheme as a system of equations specifying the transition from one layer to another. For example, the explicit Euler scheme

$$\hat{x} - x = f(x)\Delta t$$

describes such a transition and yields for $x(t + \Delta t)$ the approximate value $\hat{x}$.

Discussing the difference scheme for Eq. (1) firstly implies that this scheme approximates a differential equation. No matter how sophisticated the equations describing the transition from layer to layer are, in the vicinity of the general point x and $\Delta t = 0$, these equations must have solutions that can be expanded into a series in powers of Δt:

$$\hat{x} = x + g_1(x)\Delta t + g_2(x)\Delta t^2 + \dots$$

The exact solution can also be expanded in a series of the same form

$$x(t + \Delta t) = x(t) + f(x(t))\Delta t + \dots.$$

If in these series the terms with Δt coincide, then the difference scheme is said to approximate the differential equation. If the first s terms coincide, the approximation is said to have an order of s [4, Definition 1.2]. For example, the Euler scheme approximates the initial equation, and the order of approximation equals 1.

Note 1. Usually these power series are considered as convergent in the $\mathbb{C}$ topology, but further calculations can be preserved without change, if these series are considered as formal. The choice of topology is essential in the study of the convergence of the numerical method.

Traditionally, the main attention is paid to the problem of increasing the order of approximation and the accumulation of round-off error. In contrast with this tradition, we address the issue of using known integral manifolds.

Like many other popular difference schemes, the Euler scheme does not conserve integral manifolds. This fact means that from $x \in V$ it does not follow that $\hat{x} \in V$. Since for the exact solution $x(t + \Delta t) \in V$, the deviation of the points of an approximate solution from the manifold V can be used as an estimate for the error of the numerical method. In problems of mechanics, e.g., the problem of top rotation (Example 2), this means considerable violation of fundamental laws of mechanics, e.g., introduces "numerical" dissipation into a dissipation-free system. It is particularly annoying when the geometric constraints are affected, e.g., a sum of direction cosine squares becomes not equal to 1. The interpretation of this result is not understandable.

The idea of using conservation laws when creating numerical methods arose in the middle of the last century, and the corresponding methods were called conservative. The first successes were achieved in solving partial differential equations. As for ordinary differential equations, it was noted only in the late 1980s that in some cases, it is possible to construct difference schemes that preserve algebraic integral manifolds exactly. So, for example, according to Cooper's theorem [2, Theorem 2.2], the midpoint scheme

$$\hat{x} - x = f\left(\frac{\hat{x} + x}{2}\right) \Delta t \tag{2}$$

for system (1) preserves all integrals of motion for this system expressed by quadratic forms.

Example 3. The Jacobi elliptic functions

$$p = \operatorname{sn} t, \quad q = \operatorname{cn} t, \quad r = \operatorname{dn} t$$

are the solutions of the nonlinear system

$$\dot{p} = qr, \ \dot{q} = -pr, \ \dot{r} = -k^2 pq \tag{3}$$

with initial conditions

$$p = 0, \ q = r = 1 \text{ at } t = 0.$$

The elliptic modulus k will be fixed below and, thus, the second argument of Jacobi functions will not be indicated. This autonomous system possesses two quadratic integrals of motion

$$p^2 + q^2 = \text{const} \quad \text{and} \quad k^2 p^2 + r^2 = \text{const.} \tag{4}$$

From an analytical point of view, this system is remarkable for the property that any of its particular solutions can be represented as a ratio of two everywhere convergent series in powers of t [5]. From the point of view of the finite difference method, this system is remarkable because it can be approximated by a difference scheme (namely, the midpoint scheme (2)) that conserves these integrals exactly. We made a series of numerical experiments in Sage and were convinced that even for extremely large values of t, the amplitude of oscillations does not fall [6]. The comparison of standard solvers for the solution of the system (3) in terms of preservation of integrals of the motion is presented in the talk of Yu. A. Blinkov and V.P. Gerdt at PCA'2019 [7].

Example 4. In the problem of top rotation (Example 2), there are three quadratic integrals. In this case, it is particularly good that the law

$$\gamma^2 + \gamma'^2 + \gamma''^2 = 1$$

is preserved, so that γ, γ', and γ'' can still be interpreted as direction cosines. In the Euler–Poinsot case, the following system is separated from the initial system of 6 equations:

$$A\dot{p} + (B - C)qr = 0, \ B\dot{q} + (A - C)rp = 0, \ C\dot{r} + (B - A)pq = 0, \qquad (5)$$

where A, B, and C are the moments of inertia with respect to the principal axes. This system possesses two quadratic integrals

$$Ap^2 + Bq^2 + Cq^2 = \text{Const}, \quad A^2p^2 + B^2q^2 + C^2q^2 = \text{Const}. \qquad (6)$$

In classical courses of mechanics, it is noted that in the domain of variation of the variables p, q, and r, these two integrals define an integral curve, on which system (5) is reduced to the quadrature

$$\frac{A}{C - B} \int \frac{dp}{qr} = t + \text{Const},$$

where q and r are considered as algebraic functions of p. As is known, in this case, an elliptic integral of the first kind is obtained, the inversion of which leads to the Jacobi elliptic functions. From a point of view of the finite difference method, the midpoint scheme allows exact conservation of both these integrals.

The top rotation problem in the Euler–Poinsot case will be referred to as conservative in the sense of the following definition.

Definition 2. *A difference scheme for system* (2) *on the manifold V is called conservative, if it assigns to a general point $x \in V$ a point $\hat{x}$ that also belongs to this manifold. In this case, we will speak that the difference scheme conserves the manifold V exactly.*

Note 2. In this definition, we mean a general point x rather than any one, since as it often happens in algebraic geometry, one should eliminate singular values at which the system of equations that specify the difference scheme can become degenerate.

One can expect that conservative schemes would allow studying not only quantitative but also qualitative properties of the solutions of system (1). For instance, in Example 4, the verbal description of the exact solution and the approximate one found in this way will be identical: the angular velocity vector $\omega = (p, q, r)$ periodically oscillates with constant amplitude; two conservation laws are valid for the energy and the angular momentum. It is difficult, if possible at all, to detect qualitative differences between the obtained solutions, and only quantitative discrepancies due to the precision of the period calculation are seen at large values of t.

Within the framework of mathematical modelling, the importance of finding conservative schemes is much higher. This is because the laws of classical mechanics are formulated for infinitely small dt, although in reality this quantity should be rather large to neglect quantum mechanical effects. This idea can be traced both in Lesage and in Feynman. The problem is that the transition from a continuous equation of motion to a finite-difference one using the Euler method leads to the violation of fundamental laws of nature, including the energy conservation law. However, the use of conservative difference schemes instead of the Euler scheme will make it possible to write a difference analog of Newton's equations, for which all classical conservation laws are exactly fulfilled. Thus, not only the continuous model, but also this discrete model can be used to describe the qualitative properties of motion with equal success. Since the limit transition $dt \to 0$ in Newton's equations is an idealization, we can raise a question about which of the models is better. From this point of view, the question of the convergence of the numerical method fades into the background, together with the physically excessive reduction of the time step.

4　Explicit and Implicit Difference Schemes

Wanting to use conservative schemes in real calculations, we must not only learn how to build such schemes, but also take into account the complexity of their application. For linear differential equations, the midpoint scheme (2) is described by a system of linear algebraic equations for $\hat{x}$ and therefore, it is easy to organize the transition from layer to layer. On the contrary, in the case of nonlinear equations, this scheme is described by a system of nonlinear equations, the solution of which is the main difficulty in practical application of such schemes.

Recall that the numerical solution of systems of nonlinear equations is always associated with a number of complexities, therefore, there is no universal numerical method for solving such systems [8]. The standard way suggests using iterative methods to go from layer to layer. Of the many roots of this system, a root is chosen that is close to x for small Δt. The solution of this system is obtained numerically, usually by means of some iterative method, taking the value of x for the first approximation. At the same time, it is not possible to control the error of the iterative method due to the extreme cumbersomeness of the known estimates, and instead, the number of iterations is simply fixed.

Definition 3. *A difference scheme describing the transition from layer to layer by a system of linear algebraic equations for $\hat{x}$ is called explicit.*

Note 3. The concept of explicit and implicit schemes has developed historically and initially the question was at what point the right-hand side of the differential equation is taken. However, the schemes of interest to us are organized in a considerably more complex way, so transferring the notion of explicitness to them requires grasping the main feature that makes explicit schemes convenient for computations. Of course, from the point of view of this definition, the implicit Euler scheme for a linear differential equation is explicit, and for a nonlinear one, it is implicit. In the following, we consider nonlinear equations, so this difference is not fundamental.

Before we accept the necessity to solve a system of nonlinear equations for the transition from layer to layer, we will try to clarify what is an obstacle for combining explicitness with conservatism.

5   Explicit Conservative Difference Schemes

It is a priori unclear whether there are nonlinear systems on manifolds, for which explicit schemes can be constructed. Let us formulate this question as a problem.

Problem 1. Given an autonomous system (1) and its integral manifold V, clarify whether an explicit difference scheme exists that approximates this system and exactly preserves the integral manifold. If the answer is positive, present such a difference scheme.

First, Problem 1 assumes that the integral manifold is given. The task of finding all algebraic integrals of a given system of differential equations was formulated at the dawn of differential calculus in the correspondence between Descartes and Florimond de Beaune [10], which outstanding mathematicians of the 19th and early 20th centuries tried to solve. For modern computer algebra systems, there are several packages for finding algebraic integrals [11,12], but they all require that the user should specify an upper limit for the order of the integrals in question. It is likely that this problem is algorithmically unsolvable.

Second, the formulation of Problem 1 allows the possibility that the desired difference scheme does not exist for a given system of differential equations and a given integral manifold. Any numerical method at the dawn of its occurrence claimed that with its help it is possible to solve equally well all the problems from the subject area for which it was created. At the time of Euler, the power series method was perceived as a universal method for solving ordinary differential equations, and only in the 1860s, Lazarus Fuchs posed the problem of finding those differential equations whose general solution is representable as the ratio of two everywhere convergent power series [5]. Therefore, the modern analytic theory of differential equations studies special differential equations whose solutions are particularly well represented by power series [1]. Our formulation of Problem 1 is made by analogy with the formulation of the Fuchs problem.

6   Algebraic Background

The solution to Problem 1 is closely related to algebraic geometry. To define a difference scheme for system (1) means to specify a system of equations describing the transition from one layer to another. In the case of single-stage difference schemes, this system consists of r equations

$$g_i(x, \hat{x}, \Delta t) = 0, \quad i = 1, 2, \ldots, r,$$

the left-hand sides of which belong to $\mathbb{Q}[x, \hat{x}, \Delta t]$. For example, the explicit Euler scheme is given by the equations

$$\hat{x}_i - x_i - f_i(x_1, \ldots, x_r)\Delta t = 0, \quad i = 1, 2, \ldots, r.$$

From a geometric point of view, these equations define an algebraic correspondence between the layers, or, more precisely, a one-parameter family of such correspondences, and Δt acts as a parameter.

Definition 4. *Let V and $\hat{V}$ be affine manifolds, embedded in A_r, $\xi = (\xi_1, \ldots, \xi_r)$ and $\hat{\xi} = (\hat{\xi}_1, \ldots, \hat{\xi}_r)$ being two tuples, each with r symbolic variables. Let the system of algebraic equations*

$$g_1(\xi, \hat{\xi}) = 0, \ldots \tag{7}$$

possess the following two properties:

- *if ξ is chosen as coordinates of a general point x of the manifold V, then the system*

$$g_1(x, \hat{\xi}) = 0, \ldots \tag{8}$$

with respect to $\hat{\xi}$ has $\hat{n}$ different roots, lying on the manifold $\hat{V}$ and changing under variation of x,
- *if $\hat{\xi}$ is taken to be coordinates of a general point $\hat{x}$ of the manifold $\hat{V}$, then the system*

$$g_1(\xi, \hat{x}) = 0, \ldots \tag{9}$$

with respect to ξ possesses n different roots, lying on the manifold V and changing under variation of $\hat{x}$.

In this case the system of algebraic Eq. (7) is said to specify an algebraic correspondence of the $(n, \hat{n})$ type between the manifolds V and $\hat{V}$. The roots of system (8) are considered as the mapping of the point x on $\hat{V}$, and the roots of (9) as the mapping of the point $\hat{x}$ on V. The correspondences on the $(n, 1)$ type are referred to as rational, and the corresponcences of the $(1, 1)$ type as birational.

Note 4. For particular positions of the point x, system (8) can have a smaller number of roots. Basing on the continuity principle, in this case, it is possible to say that the roots merge or escape to infinity, which implies a projective closure of the considered manifolds.

Algebraic correspondences between curves have been a subject of investigations in algebraic geometry in the middle of the XIX century. Modern authors usually consider only birational correspondences, for which the theory constructed is much simpler thanks to the remarkable results obtained by Hurwitz [14]. Therefore, we present here a brief background on the theory of algebraic correspondence.

In algebraic geometry, curves are characterized by an integer non-negative number called the genus. Modern computer algebra systems, including Maple and Sage, can calculate this number for a given curve.

Let there be an algebraic correspondence between the curves V and $\hat{V}$ of type $(n, \hat{n})$. Let ρ be the genus of the first curve, and $\hat{\rho}$ be the genus of the second one. According to the Zeuthen formula, [13, n. 65, (12)], these numbers are related as

$$\hat{\eta} - \eta = 2n(\hat{\rho} - 1) - 2\hat{n}(\rho - 1). \tag{10}$$

Here η and $\hat{\eta}$ are non-negative integers introduced in the following way.

- A general point x on the curve V corresponds to $\hat{n}$ moving points on $\hat{V}$, and, generally speaking, there are such particular positions of the point x at which two of these points merge into one. We denote the number of such merged points on $\hat{V}$ as $\hat{\eta}$; of course, we should consider the positions at which several points merge simultaneously, as multiple ones. If $\hat{n} = 1$, then there are no such singular points, and $\hat{\eta} = 0$.
- Similarly, the point $\hat{x}$ on the curve $\hat{V}$ corresponds to n moving points on V, and there are special positions of the point $\hat{x}$ at which two of these points merge into one. The number of such merged points on V is denoted as η. If $n = 1$, then such singular points do not exist and $\eta = 0$.

For birational correspondence ($n = \hat{n} = 1$ and, thus, $\eta = \hat{\eta} = 0$) due to Zeuthen formula, we have

$$\hat{\rho} = \rho,$$

that is the genus is an invariant of birational transformations. In particular the genus of the straight line is equal to zero thus any curve which is birational equivalent to the straight line has zero genus. Fairly stronger statement: a curve admits a rational parametrization iff its genus is equal to zero. This is consequence from Lüroth's theorem [15].

Further development of the theory is connected with the use of Abelian integrals [15, 16]. At the end of XIX century the group of birational morphisms of a curve was investigated, in particular Picard proved that this group is finite for any curve of the genus $\rho > 1$, Hurwitz estimated the order of this group [14]. The investigation uses Abelian integrals of the 1st kind. The same construction was used in the works of Painlevé about integration of ordinary differential equations in finite terms [17–19]. We will use this construction for an investigation of Problem 1.

7   Difference Scheme as Algebraic Correspondence

Each one-stage difference scheme approximating system (1) and preserving the manifold V specifies an algebraic correspondence on the manifold V. This scheme is explicit when and only when only a single point $\hat{x} \in V$ corresponds to a general point $x \in V$, i.e., when the algebraic correspondence has the type $(n, 1)$ that is rational correspondence.

All the above considerations can be generalized for multistage schemes. In the case of multistage schemes, the set of variables $x, \hat{x}$, and Δt is completed with additional variables and the number of equations is appropriately increased.

For example, in the s-stage Runge–Kutta method, the additional variables $k_1, \ldots, k_s$ are interpreted as auxiliary variables of the integral curve slope. For a transition from layer to layer, the slopes are first calculated from the system of equations

$$k_i = f\left(x + (a_{i1}k_1 + \cdots + a_{is}k_s)\Delta t\right),$$

and then

$$\hat{x} = x + (b_1 k_1 + \cdots + b_s k_s)\Delta t$$

is calculated. The coefficients a_{ij} and b_j of the scheme are chosen to ensure the approximation of the initial system (1) with the maximal order. The approximation conditions lead to algebraic equations for the coefficients, which are, however, rich in rational solutions. Hence, finally the Runge–Kutta scheme with any number of stages is described by a system of equations of the form

$$g(x, k, \hat{x}, \Delta t) = 0,$$

the left-hand sides of which belong to a polynomial ring $\mathbb{Q}[x, k, \hat{x}, \Delta t]$. However, practically they sometimes use non-rational coefficients, so it seems reasonable not to exclude from consideration the case when the coefficients of the scheme belong to the algebraic closure of $\mathbb{Q}$ [20, 21].

From an algebraic point of view, auxiliary variables can be eliminated and algebraic equations describing the transition from layer to layer can be obtained. Therefore, multistage schemes can also be considered as algebraic correspondences between the layers.

We would especially like to dwell on the dependence on Δt. The Euler, Runge–Kutta, and even more exotic methods imply that the left-hand sides of equations describing the transition from layer to layer are polynomials with respect to $x, \hat{x}$, and Δt. In specific calculations, the variable Δt is always given a small numerical value, so there is no reason to limit consideration to this assumption. Below we assume that the left-hand sides of the equations defining the difference scheme are polynomials with respect to x and $\hat{x}$, the coefficients of which belong to the algebraic closure of $\mathbb{Q}[\Delta t]$ or, in short, are algebraic functions of Δt.

Note 5. The assumption of a polynomial dependence on Δt apparently imposes excess narrowing of the class of difference schemes, among which the solution of Problem 1 should be sought. Suppose, for example, that the integral manifold is a curve. In this case, the explicit scheme equation

$$\hat{x} = g(x, \Delta t)$$

for a fixed x and a variable Δt gives a rational parametrization. Therefore, the required scheme obviously does not exist, if the integral curve is not unicursal. In the case of integral manifolds of higher dimension, the existence of an explicit difference scheme will immediately give a family of unicursal curves on the manifold. This property is very rare and, therefore, we again will have to state that the desired difference scheme does not exist.

8   Autonomous System on a Curve

In the case when the considered integral manifold V has dimension 1, the Problem 1 is solved on the basis of classical results from the theory of algebraic correspondences between two curves [13, Sect. 65]. In this case, the system of differential equations reduces to Abelian quadratures, but this does not make this case trivial.

Example 5. The motion of a top fixed at the centre of gravity is described by system (5) of three differential equations, which has two quadratic integrals (see Example 4). Therefore, the integral curves in the space p, q, and r are elliptic curves defined by two Eq. (6). On an integral curve, a system of three differential equations reduces to quadrature, which describes the dependence of p, q, and r on t. This dependence is described by elliptic functions. Moreover, this example involves the whole theory of elliptic functions which is a compelling evidence in favour of its nontriviality.

If a given integral curve V admits a rational parametrization, then, by Lüroth's theorem, a birational correspondence [15] can be established between this curve and the straight line. Denote the coordinate of a point on the line as u, then the connection between $x \in V$ and u can be written using rational functions: $u = U(x)$ and $x = X(u)$. Let $x = x(t)$ be the solution of system (1) belonging to the integral curve under consideration, then $u = U(x(t))$ as a function of t is a solution of the first-order differential equation

$$\dot{u} = \frac{\partial U}{\partial x_1} f_1 + \cdots + \frac{\partial U}{\partial x_r} f_r = H(u).$$

Let us write for this equation any explicit difference scheme, e.g., the Euler scheme

$$\hat{u} = u + H(u)\Delta t$$

and go back to the variables $x = X(u)$. As a result, we get the explicit difference scheme

$$\hat{x} = X(\hat{u}) = X(u + H(u)\Delta t),$$

approximating the initial system. Thus, in the case when a given integral manifold is a curve that allows rational parametrization, the difference scheme for Problem 1 exists. This scheme is given by equations whose coefficients depend on the parameter Δt polynomially (see Remark 5). Instead of the Euler scheme, one can use any other explicit difference scheme, say, any of the explicit Runge–Kutta schemes. Therefore, there are infinitely many such schemes.

Proposition 1. *Let the autonomous system* (1) *have an integral curve that allows rational parametrization. It is always possible to approximate this system with an explicit difference scheme that preserves exactly this integral curve.*

We now proceed to the case when a given integral curve does not allow rational parametrization. An explicit difference scheme that preserves the integral curve V defines a correspondence of the type $(n, 1)$ on this curve. Therefore, we should substitute $\hat{n} = 1$ in the Zeuthen formula (10) and, therefore, $\hat{\eta} = 0$. At the next step, the manifold itself does not change, so $\hat{\rho} = \rho$. Thus, the Zeuthen formula gives

$$-\eta = 2(n - 1)(\rho - 1).$$

For $n > 1$ and $\rho > 1$, the right-hand side becomes negative, which contradicts the very definition of the number $\hat{\eta}$. Therefore, for $\rho > 1$, the number is $n = 1$ and the explicit scheme defines a birational correspondence. If such a scheme really existed, then, by giving Δx different values, we would get infinitely many birational transformations on the curve V. However, by virtue of the theorem, first indicated, probably by Picard [17], the group of birational transformations of the genus $\rho > 1$ on an algebraic curve is finite.

Proposition 2. *Let the autonomous system* (1) *have an integral curve of the genus $\rho > 1$. This system cannot be approximated by an explicit difference scheme that retains exactly this integral curve.*

It remains to consider two cases in which $\rho = 0$ and $\rho = 1$.

If $\rho = 0$, then, as is known, the curve V admits a rational parametrization, therefore, the first case is described by Proposition 1.

Let us proceed to the case when the integral curve V is of genus 1. Assume that there is an explicit scheme that approximates system (1) and preserves the integral curve V of the genus 1 [15,16]. On this curve, there is a single Abelian differential of the first kind up to a multiplicative constant. Since only one of the variables $x_1, \ldots x_r$ is independent on the curve V, it can always be written as $H(x)dx_1$, where H is a rational function on V.

Let us fix an arbitrary point o on V and consider the integral

$$\int_o^{\hat{x}} H(x)\, dx_1.$$

The derivative of this integral with respect to x_1 is a rational function on V, and the integral itself, with any choice of the upper limit, remains finite, and, therefore, is an integral of the first kind. This means that there are two constants α and β such that

$$\int_o^{\hat{x}} H(x)\, dx_1 = \alpha \int_o^{x} H(x)\, dx_1 + \beta. \tag{11}$$

These constants are independent of x, but may depend on Δt. If α really depended on Δt, then it would be an algebraic function of Δt and, therefore, would have singularities. This would allow such position $\hat{x}$ for a fixed x that the integral

$$\int_0^{\hat{x}} H(x)\, dx_1$$

would become arbitrarily large, which is impossible. For $\Delta t = 0$, ratio (11) degenerates into

$$\int_0^x H(x)\, dx_1 = \alpha(0) \int_0^x H(x)\, dx_1 + \beta(0).$$

Therefore, α is identically equal to 1. This fact allows rewriting relation (11) in the following way:

$$\int_x^{\hat{x}} H(x)\, dx_1 = \beta(\Delta t). \tag{12}$$

Let us fix the value of x and give Δt a small value. To say that the scheme approximates system (1) is the same as to write a decomposition of the form

$$\hat{x} = x + f(x)\Delta t + \mathcal{O}(\Delta t^2).$$

Substituting this expression into Eq. (12) we get

$$\beta = H(x)f_1(x)\Delta t + \mathcal{O}(\Delta t^2).$$

Thus, on the integral curve, the following equality must hold:

$$H(x)f_1(x) = \text{const}.$$

Let us summarize the proved above by the following proposition.

Proposition 3. *Let the autonomous system (1) have an integral curve of the genus $\rho = 1$. This system can be approximated by an explicit difference scheme that preserves exactly this integral curve only in the case when the integral*

$$\int \frac{dx_1}{f_1(x)}$$

is an Abelian integral of the first kind on an integral curve. In this case, the scheme sought in Problem 1 is described by the equation

$$\int_x^{\hat{x}} \frac{dx_1}{f_1} = \Delta t + \mathcal{O}(\Delta t^2). \tag{13}$$

Let us reverse the assertion of the proposition. Let the autonomous system (1) have an integral curve V of the genus $\rho = 1$ and let the integral

$$\int \frac{dx_1}{f_1(x)}$$

be an Abelian integral of the first kind on an integral curve.

We fix on V an arbitrary non-singular point o and next to it take the point $\hat{o}$, which depends on Δt in such a way that

$$\int\limits_{o}^{\hat{o}} \frac{dx_1}{f_1} = \Delta t + \mathcal{O}(\Delta t^2). \tag{14}$$

In order to find the point $\hat{o}$, it is sufficient to take the first of its coordinates equal to

$$f_1(o)\Delta t,$$

and determine the rest coordinates from the condition of belonging of the point $\hat{o}$ to the curve V. In this case, it is necessary to solve algebraic equations, therefore, the coordinates of the point $\hat{o}$ are algebraic functions of Δt. By Abel's theorem, the relation

$$\int\limits_{x}^{\hat{x}} \frac{dx_1}{f_1} = \int\limits_{o}^{\hat{o}} \frac{dx_1}{f_1}$$

allows unique definition of $\hat{x}$ as rational function of x. To find this function, we need to construct a rational function on V that has two simple poles x and $\hat{o}$ and one known zero o. Since $\hat{o}$ depends on Δt algebraically, $\hat{x}$ is a rational function of x, whose coefficients depend on Δt algebraically.

Consider a difference scheme, in which this function describes the transition from layer to layer. From equation (13), it follows that

$$\frac{\Delta x}{f_1} = \Delta t + \mathcal{O}(\Delta t^2),$$

thus, the difference scheme approximates the first of the equations of system (1) and exactly preserves the integral curve and, therefore, approximates all other equations. Therefore, the difference scheme specified in this way is the one required in Problem 1. From here we get the proposition inverse to Proposition 3.

Proposition 4. *Let the autonomous system* (1) *have an integral curve of the genus $\rho = 1$. This system can be approximated by an explicit difference scheme that preserves this integral curve exactly, if the integral*

$$\int \frac{dx_1}{f_1(x)}$$

is an Abelian integral of the first kind on an integral curve.

Note 6. Under the conditions of Proposition 4, system (1) has an integral curve V, on which the dependence of x_1 on t is described by quadrature

$$\int \frac{dx_1}{f_1(x)} = t + C.$$

Therefore, on the exact solution, the following relation is true:

$$\int\limits_{x(t)}^{x(t+\Delta t)} \frac{dx_1}{f_1} = \Delta t.$$

In other words, our difference scheme would give an exact solution if we could take $\beta = \Delta t$ in Eq. (12). Such a choice of β, however, is impossible, since the equality

$$\int\limits_{o}^{\hat{o}} \frac{dx_1}{f_1} = \Delta t$$

would cause the dependence of the coordinates of $\hat{o}$ on Δt in a transcendental way. Thus, we would obtain a difference scheme whose equations would contain Δt in a transcendental way.

We summarize the proved above as a theorem.

Theorem 1. *Let the autonomous system (1) have an integral curve of the genus ρ.*

- *If the genus is 0, then there is an infinite number of explicit difference schemes that preserve this curve exactly.*
- *If the genus is 1, then such scheme exists if and only if*

$$\int \frac{dx_1}{f_1}$$

  is an integral of the first kind on the curve V.
- *If the genus is greater than 1, then such a scheme does not exist.*

Example 6. The Jacobi elliptic functions are a solution of system (3) on the integral curve

$$p^2 + q^2 = 1 \quad \text{and} \quad k^2 p^2 + r^2 = 1.$$

This curve is of genus 1, and the system itself on this curve can be written as a quadrature

$$\int \frac{dp}{qr} = t + \text{const}$$

or

$$\int \frac{dp}{\sqrt{(1-p^2)(1-k^2 p^2)}} = t + \text{const.}$$

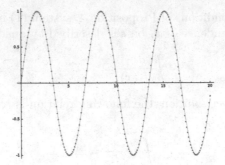

Fig. 1. Plot of $\text{sn}(t, \frac{1}{2})$, calculated using an explicit conservative difference scheme

Jacobi also noted that the integral on the right remains finite for any choice of the limits of integration, that is, according to the modern classification, it is an Abelian integral of the first kind on the elliptic curve V. Therefore, by Proposition 4, there is an explicit difference scheme that preserves this curve. The transition from layer to layer is carried out according to the formula

$$\int\limits_{(p,q,r)}^{(\hat{p},\hat{q},\hat{r})} \frac{dp}{qr} = \beta(\Delta t)$$

Using the addition theorem for elliptic functions [22, Sect. 22.8], this relation can be rewritten in algebraic form as

$$\hat{p} = \frac{p \,\text{cn}\, \lambda \,\text{dn}\, \beta - \text{sn}\, \beta \, qr}{1 - k^2 p^2 \,\text{sn}^2\, \beta}$$

$$\hat{q} = \frac{q \,\text{cn}\, \beta - \text{sn}\, \beta \,\text{dn}\, \beta \, pr}{1 - k^2 p^2 \,\text{sn}^2\, \beta}$$

and

$$\hat{r} = \frac{r \,\text{dn}\, \beta - k^2 \,\text{sn}\, \beta \,\text{cn}\, \beta \, pq}{1 - k^2 p^2 \,\text{sn}^2\, \beta}.$$

It remains to choose β so that the difference scheme approximates the original differential equation, i.e.,

$$\beta = \Delta t + \mathcal{O}(\Delta t^2),$$

and that the equations describing the transition from layer to layer depend on Δt algebraically. This can be achieved by taking

$$\text{sn}\, \beta = \Delta t, \quad \text{cn}\, \beta = \sqrt{1 - \Delta t^2}, \quad \text{dn}\, \beta = \sqrt{1 - k^2 \Delta t^2}.$$

Figure 1 presents a plot of the elliptical sine calculated using the standard means of Sage (solid line) and according to the proposed explicit scheme. The perfect coincidence of results is clearly seen, which is not surprising, since we have a difference scheme that without any connection with the finite difference method was used by Guderman to create tables of elliptic functions [23, Abh.1].

Example 7. In the case of Euler–Poinsot (Example 4), the integral

$$\int \frac{dp}{qr}$$

on an elliptic curve (6) is an integral of the first kind, therefore, by Proposition 4, there is an explicit difference scheme that preserves exactly the integral curve. The fact that among the tools for working with algebraic curves, there is no one for the construction of the main function, hinders the compilation of this scheme by means of Sage. It is convenient to carry out calculations by hand, after making the change of variables, leading to Jacobi elliptic functions.

9   Conclusion

In this paper, an autonomous system of ordinary differential equations on an algebraic manifold V was considered.

Explicit difference schemes are new and convenient for calculating approximate solutions of this system, but they divert the solution from the integral manifold. In mechanical problems, such schemes introduce parasitic numerical effects, e.g., dissipation, into the mathematical model, or even lead to results that cannot be interpreted geometrically.

Difference schemes that do not divert the solution from the manifold V are called conservative. These schemes are remarkable in that they provide us with a discrete model of a mechanical phenomenon, within which the fundamental conservation laws are exactly fulfilled. This allows the use of such models not only for quantitative, but also for qualitative research of mechanical phenomena.

Although the laws of classical mechanics are formulated for infinitely small dt, in real problems, there is always a characteristic minimum time scale Δt. Attempts to consider Newton's equations as finite-difference ones are hampered only by the fact that they lead to a non-conservative scheme. Having found a conservative scheme, we get a discrete mathematical model, the status of which is not lower than that of a continuous model. In this case, the choice of step is determined by physical considerations, and not by the convergence of the numerical method.

Unfortunately, conservative schemes are usually implicit and the transition from layer to layer is very costly. Therefore, we formulated Problem 1 about finding an explicit conservative difference scheme for a given system of differential equations and gave this problem a geometric interpretation.

For the case of a manifold of dimension 1, we gave a complete solution to this problem. This solution allows us to make several hypotheses regarding the general case.

First, it turned out that there is a purely geometric obstacle to the existence of explicit and simultaneously conservative schemes, namely, the genus of the curve V. If the genus of the curve is greater than 1, then such a scheme does not exist. In particular, for an autonomous system on a general curve whose order is greater than 3, such a scheme does not exist (Proposition 2). This means that the attempts to find a universal method for constructing an explicit conservative scheme for any system of differential equations are futile.

As a hypothesis, we note that an autonomous system on a manifold of a general form must not allow explicit conservative schemes.

Second, it turned out that an explicit conservative scheme for a system on a curve exists in two very dissimilar cases: if the genus of the integral curve is 1 and if the genus is 0. In the first case (Proposition 3), the difference scheme specifies a birational correspondence between the layers, and the autonomous system itself must be integrated in elliptic functions. In the second case (Proposition 1), there are infinitely many such schemes, but the case itself is not at all interesting, since a rational change of variables can reduce the order of the system without changing its form. In fact, this is the case of a system on affine space.

We assume that an autonomous system on a manifold admits an explicit conservative difference scheme in two cases.

First, if this variety is Abelian and the system itself is integrated in Abelian functions, an example of such a system is a top in the Kovalevskaya case (Example 2), a double pendulum [24], the Garnier system [25] or the system considered in our paper [26].

Second, if a rational replacement can reduce the original system to a system of the same kind, but of a smaller order.

Since the theory of algebraic surfaces is already sufficiently developed, we expect Problem 1 to be solved in the near future for the case of surfaces. On the other hand, there is no good package to work with algebraic surfaces, thus, the constructive solution of Problem 1 is a challenge for the computer algebra.

References

1. Goriely, A.: Integrability and Nonintegrability of Dynamical Systems. World Scientific Publishing, Singapore (2001)
2. Hairer, E., Wanner, G., Lubich, C.: Geometric Numerical Integration. Structure-Preserving Algorithms for Ordinary Differential Equations, 2nd edn. Springer, Berlin (2000). https://doi.org/10.1007/978-3-662-05018-7
3. Polubarinova-Kochina, P. Ya.: On unambiguous solutions and algebraic integrals of a problem about rotation of a gyroscope at a motionless point. In:. Chaplygin, S.A. (ed.) Dvizhenie tverdogo tela vokrug nepodvizhnoj tochki. Academy of Sciences of the USSR, Moscow-Leningrad (1940). (in Russian)
4. Hairer, E., Wanner, G., Nørsett, S.P.: Solving Ordinary Differential Equations. Nonstiff Problems, vol. 1. Springer, Berlin (2008). https://doi.org/10.1007/978-3-540-78862-1
5. Schlesinger, L.: Einführung in die Theorie der gewöhnlichen Differentialgleichungen auf Funktionentheoretischer Grundlage. De Gruyter, Berlin-Leipzig (1922)
6. Ayryan, E.A., Malykh, M.D., Sevastianov, L.A., Ying, Y.: Finite difference schemes and classical transcendental functions. In: Nikolov, G., Kolkovska, N., Georgiev, K. (eds.) NMA 2018. LNCS, vol. 11189, pp. 235–242. Springer, Cham (2019). https://doi.org/10.1007/978-3-030-10692-8_26
7. Blinkov, Yu.A., Gerdt, V.P.: On computer algebra aided numerical solution of ODE by finite difference method. In: Vassiliev, N.N. (ed.) Polynomial Computer Algebra '2019, 29–31. VVM Publication, St-Petersburg (2019)

8. Numerical Recipes (1993). https://www.numerical.recipes
9. SageMath, the Sage Mathematics Software System (Version 7.4), The Sage Developers (2016). https://www.sagemath.org
10. Descartes, R.: Geometry with the Appendix of Some Works of P. Fermat and Descartes's Correspondence. GONTI NKTP SSSR, Moscow-Leningrad (1938). (in Russian)
11. Bostan, A., Chéze, G., Cluzeau, T., Weil, J.-A.: Efficient algorithms for computing rational first integrals and Darboux polynomials of planar polynomial vector fields. Math. Comput. **85**, 1393–1425 (2016)
12. Malykh, M.D.: On integration of the first order differential equations in finite terms. J. Phys. Conf. Ser. 788 012026. (2017). https://doi.org/10.1088/1742-6596/788/1/012026
13. Zeuthen, H.G.: Lehrbuch der abzählenden Methoden der Geometrie. Teubner, Leipzig und Berlin (1914)
14. Hartshorne, R.: Algebraic Geometry. Springer, New York (1977). https://doi.org/10.1007/978-1-4757-3849-0
15. Severi, F.: Lezioni di geometria algebrica. Angelo Graghi, Padova (1908)
16. Weierstrass, K.: Mathematische Werke, 4. Mayer & Müller, Berlin (1902)
17. Painlevé, P.: Leçons sur la théorie analytique des équations différentielles, professées à Stockholm (septembre, octobre, novembre 1895) sur l'invitation de S. M. le roi de Suède et de Norwège. In: Œuvres de Painlevé, 1, Paris, CNRS (1971)
18. Umemura, H.: Birational automorphism groups and differential equations. Nagoya Math. J. **119**, 1–80 (1990)
19. Malykh, M.D.: On transcendental functions arising from integrating differential equations in finite terms. J. Math. Sci. **209**(6), 935–952 (2015). https://doi.org/10.1007/s10958-015-2539-6
20. Khashin, S.I.: A symbolic-numeric approach to the solution of the Butcher equations. Can. Appl. Math. Q. **17**(1), 555–569 (2009)
21. Khashin, S.I.: Butcher algebras for Butcher systems. Numer. Algorithms **63**(4), 679–689 (2013). https://doi.org/10.1007/s11075-012-9647-x
22. NIST Digital Library of Mathematical Functions (2018). Version 1.0.21. https://dlmf.nist.gov
23. Weierstrass, K.: Mathematische Werke, 1. Mayer & Müller, Berlin (1892)
24. Enolskii, V.Z., Pronine, M., Richter, P.H.: Double pendulum and ϑ-divisor. J. Nonlinear Sci. **13**, 157–174 (2003). https://doi.org/10.1007/s00332-002-0514-0
25. Garnier, R.: Sur une classe de systèmes différentiels abéliens déduits de la théorie des équations linéaires. Rendiconti del Circolo mat. di Palermo **43**, 155–191 (1919)
26. Malykh, M.D., Sevastianov, L.A.: On an example of a system of differential equations that are integrated in Abelian functions. J. Phys. Conf. Ser. 937 012027 (2017). https://doi.org/10.1088/1742-6596/937/1/012027

On Berlekamp–Massey and Berlekamp–Massey–Sakata Algorithms

Chenqi Mou[1,2(✉)] and Xiaolin Fan[1]

[1] LMIB–School of Mathematics and Systems Science,
Beihang University, Beijing 100191, China
{chenqi.mou,neil_fan}@buaa.edu.cn
[2] Beijing Advanced Innovation Center for Big Data and Brain Computing,
Beihang University, Beijing 100191, China

Abstract. The Berlekamp–Massey and Berlekamp–Massey–Sakata algorithms compute a minimal polynomial or polynomial set of a linearly recurring sequence or multi-dimensional array. In this paper some underlying properties of and connections between these two algorithms are clarified theoretically: a unified flow chart for both algorithms is proposed to reveal their connections; the polynomials these two algorithms maintain at each iteration are proved to be reciprocal when both algorithms are applied to the same sequence; and the uniqueness of the choices of polynomials from two critical polynomial sets in the Berlekamp–Massey–Sakata algorithm is investigated.

Keywords: Berlekamp–Massey algorithm ·
Berlekamp–Massey–Sakata algorithm · Minimal polynomial ·
Reciprocal polynomial

1 Introduction

The Berlekamp–Massey (BM) algorithm from coding theory is to find the shortest linear feedback shift register of an output sequence [1,9,10], and it can be used to decode Reed–Solomon codes by finding the error-locator polynomial. Mathematically the BM algoirthm is also an algorithm to compute the minimal polynomial of a linearly recurring sequence over a field, with successful application in solving sparse linear systems [17].

The Berlekamp–Massey–Sakata (BMS) algorithm, as one can find from its name, is generalization of the BM algorithm to the multivariate case [14–16]. This algorithm, also from coding theory, is a decoding algorithm to find the minimal polynomial set of the error locator ideal in algebraic geometry codes [5,6,11,12].

This work was partially supported by the National Natural Science Foundation of China (NSFC 11771034) and the Fundamental Research Funds for the Central Universities in China (YWF-19-BJ-J-324).

© Springer Nature Switzerland AG 2019
M. England et al. (Eds.): CASC 2019, LNCS 11661, pp. 362–376, 2019.
https://doi.org/10.1007/978-3-030-26831-2_24

The BMS algorithm draws particular interest of the community of symbolic computation due to the fact that mathematically the BMS algorithm computes a minimal Gröbner basis of the polynomial ideal defined by a linearly recurring array [5,13,15]. With this observation, the BMS algorithm has been applied to computation of Gröbner bases: it is used in the sparse FGLM algorithm to change the term orderings of Gröbner bases from the degree reverse lexicographic one to the lexicographic one [7,8]; the scalar-FGLM algoirthm is proposed in [2,4] to compute the Gröbner basis of the ideal defined by a multi-dimensional array, solving the same problem as the BMS algorithm does, by using multi-Hankel matrices. In particular, the BMS and scalar-FGLM algoirthms are compared in details in [3].

In this paper we study the BM and BMS algorithms and clarify some detailed connections and differences between these two algorithms. We first propose a unified flow chart for both the BM and BMS algorithms by introducing some notions of the latter algorithm to the former. We compare corresponding parts of the two algorithms to justify such a flow chart. In particular, we show that a case of the BMS algorithm is only meaningful when the array is multi-dimensional because of the partial order and that the two cases in the BM algorithm indeed correspond to one case in the BMS algorithm. Next we compare the BM and BMS algorithms by applying them to the same sequence and prove that in this case the polynomials they compute at each iteration are reciprocal. Then we study the numbers of polynomials in two critical sets in the BMS algorithm, which have been proved to be nonempty (see. e.g, [13, Lemma 37]) for the correctness of the algorithm, and prove that in a particular case one set consists of only one element and provide a counter-example of uniqueness of polynomials in the two sets in the other cases.

The detailed study on the BM and BMS algorithms in this paper are helpful for further understanding the BMS algorithm, which is essential at the current early research stage of this algorithm from the algebraic viewpoint. In particular, by unifying the two algorithms in terms of their underlying languages and a unified flow chart and revealing the relationships between the BM algorithm and the degenerated BMS algorithm applied to sequences, we better justify that the BMS algorithm is indeed multivariate generalization of the BM algorithm, which seems not to be adequately investigated in the literatures [15,16] to our best knowledge, though the name BMS algorithm has been widely acknowledged. The observation that two critical polynomial sets in the BMS algorithm may contain multiple polynomials indicates that the BMS algorithm may return different minimal polynomial sets for one array, for any polynomial in these two polynomial sets may be chosen for the construction. This accords with the fact that the BMS algorithm, viewed as an algorithm for computing Gröbner bases, only returns minimal Gröbner bases instead of reduced ones which are unique.

This paper is organized as follows. The notion and notations used in the BM and BMS algorithms are recalled in Sect. 2 in a comparative way. Then in Sects. 3, 4, and 5, we unify these two algorithms in one flow chart, compare them by applying them to one sequence, and study the numbers of polynomials in two critical sets for the update in the BMS algorithm respectively.

2 Preliminaries

Let $\mathbb{K}$ be a field, and $\mathbb{K}[x_1, \ldots, x_n]$ be the multivariate polynomial ring over $\mathbb{K}$ with the variables $x_1, \ldots, x_n$. We fix a variable ordering $x_1 < \cdots < x_n$ throughout this paper. For simplicity, we write $\boldsymbol{x}$ for $x_1, \ldots, x_n$ and $\mathbb{K}[\boldsymbol{x}]$ for $\mathbb{K}[x_1, \ldots, x_n]$. With the one-one correspondence between a vector $\boldsymbol{u} = (u_1, \ldots, u_n) \in \mathbb{Z}_+^n$ and a term $\boldsymbol{x^u} = x_1^{u_1} \cdots x_n^{u_n} \in \mathbb{K}[\boldsymbol{x}]$, where $\mathbb{Z}_+$ denotes the set of non-negative integers, we do not distinguish these two representations of a term in this paper.

Denote by $\mathcal{T}(\boldsymbol{x})$ the set of all the terms in $\mathbb{K}[\boldsymbol{x}]$. Next we define two orders on $\mathcal{T}(\boldsymbol{x})$ which are frequently used in this paper.

(1) $\preceq$: for any $\boldsymbol{u} = (u_1, \ldots, u_n)$, $\boldsymbol{v} = (v_1, \ldots, v_n) \in \mathcal{T}(\boldsymbol{x})$, define $\boldsymbol{u} \preceq \boldsymbol{v}$ if $u_i \leq v_i$ for $i = 1, \ldots, n$. Obviously this is a partial order defined by divisibility of terms in $\mathcal{T}(\boldsymbol{x})$.

(2) $<_t$: for any $\boldsymbol{u} = (u_1, \ldots, u_n)$, $\boldsymbol{v} = (v_1, \ldots, v_n) \in \mathcal{T}(\boldsymbol{x})$, define $\boldsymbol{u} <_t \boldsymbol{v}$ if $\sum_{i=1}^n u_i < \sum_{i=1}^n v_i$ or $\sum_{i=1}^n u_i = \sum_{i=1}^n v_i$ but there exists an integer j $(1 \leq j \leq n)$ such that $u_i = v_i$ $(i = 1, \ldots, j-1)$ and $u_j > v_j$. Those who are familiar with the theory of Gröbner bases will immediately recognize this as the degree reverse lexicographic term order (which is a total order). We will only use this term order in this paper.

Definition 1. A mapping $s : \mathbb{Z}_+ \to \mathbb{K}$ is called a *sequence* over $\mathbb{K}$. For a given integer $m \geq 1$, the restriction $s|_{\{0,1,\ldots,m-1\}}$ is called a sequence of length m over $\mathbb{K}$, and it is often written as an ordered sequence $[s(0), s(1), \ldots, s(m-1)]$.

Similarly, a mapping $E : \mathbb{Z}_+^n \to \mathbb{K}$ is called an n-dimensional *array* over $\mathbb{K}$. For a given term $\boldsymbol{u} \in \mathbb{Z}_+^n$, denote $\mathbb{Z}_+^n(\boldsymbol{u}) = \{\boldsymbol{v} \in \mathbb{Z}_+^n : \boldsymbol{v} \leq_t \boldsymbol{u}\}$. Then the restriction $E|_{\mathbb{Z}_+^n(\boldsymbol{u})}$ is called an n-dimensional array over $\mathbb{K}$ up to $\boldsymbol{u}$, denoted by $E^{\boldsymbol{u}}$.

Example 1. A 2-dimensional array over $\mathbb{F}_2$ up to $(2,1)$ is illustrated in the following picture. Please note that with respect to $<_t$, the terms smaller than $(2,1)$ are $(0,0), (1,0), (0,1), (2,0), (1,1), (0,2)$, and $(3,0)$ (Fig. 1).

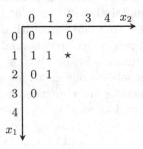

Fig. 1. A 2-dimensional array over $\mathbb{F}_2$

Definition 2. Let $F = \sum_{i=0}^{d} c_i x^i$ be a univariate polynomial in $\mathbb{K}[x]$ and $s = [s_0, s_1, \ldots, s_{m-1}]$ be a sequence of length m over $\mathbb{K}$. For an integer k $(d \leq k < m)$, define $F[s, k] := \sum_{i=0}^{d} c_i s_{k-i}$. If $k < d$ or $d \leq k$ and $F[s, k] = 0$, then F is said to be *valid* at k for s in the BM sense.

Furthermore, if F is valid at i for s in the BM sense for $i = 0, \ldots, k$, then F is said to be *valid up to* k for s in the BM sense. We denote by $V[s, k]$ the set of polynomials in $\mathbb{K}[x]$ which are valid up to k for s in the BM sense. A polynomial in $V[s, m-1]$ of the smallest degree is called a *minimal polynomial* of s in the BM sense.

For an arbitrary univariate polynomial $F \in \mathbb{K}[x]$, we denote the degree of F by $\deg(F)$; for an arbitrary multivariate polynomial $G \in \mathbb{K}[\boldsymbol{x}]$, we denote the exponent of the greatest term of G with respect to $<_t$ by $\mathrm{lead}(G)$.

Definition 3. Let $F = \sum_{v \in \Gamma} c_v \boldsymbol{x}^v \in \mathbb{K}[\boldsymbol{x}]$ with Γ a subset of $\mathbb{Z}_+^n$, and E^u be an n-dimensional array up to some term $\boldsymbol{u} \in \mathbb{Z}_+^n$. For a term $\boldsymbol{r} \in \mathbb{Z}_+^n$ $(\mathrm{lead}(F) \preceq \boldsymbol{r} \leq_t \boldsymbol{u})$, define $F[[E^u, \boldsymbol{r}]] := \sum_{v \in \Gamma} c_v E^u(\boldsymbol{v} + \boldsymbol{r} - \mathrm{lead}(F))$. If $\mathrm{lead}(F) \npreceq \boldsymbol{r}$ or $\mathrm{lead}(F) \preceq \boldsymbol{r}$ and $F[[E^u, \boldsymbol{r}]] = 0$, then F is said to be *valid* at $\boldsymbol{r}$ for E^u in the BMS sense.

Furthermore, if F is valid at $\boldsymbol{w}$ for E^u in the BMS sense for all $\boldsymbol{w}$ such that $\boldsymbol{w} \leq_t \boldsymbol{r}$, then F is said to be *valid up to* $\boldsymbol{r}$ for E^u in the BMS sense. We denote by $V[[E^u, \boldsymbol{r}]]$ the set of polynomials in $\mathbb{K}[\boldsymbol{x}]$ which are valid up to $\boldsymbol{r}$ for E^u in the BMS sense. Let $\mathcal{F} = \{F_1, \ldots, F_s\}$ be a polynomial set in $\mathbb{K}[\boldsymbol{x}]$. Then $\mathcal{F}$ is called a *minimal polynomial set* of E^u in the BMS sense if the following conditions hold:

(1) For each $i = 1, \ldots, s$, $F_i \in V[[E^u, \boldsymbol{u}]]$;
(2) There does not exist a polynomial $G \in V[[E^u, \boldsymbol{u}]]$ such that $\mathrm{lead}(G) \preceq \mathrm{lead}(F_i)$ for some $F_i \in \mathcal{F}$.

Remark 1. If one applies the BMS algorithm to a sequence s, in which case the partial order $\preceq$ and the term order $<_t$ coincide and the algorithm just degenerates to handle univariate polynomials, and compare it with the BM algorithm, then one finds that the expression of $F[s, k]$ is different from that of $F[[s, k]]$ in the sense that the coefficients of F are used in a reverse order. We have to distinguish these two expressions and this is why we add the tedious words "in the BM/BMS sense" in the above definitions.

Example 2. Let E^u be the 2-dimensional array up to $\boldsymbol{u} = (2, 1)$ in Example 1 and $F = x_2 + x_1 + 1 \in \mathbb{F}_2[x_1, x_2]$. Then

$$F[[E^u, (0, 2)]] = E[(0, 0) + (0, 2) - (0, 1)] + E[(1, 0) + (0, 2) - (0, 1)]$$
$$+ E[(0, 1) + (0, 2) - (0, 1)] = 0,$$

and thus F is valid at $(0, 2)$ for E^u in the BMS sense. One can also check that $F[[E^u, (0, 1)]] = F[[E^u, (1, 1)]] = 0$, and thus F is valid up to $(0, 2)$ in the BMS sense. Note that in the process above of checking validity of F at some term, we omit those terms which are not $\succeq \mathrm{lead}(F)$ and thus valid naturally by Definition 3.

With the terminologies above, we are now ready to describe the specifications of the two algorithms of our interest in this paper: given a sequence s of length m over $\mathbb{K}$, the BM algorithm computes a minimal polynomial of s in the BM sense; given an n-dimensional array E^u up to some term $u \in \mathbb{Z}_+^n$ over $\mathbb{K}$, the BMS algorithm computes a minimal polynomial set of E^u in the BMS sense.

Definition 4. Let $\mathcal{F}$ be a zero-dimensional polynomial set in $\mathbb{K}[x]$. Define

$$\Delta(\mathcal{F}) := \{u \in \mathbb{Z}_+^n : \operatorname{lead}(F) \not\preceq u \text{ for all } F \in \mathcal{F}\}.$$

A set $\Delta \subset \mathbb{Z}_+^n$ is called a *delta set* if for any $u \in \Delta$, we have $v \in \Delta$ for all $v \preceq u$. Obviously $\Delta(\mathcal{F})$ above is a delta set.

Let $\Delta \subset \mathbb{Z}_+^n$ be a delta set. A term $u \in \Delta$ is called an *interior corner* of Δ if there does not exist $v \in \Delta$ such that $u \preceq v$, and a term $w \notin \Delta$ is called an *exterior corner* of Δ if there does not exist $v \notin \Delta$ such that $v \preceq w$. The set of all the interior corners (exterior corners respectively) of a delta set Δ is denoted by $\operatorname{Int}(\Delta)$ ($\operatorname{Ext}(\Delta)$ respectively).

By Definition 4, interior corners of Δ correspond to maximal elements in Δ with respect to $\preceq$, and the exterior corners are the minimal elements in $T(x) \setminus \Delta$. See Fig. 2 below for illustrative delta set, interior corners, and exterior corners. For a term $v \in \mathbb{Z}_+^n$, we denote the preceding and next terms of v with respect to $<_t$ by v^- and v^+ respectively.

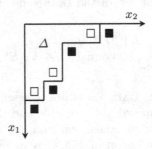

Fig. 2. Illustrative delta set (Δ), interior corners (marked with $\square$), and exterior corners (marked with $\blacksquare$)

Definition 5. Let E^u be an n-dimensional array up to some term $u \in \mathbb{Z}_+^n$ and $G \in \mathbb{K}[x]$ be a polynomial in $V[[E^u, v^-]]$ but not in $V[[E^u, v]]$. Then the *span* of G is defined to be $\operatorname{span}(G) := v - \operatorname{lead}(G)$. In this case, we denote $\delta_G := G[[E^u, v]](\neq 0)$. For a finite polynomial set $\mathcal{G} = \{G_1, \ldots, G_r\} \subset \mathbb{K}[x]$, we define

$$\operatorname{span}(\mathcal{G}) := \bigcup_{i=1}^{r} \{v \in \mathbb{Z}_+^n : v \preceq \operatorname{span}(G_i)\}.$$

One can check that $\operatorname{span}(\mathcal{G})$ for a polynomial set $\mathcal{G}$ is a delta set, and one should pay particular attention to the definition of $\operatorname{span}(\mathcal{G})$ because it is different from $\{\operatorname{span}(G) : G \in \mathcal{G}\}$.

3   BM and BMS Algorithms in a Unified Flow Chart

Next we first reproduce the pseudo codes for the BM and BMS algorithms below. The readers are referred to [6,10,13] for existing descriptions of these two algorithms. In particular, Algorithm 2 below for the BMS algorithm is based on a simplified version presented in [13].

Algorithm 1. Berlekamp–Massey Algorithm $F := \mathsf{BM}(s)$

Input: a sequence $s = [s_0, s_1, \ldots, s_{m-1}]$ of length m over $\mathbb{K}$
Output: a minimal polynomial F of s in the BM sense

1 $F := 1, G := 1, p := 1, b := 1$;
2 **for** $i = 0, \ldots, m-1$ **do**
3      $d := F[s, i]$;
4   **if** $d = 0$ **then**
5          $p := p + 1$;   [Case 1]
6   next;
7   **else**
8   **if** $2 \deg(F) \le i$ **then**
9              $T := F, F := F - \frac{dx^p G}{b}$;   [Case 2.1]
10             $G := T, b := d, p := 1$;
11   **else**
12             $F := F - \frac{dx^p G}{b}$, $p := p + 1$;   [Case 2.2]

13 **return** F

The BM algorithm maintains two polynomials F and G to update them in each new iteration when necessary, and at the end F is output as a minimal polynomial of the input sequence in the BM sense. As shown in the descriptions of the algorithm, we split Algorithm 1 into 3 cases: Case 1 corresponds to the situation when F is valid at the current iteration and thus neither F nor G changes, in Case 2.1 both F and G are updated, and in Case 2.2 only F is updated. It is noted from Lines 9 and 12 of Algorithm 1 that in both Cases 2.1 and 2.2 the updates of F share the same expression.

The BMS algorithm maintains two polynomial sets $\mathcal{F}$ and $\mathcal{G}$ to update them in each new iteration when necessary, and at the end $\mathcal{F}$ is output as a minimal polynomial set of the input n-dimensional array in the BMS sense. In the iteration handling a term v, for each $r \in \mathrm{Ext}(\mathrm{span}(\mathcal{G}^+))$, there will be a polynomial $F \in \mathcal{F}^+$ such that $\mathrm{lead}(F) = r$ (see the polynomials adjoined to $\mathcal{F}^+$ in Lines 7, 13, and 16 in Algorithm 2). As shown in the descriptions of the algorithm, we also split Algorithm 2 into 3 cases. Case 1 corresponds to the situation when all the polynomials in $\mathcal{F}$ are valid to u (namely $\mathcal{N} = \varnothing$), and thus $\mathcal{G}^+ = \mathcal{G}$ and $\mathrm{span}(\mathcal{G}^+) = \mathrm{span}(\mathcal{G})$. In this case neither $\mathcal{F}$ nor $\mathcal{G}$ changes. In both of the remaining Cases 2 and 3, $\mathcal{F}$ changes, and how a new polynomial F^+ with $\mathrm{lead}(F^+) = r$ is constructed is dependent on the condition $r \preceq v$. This is because in Case 2

Algorithm 2. Berlekamp–Massey–Sakata Algorithm $\mathcal{F} := \mathsf{BMS}(E^u)$

Input: an n-dimensional array E^u over $\mathbb{K}$ up to $u \in \mathbb{Z}_+^n$
Output: a minimal polynomial set $\mathcal{F}$ of E^u in the BMS sense

1 $\mathcal{F} := \{1\}, \mathcal{G} := \varnothing$;
2 **for** $v = 0, \ldots, u$ **do**
3 $\quad$ $\mathcal{N} := \{F \in \mathcal{F} : F[[E^u, v]] \neq 0\}$;
4 $\quad$ $\mathcal{G}^+ := \mathcal{G} \cup \mathcal{N}, \Delta := \mathrm{span}(\mathcal{G}^+), \mathcal{F}^+ := \varnothing$;
5 $\quad$ **for** $r \in \mathrm{Ext}(\Delta)$ **do**
6 $\quad\quad$ **if** $\exists F \in \mathcal{F} \setminus \mathcal{N}$ such that $\mathrm{lead}(F) = r$ **then**
7 $\quad\quad\quad$ $\mathcal{F}^+ := \mathcal{F}^+ \cup \{F\}$;   [Case 1]
8 $\quad\quad$ **else**
9 $\quad\quad\quad$ **if** $r \preceq v$ **then**
10 $\quad\quad\quad\quad$ Find $F \in \mathcal{N}$ such that $\mathrm{lead}(F) \preceq r$;   [Case 2]
11 $\quad\quad\quad\quad$ Find $G \in \mathcal{G}$ such that $\mathrm{span}(G) \succeq v - r$;
12 $\quad\quad\quad\quad$ $q := r - \mathrm{lead}(F), p := \mathrm{span}(G) - v + r$;
13 $\quad\quad\quad\quad$ $\mathcal{F}^+ := \mathcal{F}^+ \cup \{x^q F - \frac{F[[E^u, v]]}{\delta_G} x^p G\}$;
14 $\quad\quad\quad$ **else**
15 $\quad\quad\quad\quad$ Find $F \in \mathcal{N}$ such that $\mathrm{lead}(F) \preceq r$;   [Case 3]
16 $\quad\quad\quad\quad$ $\mathcal{F}^+ := \mathcal{F}^+ \cup \{x^{r - \mathrm{lead}(F)} F\}$;

17 $\quad$ $\mathcal{F} := \mathcal{F}^+$;
18 $\quad$ **if** $\Delta \neq \mathrm{span}(\mathcal{G})$ **then**
19 $\quad\quad$ $\mathcal{G} := \mathcal{G}^+$;

20 **return** $\mathcal{F}$

---

the polynomial F^+ is constructed with $G \in \mathcal{G}$ such that $\mathrm{span}(G) \succeq v - r$, and here $v - r$ only makes sense when $r \preceq v$. In Case 3 when the condition fails, the new polynomial F^+ is constructed with only F.

Comparing Algorithms 1 and 2, one can find several underlying differences between them. These differences are clarified below first, and in fact they are our motivations for the study in the sequel of this paper.

(1) In the BMS algorithm for handling n-dimensional arrays, there are indeed two underlying orders of the terms in $\mathbb{K}[x]$, namely the partial order $\preceq$ and the (total) term order $<_t$ as defined in Sect. 1. In the 1-dimensional case which the BM algorithm is for, these two orders coincide as a total order: $x^i < x^j$ if and only if $i < j$ for two integers i and j. This difference is most influential on the condition $r \preceq v$ in Line 9 in Algorithm 2 for distinguishing Cases 2 and 3. In fact, when the partial order in the n-dimensional case degenerates to the sole total order in the 1-dimensional case, this condition holds trivially for the BM algorithm. This observation indicates that Case 3 of the BMS algorithm only occurs in the n-dimensional case.

(2) As the n-dimensional generalization of the BM algorithm, what the BMS algorithm maintains are multivariate polynomial sets $\mathcal{F}$ and $\mathcal{G}$ instead of univariate polynomials F and G. From Lines 10 and 11 in Algorithm 2, one

may ask whether the polynomials F and G are uniquely determined, as in the trivial case for the BM algorithm where there is only one polynomial F or G to choose. This is studied in Sect. 5.

(3) The new polynomial set $\mathcal{F}^+$ in the BMS algorithm is constructed according to the elements in the set $\mathrm{Ext}(\Delta)$, where $\Delta = \mathrm{span}(\mathcal{G}^+)$. On the contrary, the new polynomial in the BM algorithm is constructed according to the currently handled term only. With this observation we are interested in how the delta sets $\Delta(F)$ in the BM algorithm and $\Delta = \mathrm{span}(\mathcal{G}^+)$ in the BMS algorithm change. This is partially studied below and in Sect. 4.

Lemma 1 ([10, Theorem 2]). *Let* $s = [s_0, \ldots, s_{m-1}]$ *be a sequence of length* m *over* $\mathbb{K}$ *and* $F \in \mathbb{K}[x]$ *be the polynomial in the BM algorithm applied to* s *after handling* $i - 1$, *where* $i \leq m - 1$ *is an integer. If* $F[s, i] \neq 0$, *then for the updated polynomial denoted by* F^+ *after handling* i, *we have* $\deg(F^+) = \max(\deg(F), i + 1 - \deg(F))$.

Proposition 1. *Let* $s = [s_0, \ldots, s_{m-1}]$ *be a sequence of length* m *over* $\mathbb{K}$ *and* $F, F^+ \in \mathbb{K}[x]$ *be the polynomials in the BM algorithm applied to* s *after handling* $i - 1$ *and* i *respectively, where* $i \leq m - 1$ *is an integer. Then* $\Delta(F^+) = \Delta(F)$ *if and only if* $2 \deg(F) > i$.

Proof. By Definition 4, we know that $\Delta(F) = \{0, 1, \ldots, \deg(F) - 1\}$, and thus $\Delta(F^+) = \Delta(F)$ if and only if $\deg(F^+) = \deg(F)$.

(1) If $2 \deg(F) > i$, then $2 \deg(F) \geq i + 1$, and thus $i + 1 - \deg(F) \leq \deg(F)$. By Lemma 1 we know that $\deg(F^+) = \max(\deg(F), i + 1 - \deg(F)) = \deg(F)$.

(2) Otherwise, we have $2 \deg(F) < i + 1$, and thus $i + 1 - \deg(F) > \deg(F)$. In this case $\deg(F^+) = i + 1 - \deg(F) > \deg(F)$. This ends the proof of this proposition. □

Corollary 1. *Let* $s = [s_0, \ldots, s_{m-1}]$ *be a sequence of length* m *over* $\mathbb{K}$ *and* $G, G^+ \in \mathbb{K}[x]$ *be the polynomials in the BM algorithm applied to* s *after handling* $i - 1$ *and* i *respectively. Then* $G^+ = G$ *if and only if* $\Delta(F^+) = \Delta(F)$.

Proof. This is straightforward with Proposition 1 and the observation that G changes only in Case 2.1 of Algorithm 1. □

Comparing the updates of F in Lines 9 and 12 in Algorithm 1 and in Line 13 of Algorithm 2, we can find that they share a similar form. We claim that Case 2.1 and Case 2.2 in Algorithm 1 correspond to Case 2 in Algorithm 2, with more reasons to come in the next section. With the connections and differences clarified between the BM and BMS algorithms, we are able to have a unified flow chart for both algorithm as in Fig. 3. It is worth mentioning that this unified flow chart is almost the same as that for the BMS algorithm, with different cases of the BM and BMS algorithms embedded. In this flow chart, for a set $\mathcal{S}$, the function $\mathrm{pop}(\mathcal{S})$ returns an element in $\mathcal{S}$ and then remove it from $\mathcal{S}$.

In this unified flow chart for both the BM and BMS algorithms, we use polynomial sets $\mathcal{F}$ and $\mathcal{G}$ to represent the polynomials or polynomial sets the

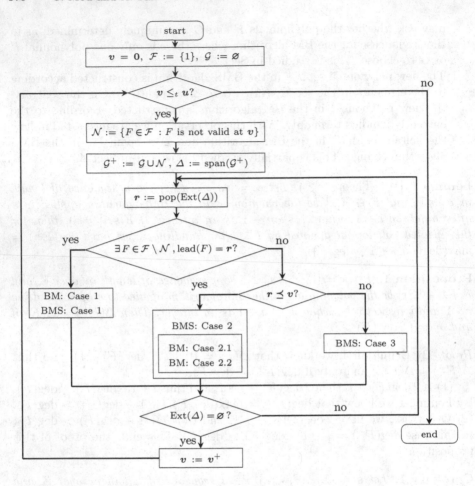

Fig. 3. One flow chart for both BM and BMS algorithms

BM or BMS algorithms maintain respectively. When it goes to the case of the BM algorithm, $\mathcal{F}$ and $\mathcal{G}$ will only contain one polynomial each. We use the n-dimensional term $\boldsymbol{u}$ to represent our last element in the sequence or array in the unified flow chart, and the term $\boldsymbol{v}$ ($\boldsymbol{v} <_t \boldsymbol{u}$) as the running handled element at each iteration. When it goes to the case of the BM algorithm, the terms $\boldsymbol{u}$ and $\boldsymbol{v}$ degenerate to $m-1$ and i ($i \leq m-1$) for a sequence of length m.

The tests of whether the polynomial F or all the polynomials in $\mathcal{F}$ are valid up to $\boldsymbol{v}$ in the BM or BMS algorithm is unified as whether $\mathcal{N} = \varnothing$ or not. Furthermore, Corollary 2 in Sect. 4 indicates that the delta set $\mathrm{span}(\mathcal{G}^+)$ is equal to $\Delta(F)$ when the unified algorithm degenerates to the case of the BM algorithm, and thus the unification with Δ is justified.

The BMS algorithm needs to traverse all the exterior corners in $\mathrm{Ext}(\Delta)$ but in the case of the BM algorithm, $\mathrm{Ext}(\Delta)$ contains a single element and thus there is only one iteration in the loop. Kept in mind that $\mathcal{F}$ in the unified flow chart contains only one polynomial in the case of the BM algorithm, it is straightforward that the test "$\exists F \in \mathcal{F} \setminus \mathcal{N}, \operatorname{lead}(F) = r$?" in the BMS algorithm corrsponds to $d = 0$ in Line 4 in Algorithm 1 for the case of the BM algorithm.

When the test "$\exists F \in \mathcal{F} \setminus \mathcal{N}, \operatorname{lead}(F) = r$" passes, the unified flow chart falls to the left path. As discussed above, both Cases 1 in the BM and BMS algorithms are for the situation when neither $\mathcal{F}$ nor $\mathcal{G}$ changes, and thus they are classified in one step in the unified flow chart. When the test does not pass, the unified flow chart falls to the right path. Then the test "$r \preceq v$?" in the unified flow chart corresponds to Line 9 in the BMS algorithm and is trivially satisfied in the BM algorithm (therefore the BM algorithm always falls to the left path after this test). This observation has been discussed above in this section.

It remains to justify the correspondence between Case 2 of the BMS algorithm and Cases 2.1 and 2.2 in the BM algorithm. Theorem 1 and Proposition 2 in Sect. 4 tell us that the BMS algorithm, when applied to a sequence, maintains polynomials which are reciprocal to those polynomials maintained in the BM algorithm applied to the same sequence, and this justifies the strong connections between the update expressions in Line 13 in Case 2 of Algorithm 2 and Line 9 in Case 2.1 and Line 12 in Case 2.2 of Algorithm 1. Furthermore, as discussed above, Cases 2.1 and 2.2 of the BM algorithm are distinguished by whether G changes or not, and by Corollary 1 we see that G changes if and only if the delta set changes. Then it is noted in Line 18 of Algorithm 2 that the same latter condition, namely whether the delta set changes, holds if and only $\mathcal{G}$ changes, and this shows that the BMS algorithm handles Cases 2.1 and 2.2 of the BM algorithm in a unified way.

4   Comparing the BM and BMS Algorithms Applied to a Sequence

In this section we study the behaviors of the BMS algorithm applied to a sequence. In other words, we study what happens when this n-dimensional algorithm degenerates into the 1-dimensional case.

We say that two univariate polynomials F and G in $\mathbb{K}[x]$ are *reciprocal* to each other if $F(x) = x^{\deg(F)} G(1/x)$, and we denote the reciprocal polynomial of $F \in \mathbb{K}[x]$ by F^r.

Let $s = [s_0, \ldots, s_{m-1}]$ be a sequence of length m over $\mathbb{K}$. Now apply both the BM and BMS algorithms to s, and after handling s_k for some k ($0 \leq k \leq m-1$), denote the polynomials F and G in the BM algorithm by F_k and G_k, and the polynomial sets $\mathcal{F} = \{F\}$ and $\mathcal{G} = \{G\}$ in the BMS algorithm by $\{\overline{F}_k\}$ and $\{\overline{G}_k\}$. The main result of this section is that F_k and $\overline{F}_k$, as well as G_k and $\overline{G}_k$, are proved to be reciprocal.

To prove these relations, we have a lemma below first. Let $s = [s_0, \ldots, s_{m-1}]$ be a sequence of length m over $\mathbb{K}$ such that $s_0 \neq 0$. For $k = 1, \ldots, m-1$, define the following polynomial sets

$$V[s(k)] = \{F \in \mathbb{K}[x] : F \in V[s, k] \text{ and } F(0) = 1\},$$
$$V[[s(k)]] = \{F \in \mathbb{K}[x] : F \in V[[s, k]] \text{ and } F \text{ is monic}\}. \tag{1}$$

Lemma 2. *Let $s = [s_0, \ldots, s_{m-1}]$ be a sequence of length m over $\mathbb{K}$ such that $s_0 \neq 0$, and the polynomial sets $V[s(k)]$ and $V[[s(k)]]$ be defined as in (1) for $k = 1, \ldots, m-1$. Then for any polynomial $F \in V[s(k)]$, its reciprocal polynomial F^r is in $V[[s(k)]]$, and vice versa.*

Proof. Write $F \in \mathbb{K}[x]$ as $F = \sum_{j=0}^{d} c_j x^j$ and $F^r = \sum_{j=0}^{d} \overline{c}_j x^j$. Since F and F^r are reciprocal, we have $c_j = \overline{c}_{d-j}$ for each $j = 0, \ldots, d$. Then for each $i = 1, \ldots, k$, by definition we have

$$F^r[[s, i]] = \sum_{j=0}^{d} \overline{c}_j s_{j+i-d} = \sum_{j=0}^{d} c_{d-j} s_{j+i-d} = \sum_{j'=0}^{d} c_{j'} s_{i-j'} = F[s, i].$$

Now it is clear that when $F \in V[s(k)]$, we have $F^r \in V[[s(k)]]$, and vice versa. $\qquad\square$

Theorem 1. *Let $s = [s_0, \ldots, s_{m-1}]$ be a sequence of length m over $\mathbb{K}$ such that $s_0 \neq 0$. Then F_k and $\overline{F}_k$ are reciprocal for each $k = 1, \ldots, m-1$.*

Proof. For each $k = 1, \ldots, m-1$, we can regard F_k and $\overline{F}_k$ as the polynomials returned by the BM and BMS algorithms applied to the truncated sequence $[s_0, \ldots, s_k]$, namely $F_k = \mathsf{BM}([s_0, \ldots, s_k])$ and $\overline{F}_k = \mathsf{BMS}([s_0, \ldots, s_k])$. Then by the specifications of the BM and BMS algorithms, we know that F_k and $\overline{F}_k$ are respectively polynomials in $V[s(k)]$ and $V[[s(k)]]$ of minimal degrees.

We first show that $\overline{F}_k$ is the unique polynomial of minimal degree in $V[[s(k)]]$. Otherwise suppose that there exists a distinct polynomial $\tilde{F} \in V[[s(k)]]$ of the same degree as $\overline{F}_k$. Then let $\tilde{F}'$ be the polynomial $\overline{F}_k - \tilde{F}$ divided by the leading coefficient. Since both $\overline{F}_k$ and $\tilde{F}$ in $V[[s(k)]]$ are monic, we know that the polynomial $\tilde{F}'$ constructed above is monic and of strictly smaller degree than $\overline{F}_k$, and one can further check that $\tilde{F}'$ also belongs to $V[[s(k)]]$. This contradicts with the fact that $\overline{F}_k$ is of the minimal degree in $V[[s(k)]]$.

Next we prove the equality $F_k^r = \overline{F}_k$ by showing that F_k^r is a polynomial in $V[[s(k)]]$ of the minimal degree. By Lemma 2 we know that $F_k^r \in V[[s(k)]]$, and clearly $\deg(F_k^r) = \deg(F_k)$. Suppose now that F_k^r is not of the minimal degree in $V[[s(k)]]$. Then $\deg(\overline{F}_k) < \deg(F_k^r) = \deg(F_k)$, and thus $\deg(\overline{F}_k^r) < \deg(F_k)$. By Lemma 2 we know that $\overline{F}_k^r \in V[s(k)]$, and this contradicts the fact that F_k is of the minimal degree in $V[s(k)]$. Since $F_k^r \in V[[s(k)]]$ is of the minimal degree and $\overline{F}_k$ is the unique polynomial of minimal degree in $V[[s(k)]]$, the equality $F_k^r = \overline{F}_k$ follows. $\qquad\square$

Remark 2. Note that the reciprocal relation does not hold for F_0 and $\overline{F}_0$ as defined in Theorem 1. These two polynomials in the BM and BMS algorithms are computed as follows. When handling s_0, it is not hard to check that the BM algorithm goes to Case 2.1 to have $F_0 = 1 - s_0 x$ and that the BMS algorithm goes to Case 3 to have $\overline{F}_0 = x$.

Corollary 2. *Let $s = [s_0, \ldots, s_{m-1}]$ be a sequence of length m over $\mathbb{K}$ such that $s_0 \neq 0$. Then $\Delta(F_k) = \overline{\Delta}_k$, where $\overline{\Delta}_k = \mathrm{span}(\{\overline{G}_k\})$.*

Proof. By Definition 4 and Theorem 1, $\text{Ext}(\Delta(F_k)) = \deg(F_k) = \deg(\overline{F}_k) = \text{Ext}(\overline{\Delta}_k)$, and the conclusion follows.   □

Proposition 2. *Let* $s = [s_0, \ldots, s_{m-1}]$ *be a sequence of length* m *over* $\mathbb{K}$ *such that* $s_0 \neq 0$. *Then* G_k *and* $\overline{G}_k$ *are reciprocal for each* $k = 1, \ldots, m-1$.

Proof. For $k = 0, \ldots, m-1$, let $\overline{\Delta}_k := \text{span}(\{\overline{G}_k\})$. With Line 9 in Algorithm 1 and Corollary 1 we know that $G_k = F_{k-1}$ if $\Delta(F_k)$ changes and $G_k = G_{k-1}$ otherwise. With Line 18 in Algorithm 2, we know that $\overline{G}_k = \overline{F}_{k-1}$ if $\overline{\Delta}_k$ changes and $\overline{G}_k = \overline{G}_{k-1}$ otherwise. Then Corollary 2 implies that G_k and $\overline{G}_k$ change at the same time. With the reciprocal relation between F_k and $\overline{F}_k$ in Theorem 1, the conclusion follows.   □

5   Updates in the BMS Algorithm

Let E^u be an n-dimensional array over $\mathbb{K}$ up to a term $u \in \mathbb{Z}_+^n$. For a term $v <_t u$, let $\mathcal{F}_{v-}$ and $\mathcal{G}_{v-}$ be the polynomial sets $\mathcal{F}$ and $\mathcal{G}$ of the BMS algorithm applied to E^u after handling v^-, and $\mathcal{N}$ and Δ be constructed in Lines 3 and 4. Then for each $r \in \text{Ext}(\Delta)$, define

$$\mathcal{N}_r := \{F \in \mathcal{N} : \text{lead}(F) \preceq r\}, \quad \mathcal{G}_r := \{G \in \mathcal{G}_{v-} : \text{span}(G) \succeq v + r\}. \quad (2)$$

By viewing Lines 9 and 16, the correctness of the BMS algorithm relies on the fact that for each r, $\mathcal{N}_r \neq \varnothing$ and $\mathcal{G}_r \neq \varnothing$, which is, for example, proved in the proof of [13, Lemma 37]. Next we investigate the number of polynomials in $\mathcal{G}_r$ and prove that there is only one polynomial in $\mathcal{G}_r$ for each $r \in \text{Ext}(\Delta)$ in a certain case.

Theorem 2. *Let* E^u *be an* n-*dimensional array over* $\mathbb{K}$ *up to a term* $u \in \mathbb{Z}_+^n$. *For a term* $v \leq_t u$, *let* Δ_v *and* Δ_{v-} *be the delta set in the BMS algorithm after handling the term* v *and* v^- *respectively, and* $\mathcal{G}_r$ *be as defined in* (2). *Then for each* $r \in \text{Ext}(\Delta_v) \setminus \text{Ext}(\Delta_{v-})$ *in Case 2 of Algorithm 2, we have* $\#\mathcal{G}_r = 1$.

Proof. With the one-one correspondences between $\mathcal{G}_{v-}$ and $\{\text{span}(G) : G \in \mathcal{G}_{v-}\}$ and between $\{\text{span}(G) : G \in \mathcal{G}_{v-}\}$ and $\text{Int}(\Delta_{v-})$, it suffices to show that there only exists one interior corner of Δ_{v-} which is $\succeq v + r$.

Since $r \in \text{Ext}(\Delta_v)$ but $r \notin \text{Ext}(\Delta_{v-})$, there exists a term w in both $\text{Ext}(\Delta_{v-})$ and $\text{Int}(\Delta_v)$ (the term which transforms from an interior corner to an exterior one after v is handled) and an integer i ($1 \leq i \leq n$) such that $r = w + e_i$, where

$$e_i = (0, \ldots, 0, \underbrace{1}_{ith}, 0, \ldots, 0).$$

Since w is in both $\text{Ext}(\Delta_{v-})$ and $\text{Int}(\Delta_v)$, we know that there exists a polynomial $F \in \mathcal{F}_{v-}$ such that F is not valid at v (namely $F \in \mathcal{N}_v$), $\text{lead}(F) = w$, and $\text{span}(F) = w$. This implies that $r = w + e_i = v - w + e_i$, and thus $v - r = w - e_i$.

We claim that the ith component of w, denoted by w_i, is not equal to 0. Otherwise the ith component of $w - e_i$ is -1, and the condition $\text{span}(G) \succeq$

$v + r = w - e_i$ in (2) is equivalent to $\text{span}(G) \succeq w$. Since w itself is in $\text{Ext}(\Delta_{v^-})$, we know that there is no interior corner of Δ_{v^-} which $\succeq w$. This contradicts with the non-emptiness of $\mathcal{G}_r$ proved in [13, Lem 37]. Now $w_i \geq 1$, and it is easy to show that $w - e_i$ is on the border of Δ_{v^-} (since $(w - e_i) + e_i = w \notin \Delta_{v^-}$), and thus there exists only one interior corner of Δ_{v^-} which $\succeq w - e_i$.   □

Next we show with the following example that the number of elements in $\mathcal{N}_r$ and that of $\mathcal{G}_r$ are not necessarily 1 in the cases other than the one discussed in Theorem 2. Let E^u be an n-dimensional array up to $u = (0, 2)$ over a field $\mathbb{K}$ as shown in Fig. 4. When the BMS algorithm is applied to E^u, the polynomial sets $\mathcal{F}$ and $\mathcal{G}$ and the exterior corners of the delta sets are recorded in Table 1.

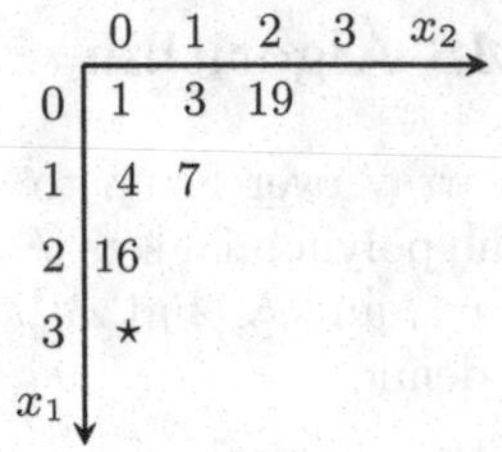

$$
\begin{array}{c|cccc}
 & 0 & 1 & 2 & 3 \quad x_2 \\
\hline
0 & 1 & 3 & 19 & \\
1 & 4 & 7 & & \\
2 & 16 & & & \\
3 & \star & & & \\
x_1 & & & &
\end{array}
$$

Fig. 4. A 2-dimensional array E^u up to $u = (0, 2)$

Table 1. BMS algorithm applied to E^u in Fig. 4

		BMS algorithm		
		$\mathcal{F} := \{1\}, \mathcal{G} := \varnothing$		
v	E_v	$\mathcal{G}$	$\text{Ext}(\Delta)$	$\mathcal{F}$
(0,0)	1	$\{1\}$	$\{(1,0),(0,1)\}$	$\{x_1, x_2\}$
(1,0)	4	$\{1\}$	$\{(1,0),(0,1)\}$	$\{F_1, x_2\}$
(0,1)	3	$\{1\}$	$\{(1,0),(0,1)\}$	$\{F_1, F_2\}$
(2,0)	16	$\{1\}$	$\{(1,0),(0,1)\}$	$\{F_1, F_2\}$
(1,1)	7	$\{F_1, F_2\}$	$\{(2,0),(1,1),(0,2)\}$	$\{F_3, F_4, F_5\}$
(0,2)	19	$\{F_1, F_2\}$	$\{(2,0),(1,1),(0,2)\}$	$\{F_3, F_4, F_6\}$

The polynomials appearing in Table 1 are listed below.

$$F_1 = x_1 - 4, \quad F_2 = x_2 - 3, \quad F_3 = x_1^2 - 4x_1,$$
$$F_4 = x_1 x_2 - 4x_2 + 5 \text{ or } x_1 x_2 - 3x_1 + 5, \quad F_5 = x_2^2 - 3x_2,$$
$$F_6 = x_2^2 - 3x_2 + 2x_1 x_2 - 6x_1 \text{ or } x_2^2 + 2x_1 x_2 - 11x_2.$$

When the BMS algorithm handles $v = (1, 1)$, we find that $F_1[[E^u, v]] = -5 \neq 0$ and $F_2[[E^u, v]] = -5 \neq 0$, and thus $\mathcal{N} = \{F_1, F_2\}$ in Line 3 of Algorithm 2.

For the exterior corner $r = (1,1) \in \mathrm{Ext}(\Delta)$, Algorithm 2 falls into Case 2 because $r = (1,1) \preceq (1,1) = v$, and one can check that $\mathrm{lead}(F_1) \preceq r$ and $\mathrm{lead}(F_2) \preceq r$. This shows that in general $\mathcal{N}_r$ defined in (2) does not necessarily contain only one element.

Next consider the term $v = (0,2)$, we find that F_3 and F_4 are valid at v but $F_5[[E^u, v]] = 10 \neq 0$, and thus $\mathcal{N} = \{F_5\}$. Note that $\mathrm{span}(F_5) = v - \mathrm{lead}(F_5) = (0,2) - (0,2) = (0,0)$, and thus Δ does not change at $(0,2)$. For the exterior corner $r = (0,2) \in \mathrm{Ext}(\Delta)$, Algorithm 2 falls into Case 2 because $r = (0,2) \preceq (0,2) = v$, and one can check that $\mathrm{span}(F_1) = (0,1) \succeq (0,0) = v - r$ and $\mathrm{span}(F_2) = (0,1) \succeq (0,0) = v - r$. This shows that when Δ in the BMS algorithm does not change, $\mathcal{G}_r$ defined in (2) does not necessarily contain only one element, and thus the condition on r in Theorem 2 is necessary.

6 Concluding Remarks

In this paper we clarify some underlying connections and differences between the BM and BMS algorithms. The main contributions of this paper lie in (1) the unification of these two algorithms in terms of their underlying languages and a unified flow chart, which is justified by proved theoretical results on these two algorithms throughout Sects. 3 and 4, and (2) identification of some differences between these two algorithms including Case 3 of the BMS algorithm, reversed coefficient orders in the definitions of valid polynomials, and the non-uniqueness of two critical polynomial sets in the BMS algorithm. These contributions are expected to provide a better understanding on the BMS algorithm against the BM one from an algebraic viewpoint.

Acknowledgments. The authors would like to thank the anonymous reviewers for their helpful suggestion which contribute to considerable improvement of this paper.

References

1. Berlekamp, E.: Nonbinary BCH decoding. IEEE Trans. Inf. Theory **14**(2), 242–242 (1968)
2. Berthomieu, J., Boyer, B., Faugère, J.C.: Linear algebra for computing Gröbner bases of linear recursive multidimensional sequences. J. Symb. Comput. **83**, 36–67 (2017)
3. Berthomieu, J., Faugère, J.C.: In-depth comparison of the Berlekamp-Massey-Sakata and the scalar-FGLM algorithms: The adaptive variants. arXiv:1806.00978 (2018, preprint)
4. Berthomieu, J., Faugère, J.C.: A polynomial-division-based algorithm for computing linear recurrence relations. In: Proceedings of ISSAC 2018, pp. 79–86. ACM Press (2018)
5. Bras-Amorós, M., O'Sullivan, M.: The correction capability of the Berlekamp-Massey-Sakata algorithm with majority voting. Appl. Algebra Eng. Commun. Comput. **17**(5), 315–335 (2006)

6. Cox, D., Little, J., O'Shea, D.: Using Algebraic Geometry, 2nd edn. Springer Verlag, New York (2005). https://doi.org/10.1007/b138611

7. Faugère, J.C., Mou, C.: Fast algorithm for change of ordering of zero-dimensional Gröbner bases with sparse multiplication matrices. In: Proceedings of ISSAC 2011, pp. 115–122. ACM, ACM Press (2011)

8. Faugère, J.C., Mou, C.: Sparse FGLM algorithms. J. Symb. Comput. **80**(3), 538–569 (2017)

9. Kaltofen, E., Yuhasz, G.: On the matrix Berlekamp-Massey algorithm. ACM Trans. Algorithms **9**(4), 33 (2013)

10. Massey, J.: Shift-register synthesis and BCH decoding. IEEE Trans. Inf. Theory **15**(1), 122–127 (1969)

11. Matsui, H., Sakata, S., Kurihara, M., Mita, S.: Systolic array architecture implementing Berlekamp-Massey-Sakata algorithm for decoding codes on a class of algebraic curves. IEEE Trans. Inf. Theory **51**(11), 3856–3871 (2005)

12. O'Sullivan, M.: New codes for the Berlekamp-Massey-Sakata algorithm. Finite Fields Appl. **7**(2), 293–317 (2001)

13. Saints, K., Heegard, C.: Algebraic-geometric codes and multidimensional cyclic codes: a unified theory and algorithms for decoding using Gröbner bases. IEEE Trans. Inf. Theory **41**(6), 1733–1751 (2002)

14. Sakata, S.: Finding a minimal set of linear recurring relations capable of generating a given finite two-dimensional array. J. Symb. Comput. **5**(3), 321–337 (1988)

15. Sakata, S.: Extension of the Berlekamp-Massey algorithm to N dimensions. Inf. Comput. **84**(2), 207–239 (1990)

16. Sakata, S.: The BM algorithm and the BMS algorithm. In: Vardy, A. (ed.) Codes, Curves, and Signals: Common Threads in Communications. SECS, vol. 485, pp. 39–52. Springer, Boston (1998). https://doi.org/10.1007/978-1-4615-5121-8_4

17. Wiedemann, D.: Solving sparse linear equations over finite fields. IEEE Trans. Inf. Theory **32**(1), 54–62 (1986)

An Algorithm for Solving a Quartic Diophantine Equation Satisfying Runge's Condition

N. N. Osipov$^{(\boxtimes)}$ and S. D. Dalinkevich

Siberian Federal University, Svobodny 79, Krasnoyarsk 660041, Russia
nnosipov@gmail.com, rgxwarer0@gmail.com

Abstract. In this paper, we suggest an implementation of elementary version of Runge's method for solving a family of diophantine equations of degree four. Moreover, the corresponding solving algorithm (in its optimized version) is implemented in the computer algebra system PARI/GP.

Keywords: Diophantine equations · Runge's method · Computer algebra systems

1 Introduction

There is a wide class of diophantine equations in two variables

$$f(x, y) = 0 \tag{1.1}$$

for which one can propose an *effective* solving method (that provides explicit upper bounds for the size of integer solutions), the so-called *Runge's method*. A description of the standard version of Runge's method can be found in the well-known monographs [4] and [10] (for more detailed proof, see [3, Ch. 4]). The original version (see old Runge's paper [9] or a modern paper [12]) is more general, below we give main theoretical result (so-called *Runge's theorem*). Despite the fact that Runge's method has been known for more than 100 years, its implementation in *computer algebra systems* (CAS) is very limited. At the same time, there is a small number of publications (see [6,8,11] and, especially, [1]) which refer to algorithmic aspects of implementation of this method (at least for some special cases) in CAS.

Assume that the polynomial $f(x, y) \in \mathbb{Z}[x, y]$ is irreducible over $\mathbb{Q}$ and let $d_0 = \max\{m, n\}$ where $m = \deg_x f(x, y)$ and $n = \deg_y f(x, y)$. If $f(x, y)$ satisfies Runge's condition (see below), then the estimate

$$\max\{|x|, |y|\} < (2d_0)^{18d_0^7} h^{12d_0^6} \tag{1.2}$$

holds for all integer solutions (x, y) of the Eq. (1.1) (see [12]). As usually, h denotes the *height* of given polynomial. This general result shows that the trivial

© Springer Nature Switzerland AG 2019
M. England et al. (Eds.): CASC 2019, LNCS 11661, pp. 377–392, 2019.
https://doi.org/10.1007/978-3-030-26831-2_25

implementation (brute force in the mentioned bounds) makes no sense in terms of the time required even in the case of d_0 small enough.

Let

$$f(x,y) = \sum_{i=0}^{m} \sum_{j=0}^{n} a_{ij} x^i y^j \tag{1.3}$$

be an irreducible polynomial in $\mathbb{Z}[x,y]$.

Runge's theorem. Assume that the Eq. (1.1) has infinitely many solutions $(x,y) \in \mathbb{Z}^2$. Then each of the following conditions holds:
   (a) $a_{in} = a_{mj} = 0$ for all $i > 0$ and $j > 0$,
   (b) $a_{ij} = 0$ for all pairs (i,j) satisfying $ni + mj > mn$,
   (c) the leading part

$$f_L(x,y) = \sum_{ni+mj=mn} a_{ij} x^i y^j$$

is a constant multiple of a power of an irreducible polynomial in $\mathbb{Z}[x,y]$,
   (d) the algebraic function $y = \Psi(x)$ defined by (1.1) has only one class of conjugate Puiseux expansions.

We say that polynomial (1.3) satisfies *Runge's condition*, if at least one of the conditions (a), (b), (c) or (d) does not hold. Runge's theorem can be reformulated in the following equivalent form: if $f(x,y)$ satisfies Runge's condition, then Eq. (1.1) has a finite set of integer solutions. In the literature, the following simplified version of this theorem is widely known. Denote by $f_d(x,y)$ the *leading homogeneous part* of polynomial (1.3), $d = \deg f(x,y)$.

Corollary. If $f_d(x,y)$ can be decomposed into a product of non-constant relatively prime polynomials in $\mathbb{Z}[x,y]$, then Eq. (1.1) has a finite set of integer solutions.

Below, the condition of Corollary will be called the *standard Runge's condition*. Under standard Runge's condition, in the case $d = 3$, a realistic (practically working) solving algorithm was proposed in [6]. This algorithm is based on the *elementary version of Runge's method* for diophantine equations of degree $d \leqslant 4$ (see [5]). In the case $d = 4$, an algorithmic implementation of elementary version of Runge's method is obtained only in some particular cases (see [8] and more recent paper [7]). It is necessary to refer to preprint [1] where it is proposed to avoid "the use of Puiseux series and algebraic coefficients" which leads to "bad" estimates (i.e., estimates of type (1.2)) for integer solutions.

The elementary version of Runge's method for diophantine equations of small degree is based on a convenient parametrization (by means of a special integer parameter) which provides enumerating possible integer solutions. As a result, the resolution of diophantine equation can be reduced to solving finitely many equations in one variable (usually, of degree two) over the integers. This idea for algorithmic implementation of Runge's method was applied in [6,7].

In our paper, we consider a family of diophantine Eq. (1.1) with the left-hand side

$$f(x, y) = (a_1 x + b_1) y^2 + (a_2 x^2 + b_2 x + c_2) y + A x^4 + B x^3 + C x^2 + D x + E. \quad (1.4)$$

By default we assume this polynomial to be irreducible in $\mathbb{Z}[x, y]$. In general case, both coefficients a_1 and A are non-zero and Runge's method can be applied because the condition (a) of Runge's theorem is violated.

In Sect. 2, we propose solving algorithm in the main case $a_1 = 1$ and $b_1 = 0$ (i.e., for Eq. (2.3), see below). This algorithm is inspired by Theorem 2.1. Technically, this algorithm differs from similar algorithms introduced in [6, 7] since it requires to resolve a number of equations in one variable of degree three. This fact must be taken into account if we want to estimate correctly the complexity of an algorithm. Therefore, we introduce an additional parameter (the so-called weight coefficient) for correct estimation of computational complexity. The weight coefficient depends on the CAS in which we plan to implement our algorithm (PARI/GP, see [13]). Further, we optimize the proposed algorithm in the same way as in [6]. The final result is represented in Theorem 2.2. At the moment, we do not know any other implementations of algorithms for solving diophantine equations of the specified type.

In Sect. 3, we give a few examples of estimating integer solutions to several diophantine equations of small degree. In the case $d = 4$, the used method does not allow the "reasonable" estimates (i.e., estimates which are close to realistic) for integer solutions, therefore, we do not give any general theorems (we refer to [6] where the reader can find relevant examples of such theorems).

In Sect. 4, we give some remarks on the obtained results. In particular, we consider different ways to construct solving algorithm for Eq. (1.1) with $f(x, y)$ of the general form (1.4). Also, we discuss a further application of the elementary version of Runge's method for diophantine equations of degree four.

2 Solving Algorithm

We begin with the case $a_1 = 0$ which is trivial in certain sense. In this case, we can improve the well-known solving algorithm (see, e.g., [8]).

2.1 The Equation $z^2 = P(x)$

In the case $a_1 = 0$ and $b_1 \neq 0$, Eq. (1.1) with polynomial (1.4) can be reduced to the equation

$$z^2 = P(x) \quad (2.1)$$

with the polynomial $P(x) \in \mathbb{Z}[x]$ which satisfies $\deg P(x) \leqslant 4$. Runge's method works for Eq. (2.1) in the case when $\deg P(x) = 4$ and the leading coefficient of $P(x)$ is a perfect square in $\mathbb{Z}$ (here, we can assume, without loss of generality, that $P(x)$ is monic). Otherwise, we need to refer to more complicated methods (see, for instance, [10]; of course, with the exception of the case $\deg P(x) \leqslant 2$ which is well studied).

We now consider Eq. (2.1) with the polynomial

$$P(x) = x^4 + ax^3 + bx^2 + cx + d.$$

A well-known algorithm for solving (2.1) with this $P(x)$ was described in [8]. Below, we refer to this algorithm as the *standard algorithm* (or *method*). Here, we propose the following alternative approach. First, we reduce Eq. (2.1) to a certain cubic diophantine equation. Next, we resolve the corresponding cubic equation using the technique from [6]. Sometimes, this approach is more effective than the standard method (for details, see Sect. 4). We demonstrate this phenomenon in the following example.

Example 2.1. Consider the equation

$$z^2 = x^4 + 8Hx^3 - 12x^2 + 4, \tag{2.2}$$

where the coefficient $H \geqslant 1$ is supposed to be rather large. Note that Eq. (2.2) was first mentioned in the short note [2]. This equation has a solution $(x, z) \in \mathbb{Z}^2$ with

$$x = 4H^3 - 2H$$

that is quite large with respect to H. At the same time, it was proved (see ibid) that the upper bound for the integer solutions (x, z) of (2.1) is

$$|x| < 26h^3$$

where h is the height of $P(x)$. Thus, Eq. (2.2) with $h = 8H$ has the biggest solution (up to a constant factor) with respect to the upper bound mentioned above. The direct computation shows that the standard solving algorithm (see [8]) needs $\approx 64H^3$ operations of taking square root for the integers with the maximal value $O(H^{12})$, and it is unexpected that Eq. (2.2) can be solved faster.

Namely, for $P(x) = x^4 + 8Hx^3 - 12x^2 + 4$, we determine

$$R(x) = x^2 + 4Hx - 8H^2 - 6 = \sqrt{P(x)} + O\left(\frac{1}{x}\right), \qquad x \to \infty.$$

Next, we introduce the new variable $w = z - R(x)$ and rewrite (2.2) in the form $F(w, x) = 0$ with the cubic polynomial

$$F(w, x) = (R(x) + w)^2 - P(x) = 2wx^2 + w^2 + 8Hwx +$$
$$+ (-16H^2 - 12)w + (-64H^3 - 48H)x + 64H^4 + 96H^2 + 32.$$

Omitting technical details, we can formulate the final result as follows. The solving algorithm from [6] which is optimized for the equation $F(w, x) = 0$ "by hands" (i.e., analytically) requires only $\approx 96H^2$ operations of taking square root for the integers with the maximal value $O(H^6)$. This may appear surprising, especially because the height of $F(w, x)$ is much larger than the height of $P(x)$.

It is easy to see that the right-hand side of (2.2) is a perfect square for

$$x \in \{0, H, 4H^3 - 2H\}.$$

Therefore, Eq. (2.2) has at least 6 integer solutions (x, y). In Table 1, we represent a certain statistical information on the number of additional (non-trivial) solutions for H taking values in the range $1 \leqslant H \leqslant 500$.

Table 1. Distribution of the number of non-trivial solutions of Eq. (2.2) in the range $1 \leqslant H \leqslant 500$.

$\#(x, z)$	$\#H$
0	393
2	85
4	20
6	1
8	1

2.2 The Main Case

Suppose that $a_1 \neq 0$. In the case $A = 0$, we obtain a cubic diophantine equation with the leading homogenous part $x(Bx^2 + a_2xy + a_1y^2)$ satisfying the standard Runge's condition. Thus, we can use the algorithmic implementation of elementary version of Runge's method proposed in [6]. Therefore, we can suppose that $A \neq 0$.

For simplicity, here we consider in detail only the particular case $a_1 = 1$, $b_1 = 0$ (the general case will be discussed briefly in Sect. 4). Then, the equation can be written as

$$xy^2 + (ax^2 + bx + c)y + Ax^4 + Bx^3 + Cx^2 + Dx + E = 0 \qquad (2.3)$$

(we use simplified notation for convenience). Also, we can suppose that $c \neq 0$ (otherwise, the possible integer values of x must be in the set of divisors of E which can be found). Assuming $x \neq 0$, consider the number

$$l = \frac{cy + E}{x}.$$

Clearly, the value of l must be integer for all the solutions $(x, y) \in \mathbb{Z}^2$ of Eq. (2.3) with $x \neq 0$. Dividing by x, we obtain

$$y^2 + (ax + b)y + Ax^3 + Bx^2 + Cx + D + l = 0.$$

This equality implies the congruence

$$y^2 + by + D + l \equiv 0 \pmod{x}$$

in the ring $\mathbb{Z}$ of integers. Next, we have

$$c^2(y^2 + by + D) \equiv c^2 D + E^2 - bcE \pmod{cy + E}$$

(here we mean the congruence in the polynomial ring $\mathbb{Z}[y]$). Taking into account that

$$cy + E \equiv 0 \pmod{x},$$

we arrive at another congruence

$$c^2 l + c^2 D + E^2 - bcE \equiv 0 \pmod{x}$$

(both congruences are in the ring $\mathbb{Z}$). Finally, we set

$$k = \frac{c^2 l + c^2 D + E^2 - bcE}{x} = \frac{c^3 y + (c^2 D + E^2 - bcE)x + c^2 E}{x^2}.$$

If (x, y) is an arbitrary integer solution of Eq. (2.3) then the value of k must be integer as well as the value of l. Thus, we obtain the following result.

Theorem 2.1. *Let* $(x, y) \in \mathbb{Z}^2$ *be a solution of Eq. (2.3) with* $x \neq 0$*. Then, the number*

$$k = \frac{c^3 y + (c^2 D + E^2 - bcE)x + c^2 E}{x^2} \tag{2.4}$$

is integer.

One can propose the following straightforward and shorter proof of Theorem 2.1 which can be obtained by computer algebra methods (i.e., using symbolic computations in a computer algebra system). Using Eq. (2.3), we find the expression for the coefficient E:

$$E = -xy^2 - (ax^2 + bx + c)y - Ax^4 - Bx^3 - Cx^2 - Dx.$$

Next, we plug it into the right-hand side of (2.4). After dividing the numerator of the fraction in (2.4) by x^2, we obtain the explicit (but rather large) expression for k as a polynomial in the ring $\mathbb{Z}[x, y]$. Hence, the value of k must be integer. In order to illustrate this method, consider the equation

$$xy^2 + (x^2 + 1)y + x^4 + 1 = 0$$

with the polynomial $f(x, y) = xy^2 + (x^2 + 1)y + x^4 + 1$. We want to prove that the number

$$k = \frac{y + x + 1}{x^2}$$

is integer for each solution $(x, y) \in \mathbb{Z}^2$ with $x \neq 0$. Indeed, using the method described above we obtain

$$\frac{y + x + 1}{x^2} = xy^4 + (2x^2 + 2)y^3 + (2x^4 + x^3 + 2x)y^2 + (2x^5 + 2x^3 - 1)y + x^7 - x^2$$

which can be viewed as an equality in the residue class ring of $\mathbb{Z}[x, y]$ modulo $f(x, y)$. Note that this representation can be simplified:

$$\frac{y + x + 1}{x^2} = y^3 + (x - 1)y^2 + (x^3 - x - 1)y - x^3 - x^2.$$

Our further reasoning is based on the following idea. It is easy to check that both explicit real solutions $y = \Psi_i(x)$ $(i = 1, 2)$ of Eq. (2.3) admit the estimate

$$\Psi_i(x) = O(|x|^{3/2}), \quad x \to \infty.$$

Hence, we have

$$\frac{c^3 \Psi_i(x) + (c^2 D + E^2 - bcE)x + c^2 E}{x^2} \to 0, \quad x \to \infty.$$

As a corollary, for any $m \geqslant 1$, there exists a number $Q = Q(m) > 0$ such that

$$\left| \frac{c^3 \Psi_i(x) + (c^2 D + E^2 - bcE)x + c^2 E}{x^2} \right| < Q(m)$$

for any x satisfying $|x| > m$ (of course, here we can use only those values of x for which $\Psi_i(x)$ are defined). Using this assertion, we can propose the following algorithm for solving Eq. (2.3) over the integers.

Solving algorithm

1. Choose $m \geqslant 1$ and compute the number $Q(m)$.
2. For all integers x satisfying $|x| \leqslant m$, solve Eq. (2.3) (as a quadratic equation in y) over the integers.
3. For all integers k with $|k| < Q(m)$, solve the system of equations

$$\begin{cases} xy^2 + (ax^2 + bx + c)y + Ax^4 + Bx^3 + Cx^2 + Dx + E = 0, \\ c^3 y + (c^2 D + E^2 - bcE)x + c^2 E - kx^2 = 0 \end{cases} \tag{2.5}$$

over the integers.

Let us consider an example in order to illustrate the proposed method.

Example 2.2. We show in detail how the equation

$$x^4 - x^2 y - xy^2 - y^2 + 1 = 0$$

can be solved over the integers (the resolution of this equation is outlined in [5]). Substituting $x - 1$ for x, we get the equation

$$xy^2 + (x^2 - 2x + 1)y - x^4 + 4x^3 - 6x^2 + 4x - 2 = 0 \tag{2.6}$$

of the form (2.3). By Theorem 2.1, the number

$$k = \frac{y + 4x - 2}{x^2}$$

must be integer for any solution $(x, y) \in \mathbb{Z}^2$ with $x \neq 0$. Eliminating y, we obtain an explicit expression for k, namely:

$$k = \frac{7x^2 - 2x - 1 \pm \sqrt{4x^5 - 15x^4 + 20x^3 - 10x^2 + 4x + 1}}{2x^3}.$$

Thus, if x satisfies $|x| > m$ then we certainly get $|k| < Q(m)$ with

$$Q(m) = \frac{7m^2 + 2m + 1 + \sqrt{4m^5 + 15m^4 + 20m^3 + 10m^2 + 4m + 1}}{2m^3}.$$

Further, we can proceed in various ways.

(1) Firstly, we can determine m_0 so that the number $Q(m_0)$ is close to 1 (which is due to the fact that $Q(m) \to 0$ as $m \to \infty$). This is reasonable since when $m = m_0$ we need to solve (mainly) only quadratic Eq. (2.3) in y over the integers. For example, taking $m_0 = 8$, we obtain $Q(m_0) < 1$. Thus, it is necessary to solve: (a) for $x \in \{0, \pm 1, \ldots, \pm 8\}$, Eq. (2.6) and, (b) for $k = 0$, system (2.5), namely

$$\begin{cases} xy^2 + (x^2 - 2x + 1)y - x^4 + 4x^3 - 6x^2 + 4x - 2 = 0, \\ y + 4x - 2 = 0. \end{cases}$$

It is easy to see that this system can be reduced to the (again) quadratic equation

$$x^2 - 16x + 12 = 0.$$

Finally, we obtain that all the solutions of Eq. (2.6) are

$$(x, y) \in \{(0, 2), (1, -1), (1, 1)\}.$$

(2) Secondly, we can find m^* such that the total number of equations needed to resolve happens to be minimal (possibly, close to being minimal) when $m = m^*$. For instance, we can take $m^* = 4$ which provides $Q(m^*) < 2$. This is somewhat better than using the previous tactics.

The first issue of the proposed method is the following: we need to determine the number $Q(m)$ as an explicit function of the so-called *control parameter* m. This can be overcome by Lemma 2.1 (see below). The second issue can be formulated as follows: how to choose the optimal value of m? More precisely, we want to minimize the *cost-function* of the form

$$\text{cost}\,(m) = 2m + 2qQ(m), \tag{2.7}$$

where the *weight coefficient* $q > 1$ can be determined by experiments in a given CAS (in our case, PARI/GP). Here, for q, we take the ratio of the complexity of resolution of quadratic equations and the complexity of resolution of algebraic system of the form (2.5) (in both cases over the integers).

Now, consider system (2.5) in detail. Eliminating y, we obtain just a cubic (with the exception of the case $k = 0$) equation with respect to x, namely

$$k^2 x^3 + K_1 x^2 + K_2 x + K_3 = 0. \tag{2.8}$$

Here, the coefficients K_j given as follows:

$$
\begin{aligned}
K_1 &= (-2c^2D - 2E^2 + 2bcE + ac^3)k + c^6A, \\
K_2 &= c^2(-2E + bc)k + c^6B + c^4D^2 + 2c^2DE^2 - 2bc^3DE - ac^5D + \\
&\quad + E^4 - 2bcE^3 - ac^3E^2 + b^2c^2E^2 + abc^4E, \\
K_3 &= c^4k + c^6C + 2c^4DE - bc^5D + 2c^2E^3 - 3bc^3E^2 + c^4(b^2 - ac)E.
\end{aligned}
\tag{2.9}
$$

Therefore, we need to determine how much harder is the problem of solving cubic equations over the integers compared to that for quadratic equations. In PARI/GP, we intend to solve both problems via the function nfroots which provides, in particular, finding all rational roots of a univariate polynomial with integer coefficients. Preliminary computer experiments with the quadratic and cubic polynomials of moderate height (up to 10^{20}) have shown that, for this purpose, one can take $q = 2$. In Sect. 4, we discuss the method of choosing q in detail.

Note that, although we can use the value $m = m_0$ with $Q(m_0)$ close to 1 (the motivation for this can be found in Example 2.2) in the algorithm, this can be disadvantageous due to the fact that m_0 may happen to be too large.

Example 2.3. Consider the equation

$$
xy^2 + (x^2 + 1)y + x^4 + H = 0
\tag{2.10}
$$

where the coefficient H is supposed to be rather large. The direct computation of $Q(m)$ based on Lemma 2.1 (see below) shows that the inequality

$$
Q(m) > \frac{|H|^2}{m}
$$

holds. Hence, if $Q(m_0) = 1$ then $m_0 > |H|^2$. On the other hand, taking $m^* = |H|$, we get $Q(m^*) \sim |H|$ as $H \to \infty$. Obviously, for Eq. (2.10), the proposed algorithm with $m = m^*$ works faster than that with $m = m_0$.

For every H, Eq. (2.10) has the trivial solution $(x, y) = (0, -H)$. A statistical information on the number of non-trivial solutions in the range $1 \leqslant H \leqslant 10^4$ is represented in Table 2.

For convenience purposes, let us introduce the notation:

$$
\begin{aligned}
Q_1 &= 2c^2D + 2E^2 - 2bcE - ac^3, & Q_6 &= -4C + 2ab, \\
Q_2 &= 2c^2E - bc^3, & Q_7 &= -4D + 2ac + b^2, \\
Q_3 &= -c^4, & Q_8 &= -4E + 2bc, \\
Q_4 &= -4A, & Q_9 &= c^2. \\
Q_5 &= -4B + a^2, &
\end{aligned}
\tag{2.11}
$$

The following technical result is necessary for an algorithmic implementation of the described method.

Table 2. Distribution of the number of non-trivial solutions of the Eq. (2.10) in the range $1 \leqslant H \leqslant 10^4$.

$\#(x,y)$	$\#H$
0	9200
1	639
2	133
3	26
4	1
5	1

Lemma 2.1. *For any $m \geqslant 1$, the number $Q(m)$ can be defined as follows:*

$$Q(m) = \frac{1}{2} \sum_{i=1}^{3} \frac{|Q_i|}{m^i} + \frac{|c|^3}{2} \left(\sum_{i=1}^{6} \frac{|Q_{i+3}|}{m^i} \right)^{1/2}, \tag{2.12}$$

where the coefficients $Q_1, \ldots, Q_9$ are given by (2.11).

Proof. The formulas (2.9) for the coefficients K_j show that Eq. (2.8) is quadratic in k. Dividing by the leading coefficient x^3 and resolving with respect to k, we obtain

$$k = \frac{1}{2} \sum_{i=1}^{3} \frac{Q_i}{x^i} \pm \frac{c^3}{2} \left(\sum_{i=1}^{6} \frac{Q_{i+3}}{x^i} \right)^{1/2}.$$

Obviously, the condition $|x| > m$ implies the required estimate $|k| < Q(m)$ with $Q(m)$ given by (2.12).

Unfortunately, the analytic expression for $Q(m)$ provided by Lemma 2.1 is too complicated to minimize the cost-function (2.7) by means of symbolic methods. Therefore, we need to focus on the reasonable estimates for cost (m^*) where m^* is a such value of m that it delivers the global minimum of cost (m). Further, the proposed solving algorithm with $m = m^*$ will be called the *optimized algorithm*. Denote by H the height of the left hand side of Eq. (2.3).

Theorem 2.2. *For the optimized algorithm, the estimate*

$$\mathrm{cost}\,(m^*) \leqslant C_1 |c|^{4/3} H \tag{2.13}$$

holds. Here $C_1 > 0$ is a constant which depends only on q.

Proof. Let $m_1 = 4|c|^{4/3} H$. Since

$$\mathrm{cost}\,(m^*) \leqslant \mathrm{cost}\,(m_1) = 2m_1 + 2qQ(m_1) = 8|c|^{4/3} H + 2qQ(m_1),$$

it is sufficient to estimate the number $Q(m_1)$. We can perform this in a straightforward manner (i.e., by estimating each of the fractions $|Q_i|/m^i$, $|Q_{i+3}|/m^i$ at

$m = m_1$ in the right-hand side of (2.12); also, we use the obvious inequality $\sqrt{\alpha_1 + \ldots + \alpha_n} \leqslant \sqrt{\alpha_1} + \ldots + \sqrt{\alpha_n}$. The extremal case is the following:

$$|c|^3 \sqrt{\frac{|Q_4|}{m_1}} \leqslant |c|^3 \sqrt{\frac{4H}{4|c|^{4/3}H}} = |c|^{7/3} = |c|^{4/3} \cdot |c| \leqslant |c|^{4/3}H.$$

As a result, we arrive at the inequality $Q(m_1) \leqslant 2|c|^{4/3}H$. Thus,

$$\mathrm{cost}\,(m_1) \leqslant (8 + 4q)|c|^{4/3}H,$$

and we can set $C_1 = 8 + 4q$.

The estimate (2.13) of complexity of the optimized algorithm in some cases occurs to be accurate (of course, up to a constant factor). For example, this is true for Eq. (2.10) because $m^* \asymp m_1 \asymp H$ and $\mathrm{cost}\,(m^*) \asymp H$ as $H \to \infty$. On the other hand, it happens that sometimes the general estimate (2.13) can be improved.

Example 2.4. For the equation

$$xy^2 + (Hx^2 + 1)y + x^4 + 1 = 0 \tag{2.14}$$

we have $m^* \asymp |H|^{1/2}$ and, consequently, $\mathrm{cost}\,(m^*) \asymp |H|^{1/2}$ as $H \to \infty$. Using the optimized algorithm, we can check that for $1 \leqslant H \leqslant 10^5$, Eq. (2.14) has no solutions $(x, y) \neq (0, -1)$, with the exception of $H = 2$ and $H = 8$ (see Example 3.5 below).

In general, the minimization of the cost-function (2.7) can be performed by a numerical method (for instance, we can use the well-known *golden-section search*). The starting (and, probably, rough) approximation $m^* \approx m_1$ proposed in the proof of Theorem 2.2 can be used as follows. Let us introduce $m_2 = tm_1$ where a constant factor $t > 1$ will be determined later. Earlier, we showed that the inequality $Q(m_1) \leqslant m_1/2$ holds. Hence, we have

$$\mathrm{cost}\,(m_2) = 2tm_1 + 2qQ(m_2) > 2tm_1 = 2m_1 + 2(t-1)m_1 \geqslant$$
$$\geqslant 2m_1 + 4(t-1)Q(m_1) \geqslant 2m_1 + 2qQ(m_1) = \mathrm{cost}\,(m_1)$$

whenever $4(t-1) \geqslant 2q$. Therefore, setting $t = q/2 + 1$, we localize m^* in the interval $[1, m_2]$. It remains to apply a numerical search algorithm in the given interval. Heuristically, this additional procedure of optimization has a small (negligible) contribution to the total computational complexity.

3   Estimates for Integer Solutions

In this section, we give a few examples of explicit bounds for integer solutions of diophantine equations of small degree satisfying Runge's condition. Usually, these bounds are supposed to be used in order to find the solutions them-selves, but the method (based on the elementary version of Runge's method) provides some estimates for solutions as an additional result (for more informa-tion, see [6]).

We start with three examples of cubic diophantine equations in order to demonstrate that the result entirely depends on the specifics of an equation.

Let H be a positive integer, C_2, C_3, etc. denote some positive absolute constants.

Example 3.1. For all the solutions (x, y) of the equation

$$x(y^2 - x^2) = Hy + 1 \tag{3.1}$$

in positive integers, we have the estimate

$$y \leqslant (H + 3)/2$$

(the elementary proof can be obtained via the technique proposed in [6]). The upper bound is achieved for any odd H since the pair $(x, y) = ((H+1)/2, (H+3)/3)$ satisfies (3.1).

Example 3.2. For all the solutions (x, y) of the equation

$$x(y^2 - x^2) = Hy \tag{3.2}$$

in positive integers, we can propose the estimate

$$y \leqslant (H + 1)^{3/4}$$

(the proof is also elementary, yet it requires some effort). The upper bound is achieved for infinitely many H since the pair $(x, y) = ((H + 1)^{1/4}, (H + 1)^{3/4})$ satisfies (3.2). This improves the expected estimate $y < C_2 H$ (see Exercise 4.15 [3]).

Example 3.3. For all the solutions (x, y) of the equation

$$x(y^2 - 2x^2) = Hy$$

in integers, the estimate

$$|x| < C_3 H^{3/2}$$

holds (see [6] for further details). There are no proved results on the accuracy of this estimate (apparently, it is achieved for infinitely many H).

For diophantine equations of degree four, the problem of estimating integer solutions is much harder. In the case of (2.3), we can hope to obtain an estimate for integer solutions (x, y) by rewriting the auxiliary Eq. (2.8) as

$$1 + \frac{K_1}{k^2 x} + \frac{K_2}{k^2 x^2} + \frac{K_3}{k^2 x^3} = 0$$

and showing that $|x|$ cannot be too large. However, this method leads to quite rough estimates which are overvalued (not achieved in reality). In order to illus-trate this fact, we consider the following three examples.

Example 3.4. For integer solutions (x, y) of Eq. (2.10), we have the estimate

$$|x| < C_4 H^2$$

which can be obtained by the above-mentioned technique. Using the optimized algorithm, we can see that this estimate is unrealistic for $1 \leqslant H \leqslant 10^4$. On the other hand, for $H = t^3 + t^2$, the pair $(x, y) = (-t^2 - t, -t^3 - t^2)$ satisfies (2.10) and for this solution, we have $|x| \sim H^{2/3}$ as $H \to \infty$. The hypothetical estimate

$$|x| < C_5 H^{2/3}$$

for non-trivial integer solutions $(x, y) \neq (0, -H)$ is confirmed by computer experiments. This estimate seems more realistic, but it is not clear how to prove it.

Example 3.5. Similarly, for integer solutions (x, y) of Eq. (2.14), we can give the estimate

$$|x| < C_5 H.$$

At the same time, computer experiments (see Example 2.4) suggest the following conjecture: Eq. (2.14) has integer solutions $(x, y) \neq (0, -1)$ if and only if $H \in \{2, 8\}$.

This conjecture is actually true, and we now outline the proof. Rewrite Eq. (2.14) in the form

$$H = -\frac{xy^2 + y + x^4 + 1}{x^2 y}.$$

From this, one can conclude that the number

$$l = \frac{y + x^4 + 1}{xy}$$

must be in $\mathbb{Z}$. The last equality can be rewritten as

$$y = \frac{x^4 + 1}{lx - 1}.$$

Since $y \in \mathbb{Z}$, the number

$$d = \frac{x^2 + l^2}{lx - 1} \tag{3.3}$$

is also in $\mathbb{Z}$. Next, eliminating l, we get the equation

$$y^2 - (dx^2 - 2)y + x^4 + 1 = 0$$

which implies

$$y = \frac{dx^2 - 2 \pm xz}{2}, \quad z = \sqrt{(d^2 - 4)x^2 - 4d} \geqslant 0.$$

Since $x \neq 0$, it follows that $z \in \mathbb{Z}$. Finally, eliminating y, we obtain

$$2H = -d(x + 1) \mp z + \frac{2 \pm z}{x}.$$

Since $H \in \mathbb{Z}$, we have $2 \pm z \equiv 0 \pmod{x}$ that yields

$$4d + 4 \equiv 0 \pmod{x}. \tag{3.4}$$

It remains to prove that congruence (3.4) and the condition $z \in \mathbb{Z}$ can be simultaneously held for finitely many pairs (x, d) at most. Thus, there are only finitely many possible values of H. More precisely, in the case of an arbitrary integer H, we conclude that

$$H \in \{-14, -9, -5, -4, -2, 0, 2, 8\}.$$

Using *Pell's equations*, we can somewhat simplify the proof. Namely, we can use the following well-known result: if a triple (x, l, d) of integers satisfies (3.3) then $d = 5$ or $d = -t$ where t is a perfect square.

Example 3.6. For integer solutions (x, y) of the equation

$$xy^2 + (Hx + 1)y + x^4 + 1 = 0, \tag{3.5}$$

we have the same rough estimate as in Example 3.5. However, Eq. (3.5) unlike Eq. (2.14) is solvable for infinitely many H. For instance, the triple

$$x = \pm\sqrt{t}(t^2 - 1), \quad y = -t^4 + t^2 - 1, \quad H = t^4 - t^2 + 1 \pm \sqrt{t}(t^3 - 2t)$$

satisfies (3.5) and $|x| \sim H^{5/8}$ as $H \to \infty$.

Note that Eq. (3.5) can be studied in the same way as Eq. (2.14). The final description of the set of all integer solutions (x, y, H) use the *Chebyshev polynomials* of the second kind.

The last two examples may look artificial, but they vividly illustrate that, in general, obtaining exact bounds for integer solutions can be very difficult.

4   Concluding Remarks

In conclusion, we comment on some obtained results and discuss further applications of the elementary version of Runge's method.

In view of Example 2.1, it is worth discussing a strategy for solving Eq. (2.1). The following seems to be reasonable. If the height of $P(x)$ is determined by the coefficient of x^3 (i.e., the other coefficients are small compared to it) then it is recommended to reduce the given equation to the corresponding cubic equation (similarly to the case of Eq. (2.2)). Otherwise, we recommend to use the standard method since this trick does not give a significant advantage (at least, the case of one-parametric equations of the type (2.2) confirms this).

Now, let us get back to the general case. Given polynomial (1.4), we can use the linear substitution $a_1 x + b_1 \to x$ that reduces the problem to solving Eq. (2.3). However, this may lead to a significant increase in the height of the polynomial $f(x, y)$ as well as in the case of cubic diophantine equations (see [6]).

It seems that a more successful way is to generalize the already available solving algorithm for Eq. (2.3) (we mean that such generalization is based on the direct analogue of Theorem 2.1). The expected estimate for complexity of the generalized algorithm (which is similar to estimate (2.13), see Theorem 2.2) will be worse than that in the case of $a_1 = 1$, $b_1 = 0$.

For optimization of solving algorithm we need to choose the weight coefficient q correctly. Now, we describe how to do this in the case when H (the height of the left-hand side of (2.3)) is moderate enough (up to 10^5) and $|c| \ll H$. Let $\widetilde{H}$ be the height of the left-hand side of (2.8). Due to (2.9) and Theorem 2.2, we can assume $\widetilde{H} \approx H^4$ to be moderate (up to 10^{20}). Then,

$$q = \frac{\texttt{time}\,(\text{quadratic}, H, M)}{\texttt{time}\,(\text{cubic}, \widetilde{H}, M)},$$

where $\texttt{time}\,(\cdot)$ is the running time for solving $M = 10^6$ randomly chosen equations of the given type. For $H = 10^5$ (and $\widetilde{H} = 10^{20}$, respectively), using the function $\texttt{nfroots}$ for finding rational roots in PARI/GP CAS, we obtain $q \approx 2$. However, in the case $c \asymp H$, we have $\widetilde{H} \gg H$, so that we recommend to increase q up to 6. In this case, the running time of the optimized algorithm will be reasonable for H up to at least 10^2.

Clearly, the results of computer experiments represented in Tables 1 and 2 should be developed further. At the moment, the running time for obtaining Table 2 is $t_1 \approx 13.5$ min and the similar table for the range $1 \leqslant H \leqslant 10^5$ requires $t_2 \approx 100 t_1$ min (by using the processor AMD Ryzen 7 2700x 3.7 GHs and 16gb RAM). Obviously, the running time can be decreased by implementing a parallel version of the proposed algorithm. Namely, the procedure of finding integer roots of a collection of univariate polynomials can be distributed between CPU threads that allows to use computer resources more efficiently, since PARI/GP CAS supports parallel programming.

It seems that the elementary version of Runge's method for $d = 4$ proposed in [5] can be implemented in the same way—at least for the polynomial $f(x, y)$ with the leading homogenous part of the form

$$f_4(x, y) = (a_1 x + b_1 y)(a_2 x^3 + b_2 x^2 y + c_2 x y^2 + d_2 y^3).$$

We expect considerably more technical aspects in such an implementation. In particular, the corresponding auxiliary equation (as an analog of (2.8)) will be more complicated, although we hope that this is not crucial.

References

1. Beukers, F., Tengely, Sz.: An implementation of Runge's method for diophantine equations. arXiv:math/0512418 [math.NT]
2. Masser, D.W.: Polynomial bounds for diophantine equations. Am. Math. Monthly **93**, 486–488 (1986)
3. Masser, D.W.: Auxiliary Polynomials in Number Theory. Cambridge University Press, Cambridge (2016)
4. Mordell, L.J.: Diophantine Equations. Academic Press Inc., London (1969)

5. Osipov, N.N.: Runge's method for the equations of fourth degree: an elementary approach. In: Matematicheskoe Prosveshchenie, Ser. 3, vol. 19, pp. 178–198. MCCME, Moscow (2015). (in Russian)

6. Osipov, N.N., Gulnova, B.V.: An algorithmic implementation of Runge's method for cubic diophantine equations. J. Sib. Fed. Univ. Math. Phys. **11**(2), 137–147 (2018)

7. Osipov, N.N., Medvedeva, M.I.: An elementary algorithm for solving a diophantine equation of degree four with Runge's condition. J. Sib. Fed. Univ. Math. Phys. **12**(3), 331–341 (2019)

8. Poulakis, D.: A simple method for solving the diophantine equation $Y^2 = X^4 + aX^3 + bX^2 + cX + d$. Elem. Math. **54**, 32–36 (1999)

9. Runge, C.: Ueber ganzzahlige Lösungen von Gleichungen zwischen zwei Veränderlichen. J. reine und angew. Math. **100**, 425–435 (1887)

10. Sprindžuk, V.G.: Classical Diophantine Equations. Springer-Verlag, New York (1993). https://doi.org/10.1007/BFb0073786

11. Tengely, S.: On the Diophantine equation $F(x) = G(y)$. Acta Arith. **110**, 185–200 (2003)

12. Walsh, P.G.: A quantitative version of Runge's theorem on diophantine equations. Acta Arith. **62**, 157–172 (1992)

13. PARI/GP Homepage. https://pari.math.u-bordeaux.fr

Old and New Nearly Optimal Polynomial Root-Finders

Victor Y. Pan[1,2(✉)]

[1] Department of Computer Science, Lehman College of the City University
of New York, Bronx, NY 10468, USA
victor.pan@lehman.cuny.edu
[2] Ph.D. Programs in Mathematics and Computer Science,
The Graduate Center of the City University of New York,
New York, NY 10036, USA
http://comet.lehman.cuny.edu/vpan/

Abstract. Univariate polynomial root-finding has been studied for four millennia and still remains the subject of intensive research. Hundreds if not thousands of efficient algorithms for this task have been proposed and analyzed. Two nearly optimal solution algorithms have been devised in 1995 and 2016, based on recursive factorization of a polynomial and subdivision iterations, respectively, but both of them are superseded in practice by Ehrlich's functional iterations. By combining factorization techniques with Ehrlich's and subdivision iterations we devise a variety of new root-finders. They match or supersede the known algorithms in terms of their estimated complexity for root-finding on the complex plane, in a disc, and in a line segment and promise to be practically competitive.

Keywords: Polynomial root-finding · Deflation ·
Polynomial factorization · Functional iterations · Subdivision ·
Real root-finding

2000 Math. Subject Classification: 65H05· 26C10· 30C15

1 Introduction

1. The Problem and Three Known Efficient Algorithms. Univariate polynomial root-finding has been the central problem of mathematics since Sumerian times (see [1,2,33,34]) and still remains the subject of intensive research due to applications to signal processing, control, financial mathematics, geometric modeling, and computer algebra (see the books [28,30], a survey [19], the recent papers [11,12,24,40,45,48], and the bibliography therein).

Hundreds if not thousands of efficient polynomial root-finders have been proposed. The algorithm of [32] and [37], extending the previous progress in

© Springer Nature Switzerland AG 2019
M. England et al. (Eds.): CASC 2019, LNCS 11661, pp. 393–411, 2019.
https://doi.org/10.1007/978-3-030-26831-2_26

[15,31,50], first computes numerical factorization of a polynomial into the product of its linear factors and then approximates the roots; it solves both tasks in nearly optimal Boolean time – almost as fast as one can access the input coefficients with the precision required for these tasks.[1]

Since 2000 the root-finder of the user's choice has been the package MPSolve,[2] implementing Ehrlich's iterations of [18], also known from their rediscovery by Aberth in 1973. In 2016, a distinct nearly optimal polynomial root-finder appeared in [11,12], based on subdivision iterations. That algorithm promises to compete with MPSolve for root-finding in a disc on the complex plain [24], but less likely for the approximation of all roots of a polynomial unless our innovations raise practical efficiency of subdivision to a much higher level.

2. New Hybrid Algorithms. We propose new hybrid root-finders seeking *synergistic combination* of the known techniques.

We first recall that [32] and [37] factorize a polynomial by splitting it into the product of two factors of comparable degrees and then recursively splitting the factors in a similar way as long as their degree exceeds 1. This advanced recursive construction turned out to be hard to implement, but its basic deflation algorithm developed by Schönhage in [50], traced back to Delves and Lyness [15], and hereafter referred to as *DLS algorithm* or *DLS deflation* can be handled quite readily. Presently we devise new hybrid root-finders by incorporating the DLS algorithm into Ehrlich's and subdivision iterations. In both cases deflation enables us to apply root-finding to smaller degree factors, possibly having fewer root clusters rather than to the input polynomial.

We also enhance the efficiency of subdivision iterations by incorporating a very fast and robust sub-algorithm of DLS deflation that computes the number of roots in a disc as the sum of the 0th powers of all roots in that disc. This is *dramatic improvement of root counting* in the papers [11,12], which listed root counting algorithm as their main novelty compared to their predecessors.

We also propose an additional simplification of real root-finding based on fast estimation of the distances of real roots from the origin.

Our hybrid root-finders are nearly optimal and can become the user's choice. Their implementation, testing and refinement are major challenges.

3. Some Extensions. Our hybrid algorithms can be readily extended to various functional iterations such as Newton's and Weierstrass's that approximate all roots of a polynomial, but we also extend our approach to nearly optimal root-finding in a disc and a line interval. In both cases non-costly removal of

---

[1] Required precision and Boolean time are smaller by a factor of d, the degree of the input polynomial, at the stage of numerical polynomial factorization, which has various important applications to modern computations, besides root-finding, e.g., to time series analysis, Wiener filtering, noise variance estimation, co-variance matrix computation, and the study of multi-channel systems (see Wilson [59], Box and Jenkins [7], Barnett [3], Demeure and Mullis [16] and [17], Van Dooren [56]).

[2] Some competition came in 2001 from the package EigenSolve of [20], but the latest version of MPSolve of [10] has combined the benefits of both packages.

the external roots by means of deflation promises substantial advantage over the customary solution of these tasks by means of application of MPSolve and subdivision to the original polynomial of higher degree.

We hope that our work will motivate further efforts towards synergistic combination of some efficient techniques well- and less- known for polynomial rootfinding (see, e.g., the little explored methods of [40] and [48]).

Devising practical and nearly optimal factorization algorithms is still a research challenge because for that task both Ehrlich's and subdivision iterations are slower by at least a factor of d than [32] and [37].

4. Variations of Deflation. With Ehrlich's iterations we can combine the DLS deflation, but also two simpler algorithms (see Sect. 4.3). One of them only involves shift and scaling of the variable and forward and inverse FFTs and allows us to represent an input polynomial with a black box for its evaluation rather than with coefficients. In [39] and [25] we alternatively combine Ehrlich's iterations with implicit deflation, which ensures preserving sparseness of an input and avoiding coefficient growth.

With subdivision iterations we combine the DLS deflation, which is highly efficient and relatively simple but has been too little (if at all) used by researchers since [37]. It has been hidden in the long paper [50], within a realm of intricate and advanced techniques for the theoretical estimation of asymptotic Boolean complexity where extremely accurate polynomial factorization is required, but some of these techniques can help enhance performance of the most popular root-finders.

Now suppose that the root set of a factor is a strongly isolated cluster of w roots of p having a small diameter. Such a cluster may appear in subdivision process and then can be readily detected. In this special case we can perform deflation at a low cost by means of shifting the origin into the cluster and then reducing the resulting polynomial $q(x)$ modulo x^{w+1} (see Sect. 5.4).

5. Organization of the Paper. We state four variations of the main rootfinding problem in Sect. 2 and deduce lower bounds on their Boolean complexity in Sect. 3. We cover computation of a factor with root set in a fixed disc in Sect. 5 and our hybrids of deflation with functional iterations in Sect. 4 and with subdivision in Sect. 6. In Sect. 7 we devise a new nearly optimal polynomial root-finder on a line segment. We refer the reader to [42, Appendix] for *concise exposition* of factorization algorithms of [50] by Schönhage and [26] by Kirrinnis (67 pages) and for some other auxiliary and complementary algorithms and techniques for polynomial root-finding.

2   Four Fundamental Computational Problems

Problem 1. *Univariate Polynomial Root-finding.* Given a positive b and the coefficients $p_0, p_1, \ldots, p_d$ of a univariate polynomial $p(x)$,

$$p(x) = \sum_{i=0}^{d} p_i x^i = p_d \prod_{j=1}^{d} (x - x_j), \ p_d \neq 0. \tag{1}$$

approximate all d roots[3] $x_1, \ldots, x_d$ within the error bound $1/2^b$ provided that $\max_{j=0}^{d} |x_j| \leq 1$. We can ensure the latter customary assumption at a dominated computational cost by first approximating the root radius $r_1 = \max_{j=1}^{d} |x_j|$ and then scaling the variable x (cf., e.g., [33]).

Before proceeding any further we recall some **Basic Definitions.**

- Hereafter we freely denote polynomials $p(x)$, $t(x) = \sum_i t_i x^i$, $u(x) = \sum_i u_i x^i$ etc. by p, t, u etc. unless this can cause confusion.
- We use the norm $|u| = \sum_i |u_i|$ for $u = \sum_i u_i x^i$.
- $d_u := \deg(u)$ denotes the degree of a polynomial u; in particular $d_p = d$.
- ϵ-*cluster* of roots of p is a root set lying in a disc of radius ϵ; in particular a 0-cluster of m roots of p is its root of multiplicity m.

Problem 2. *Approximate Factorization of a Polynomial.* Given a positive b' and the coefficients $p_0, p_1, \ldots, p_d$ of a polynomial $p = p(x)$ of (1), compute $2d$ complex numbers u_j, v_j for $j = 1, \ldots, d$ such that

$$|p - \prod_{j=1}^{d}(u_j x - v_j)| \leq 2^{-b'} |p|. \tag{2}$$

Problem 3. *Polynomial root-finding in a disc.* This is Problem 1 restricted to root-finding in a disc on the complex plain for a polynomial p that has no roots lying outside the disc but close to it.

Problem 4. *Polynomial root-finding in a line segment.* This is Problem 1 restricted to root-finding in a line segment for a polynomial p that has no roots lying outside the segment but close to it.

The above concept "close" is quantified in Definition 1 in the case of Problem 3 and is extended to Problem 4 via its reduction to Problem 3 in Sect. 1.

Remark 1. It is not easy to optimize *working precision* for the solution of Problems 1–4 a priori, but we nearly optimize it *by action*, that is, by applying the solution algorithms with recursively doubled or halved precision and monitoring the results (see Sect. 4.2 and recall similar policies in [5,10,12,45]).

Remark 2. It is customary to reduce Problems 3 and 4 to root-finding in the unit disc

$$D(0,1) := \{x : |x| < 1\}$$

and unit segment

$$S[-1,1] := \{x : -1 \leq x \leq 1\}$$

by means of shifting and scaling the variable. Then the working precision and Boolean cost grow but within the nearly optimal bounds.

---

[3] We count m times a root of multiplicity m.

3   Boolean Complexity: Lower Estimates and Nearly Optimal Upper Bounds

Proposition 1. *The solution of Problem 2 involves at least db' bits of memory and at least as many Boolean (bit-wise) operations.*

Proof. The solution of Problem 2 is given by the $2d$ coefficients u_j and v_j of the d linear factors $u_j x - v_j$ of p for $j = 1, \ldots, d$. Let $u_j = 1$ and $1/2 < |v_j| < 1$ for all j. Then each v_j must be represented with b' bits and hence all v_j must be represented with db' bits in order to satisfy (2). A Boolean operation outputs a single bit, and so we need at least db' operations in order to output db' bits. $\qquad\blacksquare$

Next we bound from below the Boolean complexity of Problems 1, 3 and 4.

Lemma 1. *Let $p(x) = (x - x_1)^m f(x)$ for a polynomial $f(x)$ and a positive integer m. Fix a positive b. Then the polynomial $p_j(x) = p(x) + 2^{(j-m)b}(x - x_1)^j f(x)$ has $m - j$ roots $x_1 + \omega_{m-j}^i 2^{-b}$ for $i = 0, \ldots, m - j - 1$ and $\omega_{m-j} = \exp(2\pi i/(m-j))$ denoting a primitive $(m-j)$th root of unity, such that $\omega_{m-j}^{m-j} = 1$, $\omega_{m-j}^i \neq 1$ for $0 < i < m - j$.*

Proof. Observe that $p_j(x) = ((x - x_1)^{m-j} + 2^{(j-m)b})(x - x_1)^j f(x)$ and consider the roots of the factor $(x - x_1)^{m-j} + 2^{(j-m)b}$. $\qquad\blacksquare$

Corollary 1. *Under the assumption of Lemma 1 write $f := \lceil \log_2 |f| \rceil$ and $g := \sum_{j=1}^{m-1} \lceil \log_2 |g_j| \rceil$. Then one must process at least*

$$\mathcal{B}_p = \left(d - m + 1 + \frac{m-1}{2}\right) mb - f - g \qquad (3)$$

bits of the coefficients of p and must perform at least $\mathcal{B}_p/2$ Boolean operations in order to approximate the m-multiple root x_1 of p within $1/2^b$.

Proof. By virtue of Lemma 1 the perturbation of the coefficients $p_0, \ldots, p_{d-m}$ of $p(x)$ by $|f|/2^{mb}$ turns the $(m - j)$-multiple root x_1 of the factor $(x - x_1)^{m-j}$ of $p(x)$ into $m - j$ simple roots $p_j(x)$, all lying at the distance $1/2^b$ from x_1. Therefore, one must access at least $(d - m + 1)mb - f$ bits of the coefficients $p_0, \ldots, p_{d-m}$ of p in order to approximate the root x_1 within $1/2^b$.

Now represent the same polynomial $p(x)$ as $(x - x_1)^{m-j} g_j(x)$ for $g_j(x) = (x - x_1)^j f(x)$ and $j = 1, \ldots, m - 1$. Apply Lemma 1 for m replaced by $m - j$ and for $f(x)$ replaced by $g_j(x)$ and deduce that a perturbation of the coefficient p_{d-m+j} of p by $|g_j|/2^{(m-j)b}$ turns the j-multiple root x_1 of $g_j(x) = (x - x_1)^j f(x)$ into j simple roots, all lying at the distance $1/2^b$ from x_1. Therefore, one must access at least $\sum_{j=1}^{m-1}((m-j)b - g = \frac{m-1}{2}mb - g$ bits of the coefficients $p_{d-m+1}, \ldots, p_{d-1}$ in order to approximate the root x_1 within $1/2^b$ Sum the bounds $(d - m + 1)mb - f$ and $\frac{m-1}{2}mb - g$ and arrive at the bound (3) on the overall number $\mathcal{B}_p$ bits; at least $\mathcal{B}_p/2$ Boolean operations must be used in order to access these bits – at least one operation per each pair of bits. $\qquad\blacksquare$

Let us specify bound (3) in two cases. (i) If $m = d$, $f(x) = 1$, and $|x_1| \leq 0.5/d$, then $f = 0$, $|g_j| \leq 2$ for all j, $g \leq d - 1$, and

$$\mathcal{B}_p \geq (d + 1)db/2 - d + 1. \tag{4}$$

(ii) If x_1 is a simple root, well-isolated from the other roots of p, then substitute $m = 1$ and $g = 0$ into Eq. (3), thus turning it into $\mathcal{B}_p = db - f$ for $f(x) = p(x)/(x - x_1)$ such that $|f| \leq d|p|$. This implies that

$$\mathcal{B}_p \geq (b - |p|)d.$$

Remark 3. Corollary 1 defines lower bounds on the Boolean complexity of Problems 1, 3, and 4 as long as an input polynomial p has an m-multiple root in the complex plain, a disc, and a segment, respectively. One can extend all these bounds to the case where a polynomial has an ϵ-cluster of m roots for a sufficiently small positive ϵ rather than an m-multiple root.

The algorithm of [32] and [37] solves Problem 2 by using $\tilde{O}(db')$ bits and Boolean operations.[4] This Boolean cost bound is within a poly-logarithmic factor from the information-theoretic lower bound db' of Proposition 1. Based on [52, Theorem 2.7] one can extend this estimate to the solution of Problems 1, 3 and 4 at a Boolean cost in $\tilde{O}(d^2b)$, which is also nearly optimal by virtue of (4), and to nearly optimal solution of the problem of polynomial root isolation (see [42, Corollaries D.1 and D.2]).

4   Ehrlich's Iterations and Deflation

4.1   Ehrlich's Iterations and Their Super-Linear Convergence

The papers [5] and [10] present two distinct versions of MPSolve based on two distinct implementations of Ehrlich's functional iterations.

[5] applies original Ehrlich's iterations by updating current approximations z_i to all or selected roots x_i as follows:

$$z_i \leftarrow z_i - E_{p,i}(z_i), \quad i = 1, \ldots, d, \tag{5}$$

$$E_{p,i}(x) = 0 \text{ if } p(x) = 0; \quad \frac{1}{E_{p,i}(x)} = \frac{1}{N_p(x)} - \sum_{j=1, j \neq i}^{d} \frac{1}{x - z_j} \text{ otherwise}, \tag{6}$$

$$N_p(x) = p(x)/p'(x). \tag{7}$$

---

[4] Here and hereafter we write $\tilde{O}(s)$ for $O(s)$ defined up to a poly-logarithmic factor in s.

[10] modifies these iterations by replacing polynomial equation $p(x) = 0$ by an equivalent rational *secular equation*[5]

$$S(x) := \sum_{j=1}^{d} \frac{v_j}{x - z_j} - 1 = 0 \tag{8}$$

where $z_j \approx x_j$ and $v_j = \frac{p(z_j)l(x)}{x-z_j}$ for $l(x) = \prod_{j=1}^{d}(x - z_j)$ and $j = 1, \ldots, d$.

Cubic convergence of these iterations simultaneously to all roots of a polynomial has been proved locally, near the roots, but under some standard choices of initial approximations very fast global convergence to all roots, right from the start, has been consistently observed in all decades–long applications of the iterations worldwide.

4.2 Precision Management

The condition number of a root defines computational precision sufficient in order to ensure approximation within a fixed relative output error bound (see the relevant estimates in [5] and [10]). The value of the condition is not known a priori, however, and MPSolve adopts the following policy: at first apply Ehrlich's iterations with a fixed low precision (e.g., the IEEE double precision of 53 bits) and then recursively double it until all roots are approximated within a selected error tolerance. More precisely, MPSolve updates approximations only until they are close enough in order to satisfy a fixed stopping criterion, verified at a low computational cost. We call such roots *tame* and the remaining roots *wild*. MPSolve stops applying Ehrlich's iterations to a root when it is tamed but keep applying them to the wilde roots – with recursive doubling of the working precision. When a root is tamed this precision is optimal up to at most a factor of two. Recall from [50] and [52, Section 2.7] that working precision does not need to exceed the output precision b by more than a factor of d, and so at most $O(\log(db))$ steps of doubling precision are sufficient. This natural policy has been proposed and elaborated upon in [5] and [10], greatly improving the efficiency of the previous implementations of functional iterations for polynomial root-finding.

4.3 Ehrlich's Iterations with Deflation

Suppose that MPsolve seeks w wild roots $x_1, \ldots, x_w$ of p and perform ITER$_p$ Ehrlich's iterations with a working precision $\bar{b}$. This involves $O(dw\, \text{ITER}_p)$ arithmetic operations performed at the Boolean cost $O(dw\mu(\bar{b})\, \text{ITER}_p)$.

We propose a modification where we first (a) *deflate* p by computing its factor $f(x) = \prod_{j=1}^{w}(x - x_j)$ and then (b) apply to this factor ITER$_f$ Ehrlich's iterations, at both stages (a) and (b) using the same working precision $\bar{b}$.

---

[5] The paper [10] elaborates upon expression of Ehrlich's iterations via secular equation, shows significant numerical benefits of root-finding by using this expression, and traces the previous study of this approach back to [6].

At stage (b) the Boolean cost bound $O(dw\mu(\bar{b}))$ decreases by a factor of d/w at the price of adding the cost of deflation at stage (a).[6]

Each of the three algorithms of the next subsection performs deflation at a Boolean cost in $O(dw\mu(\bar{b}))$, favorably compared to the above cost bound $O(dw\mu(\bar{b}) \text{ ITER}_p)$. This comparison suggests that using deflation is competitive in terms of the Boolean cost and becomes more favorable as ITER_p grows.

Subsequent Ehrlich's iterations may in turn tame part of the roots of the polynomial $f(x)$, and then we can deflate $f(x)$. We can do this recursively, computing a sequence of factors $f_i(x)$ for $i = 1, 2, \ldots, t$ where $f_1 = f$ and f_{i+1} is a smaller degree factor of f_i for all i.[7] The algorithm would stay nearly optimal overall if we perform deflation poly-logarithmic number of times t, e.g., if we delay deflation of the current factor until $\deg(f_{i+1}) \leq \beta \deg(f_i)$ for a fixed constant[8] $\beta < 1$ and for all i. In the following example computational cost becomes too high if we perform deflation $d - 1$ times but stays nearly optimal if we properly delay deflation.

Example 1. Let $p = \prod_{j=1}^{d}(x - 1 + 1/2^j)$ for a large integer d. In this case the roots $1 - 1/2^j$ are stronger isolated and better conditioned for smaller j, and so Ehrlich's iterations may peel out one such root of p at a time. Then we would $d - 1$ times approximate polynomials of the form $p_i := f_i(x) \prod_{j=1}^{i}(x - 1 + 1/2^j)$ for some polynomials $f_i(x)$, $i = 1, \ldots, d - 1$. For each i the polynomial $f_i(x)$ is a factor of p of degree $d - i$ sharing a cluster of at least $d - i$ roots with p, and so approximation of such a factor involves at least $b_i' d/2$ bits and at least $b_i' d/4$ Boolean operations. Now consider just the polynomials f_i for $i = 1, \ldots, \lfloor d/2 \rfloor$. Each of them shares a cluster of at least $d/2$ roots with the polynomial p, and so we must choose $b_i' \geq bd/2$ (see Corollary 1). Hence approximation of all these factors requires at least order of bd^3 bits and Boolean operations, but we can decrease this large bound to nearly optimal level if we skip deflation at the ith step unless $\deg(f_i(x))/\deg(f_{i+1}(x)) \geq \gamma$ for a fixed $\gamma > 1$, e.g., $\gamma = 2$.

4.4 Deflation Algorithms for Ehrlich's Iterations

Recipe 1. Fix $\rho \geq 2 \max_{j=1}^{w} |x_j|$ and an integer q such that $2^{q-1} \leq w < 2^q$, write $z_j = \rho \exp(2\pi \mathbf{i}/2^q)$ for $j = 0, 1, \ldots, 2^q - 1$, and compute (i) $p(z_j)$ for all j, (ii) $f(z_j) = p(z_j) - \prod_{g=w+1}^{d}(z_j - x_g)$ for all j, and (iii) the coefficients of $f(x)$.

---

[6] Furthermore we may have $\text{ITER}_f < \text{ITER}_p$ because of the decrease of the maximal distance between a pair of roots and of the number and sizes of root clusters in the transition from p to the polynomial $f(x)$ of a smaller degree w.

[7] [50] supplies estimates for the working precision in such a recursive process, which ensure the bound $1/2^b$ on the errors of the output approximations to the roots of p.

[8] The wild roots are much less numerous than the tame roots in a typical partition of a root set observed in Ehrlich's, Weierstrass's and other functional iterations that simultaneously approximate all roots of p as well as in Newton's iteration in [54]. Consequently the coefficient growth and the loss of sparseness are not dramatic in the transition to the factors defined by the wild roots.

Besides scaling the variable x, we perform $(d - w)2^q$ arithmetic operations at stage (ii), $\lceil d/w \rceil$ FFTs on 2^q points at stage (i), and a single inverse FFT on 2^q points; overall we need $O(dw\mu(\bar{b}))$ Boolean operations [26,46]. This cost bound can be verified for the following two recipes as well.

Recipe 2. Compute at first the values of the polynomial $f(x)$ at scaled roots of unity (as in Recipe 1), then the power sums of its roots, and finally its coefficients, (cf. [50, Section 12] or [42, Appendices A and B]).

Recipe 3 [53]. Compute the power sums $s_k = \sum_{j=1}^{d} x_j^k$, $k = 0, 1, \ldots$, of the roots of p by applying Newton's identities (cf., e.g., [36, Equations (2.5.4) and (2.5.5)]). Then by subtracting the powers of all tame roots compute the power sums of the roots of the polynomial $f(x)$. Finally recover its coefficients by applying Newton's identities.

Recipe 3 involves the coefficients of p, while Recipes 1 and 2 as well as Ehrlich's, Weierstrass's and Newton's iterations can be applied to a polynomial p given just by a subroutine for its evaluation, which is an advantage when, say, the polynomial is presented in a compressed form or in Bernstein basis.

4.5   Extension to Other Functional Iterations

The recipes of doubling the working precision and consequently of partitioning the roots into tame and wild ones and our recipes for deflation and its analysis can be extended to Weierstrass's [57], Werner's [60], various other functional iterations for simultaneous approximation of all roots of p [28, Chap. 4], and Newton's iterations applied to the approximation of all roots of p. E.g., Schleicher and Stoll in [54] apply Newton's iterations to the approximation of all roots of a polynomial of degree $d = 2^{17}$ and arrive at $w \approx d/1000$ wild roots.

4.6   Boolean Complexity of Problems 1–4 with and Without MPSolve

Let us compare Boolean complexity of the solution of Problems 1–4 by using MPSolve versus the algorithms of [11,32,37], and [12].

Empirically Ehrlich's iterations in MPSolve have simpler structure and smaller overhead in comparison with the algorithms of the latter papers, but unlike them have only empirical support for being nearly optimal. Moreover super-linear convergence of Ehrlich's iterations has only been observed for simultaneous approximation of all roots, and so these iterations solve Problems 3 and 4 of root-finding in a disc and on a line interval about as fast and as slow as Problem 1 of root-finding on the complex plain, while the nearly optimal cost of root-finding by the algorithms of [11,32,37], and [12] decreases at least proportionally to the number of roots in the input domain. As we have already said in the introduction, MPSolve, [11] and [12] can solve Problem 2 of factorization of p within the same Boolean cost bound as Problem 1, whereas the algorithm of [32] and [37] solves Problem 2 faster by a factor of d, reaching a nearly optimal Boolean cost bound.

5   Approximation of a Factor of a Polynomial

In Sect. 4.4 we approximated a factor $f(x)$ by first computing its values on a fixed circle or the power sums of its roots. These computations were inexpensive in applications to Ehrlich's iterations and were readily combined with them. In applications to subdivision iterations and to root-finding in the line interval the power sum techniques of the DLS deflation also work but require more elaboration. We begin that elaboration in this section, complete it in [42, Appendices A and B], and apply it in Sects. 6 and 7.

5.1   Isolation of a Domain and Its Boundary

The following definition covers quite a general class of domains on the complex plain, although in this paper we only apply it to the unit disc $D(0,1)$.

Definition 1. Isolation of a domain and its boundary. *Let a domain D on the complex plain allow its dilation from a fixed center. Then this domain has an* isolation ratio *at least θ and is θ-isolated for a polynomial p and real $\theta > 1$ if the root set of p in the domain D is invariant in the θ-dilation of D. The boundary of such a domain D has an* isolation ratio *at least θ and is θ-isolated if the root set of p in D stays invariant in both θ- and $1/\theta$-dilation of the domain.*

5.2   Approximation of a Factor with Root Set in an Isolated Disc

Problem 5: *Approximation of a factor with the root set in an isolated disc.*
    INPUT: a polynomial p of (1), a complex number c, a positive number r, and $\theta = 1 + g/\log^h(d)$, a positive constant g and a real constant h such that the disc $D(c,r) = \{z : |z - c| < r\}$ on the complex plain is θ-isolated.
    OUTPUT: the number w of roots of p in the disc $D(c,r)$ and a monic factor f of p whose root set is precisely the set of the roots of p that lie in that disc.

Dual **Problem 5a** of the approximation of a factor p/f whose root set lies outside an isolated disc D is equivalent to Problem 5 of the approximation of the factor of the reverse polynomial $p_{\mathrm{rev}}(x) := x^d p(1/x) = \sum_{i=0}^{d} p_{d-i} x^i$ with root set in D. [Notice that $p_{\mathrm{rev}}(x) = p_0 \prod_{j=1}^{d}(x - 1/x_j)$ if $p_0 \neq 0$.] Accordingly we can re-use the algorithms for Problem 5 in order to solve Problem 5a.

If we are given a factor f of p, we can also solve Problem 5a by means of any of the algorithms of [4, 43, 44, 51, Section 3], and [26] for approximate polynomial division.

Alternatively (cf. Recipe 3 in Section 4.4) we can first compute (i) the sums of the ith powers of the roots of p and f for $i = 0, 1, \ldots, d - w$, by applying Newton's identities or the algorithm of [50, Section 12], then (ii) the sums of the ith powers of the roots of the polynomial p/f for $i = 0, 1, \ldots, d - w$, by subtracting the $d - w + 1$ power sums of the roots of f from those of p, and finally (iii) the coefficients of p/f, by applying either Newton's identities or the algorithm [50, Section 12].

5.3 DLS Deflation: Outline and Complexity

Here is a very brief outline of the DLS algorithm, elaborated upon in [42, Appendices A and B] or [50, Section 12]. Given a polynomial p of degree d the DLS algorithm approximates its factor having root set in a unit disc θ-isolated from the other roots of p. The algorithm first computes $2q$ values of the factor at the $2q$th roots of unity, ensuring error bound roughly d/θ^q, then the power sums of its roots, and finally its coefficients.

The algorithm solves Problem 5 within the same asymptotic Boolean cost bound, $O(d\log(d)\mu(\bar{b}'))$, as in the special case of Sect. 4.4. [42, Corollary A.2] enables extension of the solution and its complexity estimates to the approximation of a factor of p with a root set on the θ-isolated unit circle $C(0,1)$ for $\theta = 1 + g/\log^h(d)$, a positive constant g and a real constant h. At the Boolean cost bound in $O(d\log(d)\mu(\bar{b}'))$ this reduces Problem 3 for a polynomial p of degree d to Problem 1 for a polynomial f of a degree $d_f < d$.

5.4 Deflation in the Case of an Isolated Cluster of a Small Number of Roots

Suppose that all d roots of p lie in the unit disc $D(0,1)$ and that in a subdivision iteration we observe that w roots of p form a cluster strongly isolated from the other roots of p. Let a small disc $D(c,r)$ cover the cluster,1 let $q(x) = p(x-c) = \sum_{j=0}^d q_j x^j$, let

$$\tilde{f}(x) = \sum_{j=0}^w q_j x^j = q(x) \mod x^{w+1} \tag{9}$$

be the sum of the $w + 1$ trailing terms of the polynomial $q(x)$, and consider this sum an approximation of the factor $f(x)$ of $q(x)$ whose root set is made up of the roots of p lying in the cluster, as this was proposed in [27, Section 3.2] for real root-finding. Such disc $D(c,r)$ can be found as a cost-free by-product of subdivision iterations; then we obtain $\tilde{f}(x)$ by means of shifting the variable x.

Let $w \ll d$, that is, let the cluster size w be small. Let is be also well-isolated and estimate the norm $|\tilde{f} - f|$. Write $f := \sum_{i=0}^w f_i x^i$, $g := q/f = \sum_{i=0}^{d-w} g_i x^i$; let $f_w = g_0 = 1$.Then

$$\tilde{f}(x) - f(x) = f(x) \sum_{j=1}^w g_j x^j \mod x^{w+1}. \tag{10}$$

Clearly $\tilde{f}(x) = f(x)$ if $c = 0$ and $f(x) = x^w$. Consider the special case where all roots of f lie in a small r-neighborhood of 0 such that $rw \le 1/2$. Then we readily verify that $|f(x) - x^w| \le |(x-r)^w - x^w| \le 2rw$ and $|\tilde{f}(x) - x^w| \le 2(d-w)^w rw$, and so $|\tilde{f} - f| \le 2((d-w)^w + 1)rw$. This bound can be sufficient in some applications where w is a small positive integer, although the bound is by far not as strong as what we can obtain by applying the algorithms of [42, Part I of the Appendix].

A root-squaring iteration [21] lifts the isolation of the cluster or equivalently squares the radius of the disc. We can apply such lifting recursively, although limited number of times, within $O(\log(\log(d)))$, because of numerical problems. Having approximated the lifted roots in the cluster, we can recover the associated roots of p by applying the descending process of [32] and [37], at dominated overall Boolean cost at both lifting and descending stages.

6   Subdivision Iterations with Deflation

1. Background on Subdivision. Subdivision iterations extend the classical bisection iterations from root-finding on a line to polynomial root-finding in the complex plain. Under the name of *Quad-tree Construction* these iterative algorithms have been studied in [22,23,49], and [35] and extensively used in Computational Geometry. The algorithms have been introduced by Herman Weyl in [58] and advanced in [22,23,49], and [35]; under the name of subdivision Becker et al. modified them in [11] and [12].[9] Let us briefly recall subdivision algorithms for Problem 1; they are similar for Problem 3.

At the beginning of a subdivision (quad-tree) iteration all the d roots of p are covered by at most $4d$ congruent *suspect squares* on the complex plain that have horizontal and vertical edges, all of the same length. The iteration outputs a similar cover of all d roots of p with a new set of at most $4d$ suspect squares whose edge length is halved. Hence the centers of the suspect squares approximate the root set with an error bound linearly converging to 0.

At every iteration suspect squares form connected components. Given $s > 1$ components, embed them into the minimal discs $D_i = D(c_i, R_i)$, $i = 1, \ldots, s$; they become well-isolated from each other in $O(\log(d))$ iterations. For $i = 1, \ldots, s$ cover the root sets of p in the discs D_i by the minimal discs $D_i' = D(c_i', R_i')$. For some i the distances $|c_i - c_i'|$ may greatly exceed R_i', and then linear convergence of subdivision iterations to roots lying in the disc D_i' can be too slow in order to support root-finding in nearly optimal Boolean time.

The algorithms of [11,35,49] and [12] yield super-linear convergence at those stages. [49] and [35] apply Newton's iterations, whose convergence to a disc D_i' is quadratic right from the start if they begin in its θ-dilation $D(c_i', \theta R_i')$ and if the disc D_i' is θ-isolated for a sufficiently large θ. Tilli in [55] proves that it is sufficient if $\theta \geq 3d - 3$, which improves the earlier estimate $\theta \geq 5d^2$ of [49]. [11] and [12] achieve super-linear convergence to the roots by applying Pellet's theorem for root-counting in a disc. This was the main algorithmic innovation of [11] and [12] versus [49] and [35]; the papers [11] and [12] have also extended to the complex plain the QIR iterations, proposed by Abbott for a line segment, and then laboriously estimated the Boolean cost of the resulting algorithm.

2. Our Alternatives: Outline. We deviate from the algorithms of [11] and [12] in two ways.

---

[9] The algorithms of [11] and [12] are quite similar to one another.

(i) We replace the counting algorithm of [11] and [12] with a distinct method, which is faster, more robust, and can be applied where a polynomial p is defined by a black box subroutine for its evaluation rather than by its coefficients.

(ii) If the algorithm of [11] and [12] encounters an isolated component containing d_f roots of p with sufficiently small d_f (we either monitor this explicitly or detect by action), we cover that component with an isolated disc containing these d_f roots and then approximate them by solving Problem 3.

3. Our Alternative Counting. We fix a sufficiently large integer q, let ω denote a primitive qth root of unity, and approximate the number s of the roots of p in the θ-isolated unit disc $D(0,1) = \{x : |x| = 1\}$ as follows:

$$s \approx s^* = \frac{1}{q} \sum_{g=0}^{q-1} \omega^g \frac{p'(\omega^g)}{p(\omega^g)}, \quad \omega = \exp(2\pi\sqrt{-1}/q). \tag{11}$$

By virtue of [42, Theorem 12], extracted from [50], $|s-s^*| < 1/2$ if $2d+1 < \theta^q$. For example, if $\theta = 2$, then choosing any $q \geq 11$ is sufficient where $d = 1,000$ and choosing any $q \geq 21$ is sufficient where $d = 1,000,000$.

We obtain s^* of (11) by evaluating both $p(x)$ and $p'(x)$ at the qth roots of unity (which means performing discrete Fourier transform at q points twice) and in addition performing discrete Fourier transform at q points once again. We can perform Fourier transforms by applying FFT if we choose q being the power of 2.

We can extend our recipe to any sufficiently well isolated disc D on complex plain by means of shifting and scaling the variable x (see Remark 2), but alternatively we can just evaluate $p(x)$ and $p'(x)$ at q equally spaced points on the boundary circle of the disc D. Instead of two FFTs we can apply the so called Horner's algorithm $2q$ times or the algorithms of [29,38], or [41] for fast multipoint polynomial evaluation.

4. Root-Finding in an Isolated dDisc (Problem 3). As soon as subdivision defines a well-isolated disc D_i' containing a positive but reasonably small number d_f of the roots of a polynomial p we approximate its factor $f = f(x)$ having degree d_f and having all its roots in that disc (cf. Sect. 4.3); then we approximate all the d_f roots of the factor by applying the subdivision iterations or MPSolve.

By applying a subdivision algorithm to f rather than p, we approximate the roots of f at the cost that decreases by at least the factor of d/w, because of the decrease of the degree, but possibly more than that if we get rid of some root clusters in the transition from p to f.

Empirically we may additionally benefit from shifting to MPSolve rather than continuing subdivision iterations unless our simplification of subdivision makes it competitive with or even superior to MPSolve.

By applying the DLS algorithm we approximate the factor f within the same asymptotic estimates for the Boolean cost as we deduced for deflation algorithms of Sect. 4.3.[10]

Then again we can apply deflation of p repeatedly for a number of discs D_i' and can recursively extend it to deflation of the computed factors $f(x)$, together with the policy of delaying the deflation until we decrease the number of remaining roots below a fixed bound.[11] The estimates of [42, Appendices C.2 and D] for the working precision in this deflation ensure the upper bound of $1/2^b$ on the errors of the output approximation to the roots.

7   Root-Finding on a Line Segment with Deflation

The algorithms of [11,27,32,37,45] and [12] solve Problem 4 in nearly optimal Boolean time. Next we reduce Problem 4 on the unit segment $S[-1,1] = \{x : -1 \leq x \leq 1\}$ to the special case where all roots of p lie in that segment. In this case the algorithms of [8,13], and [14] are also nearly optimal, but application of MPSolve or [27], complemented with the policy of recursive doubling of working precision, may have even better chances to become the method of user's choice, particularly if its efficiency is enhanced by incorporating the initial approximation of the real roots by means of the simple algorithm of [47], which we recall at the end of this section.

Next we approximate the factor of $p(x)$ whose root set is precisely the set of the roots of p restricted to the segment $S[-1,1]$. The algorithm of [42, Appendices A and B] can be applied for the deflation over any convex domain of the complex plain (see [42, Remark 15]), but in the case of a disc its output approximation to the factor is much closer than for general convex domain (at the same computational cost). Thus we obtain better output approximation by means of reducing Problem 4 to Problem 3.

Towards this goal we first recall the two-to-one Zhukovsky function $z = J(x)$, which maps the unit circle $C(0,1)$ onto the unit segment $S[-1,1]$, and its one-to-two inverse:

$$z = J(x) := \frac{1}{2}\left(x + \frac{1}{x}\right); \quad x = J^{-1}(z) := z \pm \sqrt{z^2 - 1}. \tag{12}$$

Here x and z are complex variables. Now perform the following steps:

1. Compute the polynomial $s(z) := x^d p(x)p(1/x)$ of degree $2d$ by applying [9, Algorithm 2.1], based on the evaluation of the polynomials $p(x)$ and $x^d p(1/x)$

---

[10] The DLS algorithms (cf. [42, Appendices A and B]) approximates a factor f at a nearly optimal Boolean cost if the disc D_i' is θ-isolated for isolation ratio $\theta = 1 + g/\log^h(d)$, a positive constant g and a real constant h. Such an isolation ratio is smaller than those required in [11,35,49] and [12] and thus can be ensured by means of performing fewer subdivision steps.

[11] Then again with such a delay we bound the overall cost of all deflation steps (cf. Example 1), avoid coefficient growth and do not lose sparseness.

at Chebyshev points and the interpolation to $s(z)$ at roots of unity. Recall that the set of the roots of $p(x)$ lying in the segment $S[-1, 1]$ is well-isolated and observe (see Remark 5) that it is mapped in one-to-two mapping (12) into a well-isolated set of the roots of $s(z)$ lying on the unit circle $C(0, 1)$.

2. Let $g(z)$ denote the monic factor of the polynomial $s(z)$ with the root set made up of the roots of $s(z)$ lying on the unit circle $C(0, 1)$ and such that $\deg(g(z)) = \deg(f)$. By applying the algorithm of [50, Section 12] (cf. [42, Corollary A.2]) approximate the power sums of the roots of the polynomial $g(z)$.

3. By applying Newton's identities of the algorithm of [50, Section 13] (cf. [42, Solution 2 of Appendix B]) approximate the coefficients of $g(z)$.

4. Compute the polynomial $h(x) := x^{2a} g(\frac{1}{2}(x + \frac{1}{x}))$ of degree $4 \deg(f)$ in x. Its root set is made up of the roots of the polynomial p lying in the segment $S[-1, 1]$ and of their reciprocals; in the transition to $h(x)$ the multiplicity of the roots of p either grows 4-fold (for the roots 1 and -1 if they are the roots of p) or is doubled, for all other roots.

5. By applying the algorithm of [14] approximate all roots of the polynomial $h(x)$.

6. Among them identify and output $\deg(f)$ roots that lie in the segment $S[-1, 1]$; they are precisely the roots of $p(x)$.

Remark 4. We can simplify stage 5 by replacing the polynomial $h(x)$ with its half-degree square root $j(x) := x^a f(x) f(1/x)$ at stage 5, but further study is needed to find out whether and how much this could decrease the overall computational cost.

Remark 5. Represent complex numbers as $z := u + iv$. Then Zhukovsky's map transforms a circle $C(0, \rho)$ for $\rho \neq 1$ into the ellipse $E(0, \rho)$ whose points (u, v) satisfy the following equation,

$$\frac{u^2}{s^2} + \frac{v^2}{t^2} = 1 \text{ for } s = \frac{1}{2}\left(\rho + \frac{1}{\rho}\right), \quad t = \frac{1}{2}\left(\rho - \frac{1}{\rho}\right).$$

Consequently it transforms the annulus $A(0, 1/\theta, \theta)$ into the domain bounded by the ellipses $E(0, 1/\theta)$ and $E(0, \theta)$, so the circle $C(0, 1)$ is θ-isolated if and only if no roots of p lie in the latter domain.

We conclude this section with recalling an efficient algorithm of [47] for computing crude initial approximations to real roots. Extensive tests in [47] showed particular efficiency of this algorithm for the approximation of the real roots of p that are sufficiently well-isolated from the other roots.

Theorem 1. *See [50, Corollary 14.3]. Assume that we are given a polynomial $p = p(x)$ of (1) and a pair of real constants $c > 0$ and h. Write $\theta = 1 + c/d^h$ and $r_j := |x_j|$ for $j = 1, \ldots, d$. (r_j are said to be the root radii for p.) Then, within the Boolean cost bound $O_B(d^2 \log^2(d))$, one can compute approximations $\tilde{r}_j$ to all root radii r_j such that $1/\theta \leq \tilde{r}_j/r_j \leq \theta$ for $j = 1, \ldots, d$, provided that $\lg(\frac{1}{\theta-1}) = O(\lg(d))$, that is, $|\tilde{r}_j/r_j - 1| \leq c/d^h$.*

Apply the algorithm supporting this theorem and compute d narrow annuli covering all roots of p. Their intersection with real line defines at most $2d$ small segments that contain all real roots of p. Then we weed out the extraneous empty segments containing no roots of p and obtain close approximations to all real roots, in particular to those lying in the segment $S[-1, 1]$. See [47] for further details.

Acknowledgements. This research has been supported by NSF Grants CCF–1563942 and CCF–1733834 and PSC CUNY Award 69813 00 48.

References

1. Bell, E.T.: The Development of Mathematics. McGraw-Hill, New York (1940)
2. Boyer, C.A.: A History of Mathematics. Wiley, New York (1968)
3. Barnett, S.: Polynomial and Linear Control Systems. Marcel Dekker, New York (1983)
4. Bini, D.A.: Parallel solution of certain Toeplitz linear systems. SIAM J. Comput. **13**(2), 268–279 (1984)
5. Bini, D.A., Fiorentino, G.: Design, Analysis, and Implementation of a Multiprecision Polynomial Rootfinder. Numer. Algorithms **23**, 127–173 (2000)
6. Bini, D.A., Gemignani, L., Pan, V.Y.: Inverse power and Durand/Kerner iteration for univariate polynomial root-finding. Comput. Math. Appl. **47**(2/3), 447–459 (2004)
7. Box, G.E.P., Jenkins, G.M.: Time Series Analysis: Forecasting and Control. Holden-Day, San Francisco (1976)
8. Bini, D., Pan, V.Y.: Computing matrix eigenvalues and polynomial zeros where the output is real. SIAM J. Comput. 27(4), 1099–1115 (1998). Proc. version. In: SODA 1991, pp. 384–393. ACM Press, NY, and SIAM Publ., Philadelphia (1991)
9. Bini, D., Pan, V.Y.: Graeffe's, Chebyshev, and Cardinal's processes for splitting a polynomial into factors. J. Complex. **12**, 492–511 (1996)
10. Bini, D.A., Robol, L.: Solving secular and polynomial equations: a multiprecision algorithm. J. Comput. Appl. Math. **272**, 276–292 (2014)
11. Becker, R., Sagraloff, M., Sharma, V., Xu, J., Yap, C.: Complexity analysis of root clustering for a complex polynomial. In: International Symposium on Symbolic and Algebraic Computation (ISSAC 2016), pp. 71–78. ACM Press, New York (2016)
12. Becker, R., Sagraloff, M., Sharma, V., Yap, C.: A near-optimal subdivision algorithm for complex root isolation based on the Pellet test and Newton iteration. J. Symb. Comput. **86**, 51–96 (2018)
13. Ben-Or, M., Tiwari, P.: Simple algorithms for approximating all roots of a polynomial with real roots. J. Complex. **6**(4), 417–442 (1990)
14. Du, Q., Jin, M., Li, T.Y., Zeng, Z.: The quasi-Laguerre iteration. Math. Comput. **66**(217), 345–361 (1997)
15. Delves, L.M., Lyness, J.N.: A numerical method for locating the zeros of an analytic function. Math. Comput. **21**, 543–560 (1967)
16. Demeure, C.J., Mullis, C.T.: The Euclid algorithm and the fast computation of cross-covariance and autocovariance sequences. IEEE Trans. Acoust. Speech Signal Process. **37**, 545–552 (1989)

17. Demeure, C.J., Mullis, C.T.: A Newton-Raphson method for moving-average spectral factorization using the Euclid algorithm. IEEE Trans. Acoust. Speech Signal Process. **38**, 1697–1709 (1990)
18. Ehrlich, L.W.: A modified Newton method for polynomials. Commun. ACM **10**, 107–108 (1967)
19. Emiris, I.Z., Pan, V.Y., Tsigaridas, E.: Algebraic algorithms. In: Tucker, A.B., Gonzales, T., Diaz-Herrera, J.L. (eds.) Computing Handbook. Computer Science and Software Engineering, 3rd edn., vol. I, Chap. 10, pp. 10-1–10-40. Taylor and Francis Group (2014)
20. Fortune, S.: J. Symbol. Comput. **33**(5), 627–646 (2002). Proc. version in Proc. Intern. Symp. on Symbolic and Algebraic Computation An Iterated Eigenvalue Algorithm for Approximating Roots of Univariate Polynomials, (ISSAC 2001), 121–128, ACM Press, New York (2001)
21. Householder, A.S.: Dandelin, Lobachevskii, or Graeffe? Amer. Math. Mon. **66**, 464–466 (1959)
22. Henrici, P.: Applied and computational complex analysis. In: Power Series, Integration, Conformal Mapping, Location of Zeros, vol. 1. Wiley, New York (1974)
23. Henrici, P., Gargantini, I.: Uniformly convergent algorithms for the simultaneous approximation of all zeros of a polynomial. In: Dejon, B., Henrici, P. (eds.) Constructive Aspects of the Fundamental Theorem of Algebra. Wiley, New York (1969)
24. Imbach, R., Pan, V.Y., Yap, C.: Implementation of a near-optimal complex root clustering algorithm. In: Davenport, J.H., Kauers, M., Labahn, G., Urban, J. (eds.) ICMS 2018. LNCS, vol. 10931, pp. 235–244. Springer, Cham (2018). https://doi.org/10.1007/978-3-319-96418-8_28
25. Inbach, R., Pan, V.Y., Yap, C., Kotsireas, I.S., Zaderman, V.: Root-finding with implicit deflation. In: Proceedings CASC 2019. arxiv:1606.01396. Accepted 21 May 2019
26. Kirrinnis, P.: Polynomial factorization and partial fraction decomposition by simultaneous Newton's iteration. J. Complex. **14**, 378–444 (1998)
27. Kobel, A., Rouillier, F., Sagraloff, M.: Computing real roots of real polynomials ... and now for real! In: The International Symposium on Symbolic and Algebraic Computation (ISSAC 2016), pp. 301–310. ACM Press, New York (2016)
28. McNamee, J.M.: Numerical Methods for Roots of Polynomials, Part I, p. XIX+354. Elsevier, Amsterdam (2007)
29. Moenck, R., Borodin, A.: Fast modular transforms via division., In: Proceedings of 13th Annual Symposium on Switching and Automata Theory (SWAT 1972), pp. 90–96. IEEE Computer Society Press (1972)
30. McNamee, J.M., Pan, V.Y.: Numerical Methods for Roots of Polynomials, Part II, p. XXI+728. Elsevier, Amsterdam (2013)
31. Neff, C.A., Reif, J.H.: An $o(n^{1+\epsilon})$ algorithm for the complex root problem. In: Proceedings 35th Annual Symposium on Foundations of Computer Science (FOCS 1994), pp. 540–547. IEEE Computer Society Press (1994)
32. Pan, V.Y.: Optimal (up to polylog factors) sequential and parallel algorithms for approximating complex polynomial zeros. In: Proceedings of 27th Annual ACM Symposium on Theory of Computing (STOC 1995), pp. 741–750. ACM Press, New York (1995)
33. Pan, V.Y.: Solving a polynomial equation: some history and recent progress. SIAM Rev. **39**(2), 187–220 (1997)
34. Pan, V.Y.: Solving polynomials with computers. Am. Sci. **86**, 62–69 (1998)

35. Pan, V.Y.: Approximation of complex polynomial zeros: modified quadtree (Weyl's) construction and improved Newton's iteration. J. Complex. **16**(1), 213–264 (2000)

36. Pan, V.Y.: Structured Matrices and Polynomials: Unified Superfast Algorithms. Birkhäuser/Springer, Boston/New York (2001). https://doi.org/10.1007/978-1-4612-0129-8

37. Pan, V.Y.: Univariate polynomials: nearly optimal algorithms for factorization and rootfinding. J. Symb. Comput. **33**(5), 701–733 (2002)

38. Pan, V.Y.: Transformations of matrix structures work again. Linear Algebra Appl. **465**, 1–32 (2015)

39. Pan, V.Y.: Root-finding with Implicit Deflation. arXiv:1606.01396, Accepted 4 June 2016

40. Pan, V.Y.: Simple and nearly optimal polynomial root-finding by means of root radii approximation. In: Kotsireas, I., Martinez-Moro, E. (eds.) ACA 2015. SPMS, vol. 198, pp. 329–340. Springer, Cham (2017). https://doi.org/10.1007/978-3-319-56932-1_23

41. Pan, V.Y.: Fast approximate computations with Cauchy matrices and polynomials. Math. Comput. **86**, 2799–2826 (2017)

42. Pan, V.Y.: Old and new nearly optimal polynomial root-finders, In: Proceedings of CASC (2019). Also arxiv: 1805.12042 May 2019

43. Pan, V.Y., Sadikou, A., Landowne, E.: Univariate polynomial division with a remainder by means of evaluation and interpolation. In: Proceedings of 3rd IEEE Symposium on Parallel and Distributed Processing, pp. 212–217. IEEE Computer Society Press, Los Alamitos (1991)

44. Pan, V.Y., Sadikou, A., Landowne, E.: Polynomial division with a remainder by means of evaluation and interpolation. Inform. Process. Lett. **44**, 149–153 (1992)

45. Pan, V.Y., Tsigaridas, E.P.: Nearly optimal refinement of real roots of a univariate polynomial. J. Symb. Comput. **74**, 181–204 (2016). Proceedings version. In: Kauers, M. (ed.) Proc. ISSAC 2013, pp. 299–306. ACM Press, New York (2013)

46. Pan, V.Y., Tsigaridas, E.P.: Nearly optimal computations with structured matrices. Theor. Comput. Sci. **681**, 117–137 (2017)

47. Pan, V.Y., Zhao, L.: Real root isolation by means of root radii approximation. In: Gerdt, V.P., Koepf, W., Seiler, W.M., Vorozhtsov, E.V. (eds.) CASC 2015. LNCS, vol. 9301, pp. 349–360. Springer, Heidelberg (2015). https://doi.org/10.1007/978-3-319-24021-3_26

48. Pan, V.Y., Zhao, L.: Real polynomial root-finding by means of matrix and polynomial iterations. Theor. Comput. Sci. **681**, 101–116 (2017)

49. Renegar, J.: On the worst-case arithmetic complexity of approximating zeros of polynomials. J. Complex. **3**(2), 90–113 (1987)

50. Schönhage, A.: The Fundamental Theorem of Algebra in Terms of Computational Complexity. Department of Mathematics. University of Tübingen, Tübingen, Germany (1982)

51. Schönhage, A.: Asymptotically fast algorithms for the numerical muitiplication and division of polynomials with complex coefficients. In: Calmet, J. (ed.) EUROCAM 1982. LNCS, vol. 144, pp. 3–15. Springer, Heidelberg (1982). https://doi.org/10.1007/3-540-11607-9_1

52. Schönhage, A.: Quasi GCD computations. J. Complex. **1**, 118–137 (1985)

53. Schleicher, D.: Private communication (2018)

54. Schleicher, D., Stoll, R.: Newton's method in practice: finding all roots of polynomials of degree one million efficiently. Theor. Comput. Sci. **681**, 146–166 (2017)

55. Tilli, P.: Convergence conditions of some methods for the simultaneous computation of polynomial zeros. Calcolo **35**, 3–15 (1998)
56. Van Dooren, P.: Some numerical challenges in control theory. In: Van Dooren, P., Wyman, B. (eds.) Linear Algebra for Control Theory. The IMA Volumes in Mathematics and its Applications, vol. 62. Springer, New York (1994). https://doi.org/10.1007/978-1-4613-8419-9_12
57. Weierstrass, K.: Neuer Beweis des Fundamentalsatzes der Algebra. Mathematische Werke, Bd, vol. III, pp. 251–269. Mayer und Mueller, Berlin (1903)
58. Weyl, H.: Randbemerkungen zu Hauptproblemen der Mathematik. II. Fundamentalsatz der Algebra und Grundlagen der Mathematik. Math. Z. **20**, 131–151 (1924)
59. Wilson, G.T.: Factorization of the covariance generating function of a pure moving-average. SIAM J. Numer. Anal. **6**, 1–7 (1969)
60. Werner, W.: Some improvements of classical iterative methods for the solution of nonlinear equations. In: Allgower, E.L., Glashoff, K., Peitgen, H.-O. (eds.) Numerical Solution of Nonlinear Equations. LNM, vol. 878, pp. 426–440. Springer, Heidelberg (1981). https://doi.org/10.1007/BFb0090691

Symbolic-Numeric Implementation of the Four Potential Method for Calculating Normal Modes: An Example of Square Electromagnetic Waveguide with Rectangular Insert

A. A. Tiutiunnik[1], D. V. Divakov[1]([✉]), M. D. Malykh[1], and L. A. Sevastianov[1,2]

[1] Department of Applied Probability and Informatics,
Peoples' Friendship University of Russia (RUDN University), 6 Miklukho-Maklaya St,
Moscow 117198, Russia
{tyutyunnik-aa,divakov-dv,malykh-md,sevastianov-la}@rudn.ru
[2] Joint Institute for Nuclear Research (Dubna), Joliot-Curie, 6, Dubna,
Moscow Region 141980, Russia

Abstract. In this paper, the Maple computer algebra system is used to construct a symbolic-numeric implementation of the method for calculating normal modes of square closed waveguides in a vector formulation. The method earlier proposed by Malykh et al. [M.D. Malykh, L.A. Sevastianov, A.A. Tiutiunnik, N.E. Nikolaev. On the representation of electromagnetic fields in closed waveguides using four scalar potentials // Journal of Electromagnetic Waves and Applications, 32 (7), 886–898 (2018)] will be referred to as the method of four potentials. The Maple system is used at all stages of treating the system of differential equations for four potentials: the generation of the Galerkin basis, the substitution of approximate solution into the system under study, the formulation of a computational problem, and its approximate solution.

Thanks to the symbolic-numeric implementation of the method, it is possible to carry out calculations for a large number of basis functions of the Galerkin decomposition with reasonable computation time and then to investigate the convergence of the method and verify it, which is done in the present paper, too.

Keywords: Computer algebra system · Symbolic-numeric method · Waveguide · Maxwell equations · Four potentials method · Normal modes · Incomplete galerkin method

The publication has been prepared with the support of the "RUDN University Program 5-100" and funded by RFBR according to the research projects Nos. 18-07-00567 and 18-51-18005.

M. England et al. (Eds.): CASC 2019, LNCS 11661, pp. 412–429, 2019.
https://doi.org/10.1007/978-3-030-26831-2_27

1   Introduction

Computer algebra systems have recently been actively used to construct symbolic-numeric methods for approximate solution of computational problems of mathematical physics [3–5], electrodynamics [1,2,6], mechanics [7–9], and fluid simulations too [10–13].

Particularly efficient is the use of a computer algebra system in the application of methods based on Galerkin and Kantorovich decompositions. Approximate solutions within the framework of these methods are represented as partial sums of a series over some complete system of basis functions. Approximate solutions are substituted into the problem under consideration and the resulting relations are projected onto some finite-dimensional subspace of the function space, resulting in the formulation of an approximate problem.

When computerising the methods of Galerkin and Kantorovich [14,15], it becomes necessary to apply projection relations, including integrals (often from rapidly oscillating functions) (see [3,4]). If the basic functions of the Galerkin and Kantorovich methods are known in symbolic form (for example, polynomials, trigonometric functions, etc.), then it is convenient to set up an approximate problem in a computer algebra system.

In this paper, we have developed a symbolic-numeric implementation of the four potential method [1] for computing the normal modes of a rectangular closed waveguides in vector formulation using the Maple computer algebra system.

2   Preliminary Information: Four Potential Method

In the framework of the four potential method [1,2], a waveguide of constant cross section S with ideally conducting walls is considered. The axis Oz with unit vector e_z is directed along the axis of the waveguide; the normal to the side wall will be denoted by n, and the tangent vector perpendicular to both e_z and n by τ (Fig. 1).

In a closed waveguide, the components of the electromagnetic field E_z and H_z are expressed in terms of $E_\perp = (E_x, \ E_y)^T$ and $H_\perp = (H_x, \ H_y)^T$ [16–18], therefore, within the framework of the method, the electromagnetic field can be represented as $E_\perp = \nabla u_e + \nabla' v_e$, $H_\perp = \nabla v_h + \nabla' u_h$, where $u_{e,h}$ and $v_{e,h}$ are scalar functions that satisfy the Dirichlet and Neumann conditions at the waveguide boundary. The Maxwell equations in this representation are reduced to a system of differential equations for four potentials [1]:

$$\begin{cases} \nabla'\left(\partial_z v_h - \frac{1}{ik_0\mu}\Delta v_e\right) + ik_0\varepsilon\nabla'v_e - \nabla\partial_z u_h + ik_0\varepsilon\nabla u_e = 0, \\ \nabla'\left(\partial_z u_e + \frac{1}{ik_0\epsilon}\Delta u_h\right) - ik_0\mu\nabla'u_h - \nabla\partial_z v_e - ik_0\mu\nabla v_h = 0, \end{cases} \quad (1)$$

where ε and μ are permittivity and permeability, k_0 is the wavenumber, i is the imaginary unit, $\nabla = (\partial/\partial x, \partial/\partial y)^T$, $\nabla' = (-\partial/\partial y, \partial/\partial x)^T$. The boundary conditions for the potentials are as follows [1]:

$$u_e|_{\partial S} = u_h|_{\partial S} = 0, \ \ \nabla v_e \cdot n|_{\partial S} = \nabla v_h \cdot n|_{\partial S} = 0. \quad (2)$$

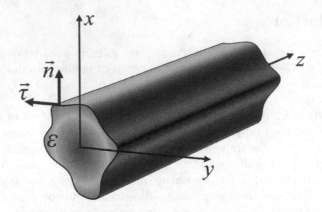

Fig. 1. Waveguide of constant section S

The Dirichlet and Neumann boundary conditions (2) that are classical for mathematical physics [19, 20] often allow constructing an approximate solution applying the Galerkin method with, e.g., the eigenfunctions of the Laplace operator with appropriate boundary conditions as the basis. In the case of a complex cross section geometry, it is more convenient to construct an approximate solution by means of the finite element method [3, 21–26]. However, in the case of simple geometry that allows separation of variables, e.g., a rectangular cross section, an approximate solution can be constructed using the Galerkin method.

To test the performance of the method, we will first consider waveguides of square cross section with an arbitrary rectangular insert. In this case, we will construct an approximate solution using the incomplete Galerkin method [27, 28] and then investigate the numerically obtained solutions.

3   Setting a Computational Problem

We first consider a waveguide homogeneous along the z-axis with square cross section $S = \{x \in [0; 1], y \in [0; 1]\}$.

Let the permittivity and permeability ε, μ do not change along the z-axis on a certain segment $a < z < b$. We decompose the approximate solution in a finite number of linearly independent functions satisfying the boundary conditions (2). A suitable basis in this case is the basis $\{\varphi_j\}_{j=1}^{4N^2-2}$ composed of eigenfunctions of the Laplace operator with Dirichlet conditions and Neumann conditions, namely [19] (Fig. 2):

$$\{\varphi_j\}_{j=1}^{N^2-1} = (\psi_{nm}(x,y), 0, 0, 0)^T,$$

$$\{\varphi_j\}_{j=N^2}^{2N^2-1} = (0, \psi_{nm}(x,y), 0, 0)^T, \tag{3}$$

$$m, n = 0, \ldots, N-1, m+n > 0,$$

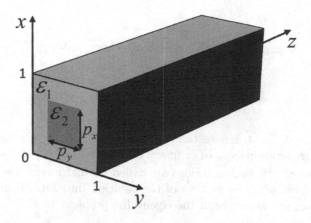

Fig. 2. The geometry of a square waveguide with an arbitrary rectangular insert having the width p_y and the height p_x

$$\{\varphi_j\}_{j=2N^2}^{3N^2-2} = (0, 0, \phi_{nm}(x, y), 0)^T,$$

$$\{\varphi_j\}_{j=3N^2-1}^{4N^2-2} = (0, 0, 0, \phi_{nm}(x, y))^T, \qquad (4)$$

$$m, n = 1, \ldots, N,$$

where $\psi_{nm}(x, y) = \cos(\pi nx)\cos(\pi my)$, $\phi_{nm}(x, y) = \sin(\pi nx)\sin(\pi my)$ are eigenfunctions of the Neumann and Dirichlet problems for the two-dimensional Laplace operator in the square S. An approximate solution of the system will be constructed by analogy with the incomplete Galerkin method [27,28] in the form of expansion

$$(v_h, v_e, u_e, u_h)^T = \sum_{j=1}^{4N^2-2} w_j(z)\,\varphi_j(x, y). \qquad (5)$$

After substituting the approximate solution into the equations of the system and applying the projection scheme of the Galerkin method, we arrive at a system of ordinary differential equations for the vector of the desired coefficient functions $w = (w_1, w_2, \ldots, w_{4N^2-2})^T$:

$$Bw' + ik_0Aw + (ik_0)^{-1}Cw = 0, \qquad (6)$$

where A, B, and C are square matrices, whose elements are double integrals over the waveguide section S:

$$a_{ij} = \iint\limits_S \varepsilon\,(\nabla'\varphi_{j2} + \nabla\varphi_{j3}) \cdot (\nabla'\varphi_{i1} - \nabla\varphi_{i4})\,dxdy+$$

$$+ \iint\limits_S \mu\,(\nabla'\varphi_{j4} + \nabla\varphi_{j1}) \cdot (\nabla'\varphi_{i2} - \nabla\varphi_{i3})\,dxdy, \qquad (7)$$

$$b_{ij} = \iint\limits_{S} (\nabla'\varphi_{j1} \cdot \nabla'\varphi_{i1} + \nabla'\varphi_{j3} \cdot \nabla'\varphi_{i3} + \nabla\varphi_{j2} \cdot \nabla\varphi_{i2} + \nabla\varphi_{j4} \cdot \nabla\varphi_{i4}) \, dx dy, \quad (8)$$

$$c_{ij} = \iint\limits_{S} (\Delta\varphi_{j2}\Delta\varphi_{i1}/\mu - \Delta\varphi_{j4}\Delta\varphi_{i3}/\varepsilon) \, dx dy, \quad (9)$$

where φ_{jn}, $n = 1, \ldots, 4$ denote the nth component of the jth vector function φ_j. The homogeneous system of ordinary differential Eq. (6) has solutions that depend on z as $e^{ik_0\beta z}$; such solutions are called normal waveguide modes. Substituting the form $w(z) = \psi e^{ik_0\beta z}$ of the solution into Eq. (6) and cancelling the nonzero factors, we arrive at the eigenvalue problem

$$K\psi = \beta\psi, \quad (10)$$

where the matrix K is expressed through the matrices A, B, and C as $K = B^{-1}(A - k_0^{-2}C)$. The eigenvalue problem (10) must then be solved numerically. The desired eigenvalue β entering the exponent of the solution is called the mode phase constant; it determines the mode propagation behaviour. Modes with real $\beta > 0$ are waves traveling along the axis of the waveguide in the positive direction, modes with $\beta < 0$ propagate in the negative direction. Modes with real β are called propagating modes. Modes with imaginary β exponentially decrease in one of the directions and are called evanescent modes.

4   Symbolic-Numeric Algorithm for the Approximate Calculation of Normal Modes

4.1   Introductory Notes

The calculation of the elements of the matrices A, B, and C using the formulas (7)–(9) is associated with certain difficulties. The basis (3)–(4) chosen by us consists of four-component vector functions, each of whose nonzero components is an eigenfunction of the Dirichlet or Neumann problem for the Laplace operator in the domain $S = \{x \in [0; 1], y \in [0; 1]\}$.

The complexity of the calculations in the framework of the proposed method is easily illustrated by a simple example. Suppose we want to describe the modes that are characterised by no more than five oscillations on each of the axes Ox and Oy. In this case, for $\psi_{nm}(x, y) = \cos(\pi n x)\cos(\pi m y)$ you must keep at least $n = \overline{1, 5}$, $m = \overline{1, 5}$, i.e., at least 25 eigenfunctions. Moreover, since in the vector basis (3)–(4), the functions are four-component, the basis will consist of $25 \times 4 = 100$ functions. In this case, the matrix K to be considered will have a dimension of 100×100 (that is, consist of 10^4 elements). In more realistic calculations, the dimension of the matrix K under study will be greater by an order of magnitude.

Note also that the matrix $K = B^{-1}\left(A - k_0^{-2}C\right)$ is defined by three matrices A, B, C and each of the elements of each of the three matrices is a double integral over the cross section S of the waveguide (see (7)–(9)). Summarizing the above considerations, only for calculating the matrix K it is necessary to find at least 3×10^4 double integrals. Obviously, the numerical calculation of such a number of integrals of oscillating functions with acceptable accuracy does not allow even the formation of the matrix K in reasonable time. The study of the method requires the calculation of the matrix K for a different number of basis functions, generally speaking, for the largest possible one, with which the calculation is performed during reasonable time.

4.2 Symbolic-Numeric Algorithm

The formulation of the computational problem (10) consists in calculating the elements of the matrix K, for which it is first necessary to generate basic functions $\{\varphi_j\}$ using the built-in tools of the Maple system [29]. The details of symbolic algorithm are presented in Appendix.

Numerical calculations consist in finding the eigenvalues and eigenvectors of the matrix K using built-in Maple numerical tools. The eigenvalues β_j of the matrix K determine the phase constants for the normal modes of the waveguide in question, and the eigenvectors ψ_j determine the coefficients of expansion of the jth mode in the basis (3)–(4).

Numerical calculations are also implemented using Maple built-in numerical methods (using LAPACK library [30,31] with QR factorization [32,33]). Eigenvalues and eigenvectors are calculated using the function Eigenvectors from the LinearAlgebra library [29] with high accuracy that can be evaluated by Van Loan formulas [34,35]. The described algorithm is implemented in Maple and is available at https://bitbucket.org/AnastasiiaTyu/waveguide/downloads/.

The implemented algorithm is used to conduct numerical experiments aimed at verifying the efficiency of the method and verification.

5 Numerical Calculation of Normal Modes in a Square Waveguide with a Rectangular Insert

Consider a square-cross-section waveguide of unit width and height made from a material with a dielectric constant $\varepsilon_1 = 2$ with an insert having the width p_x and the height p_y, made from a more optically dense material with $\varepsilon_2 = 3$. Let us calculate the eigenvalues (phase constants) and normal modes for the configuration $p_x = p_y = 0.5$. In this case, the exact phase constants are not known analytically.

As a result of the calculation with $N = 12$ (which corresponds to the dimension of the matrix $M \times M$ with $M = 574$), we obtained phase constants, the first ten of which are given below:

$$\beta_1 = 1.43296289171709,$$
$$\beta_2 = 1.42946771606211,$$
$$\beta_3 = 1.23553335397222,$$
$$\beta_4 = 1.16064902183147,$$
$$\beta_5 = 0.840564995190615,$$
$$\beta_6 = 0.811608008252520,$$
$$\beta_7 = 0.663953243441277,$$
$$\beta_8 = 0.655779874680068,$$
$$\beta_9 = 0.399854493680243,$$
$$\beta_{10} = 0.396923647592231.$$

(11)

Now let us consider the computed eigenfunctions corresponding to a few first phase constants.

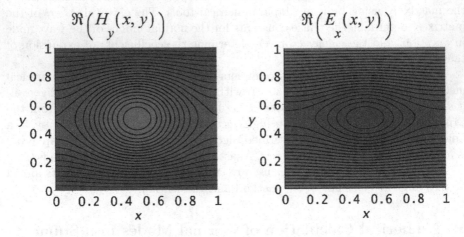

Fig. 3. Electromagnetic field components corresponding to $\beta_1 \approx 1.4330$

Figure 3 shows the level lines corresponding to the first mode of the waveguide with the phase constant $\beta_1 \approx 1.4330$. $\Re(H_y(x,y)) = \mathrm{Re}(H_y(x,y))$ and $\Re(E_x(x,y)) = \mathrm{Re}(E_x(x,y))$ are real parts of the components H_y and E_x. The components of the electromagnetic field corresponding to the TE-mode, namely E_y and H_x, are negligible in comparison with the components H_y, E_x shown in Fig. 3 and, therefore, are not presented. From the type of components H_y and E_x shown in the figure, it can be noted that the field for both components is concentrated in the region corresponding to the insert of a more optically dense material. Such behaviour is illustrated by the well-known optical principle of radiation propagation in waveguides.

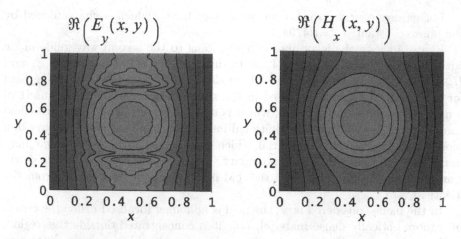

Fig. 4. Electromagnetic field components corresponding to $\beta_2 \approx 1.4294$

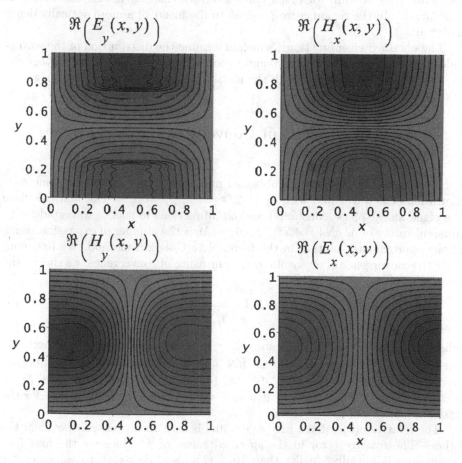

Fig. 5. Electromagnetic field components corresponding to $\beta_3 \approx 1.2355$

Let us now consider the second waveguide mode, which is characterized by the phase constant $\beta_2 \approx 1.4294$.

Figure 4 shows the level lines corresponding to the second waveguide mode with the phase constant $\beta_2 \approx 1.4294$. In this case, the field components E_y and H_x , i.e., the components of the TE mode, are different from zero. The field for this mode is also concentrated in the region corresponding to the insert of a more optically dense material, which is to be expected. In addition, the field component E_y exhibits small additional intensity peaks near the boundary of a more optically dense insert material, which may be due to the presence of a jump in the dielectric constant at the boundary. The next mode with $\beta_3 \approx 1.2355$ will contain both the TE-mode and the TM-mode components different from the identical zero.

In the mode considered now, the field is no longer localised inside the region of a more optically dense material, now it is concentrated outside this region. Such modes also exist in waveguides and, as a rule, when solving diffraction problems, they account for much lower intensity than for modes whose field is concentrated in the region corresponding to the insert of a more optically dense material.

The obtained eigenfunctions, which determine the distribution of the waveguide modes in the region corresponding to the cross section of the waveguide, are in qualitative agreement with the physical concept of radiation propagation in a waveguide (Fig. 5).

6   Numerical Analysis of Convergence of Eigenvalues and Eigenfunctions

Let us now analyse the results obtained numerically. Let us consider one configuration of a waveguide with $\varepsilon_1 = 2$, $\varepsilon_2 = 3$, $p_x = p_y = 0.5$, and calculate the eigenvalues (phase constants) and eigenfunctions in such a waveguide with different values of $M(N) = 4N^2 - 2$, where M is the number of expansion terms of the approximate solution in the incomplete Galerkin method. We first consider the convergence of eigenvalues. As a measure of convergence, we choose the relative error

$$\delta_M(\beta_j) = \left| \frac{\beta_j^{M(N)} - \beta_j^{M(N+1)}}{\beta_j^{M(N+1)}} \right|, \tag{12}$$

where $\beta_j^{M(N)}$ and $\beta_j^{M(N+1)}$ denote the jth eigenvalue obtained numerically with $M = 4N^2 - 2$ and $M = 4(N+1)^2 - 2$, respectively. Calculations were carried out for $N = 10, 12, 14, 16, 18$, which corresponds to $M = 398, 574, 782, 1022, 1294$. Below in Fig. 6, we illustrate the convergence for the first four eigenvalues.

The relative error of the first eigenvalue is seen to decrease faster than the others. The relative error in the specified range of M values for the first four eigenvalues is initially smaller than 10^{-2} (1%) and decreases to values of the order of $5 \cdot 10^{-5}$, and for the first eigenvalue, the relative error is even smaller,

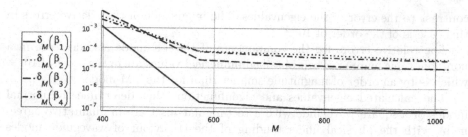

Fig. 6. Plots of relative error $\delta_M(\beta_j)$ for $j = 1, 2, 3, 4$

of the order of 10^{-7}. Thus, the convergence of eigenvalues takes place, and the order of magnitude of the relative error indicates a sufficient accuracy of the calculations.

Now let us proceed to consider the relative error convergence for the eigenfunction corresponding to β_4. The first three eigenfunctions are typically calculated with higher accuracy, so for consistency, we will examine the convergence for the relative error of the fourth eigenfunction.

As a measure of the convergence of the component, we take the relative error

$$\delta_M(E_x) = \int_0^1 \int_0^1 \left| E_x^{M(N)} - E_x^{M(N+1)} \right|^2 dxdy, \qquad (13)$$

where $E_x^{M(N)}$ denotes the component E_x $\left(\int_0^1 \int_0^1 \left| E_x^{M(N)} \right|^2 dxdy = 1 \right)$ obtained numerically and normalized to one with $M = 4N^2 - 2$, and $E_x^{M(N+1)}$ is the normalized component E_x with $M = 4(N+1)^2 - 2$. For the remaining components, the relative error is determined in a similar way (Fig. 7).

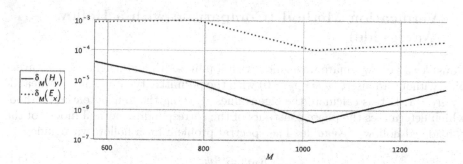

Fig. 7. Plots of the relative error $\delta_M(H_y)$, $\delta_M(E_x)$ of the components of the TM mode for the fourth mode

Considering the plots of the relative errors for the components of the TM mode, one can note that in this case, the convergence is not monotonic, in contrast to the error of the eigenvalues. The magnitude of the relative errors in this case is of the order of 10^{-4}.

The relative errors for the components of the TE mode also decrease in a nonmonotonic manner. Their obtained numerical values are of the order of 10^{-5}, which is by an order of magnitude smaller than for the TM mode.

The calculated eigenvalues and eigenfunctions that determine the normal modes of a square waveguide with a rectangular inset are in qualitative agreement with the physical understanding of the behaviour of waveguide modes propagating in a closed waveguide. Numerical estimates of the relative error also confirm the numerical stability of the calculated values, which indirectly confirms the adequacy of the results obtained.

To test the performance of the method, we also compare the numerical results of our method with the results known analytically in a hollow waveguide model (Fig. 8).

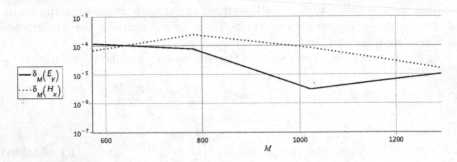

Fig. 8. Plots of the relative error $\delta_M\left(E_y\right), \delta_M\left(H_x\right)$ of the fourth TE mode components

7   Verification Method (Comparison with a Hollow Waveguide)

Consider a hollow square waveguide with a unit width and height, i.e., a waveguide without an insert ($p_x = p_y = 0$) with the permittivity $\varepsilon_1 = 1$ and permeability $\mu_1 = 1$. Let us calculate the eigenvalues β_j using the four potential method, which determines the phase constants of the corresponding normal modes of the considered hollow waveguide. The spectral problem for a hollow waveguide is

$$K_0\psi = \beta\psi, \tag{14}$$

where, $K_0 = B^{-1}\left(A - k_0^{-2}C\right)$, the elements of the matrix B are determined by Eq. (8), and the elements of the matrices A and C are defined as

$$a_{ij}^0 = \iint\limits_S (\nabla'\varphi_{j2} + \nabla\varphi_{j3}) \cdot (\nabla'\varphi_{i1} - \nabla\varphi_{i4})\, dxdy$$

$$+ \iint\limits_S (\nabla'\varphi_{j4} + \nabla\varphi_{j1}) \cdot (\nabla'\varphi_{i2} - \nabla\varphi_{i3})\, dxdy, \qquad (15)$$

$$c_{ij}^0 = \iint\limits_S (\Delta\varphi_{j2}\Delta\varphi_{i1} - \Delta\varphi_{j2}\Delta\varphi_{i3})\, dxdy.$$

Taking into account the fact that the quantities ε_1, μ_1 are constants, in the integrals of Eq. (15) the integrands are simplified, thus simplifying the structure of matrices A and C.

Note that in the hollow waveguide, the structure of the matrices is simplified, the matrix K_0 has only two diagonals differing from the identical zero.

Next, we calculate the eigenvalues of the matrix K_0 using the four potential method and compare the eigenvalues obtained as a result of a numerical calculation with the analytically known eigenvalues, expressed as $\beta_{1,2} = \pm\sqrt{1 - \pi^2/k_0^2}$, $\beta_{3,4} = \pm\sqrt{1 - 2\pi^2/k_0^2}$. The eigenvalues obtained in the framework of the proposed method will be denoted $\beta_j^{M(N)}$, where the parameter $M(N) = 4N^2 - 2$ is the number of eigenfunctions to be held in the expansion of the approximate solution. Next, we present the absolute and relative errors for the calculated eigenvalues with $N = 3, 4, 5, 6, 7$, which correspond to the dimensions $M = 34, 62, 98, 142, 194$ of the matrices:

$$\left|\beta_j^{M(N)} - \beta_j\right| < 2.82 \cdot 10^{-11},$$

$$\left|\beta_j^{M(N)} - \beta_j\right| / |\beta_j| < 3.62 \cdot 10^{-11}, \qquad (16)$$

$$j = 1, 2, 3, 4.$$

The error values obtained in (16) demonstrate the high accuracy of approximate calculations of the eigenvalues β_j that determine the phase constants of the corresponding normal modes in a hollow waveguide.

8   Conclusion

The proposed approach, based on the introduction of potentials, allows adequate description of the normal modes in closed waveguides having square cross section.

The symbolic-numeric implementation of the method allows reduction of the number of calculations needed to find double integrals from rapidly oscillating functions, due to the use of symbolic calculations. The use of the Maple computer algebra system made it possible to perform calculations with matrices having the dimension 1200×1200 in an acceptable computation time, making it possible to investigate the convergence of the method.

The validity of the four potential method within the frameworks of this study was verified by qualitative and numerical analysis of the resulting eigenvalues and eigenfunctions, as well as by comparison with known results for closed waveguides:

- for closed waveguides, the convergence of eigenvalues (phase constants of waveguide modes) and eigenfunctions was investigated, the obtained orders of the relative error are $5 \cdot 10^{-5}$ for eigenvalues and 10^{-4} for eigenfunctions;
- in the example of a model structure (a hollow waveguide), for which the exact eigenvalues are known in analytical form, the proposed method yields an error of the order of 10^{-11};
- the qualitative behaviour of the waveguide modes in the considered closed waveguides agrees with its physical understanding.

Appendix. Symbolic Algorithm

The integral expressions (7)–(9) contain three differential operators: the Laplace operator $\Delta = \partial^2/\partial x^2 + \partial^2/\partial y^2$ and two vector operators $\nabla = (\partial/\partial x, \partial/\partial y)^T, \nabla' = (-\partial/\partial y, \partial/\partial x)^T$, which we define in Maple:

```
D0 := proc (f)
return diff(f, x$2) + diff(f, y$2)
end proc;

D1 := proc (f)
return VectorCalculus[Gradient](f, [x, y])
end proc;

D2 := proc (f)
local v;
v := VectorCalculus[Gradient](f, [x, y]);
return Vector([-v[2], v[1]])
end proc;
```

Using the introduced operators, we consider the calculation of the matrix elements of the matrix C. The matrix element c_{ij} is determined by the expression

$$c_{ij} = c\left(\varphi_i, \varphi_j\right) = \iint_S \left(\Delta\varphi_{j2}\Delta\varphi_{i1}/\mu - \Delta\varphi_{j4}\Delta\varphi_{i3}/\varepsilon\right) dxdy \qquad (17)$$

From Eq. (17) it is obvious that if $\varphi_{j2} = \varphi_{j4} \equiv 0$ or, e.g., $\varphi_{i1} = \varphi_{i3} \equiv 0$, then $c_{ij} = 0$. Therefore, ignoring knowingly zero results and using the built-in tools of Maple, one can calculate only non-zero blocks of the matrix.

Here we calculate the integrands for non-zero blocks of the matrix:

```
vv1 := cos(Pi*iI1*x)*cos(Pi*jJ1*y);
vv2 := cos(Pi*iI2*x)*cos(Pi*jJ2*y);
uu1 := sin(Pi*iI1*x)*sin(Pi*jJ1*y);
uu2 := sin(Pi*iI2*x)*sin(Pi*jJ2*y);
Dvv := D0(vv1)*D0(vv2);
Duu := D0(uu1)*D0(uu2)
```

Particular attention is paid to the method of reducing the number of calculated integrals. In fact, the integrand (17) includes $\Delta\psi_{nm}$ and $\Delta\phi_{n'm'}$, where $\psi_{nm}(x,y) = \cos(\pi nx)\cos(\pi my)$, $\phi_{nm}(x,y) = \sin(\pi nx)\sin(\pi my)$. Using the fact that the components of the basis functions ψ_{nm} and ϕ_{nm} are known in the symbolic form, the integral (17) can be calculated once, keeping the subscripts n, m, n', m' in symbolic form:

```
intDvv  := int(Dvv, [x = 0 .. 1, y = 0 .. 1]);
intDuu1 := int(Duu, [x = 0 .. 1, y = 0 .. 1]);
intDuu2 := int(Duu, [x = (1-ppx)*(1/2) .. (1+ppx)*(1/2),
y = (1-ppy)*(1/2) .. (1+ppy)*(1/2)]);
intDuu3 := intDuu1/eps1+(1/eps2-1/eps1)*intDuu2
```

Further, the specific elements of nonzero blocks of matrices are calculated by the replacing the symbols n, m, n', m' with specific numbers in the symbolically calculated integrals:

```
cEPS3 := Matrix(n^2, n^2, (i, j)→ subs({iI1 =
   indxU[i][1],
iI2 = indxU[j][1]+eps, jJ1 = indxU[i][2],
jJ2 = indxU[j][2]+eps}, intDuu3));

cMU := Matrix(n^2-1, n^2-1, (i, j)→ subs({iI1 =
   indxV[i][1],
iI2 = indxV[j][1]-eps, jJ1 = indxV[i][2],
jJ2 = indxV[j][2]+eps}, intDvv));

for i to n^2 do
    for j to n^2 do
        Cc[n^2+i-1, 3*n^2+j-2] := -cEPS3[i, j]
    end do;
end do;
for i to n^2-1 do
    for j to n^2-1 do
        Cc[i, 2*n^2+j-1] := cMU[i, j]
    end do;
end do;
Ccf := evalf(Cc);
return Ccf
```

The calculation of B matrix elements is much simpler due to the orthogonality of the basis (3)–(4) used to decompose the approximate solution. Excluding zero elements of the matrix B, we obtain a matrix with nonzero elements on the main diagonal. As a result, it appears necessary to calculate only the diagonal elements (Fig. 9):

```
vv := cos(Pi*iI*x)*cos(Pi*jJ*y);
vv01 := cos(Pi*jJ*y);
vv10 := cos(Pi*iI*x);
uu := sin(Pi*iI*x)*sin(Pi*jJ*y);

D2D2v := D2(vv).D2(vv);
D1D1v := D1(vv).D1(vv);
D2D2u := D2(uu).D2(uu);
D1D1u := D1(uu).D1(uu);

intD2D2v01 := int(D2(vv01).D2(vv01), [x = 0 .. 1, y = 0 .. 1]);
intD2D2v10 := int(D2(vv10).D2(vv10), [x = 0 .. 1, y = 0 .. 1]);
intD1D1v01 := int(D1(vv01).D1(vv01), [x = 0 .. 1, y = 0 .. 1]);
intD1D1v10 := int(D1(vv10).D1(vv10), [x = 0 .. 1, y = 0 .. 1]);

intD2D2v := int(D2D2v, [x = 0 .. 1, y = 0 .. 1]);
intD2D2u := int(D2D2u, [x = 0 .. 1, y = 0 .. 1]);
intD1D1v := int(D1D1v, [x = 0 .. 1, y = 0 .. 1]);
intD1D1u := int(D1D1u, [x = 0 .. 1, y = 0 .. 1])
```

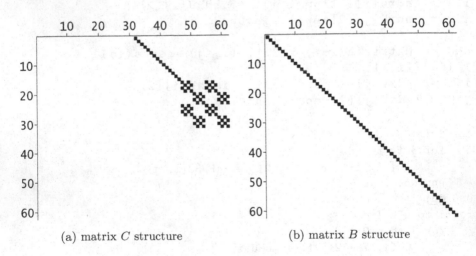

(a) matrix C structure   (b) matrix B structure

Fig. 9. Matrices C and B ($M \times M$) structures for $M = 62$: black squares correspond to non-zero elements of the matrix, and white squares correspond to the identical zero

The structure of the matrix A is more difficult, since it contains significantly greater number of nonzero blocks.

```
vv1 := cos(Pi*iI1*x)*cos(Pi*jJ1*y);
vv2 := cos(Pi*iI2*x)*cos(Pi*jJ2*y);
uu1 := sin(Pi*iI1*x)*sin(Pi*jJ1*y);
uu2 := sin(Pi*iI2*x)*sin(Pi*jJ2*y);

a12 := D2(vv1).D2(vv2);
a13 := D2(vv1).D1(uu2);
a21 := D1(vv1).D1(vv2);
a34 := -D2(uu1).D2(uu2);
a42 := -D1(uu1).D2(vv2);
a43 := -D1(uu1).D1(uu2)
```

However, the principle of calculation is similar to that already described, i.e., the integration is carried out once for a whole class of integrals containing ψ_{nm} and $\phi_{n'm'}$ in the integrand (Fig. 10).

After calculating the matrices A, B, and C, the matrix K is determined.

```
KN := BN.(AN-evalf(1.0/k0^2)*CN)
```

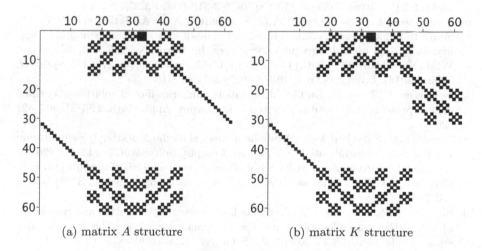

(a) matrix A structure   (b) matrix K structure

Fig. 10. Matrices A and K ($M \times M$) structures for $M = 62$: black squares correspond to non-zero elements of the matrix, and white squares correspond to the identical zero

The described symbolic transformations aimed, firstly, at calculating only non-zero blocks of matrices, secondly, at reducing the number of integrations, made it possible to avoid multiple repetitions of the time-consuming procedure of numerical integration, and immediately proceed to numerical calculations using the Maple algorithm.

References

1. Malykh, M.D., Sevastianov, L.A., Tiutiunnik, A.A., Nikolaev, N.E.: On the representation of electromagnetic fields in closed waveguides using four scalar potentials. J. Electromagn. Waves Appl. **32**(7), 886–898 (2018)
2. Divakov, D.V., Lovetskiy, K.P., Malykh, M.D., Tiutiunnik, A.A.: The application of Helmholtz decomposition method to investigation of multicore fibers and their application in next-generation communications systems. Commun. Comput. Inf. Sci. **919**, 469–480 (2018)
3. Gusev, A.A., et al.: Symbolic-numerical algorithms for solving the parametric self-adjoint 2D elliptic boundary-value problem using high-accuracy finite element method. In: Gerdt, V.P., Koepf, W., Seiler, W.M., Vorozhtsov, E.V. (eds.) CASC 2017. LNCS, vol. 10490, pp. 151–166. Springer, Cham (2017). https://doi.org/10.1007/978-3-319-66320-3_12
4. Gusev, A.A., et al.: Symbolic-numerical algorithm for generating interpolation multivariate hermite polynomials of high-accuracy finite element method. In: Gerdt, V.P., Koepf, W., Seiler, W.M., Vorozhtsov, E.V. (eds.) CASC 2017. LNCS, vol. 10490, pp. 134–150. Springer, Cham (2017). https://doi.org/10.1007/978-3-319-66320-3_11
5. Shapeev, V.P., Vorozhtsov, E.V.: The method of collocations and least residuals combining the integral form of collocation equations and the matching differential relations at the solution of pdes. In: Gerdt, V.P., Koepf, W., Seiler, W.M., Vorozhtsov, E.V. (eds.) CASC 2017. LNCS, vol. 10490, pp. 346–361. Springer, Cham (2017). https://doi.org/10.1007/978-3-319-66320-3_25
6. Sevastyanov, L.A., Sevastyanov, A.L., Tyutyunnik, A.A.: Analytical calculations in maple to implement the method of adiabatic modes for modelling smoothly irregular integrated optical waveguide structures. In: Gerdt, V.P., Koepf, W., Seiler, W.M., Vorozhtsov, E.V. (eds.) CASC 2014. LNCS, vol. 8660, pp. 419–431. Springer, Cham (2014). https://doi.org/10.1007/978-3-319-10515-4_30
7. Bertolazzi, E., Biral, F., Da Lio, M.: Symbolic-numeric efficient solution of optimal control problems for multibody systems. J. Comput. Appl. Math. **185**(2), 404–421 (2006)
8. Gutnik, S.A., Sarychev, V.A.: Symbolic-numerical methods of studying equilibrium positions of a gyrostat satellite. Program. Comput. Softw. **40**(3), 143–150 (2014)
9. Budzko, D.A., Prokopenya, A.N.: Symbolic-numerical analysis of Equilibrium solutions in a restricted four-body problem. Program. Comput. Softw. **36**(2), 68–74 (2010)
10. Shapeev, V.P., Vorozhtsov, E.V.: Symbolic-numerical optimization and realization of the method of collocations and least residuals for solving the Navier-stokes equations. In: Gerdt, V.P., Koepf, W., Seiler, W.M., Vorozhtsov, E.V. (eds.) CASC 2016. LNCS, vol. 9890, pp. 473–488. Springer, Cham (2016)
11. Semin, L., Shapeev, V.: Constructing the numerical method for Navier-stokes equations using computer algebra system. In: Ganzha, V.G., Mayr, E.W., Vorozhtsov, E.V. (eds.) CASC 2005. LNCS, vol. 3718, pp. 367–378. Springer, Berlin (2005)
12. Shapeev, V.P., Vorozhtsov, E.V.: CAS application to the construction of the collocations and least residuals method for the solution of 3D Navier-stokes equations. In: Gerdt, V.P., Koepf, W., Mayr, E.W., Vorozhtsov, E.V. (eds.) CASC 2013. LNCS, vol. 8136, pp. 381–392. Springer, Cham (2013)

13. Shapeev, V.P., Vorozhtsov, E.V.: Symbolic-numeric implementation of the method of collocations and least squares for 3D Navier-stokes equations. In: Gerdt, V.P., Koepf, W., Mayr, E.W., Vorozhtsov, E.V. (eds.) CASC 2012. LNCS, vol. 7442, pp. 321–333. Springer, Heidelberg (2012)
14. Kantorovich, L.V., Krylov, V.I.: Approximate Methods of Higher Analysis. Wiley, New York (1964)
15. Fletcher, C.A.J.: Computational Galerkin Methods. Springer-Verlag, Heidelberg (1984)
16. Adams, M.J.: An Introduction to Optical Waveguides. Wiley, New York (1981)
17. Marcuse, D.: Light Transmission Optics. Van Nostrand, New York (1974)
18. Tamir, T.: Guided-Wave Optoelectronics. Springer-Verlag, Berlin (1990)
19. Ladyzhenskaya, O.A.: The Boundary Value Problems of Mathematical Physics. Springer, Heidelberg (1985)
20. Hellwig, G.: Differential Operators of Mathematical Physics. Addison-Wesley, MA (1967)
21. Bathe, K.J.: Finite Element Procedures in Engineering Analysis. Prentice Hall, Englewood Cliffs (1982)
22. Ciarlet, P.: The Finite Element Method for Elliptic Problems. North Holland Publishing Company, Amsterdam (1978)
23. Strang, G., Fix, G.J.: An Analysis of the Finite Element Method. Prentice-Hall, Englewood Cliffs (1973)
24. Bogolyubov, A.N., Mukhartova, Yu.V., Gao, J., Bogolyubov, N.A.: Mathematical modeling of plane chiral waveguide using mixed finite elements. In: Progress in Electromagnetics Research Symposium, pp. 1216–1219 (2012)
25. Bogolyubov, A.N., Mukhartova, Y.V., Gao, T.: Calculation of a parallel-plate waveguide with a chiral insert by the mixed finite element method. Math. Models Compu. Simul. 5(5), 416–428 (2013)
26. Mukhartova, Y.V., Mongush, O.O., Bogolyubov, A.N.: Application of the finite-element method for solving a spectral problem in a waveguide with piecewise constant bi-isotropic filling. J. Commun. Technol. Electron. 62(1), 1–13 (2017)
27. Sveshnikov, A.G.: The basis for a method of calculating irregular waveguides. Comput. Math. Math. Phys. 3(1), 170–179 (1963)
28. Sveshnikov, A.G.: A substantiation of a method for computing the propagation of electromagnetic oscillations in irregular waveguides. Comput. Math. Math. Phys. 3(2), 314–326 (1963)
29. Mathematics-based software and services for education, engineering, and research. https://www.maplesoft.com/
30. Anderson, E., et al.: LAPACK Users' Guide, 3rd edn. SIAM, Philadelphia (1999). http://www.netlib.org/lapack/lug
31. LAPACK Users' Guide Release. http://www.netlib.org/lapack/lug/node93.html
32. Bellman, R.: Introduction to Matrix Analysis. McGraw-Hill, New York (1960)
33. Kressner, D.: Numerical Methods for General and Structured Eigenvalue Problems. Springer, Berlin (2006)
34. Golub, G.H., Van Loan, C.F.: Matrix Computations, 3rd edn. Johns Hopkins University Press, Baltimore (1996)
35. Van Loan, C.: On estimating the condition of eigenvalues and eigenvectors. Linear Algebra Appl. 88–89, 715–732 (1987)

A Divergence-Free Method for Solving the Incompressible Navier–Stokes Equations on Non-uniform Grids and Its Symbolic-Numeric Implementation

Evgenii V. Vorozhtsov[1]([✉]) and Vasily P. Shapeev[1,2]

[1] Khristianovich Institute of Theoretical and Applied Mechanics of the Siberian Branch of the Russian Academy of Sciences, Novosibirsk 630090, Russia
{vorozh,shapeev}@itam.nsc.ru
[2] Novosibirsk National Research University, Novosibirsk 630090, Russia

Abstract. To increase the accuracy of computations by the method of collocations and least squares (CLS) a generalization of this method is proposed for the case of a non-uniform logically rectangular grid. The main work formulas of the CLS method on non-uniform grid, including the formulas implementing the prolongation operator on a non-uniform grid at the use of a multigrid complex are obtained with the aid of the computer algebra system (CAS) *Mathematica*. The proposed method has been applied for the numerical solution of two-dimensional stationary Navier–Stokes equations governing the laminar flows of viscous incompressible fluids. On a smooth test solution, the application of a non-uniform grid has enabled a 47-fold reduction of the solution error in comparison with the uniform grid case. At the solution of the problem involving singularities – the lid-driven cavity flow – the error of the solution obtained by the CLS method was reduced by the factors from 2.65 to 3.05 depending on the Reynolds number value.

Keywords: Non-uniform grids · Logically rectangular grids · Navier–Stokes equations · Krylov subspaces · Multigrid · Preconditioners · Method of collocations and least squares

1 Introduction

At present, finite difference methods, finite element methods, and finite volume methods have gained widespread acceptance at the numerical solution of the Navier–Stokes equations governing viscous incompressible fluid flows. One can combine efficiently all these methods with spatial grids of various types, for example, with structured curvilinear grids. The numerical solution of partial

The research was carried out within the framework of the Program of Fundamental Scientific Research of the state academies of sciences in 2013–2020 (projects Nos. AAAA-A17-117030610134-9 and AAAA-A17-117030610136-3).

differential equations of fluid dynamics in physical regions with *complex geometry* is the main goal of the application of curvilinear grids as well as of the grids of other types. Such grids simplify the realization of boundary conditions on curved boundaries [7,17]. Besides, their application enables one to increase the numerical solution accuracy in spatial subregions near the boundaries in the cases of large solution gradients at the expense of grid refinement near the boundaries.

In many applications, it is possible to transform the physical region with curved boundaries with the aid of a one-to-one mapping $x_1 = \phi_1(\zeta_1, \zeta_2)$, $x_2 = \varphi_2(\zeta_1, \zeta_2)$ onto a rectangle or a square, where x_1 and x_2 are the original Cartesian physical coordinates, ζ_1 and ζ_2 are the curvilinear coordinates. Such grids are called *logically rectangular* [7]. In the present work, we restrict ourselves to the case of logically rectangular grids. We also mention here such well-known types of spatial grids as composite grids [17], structured grids [1], unstructured grids [1,3], multi-block grids [17], and chimera grids [17].

The method of collocations and least squares (CLS) of the numerical solution of boundary-value problems for partial differential equations was developed recently [6,12–16]. In these works, the CLS method was applied mainly on uniform rectangular grids.

The CLS method is a projection-grid method. The local approximate solution of the differential problem is sought therein in each cell of difference grid in the form of a linear combination of the basis elements of some functional space. The space of polynomials is mainly used as the latter in view of certain convenience. In the CLS method, the system of collocation equations was augmented by linear matching conditions for the local solution in each cell with the local solutions taken in all neighboring cells. As a result, an overdetermined SLAE was obtained. As the subsequent studies [14] showed, this technique enables a considerable reduction of the SLAE condition number. Its solution is sought in the CLS method from the requirement of minimizing the residual functional of equations of the problem on its numerical solution.

It was proposed in [14] to use a version of the weighted method of least squares, in which two weight parameters ξ and η were introduced. The parameter ξ was introduced as a factor in the both sides of the momentum equations, and the parameter η was introduced in the left- and right-hand sides of the matching condition as a factor affecting only the velocity component normal to the cell boundary. After that, the problem of minimizing the condition number as a function of parameters ξ and η was posed and solved numerically in [14]. The optimal values ξ_{opt} and η_{opt} of these parameters were found for each specific problem to be solved. It has turned out that these values depend on the problem under consideration, in particular, on the Reynolds number. The use of found values ξ_{opt} and η_{opt} has enabled a successful solution of all those test problems, which were considered in [14,15]. Thus, the condition number of a SLAE from which the approximate solution of the problem is found is a key criterion of the quality of the CLS method.

Because of the combination of the method of collocations with a "strong" requirement of minimizing the functional of the discrete problem solution residual, its properties (smoothness, accuracy) are improved in comparison with the solutions obtained by a simple method of collocations. The CLS method indeed possesses also other improved properties in comparison with the method of collocations and LSFEM: (i) this method ensures the exact satisfaction of the mass conservation law owing to the use of a solenoidal basis; (ii) there is no problem of the pressure coupling with the velocity vector because the pressure is calculated in the CLS method *simultaneously* with the velocity vector components. Furthermore, the CLS method enables an efficient solution of the problems for elliptic, parabolic, and hyperbolic PDEs on various adaptive grids with rectangular and triangular cells with the use of graphs for their ordering in the process of computation; the overview of corresponding works is available in [11, 14].

To achieve a higher acceleration we consider in the given work a combined application of three techniques for accelerating the iteration process of problem solution: the preconditioner, the operation of prolongation on a multigrid complex, which is a constituent part of the Fedorenko method [4], and the Krylov method [8]. These techniques of the iteration process acceleration are employed in practice both separately and in combination [14].

It was shown previously in [6] that it is possible to increase the accuracy of the numerical solution obtained by the CLS method by using the polynomials of higher degrees. In this connection, we apply in the present work a second-degree polynomial also for the pressure approximation. In this case, there are 18 basis functions in total in the chosen space. Since the coefficients are constant in the continuity equation, it is easy to satisfy it at the expense of the choice of basis polynomials. It is not difficult to find that to this end, they must satisfy three linear relations. From the original 18 basis vectors, only 15 vectors will finally remain independent. Their set can be termed a solenoidal basis because each basis vector is divergence-free. Therefore, the continuity equation and, consequently, the mass conservation law is satisfied by the numerical solution of the problem in the entire computational region.

In the work [6], the solution of the lid-driven cavity problem was carried out on a composite grid, which has enabled a significant increase of the numerical accuracy. However, the implementation of the CLS method on a composite grid is related to a significant complication of its program realization: it is necessary to remember the way to each location of the grid fragmentation with the aid of quadtrees. This needs an additional computer memory and increases the total CPU time of the problem solution. It is easy to see that from the viewpoint of the algorithmic complexity, the CLS method on a non-uniform logically rectangular grid takes an intermediate position between the CLS method on a uniform grid and the CLS method on a composite grid because the first method does not need the application of quadtrees. In the available literature on the CLS method, there are no works devoted to the questions of the investigation of the influence of the application of a non-uniform logically rectangular grid in the CLS method on the accuracy of the results obtained by this method. The given work just bridges this gap.

The purpose of the investigations described in the given paper is an increase in the accuracy and convergence rate of a new version of the CLS method for numerical computations of two-dimensional stationary laminar flows of viscous incompressible fluids. To this end, the following new elements are proposed:

1°. A new version of the CLS method—the CLS method on a logically rectangular non-uniform grid.

2°. The analytic investigation of the two-parameter preconditioner for the CLS method on non-uniform grid for a more rapid determination of its optimal parameters.

3°. It is shown how one can realize the boundary condition at the channel outlet in different versions of the CLS method without using the rows of external fictitious cells. This substantial simplification of the boundary condition procedure at the open outlet boundary is a fundamental advantage of the CLS method over the finite difference and finite volume methods.

2   Description of the CLS Method for Numerical Solution of the 2D Navier–Stokes Equations on Non-uniform Grids

Consider a boundary-value problem for the system of stationary Navier–Stokes equations

$$(\mathbf{V} \cdot \nabla)\mathbf{V} + \nabla p = \frac{1}{\text{Re}}\Delta\mathbf{V}, \tag{1}$$

$$\text{div}\,\mathbf{V} = 0, \quad (x_1, x_2) \in \Omega, \tag{2}$$

$$\mathbf{V}\big|_{\partial\Omega} = \mathbf{g}, \tag{3}$$

which describe the flow of a viscous non-heat-conducting incompressible fluid in the region Ω with the boundary $\partial\Omega$. In Eq. (1), x_1, x_2 are the Cartesian spatial coordinates, $\mathbf{V} = (v_1(x_1, x_2), v_2(x_1, x_2))$ is the velocity vector; $p = p(x_1, x_2)$ is the pressure, Re is the Reynolds number, $\Delta = \frac{\partial^2}{\partial x_1^2} + \frac{\partial^2}{\partial x_2^2}$, $(\mathbf{V}\cdot\nabla) = v_1\frac{\partial}{\partial x_1} + v_2\frac{\partial}{\partial x_2}$. The system of three Eqs. (1) and (2) is solved under the Dirichlet boundary conditions (3), where $\mathbf{g} = \mathbf{g}(x_1, x_2) = (g_1, g_2)$ is a given vector function. The pressure is determined from (1) and (2) with the accuracy up to a constant. We will choose this constant in the following in such a way that the following condition is satisfied:

$$\int_\Omega p\,dx_1 dx_2 = 0, \tag{4}$$

which is valid in the absence of sources and sinks in the region Ω [6]. As the solution region the rectangular region is considered in the present work:

$$\Omega = \{(x_1, x_2)| \ 0 \le x_1 \le L, \quad 0 \le x_2 \le H\}, \tag{5}$$

where $L > 0$ and $H > 0$ are the given lengths of the sides of region (5) along the axes x_1 and x_2, respectively. The quantity H was used at the solution of fluid dynamics tasks as a reference length at the nondimensionalization of variables, and it enters in a natural way the definition of the Reynolds number Re in (1).

In the problem (1)–(4), region (5) is covered by a non-uniform grid. Denote by $(x_{1,i}, x_{2,j})$ a node of this grid, where the quantities $(x_{1,i})$ and $(x_{2,j})$ satisfy the following inequalities: $0 = x_{1,1} < x_{1,2} < \ldots < x_{1,I+1} = L$, $0 = x_{2,1} < x_{2,2} < \ldots < x_{2,J+1} = H$. Introduce the cell Ω_{ij} of the numerical grid in the (x_1, x_2) plane as the following set of points: $\Omega_{ij} = \{(x_1, x_2)|\; x_{1,i} \leq x_1 \leq x_{1,i+1},\;\; x_{2,j} \leq x_2 \leq x_{2,j+1}\}$, $i = 1, \ldots, I;\; j = 1, \ldots, J$.

By analogy with [15] let us introduce the local coordinates y_1, y_2 in each cell Ω_{ij} by the following formulas: $y_1 = (x_1 - x_{1c,i})/h_{1i}$, $y_2 = (x_2 - x_{2c,j})/h_{2j}$, where $(x_{1c,i}, x_{2c,j})$ are the coordinates of the geometric center of the cell Ω_{ij}, $h_{1i} = \frac{1}{2}(x_{1,i+1} - x_{1,i})$, $h_{2j} = \frac{1}{2}(x_{2,j+1} - x_{2,j})$, $i = 1, \ldots, I;\; j = 1, \ldots, J$. that is h_{1i}, h_{2j} are the halved sizes of the cell Ω_{ij} along the axes x_1 and x_2, respectively. Consequently, the local coordinates vary in the interval $y_m \in [-1, 1]$, $m = 1, 2$. Introduce the notations $\mathbf{u}(y_1, y_2) = (u_1, u_2) = \mathbf{V}(h_{1i}y_1 + x_{1c,i}, h_{2j}y_2 + x_{2c,j})$, $q(y_1, y_2) = p(h_{1i}y_1 + x_{1c,i}, h_{2j}y_2 + x_{2c,j})$. As a result of this substitution of variables, the Navier–Stokes equations take the following form:

$$r_{ij}^2 \frac{\partial^2 u_1}{\partial y_1^2} + \frac{\partial^2 u_1}{\partial y_2^2} - \mathrm{Re} h_{2j}\left(r_{ij} u_1 \frac{\partial u_1}{\partial y_1} + u_2 \frac{\partial u_1}{\partial y_2} + r_{ij}\frac{\partial q}{\partial y_1}\right) = 0, \qquad (6)$$

$$r_{ij}^2 \frac{\partial^2 u_2}{\partial y_1^2} + \frac{\partial^2 u_2}{\partial y_2^2} - \mathrm{Re} h_{2j}\left(r_{ij} u_1 \frac{\partial u_2}{\partial y_1} + u_2 \frac{\partial u_2}{\partial y_2} + \frac{\partial q}{\partial y_2}\right) = 0, \qquad (7)$$

$$r_{ij}\frac{\partial u_1}{\partial y_1} + \frac{\partial u_2}{\partial y_2} = 0. \qquad (8)$$

In Eqs. (6)–(8), r_{ij} is the cell aspect ratio, $r_{ij} = \frac{h_{2j}}{h_{1i}}$. The value of the quantity r_{ij} is generally different for each cell Ω_{ij} of a non-uniform logically rectangular grid. In the particular case of a square grid, $r_{ij} = 1\; \forall i, j$.

One can see from (8) that it is necessary to introduce the quantity r_{ij} in some vectors of the basis of 15 vectors to ensure the solenoidal property of the basis. For the case of a square grid, the form of basis vectors $\varphi_1, \ldots, \varphi_{15}$ was presented previously in [15]. The needed changes in this basis for the case of a non-uniform grid were found by the method of indeterminate coefficients. It has turned out that it is sufficient to introduce the quantity r_{ij} only in the basis vectors $\varphi_2, \varphi_4, \varphi_5$. The final form of 15 basis vectors is presented in Table 1.

Let us perform the Newton linearization of Eqs. (6) and (7):

$$\xi\left[r_{ij}^2 \frac{\partial^2 u_1^{s+1}}{\partial y_1^2} + \frac{\partial^2 u_1^{s+1}}{\partial y_2^2} - \mathrm{Re} h_{2j}\left(r_{ij}\left(u_1^{s+1}\frac{\partial u_1^s}{\partial y_1} + u_1^s \frac{\partial u_1^{s+1}}{\partial y_1}\right) + u_2^{s+1}\frac{\partial u_1^s}{\partial y_2} \right.\right.$$
$$\left.\left. + u_2^s \frac{\partial u_1^{s+1}}{\partial y_2} + r_{ij}\frac{\partial q^{s+1}}{\partial y_1}\right)\right] = -\xi\left(u_1^s \frac{\partial u_1^s}{\partial y_1} + u_2^s \frac{\partial u_1^s}{\partial y_2}\right), \qquad (9)$$

$$\xi\left[r_{ij}^2 \frac{\partial^2 u_2^{s+1}}{\partial y_1^2} + \frac{\partial^2 u_2^{s+1}}{\partial y_2^2} - \mathrm{Re} h_{2j}\left(r_{ij}\left(u_1^{s+1}\frac{\partial u_2^s}{\partial y_1} + u_1^s \frac{\partial u_2^{s+1}}{\partial y_1}\right) + u_2^{s+1}\frac{\partial u_2^s}{\partial y_2} \right.\right.$$
$$\left.\left. + u_2^s \frac{\partial u_2^{s+1}}{\partial y_2} + \frac{\partial q^{s+1}}{\partial y_2}\right)\right] = -\xi\left(u_1^s \frac{\partial u_2^s}{\partial y_1} + u_2^s \frac{\partial u_2^s}{\partial y_2}\right), \qquad (10)$$

Table 1. The form of basis functions φ_l

l	1	2	3	4	5	6	7	8	9	10	11	12	13	14	15
φ_l	1	$\frac{y_1}{r_{ij}}$	y_2	$\frac{y_1^2}{r_{ij}}$	$-2\frac{y_1 y_2}{r_{ij}}$	y_2^2	0	0	0	0	0	0	0	0	0
	0	$-y_2$	0	$-2y_1 y_2$	y_2^2	0	1	y_1	y_1^2	0	0	0	0	0	0
	0	0	0	0	0	0	0	0	0	1	y_1	y_2	y_1^2	$y_1 y_2$	y_2^2

where s is the number of the iteration over the nonlinearity, $s = 0, 1, 2, \ldots$, u_1^s, u_1^s, q^s is the known approximation to the solution at the sth iteration starting from the chosen initial guess with index $s = 0$. Here, as in [14,15], the user-specified parameter ξ has been introduced for controlling the magnitude of the condition number of the overdetermined system of linear algebraic equations (SLAE), which must be solved in each cell Ω_{ij}.

The approximate solution in each cell $\Omega_{i,j}$ is sought in the form of a linear combination of the basis vector functions φ_l:

$$
(u_1^s, u_2^s, p^s)^T = \sum_{l=1}^{15} b_{i,j,l}^s \varphi_l, \tag{11}
$$

where the superscript T denotes the transposition operation. In the given work, the second-degree polynomials in variables y_1, y_2 are used for the approximation of both velocity components and pressure. There are eighteen basis functions in total in the chosen space. Since the coefficients are constant in the continuity equation, which has a simple form, it is easy to satisfy it at the expense of the choice of basis polynomials φ_l. It is not difficult to find that it is required to this end that they satisfy three linear relations. There will finally remain only fifteen independent basis polynomials from the original eighteen ones. They are presented in Table 1. One can term their set a solenoidal basis because $r_{ij} \cdot (\partial\varphi_{l,1}/\partial y_1) + \partial\varphi_{l,2}/\partial y_2 = 0$, $l = 1, \ldots, 15$. In such a basis, continuity equation is satisfied identically by the discrete problem solution in each cell.

The CLS method differs significantly from other methods of solving the boundary-value problems for differential equations in that the solution of the discrete problem is determined by the solution of an overdetermined SLAE. It is required here that the minimum of the residual functional of its equations is reached on the pseudo solution of the corresponding SLAE. To write the discrete problem equations, which determine the solution in the cell $\Omega_{i,j}$, we specify the collocation points therein. In the given work, three versions of the specification of the collocation point coordinates have been implemented. Although a non-uniform grid is employed, in local variables, each cell represents a square with the side length equal to 2. Therefore, in such a cell, the collocation points for the different values of numbers N_c in each cell were specified in the same way as in the case of uniform grid, see further details in [14,15].

Substituting (11) as well as the numerical values of the coordinates of each collocation point in (9), (10) we obtain $2N_c$ collocation equations of the discrete problem:

$$\sum_{m=1}^{15} a_{\nu,m}^{(1)} \cdot b_m^{s+1} = f_\nu^s, \quad \nu = 1, \ldots, 2N_c. \tag{12}$$

They are linear algebraic equations for the sought quantities b_m^{s+1}.

Let us augment the system of equations of the approximate problem in the Ω_{ij} cell by the conditions of matching with the solutions of the discrete problem, which are taken in all cells adhering to the given cell. We write them at separate points (matching points) on the sides of the $\Omega_{i,j}$ cell, which are common with its neighboring cells. In the given case, the matching conditions are the conditions of a smooth conjugation of the discrete problem solution at matching points. Their form on the non-uniform grid somewhat differs from the matching conditions on the uniform grid. As an example, let us consider the face $y_1 = +1$ of a cell. On this face, the matching conditions are as follows:

$$h_{1i}\frac{\partial[(u^+)^n]^{s+1}}{\partial n} + \eta[(u^+)^n]^{s+1} = h_{1,i+1}\frac{\partial[(u^-)^n]^*}{\partial n} + \eta[(u^-)^n]^*, \tag{13}$$

$$h_{1i}\frac{\partial[(u^+)^\tau]^{s+1}}{\partial n} + [(u^+)^\tau]^s = h_{1,i+1}\frac{\partial[(u^-)^\tau]^*}{\partial n} + [(u^-)^\tau]^*, \tag{14}$$

$$(q^+)^{s+1} = (q^-)^*. \tag{15}$$

Here $h_{1i}\frac{\partial}{\partial n} = n_1\frac{\partial}{\partial y_1} + \frac{n_2}{r_{ij}}\frac{\partial}{\partial y_2}$, $n = (n_1, n_2)$ is the external normal to the side of the cell $\Omega_{i,j}$, $(\cdot)^n$, $(\cdot)^\tau$ are the normal and tangent components of the velocity vector with respect to the cell side, u^+, u^- are the limits of the function u as its arguments tend to the matching point from inside and outside the Ω_{ij} cell. The superscript * by the quantities in the right-hand sides of relations (13)–(15) reflects the fact that in the right-hand sides, one employs either the quantities at the sth iteration or the quantities at the $(s+1)$th iteration, which have already been computed by the moment of the use of the given matching conditions.

The user-specified parameter η has been introduced here as in [14,15] for the purpose of controlling the magnitude of the condition number of a SLAE, which must be solved in each cell Ω_{ij}. On the remaining three faces of the cell, the matching conditions for the velocity vector components are written down similarly, therefore, they are not presented here for the sake of brevity.

For the uniqueness of the pressure determination in the solution, we either specify its value at a single point of the region or approximate condition (4) by the formula

$$\frac{1}{h_{1i}}\left(\int_{\Omega_{i,j}} q\,dx_1 dx_2\right) = \frac{1}{h_{1i}}\left(-I^* + \int_{\Omega_{i,j}} q^* dx_1 dx_2\right). \tag{16}$$

Here I^* is the integral over the entire region, which is computed as a sum of the integrals over each cell at the foregoing iteration, q^* is the pressure in a cell from the foregoing iteration.

Note that in conditions (13)–(14), in the form of the systems of equations determining the solution pieces in two neighboring cells, the directions of external normals to these cells on their common side are opposite. Therefore, in formulas (13)–(14), the first items will enter with one sign for one of these cells, and in the neighboring cell, the sign will be opposite. Owing to this, Eqs. (13)–(14) do not repeat in the total SLAE, which determines the numerical solution of the global problem and involves all equations written for each cell.

The points at which Eqs. (13)–(15) are written are termed the matching points. As in the case of collocation points, the number of matching points and their location on each cell side may be different in different specific versions of the method. Denote by N_m the number of matching points for the velocity vector components on the sides of each cell. The coordinates of these points in variables y_1, y_2 were specified in each cell of a non-uniform grid in the same way as in the case of a uniform grid, see [14,15] for further details. Using the solution representation (11), let us substitute the coordinates of matching points into each of three matching conditions (13)–(15). We obtain $2N_m$ linear algebraic equations for velocity components from the first two conditions.

After substituting representation (11) into (15) one obtains four linear algebraic (matching) equations with one condition on each of the four cell sides.

For the uniqueness of the pressure determination in the solution, we either specify its value at a single point of the region (in the given work, at the vertex of the cell $\Omega_{1,1}$) or use condition (16). In test solutions, the exact pressure value at this point was taken. Thus, the matching conditions for the velocity vector components and pressure yield $2N_m + 4 + \delta_i^1 \delta_j^1$ linear algebraic equations in the total for the unknown $b_{i,j,l}$ in each cell (i,j), where δ_i^j is the Kronecker symbol, $\delta_i^j = 1$ at $i = j$ and $\delta_i^j = 0$ at $i \neq j$.

If the cell side coincides with the boundary of region Ω, then the boundary conditions are written at the corresponding points instead of the matching conditions for the discrete problem solution: $u_m = g_m$, $m = 1, 2$.

Uniting the equations of collocations, matching, and the equations obtained form the boundary conditions, if the cell Ω_{ij} is the boundary cell, we obtain in each cell a SLAE of the form

$$A_{i,j} \cdot \boldsymbol{X}_{i,j}^{s+1} = \boldsymbol{f}_{i,j}^{s,s+1}, \tag{17}$$

where $\boldsymbol{X}_{i,j}^{s+1} = (b_{i,j,1}^{s+1}, \ldots, b_{i,j,15}^{s+1})^T$. The solution of system (17) determines locally in the given cell $\Omega_{i,j}$ the solution of the global discrete problem as its piece matched with neighboring pieces. The matrix $A_{i,j}$ contains $2N_c + 2N_m + 4 + \delta_i^1 \delta_j^1$ rows and 15 columns. In the versions studied in the present work, system (17) is overdetermined.

All 100% of the coefficients of all equations of SLAE (17) were derived on computer in Fortran form by using symbolic computations with the computer algebra system (CAS) *Mathematica* [18]. At the obtaining of the final form of the formulas for the coefficients of the equations, it is useful to perform the simplifications of the arithmetic expressions of polynomial form to reduce the number of the arithmetic operations needed for their numerical computation. To this end, we employed standard functions of the *Mathematica* system, such as `Simplify` and `FullSimplify` for the simplification of complex symbolic expressions arising at the symbolic stages of the construction of the formulas of

the method. Their application enabled a two-three-fold reduction of the length
of polynomial expressions.

One global $(s+1)$th iteration meant that all the cells were considered sequentially in the computational region Ω. In each cell, SLAE (17) was solved by
the orthogonal method (of Householder), and the values known at the solution
construction at the $(s + 1)$th iteration were taken in the right-hand sides of
Eqs. (13)–(15) as the u^- and q^- in a given cell.

3   Two-Parameter Preconditioner for the CLS Method

It is necessary to solve in each cell Ω_{ij} the SLAE of the form (17). Let us omit
in (17) the superscripts and subscripts for the sake of brevity:

$$AX = f. \tag{18}$$

As was mentioned in Sect. 2, the matrix A has the size $N \times 15$, where $N > 15$
that is the matrix A is rectangular. As is known, the condition number of a
rectangular matrix A is computed by the formula $\kappa(A) = \sqrt{\kappa(A^T A)}$, where the
superscript T by A denotes the transposition operation and it is assumed that
the matrix $A_1 = A^T A$ is non-singular.

The condition number of the matrix gives an universal estimate of the relative
error of the solution of system (18). In the cases of applying the CLS method for
the numerical solution of nonlinear partial differential equations, the right-hand
side f includes the solution values from the foregoing iteration. These values
have the error $O(h_{1i}^l) + O(h_{2j}^l)$, where $l \geq 1$ is the accuracy order of the method
(it is assumed that $r_{ij} = O(1)$).

The standard condition number of a nonsingular matrix A_1 is defined as

$$\kappa(A_1) =\| A_1 \| \cdot \| A_1^{-1} \|, \tag{19}$$

where $\| \cdot \|$ is a matrix norm. In this section, we investigate the preconditioner
depending on two parameters. The parameter ξ has been introduced above by
multiplying by ξ the both sides of the linearized Navier–Stokes equations, see
(9) and (10). The parameter η has been introduced in the matching condition
(13). These parameters are chosen by a numerical solution of the problem of
minimizing the quantity $\kappa(A)$.

As a result of the substitution of collocation points in Eqs. (9) and (10) one
obtains $2N_c$ collocation equations.

The pressure enters the momentum Eqs. (6) and (7) only in the form of the
derivatives $\partial q/\partial y_1$ and $\partial q/\partial y_2$, therefore, the coefficient affecting b_{10} in matrix
A_{col} is equal to zero. As a result of this, the matrix A_{col} is singular. It was
shown in [14] that the inclusion in the matrix A of a row corresponding to
approximation (16) of the pressure integral reverts this matrix into a matrix of
the full rank.

Denote by A_{mat} a matrix corresponding to the matching conditions (13),
(14), and (15). The entire matrix A of the overdetermined SLAE can then be

written as $A = (\tilde{A}_{col}, A_{mat})^T$. At the given numerical values of half-steps h_{1i} and h_{2j}, the entries of the matrix A depend on ξ and η. Since the matrix A_1 is symmetric, one can calculate its condition number in the Euclidean norm by the formula $\kappa(A_1) = \lambda_{max}/\lambda_{min}$, where $\lambda_{max} = \max(|\lambda_1|, \ldots, |\lambda_{15}|)$, $\lambda_{min} = \min(|\lambda_1|, \ldots, |\lambda_{15}|)$, $\lambda_1, \ldots, \lambda_{15}$ are the eigenvalues of the matrix A_1. At the use of these formulas, there is no need in computing the inverse matrix A_1^{-1}.

The characteristic equation corresponding to the CLS method under consideration is a 15-degree algebraic equation. It is, therefore, impossible to obtain in a closed analytic form the expressions for the eigenvalues of the corresponding matrix A_1. We nevertheless show in the following that it is possible to obtain information about some properties of the matrix A_1 and, consequently, about the properties of the two-parameter preconditioner under consideration by studying the analytic expressions for its entries. Let us write in analytic form these expressions for the case of an internal cell (i, j). The matrix A_1 is symmetric, therefore, it is sufficient to present the expressions for the entries lying in the upper triangular part of this matrix. The size of the matrix A_1 is equal to $m_b \times m_b$, where $m_b = 15$ for the version of the CLS method, which is considered here. The size of the upper triangular part of the matrix A_1 is equal to $m_b \cdot (m_b + 1)/2$ so that this matrix part contains 120 entries at $m_b = 15$. Denote the entries of this matrix by $\beta_{i,j}$, $i, j = 1, \ldots, m_b$. It is assumed that the number of collocation points $N_c = 8$ in the cell, therefore, the total number of collocation equations is $2N_c = 16$. In addition, four rows have been included in the matrix A, which correspond to four matching conditions for the pressure (at the center of each cell face, there is one point for the matching condition) and 16 rows corresponding to the matching conditions for the velocity vector components (which are set at two points on each of the four faces). Finally, one more row accounts for the integral condition for the pressure (16).

We present at first the expressions for those entries $\beta_{i,j}$, which involve the parameter η and/or the grid half-step h_{2j}. The half-step h_{1i} does not enter these elements for the reason that it has been introduced by analogy with [6] as a denominator in the both sides of equality (16), therefore, it is cancelled mutually with this half-step entering the numerators in the both sides of (16).

$$\beta_{\mu,\nu} = \xi^2 \cdot \left(\sum_{k=0}^{6} a_{10+k,\mu} a_{11+k,\nu} + \sum_{k=1}^{9} a_{k,\mu} a_{k,\nu} \right) + P(\eta, h_{2j}, \mu, \nu), \qquad (20)$$

where

$$(\mu, \nu) = (1,1), (1,2), (1,4), (1,6), (2,2), (2,4), (2,5), (2,6), (2,7), (2,9),$$
$$(3,3), (3,5), (4,4), (4,6), (4,8), (5,5), (5,7), (5,9), (6,6), (7,7), (7,9),$$
$$(8,8), (9,9), (11,11), (12,12), (13,13), (15,15),$$
$$P(\eta, h_{2j}, \mu, \nu) = 4\delta_\mu^1 \delta_\nu^1 + \delta_\mu^1 \delta_\nu^2 \cdot 4\eta + \delta_\mu^1 \delta_\nu^4 \cdot (1 + 4\eta^2) + \delta_\mu^1 \delta_\nu^6 \cdot (12 + \eta^2)$$
$$+ \delta_\mu^2 \delta_\nu^2 \cdot (10 + 8\eta^2) + \delta_\mu^2 \delta_\nu^4 \cdot 12\eta - \delta_\mu^2 \delta_\nu^5 \cdot 12\eta + \delta_\mu^2 \delta_\nu^6 \cdot \eta - \delta_\mu^2 \delta_\nu^7 \cdot 4\eta$$
$$+ \delta_\mu^2 \delta_\nu^9 \cdot \eta + \delta_\mu^3 \delta_\nu^3 \cdot (16 + \eta^2) - \delta_\mu^3 \delta_\nu^5 \cdot 2\eta + \delta_\mu^4 \delta_\nu^4 \cdot \left(\frac{145}{4} + 8\eta^2 \right)$$

$$+\delta_\mu^4\delta_\nu^6\cdot(3+\eta^2)-\delta_\mu^4\delta_\nu^8\cdot2\eta+\delta_\mu^5\delta_\nu^5\cdot\left(\frac{145}{4}+8\eta^2\right)+\delta_\mu^6\delta_\nu^6\cdot\left(\frac{\eta^2}{4}+36\right)$$

$$+\delta_\mu^7\delta_\nu^7\cdot(4+4\eta^2)\delta_\mu^7\delta_\nu^9\cdot(12+\eta^2)+\delta_\mu^8\delta_\nu^8\cdot(16+\eta^2)+\delta_\mu^9\delta_\nu^9\cdot\left(36+\frac{\eta^2}{4}\right)$$

$$+2\delta_\mu^{11}\delta_\nu^{11}+4\delta_\mu^{12}\delta_\nu^{12}+\delta_\mu^{13}\delta_\nu^{13}\left(\frac{h_{2j}^2}{9}+2\right)+\delta_\mu^{15}\delta_\nu^{15}\left(\frac{h_{2j}^2}{9}+4\right).$$

One can write the remaining entries of the matrix A_1 in the form

$$\beta_{i,j}=\xi^2\cdot\sum_{m=1}^{2N_c}\sum_{p=1}^{2N_c}a_{m,i}\cdot a_{p,j},\quad i,j=1,\ldots,m_b. \tag{21}$$

The quantities $a_{m,l}$, $m=1,\ldots,2N_c$, $l=1,\ldots,m_b$, which enter (20) and (21), are the coefficients of the collocation equations obtained from Eqs. (9) and (10) at the substitution of expansions (11) into them, of the values $\xi=1$, and of the values of local coordinates y_1 and y_2 of collocation points. According to (9) and (10), these coefficients depend on the solution obtained at the foregoing iteration and on the Reynolds number. An analysis of expressions (20) and (21) leads to the following conclusions.

1. The parameter ξ enters the quantities $\beta_{i,j}$ only as a factor of the form ξ^2. It follows from here that the surface $\kappa=\kappa(\xi,\eta)$ is symmetric with respect to the η axis. This enables one to restrict oneself to the search for the optimal value of the parameter ξ only in the half-plane $\xi>0$.

2. The expressions involving the parameter η enter the elements $\beta_{i,j}$ as the additive items, and there are both the first and the second powers of this parameter. It is, therefore, clear that the surface $\kappa=\kappa(\xi,\eta)$ is neither even nor odd function of the parameter η.

3. The half-step h_{2j} enters the formulas for $\beta_{i,j}$ only additively and only as h_{2j}^2. In many fluid dynamics problems, the computational region sizes in the plane of dimensionless spatial coordinates are usually the quantities of the order $O(1)$. Besides, one must use as a rule a grid, which has at least 10 cells in each spatial direction to ensure an acceptable accuracy of the CLS method under consideration. Therefore, the above half-step is less than unity in its magnitude, and then $h_{2j}^2\ll1$. At the same time, the remaining expressions entering $\beta_{i,j}$ are the quantities of the order $O(1)$. It follows from here that the condition number weakly depends on the magnitude of half-steps at $h_{2j}<1$. We emphasize that this conclusion is general and is independent of an applied problem. One can use this circumstance efficiently at a search for optimal values of parameters ξ and η ensuring the minimum of the condition number of the matrix A_1: it is sufficient to find these optimal values by using a numerical solution found by the CLS method on a relatively crude grid. One can then use the optimal ξ and η in the computations on grids having much smaller steps.

Because the entries of the matrix $\tilde{A}_{col}$ depend on the solution obtained at the foregoing iteration, it is necessary to carry out a further investigation of the condition number properties on a given grid at the solution of a specific problem. The corresponding numerical experiments are described below in Sect. 6.

The search of the minimum of function $\kappa(\xi_{opt}, \eta_{opt})$ was carried out in [14] by the method of uniform search with a variable step [9] for the case of a CLS method with 12 basis vectors. The value $\kappa(\xi_{opt}, \eta_{opt})$ typically satisfied the inequalities $3 < \kappa(\xi_{opt}, \eta_{opt}) < 10$ at the point of its minimum. Besides, it was found that the optimal values ξ_{opt} and η_{opt} depended weakly on the location of a specific cell in the spatial grid, at least in the cases of those test and benchmark problems, which were considered in [14]. It was shown further that a reduction of N_c affects more significantly the value ξ_{opt} than the value η_{opt}.

4   Convergence Acceleration Algorithm Based on Krylov's Subspaces

To accelerate the convergence of the iterations used for the approximate solution construction we have used in the new version of the CLS method, which is discussed in the present paper, a new variant of the well-known method [8] based on Krylov's subspaces, which was previously presented in detail in [12, 15]. We present in the following a very brief description of the corresponding algorithm. Let the SLAE have the form $X = TX + f$, where the vector X is the sought solution, T is a square matrix, and f is a column vector. Let the matrix T have a full rank, and let the following iteration process converge: $X^{n+1} = TX^n + f$, $n = 0, 1, \ldots$, in which X^n is the approximation for the solution at the nth iteration. By the definition, $r^n = TX^n + f - X^n = X^{n+1} - X^n$ is the residual of equations $X = TX + f$, and it is not difficult to obtain the following relation from the above formulas: $r^{n+1} = T r^n$. Let us assume that $k + 1$ iterations have been made starting from some initial guess X^0 that is the quantities X^1, $X^2, \ldots$, X^{k+1} and r^0, $r^1, \ldots$, r^k have been computed. The value X^{k+1} is then refined by the formula $X^{*k+1} = X^{k+1} + Y^{k+1}$. One employs the correction of the form $Y^{k+1} = \sum_{i=1}^{k} \alpha_i \, r^i$ with indefinite coefficients $\alpha_1, \ldots, \alpha_k$ that are found from the condition of the minimization of the residual functional $\Phi(\alpha_1, \ldots, \alpha_k) = \| X^{*k+1} - TX^{*k+1} - f \|_2^2$, which arises at the substitution of X^{*k+1} into the system $X = TX + f$. Here $\|u\|_2$ is the Euclidean norm of the vector u. The refined vector of the $k + 1$th approximation X^{*k+1} is used as the initial approximation for further continuation of the sequence of iterations.

5   Convergence Acceleration with the Aid of the Multigrid Algorithm

The main idea of multigrid is the selective damping of the error harmonics [4, 12]. The questions of the realization of these algorithms on non-uniform logically rectangular grids in the contexts of finite difference methods and finite volume methods were discussed in detail in the works [3] and [1], respectively.

In the CLS method, as in other methods, the number of iterations necessary for reaching the given accuracy of the approximation to the solution depends on the initial guess. As a technique for obtaining a good initial guess for the iterations on the finest grid among the grids used in a multigrid complex we have applied the prolongation operations along the ascending branch of the V-cycle—the computations on a sequence of refining grids. The passage from a coarser grid to a finer grid is made with the aid of the prolongation operators. Let us illustrate the algorithm of the prolongation operation by the example of the velocity component $u_1(y_1, y_2, b_1, \ldots, b_{15})$. Let h_{1i} and h_{2j} be the half-steps of the coarse grid.

Step 1. Let X_1 and X_2 be the global coordinates of the coarse grid cell center. We make the following substitutions into the polynomial expression for u_1:

$$y_1 = (x_1 - X_1)/h_{1i}, \quad y_2 = (x_2 - X_2)/h_{2j}. \tag{22}$$

As a result, we obtain the polynomial

$$U_1(x_1, x_2, b_1, \ldots, b_{15}) = u_1\left(\frac{x_1-X_1}{h_{1i}}, \frac{x_2-X_2}{h_{2j}}, b_1, \ldots, b_{15}\right). \tag{23}$$

Step 2. Let $(\tilde{X}_1, \tilde{X}_2)$ be the global coordinates of the center of any of the four cells of the fine grid, which lie in the coarse grid cell. We make in (23) the substitution $x_1 = \tilde{X}_1 + \tilde{y}_1 \cdot h_{1,i_2}^{(f)}$, $x_2 = \tilde{X}_2 + \tilde{y}_2 \cdot h_{2,j_2}^{(f)}$, where (i_2, j_2) are the indices of a fine grid cell, $h_{1,i_2}^{(f)}$ and $h_{2,j_2}^{(f)}$ are the halved sizes of this cell along the x_1 and x_2 axis, respectively. As a result, we obtain the second-degree polynomial $\tilde{U}_1 = P(\tilde{y}_1, \tilde{y}_2, \tilde{b}_1, \ldots, \tilde{b}_{15})$ in variables $\tilde{y}_1, \tilde{y}_2$ with coefficients $\tilde{b}_1, \ldots, \tilde{b}_{15}$. After the collection of terms of similar structure it turns out that the coordinates X_1, X_2 and $\tilde{X}_1, \tilde{X}_2$ enter $\tilde{b}_l$ $(l = 1, \ldots, 15)$ only in the form of combinations $\delta x_1 = (X_1 - \tilde{X}_1)/h_{1i}$, $\delta x_2 = (X_2 - \tilde{X}_2)/h_{2j}$. According to (22), the quantities $-\delta x_1 = (\tilde{X}_1 - X_1)/h_{1i}$, $-\delta x_2 = (\tilde{X}_2 - X_2)/h_{2j}$ are the local coordinates of the fine grid cell center in the coarse grid cell.

Let us present the expressions for coefficients $\tilde{b}_j$ $(j = 1, \ldots, 15)$ of the solution representation in a fine grid cell with the half-steps $h_{1,i_2}^{(f)}$, $h_{2,j_2}^{(f)}$ in terms of the coefficients $b_1, \ldots, b_{15}$ of the solution representation in a cell with the half-steps h_{1i}, h_{2j}:

$$\tilde{b}_1 = b_1 - b_3\delta x_2 + b_6\delta x_2^2 + \delta x_1(-b_2 + b_4\delta x_1 - 2b_5\delta x_2)/r_{ij},$$
$$\tilde{b}_2 = \sigma_1(T_1 + b_5\delta x_2)/r_{ij}, \quad \tilde{b}_3 = \sigma_2[b_3 + 2(b_5\delta x_1/r_{ij} - b_6\delta x_2)],$$
$$\tilde{b}_4 = \sigma_1^2 b_4/r_{ij}, \quad \tilde{b}_5 = \sigma_1\sigma_2 b_5/r_{ij}, \quad \tilde{b}_6 = \sigma_2^2 b_6,$$
$$\tilde{b}_7 = b_7 - \delta x_1(b_8 - b_9\delta x_1) + \delta x_2 T_1, \quad \tilde{b}_8 = \sigma_1(b_8 - 2b_9\delta x_1 + 2b_4\delta x_2),$$
$$\tilde{b}_9 = \sigma_1^2 b_9, \quad \tilde{b}_{10} = b_{10} - \delta x_1 T_2 - \delta x_2(b_{12} - b_{15}\delta x_2), \quad \tilde{b}_{11} = \sigma_1(T_2 - b_{13}\delta x_1),$$
$$\tilde{b}_{12} = \sigma_2(b_{12} - b_{14}\delta x_1 - 2b_{15}\delta x_2), \quad \tilde{b}_{13} = \sigma_1^2 b_{13}, \quad \tilde{b}_{14} = \sigma_1\sigma_2 b_{14}, \quad \tilde{b}_{15} = \sigma_2^2 b_{15},$$

where $\sigma_1 = h_{1,i_2}^{(f)}/h_{1i}$, $\sigma_2 = h_{2,j_2}^{(f)}/h_{2j}$, $r_{ij} = h_{2j}/h_{1i}$, $T_1 = b_2 - 2b_4\delta x_1 + b_5\delta x_2$, $T_2 = b_{11} - b_{13}\delta x_1 - b_{14}\delta x_2$. Note that the above expressions for $\tilde{b}_1, \ldots, \tilde{b}_{15}$ coincide with the expressions presented in [15] in the particular case of a square grid (in this case, $r_{ij} = 1 \ \forall \ i, j, \ \sigma_1 = \sigma_2 = 1/2$).

We present in the following a fragment of a program in the language of CAS *Mathematica*, which finds the analytic expressions of coefficients $\tilde{b}_1, \ldots, \tilde{b}_5$. The remaining 10 coefficients are found in a similar way.

```
u1=b1 + (b2 y1)/rij + (b4 y1^2)/rij + b3 y2 - (2 b5 y1 y2)/rij + b6 y2^2;
ru1 = {y1 -> (x1 - xg1)/h1i, y2 -> (x2 - xg2)/h2j};
u1x1 = u1 /. ru1; ry0 = {y1 -> 0, y2 -> 0};
ru12 = {x1 -> xn1 + y1*h1f, x2 -> xn2 + y2*h2f};
u1x2 = Expand[u1x1 /. ru12]; b1n = Simplify[u1x2 /. ry0];
ru19 = {xg1 -> xn1 + h1i*dx1, xg2 -> xn2 + h2j*dx2};
b1n = FullSimplify[Expand[b1n /. ru19]];
  b2n = Simplify[Coefficient[u1x2, y1]]; b2n = b2n /. {y2 -> 0};
  b2n = Expand[b2n /. ru19]; ru13 = {h1f -> rat1*h1i, h2f -> rat2*h2j};
  b2n = Simplify[b2n /. h1f -> rat1*h1i];
  b3n = Coefficient[u1x2, y2]; b3n = Expand[b3n /. ru19];
  b3n = b3n /. y1 -> 0; b3n = b3n /. ru13;
    b4n = Simplify[Coefficient[u1x2, y1^2]];
    b4n = Expand[b4n /. ru19]; b4n = Simplify[b4n /. ru13];
  b5n = Coefficient[u1x2, y1*y2];
  b5n = -b5n/2; b5n = Simplify[b5n /. ru13];
```

Here b1n = $\tilde{b}_1, \ldots$, b5n = $\tilde{b}_5$, dx1 = δx_1, dx2 = δx_2, (xn1, xn2) = $(\tilde{X}_1, \tilde{X}_2)$, (xg1, xg2) = (X_1, X_2), rat1 = σ_1, rat2 = σ_2, h1f = $h_{1,i_2}^{(f)}$, h2f = $h_{2,j_2}^{(f)}$.

6   Results of Numerical Experiments

6.1   Testing

Consider the following exact solution of the Navier–Stokes Eq. (1) [15]:

$$u_1 = \frac{-2(1+x_1)}{(1+x_1)^2 + (1+x_2)^2}, \quad u_2 = \frac{2(1+x_1)}{(1+x_1)^2 + (1+x_2)^2},$$

$$p = -\frac{2}{(1+x_1)^2 + (1+x_2)^2}, \quad 0 \le x_1, x_2 \le 1. \qquad (24)$$

Note that the functions $u_1(x_1, x_2)$ and $u_2(x_1, x_2)$ describe the divergence-free velocity field. Furthermore,

$$\int_0^1 \int_0^1 p \, dx_1 dx_2 = 4G - \pi \ln 2 - 2i \left[\text{Li}_2 \left(-\frac{i}{2} \right) - \text{Li}_2 \left(\frac{i}{2} \right) \right]$$
$$\approx -0.46261314677281549872,$$

where $i = \sqrt{-1}$, G is the Catalan's constant [18], $G \approx 0.91596559417721901505$, $\text{Li}_2(z)$ is the polylogarithmic function. To ensure the satisfaction of Eq. (4) with an error not exceeding the error of machine computations, the pressure p in (4) was replaced with the quantity $\bar{p} = p + 0.4626131467728155$.

The zero initial guess was set for quantities u_i and p.

The non-uniform grid was generated in intervals $0 \leq x_1 \leq L$ and $0 \leq x_2 \leq H$ by the algorithm presented in [17] and using the function $\sinh(\cdot)$. Let us very briefly describe the grid generation by the example of the grid in the interval $0 \leq x_1 \leq L$. Let δx_1 be the uniform grid step in this interval that is $\delta x_1 = L/I$. This step is used as a reference step for specifying the grid step near the boundaries of the computational region: $x_{1,2} - x_{1,1} = c_L \delta x_1$, $x_{1,I+1} - x_{1,I} = c_R \delta x_1$. After that, the abscissa $x_{1,k}$ of any grid node is calculated by a nonlinear mapping $x_{1,k} = L \cdot \phi(k, c_L, c_R, \sigma)$, the form of $\phi(\cdot)$ is given in [17]. The parameter σ is found by the numerical solution of the following transcendental equation: $\sinh(\sigma)/\sigma = (c_L c_R)^{-0.5}$. This equation was solved by the bisection method, the tolerance for root finding was set to 10^{-14}.

The step sizes near the boundaries $x_2 = 0$ and $x_2 = H$ are set similarly: $x_{2,2} - x_{2,1} = c_{bot}\delta x_2$, $x_{2,J+1} - x_{2,J} = c_{top}\delta x_2$, where $\delta x_2 = H/J$. The quantities c_L, c_R, c_{bot}, and c_{top} are termed the stretching factors. If one sets $c_{top} < 1$, then one obtains a non-uniform grid, which clusters near the boundary $x_2 = H$.

The formulas for computing the root-mean-square errors of the numerical solution obtained on a non-uniform grid are as follows:

$$\delta \mathbf{u}(M) = \left[\frac{1}{2HL} \sum_{i=1}^{M} \sum_{j=1}^{M} \sum_{\nu=1}^{2} (u_{\nu,i,j} - u_{\nu,i,j}^{ex})^2 2h_{1i} 2h_{2j} \right]^{\frac{1}{2}},$$

$$\delta p(M) = \left[\frac{1}{HL} \sum_{i=1}^{M} \sum_{j=1}^{M} (p_{i,j} - p_{i,j}^{ex})^2 2h_{1i} 2h_{2j} \right]^{\frac{1}{2}},$$

where $M = I = J$ is the number of cells along each coordinate direction, $\mathbf{u}_{i,j}^{ex}$ and $p_{i,j}^{ex}$ are the velocity vector and the pressure according to the exact solution (24). The quantities $\mathbf{u}_{i,j}$ and $p_{i,j}$ denote the numerical solution obtained by the CLS method described above.

We will compute the convergence orders ν_u and ν_p from the numerical solutions for the velocity vector $\mathbf{u}$ and the pressure p by the formulas known in numerical analysis [12,13]. Let $b_{i,j,l}^s$, $s = 0, 1, \ldots$ be the value of the coefficient $b_{i,j,l}$ in (11) at the sth iteration. The following condition was used for termination of the iterations: $\delta b^{s+1} < \varepsilon$, where $\delta b^{s+1} = \max_{i,j} \left(\max_{1 \leq l \leq 12} \left| b_{i,j,l}^{s+1} - b_{i,j,l}^{s} \right| \right)$, and $\varepsilon < h^2$ is a small positive quantity. We will call the quantity δb^{s+1} the pseudo-error of the approximate solution.

Table 2 presents the results of numerical experiments, in which only two of the above-described techniques for convergence acceleration were used: the two-parameter preconditioner and the Krylov subspace method. The Reynolds number Re = 1000, $L = H = 1$ in (5). The satisfaction of the inequality $\delta b^n < 10^{-9}$ was the criterion for termination of the computations by the CLS method. Equation (16) was incorporated into the overdetermined SLAE (17). In the process of iterations by the CLS method, the absolute value of integral (4) dropped from the value of the order 10^{-3} to the value of the order $10^{-12} - 10^{-13}$ that is the value of the order of the machine roundoff errors at the computations with double precision by the Fortran code. This can serve one of the criteria for the correctness of the program implementation of the CLS method presented above.

Table 2. The errors $\delta \mathbf{u}$, δp and their convergence orders ν_u and ν_p on a sequence of non-uniform grids, Re $=1000$, $L = H = 1$, $N_c = 8$, $c_{2top} = 0.5$, $c_{2bot} = 1.0$, $c_{1L} = c_{1R} = 1$, $\eta = \eta_{\mathrm{opt}} = 1.745$

M	$\delta \mathbf{u}$	δp	ν_u	ν_p
10	6.910E−05	1.267E−04		
20	1.781E−05	3.454E−04	1.96	−1.45
40	4.817E−06	6.598E−04	1.89	−0.93
80	8.547E−07	4.880E−04	2.15	0.44

A comparison of Table 2 with a similar Table 2 from [15] leads to the following conclusions. The error $\delta \mathbf{u}$ obtained at the use of a non-uniform grid is by 1–2 decimal orders less than in the uniform grid case. The error δp on relatively crude grids of $10^2, 20^2$, and 40^2 cells has dropped by one decimal order. However, it has proved to be somewhat higher in comparison with the uniform grid case on the non-uniform grid of 80^2 cells.

6.2   Lid-Driven Cavity Flow

A smooth test problem was used in the foregoing subsection for the verification of the proposed version of the CLS method. For the control and verification of the results of numerical experiments we have carried out here also a comparison of typical quantities obtained in numerical experiments on the solution of the benchmark problem of the lid-driven cavity flow. The results of its solution of increasing accuracy are published by different researchers during the last thirty years in detailed tables. The most accurate solutions of this problem available at present [2,6,10] may be used for elucidating the accuracy of the new and modified existing numerical methods for solving the Navier–Stokes equations.

In the 2D driven cavity problem, the computational region is the cavity, which is a square (5) with side $L = H = 1$, the coordinate origin lies in its left lower corner. The upper lid of the cavity moves with unit velocity in dimensionless variables in the positive direction of the Ox_1 axis. The other sides of cavity (5) are at rest. The no-slip conditions are specified on all sides: $v_1 = 1$, $v_2 = 0$ at $x_2 = 1$ and $v_m = 0$, $m = 1, 2$ on the remaining sides.

The lid-driven cavity flow has the singularities in the region upper corners. Their influence on the numerical solution accuracy enhances with increasing Reynolds number. Therefore, at high Reynolds numbers, it is necessary to apply adaptive grids for obtaining a more accurate solution: the grids with finer cells in the neighborhood of singularities.

It has turned out that the most accurate result is obtained at Re = 100 when the grid along the x_1 axis is uniform, and along the x_2 axis, it is clustered near the upper boundary. The application of such a non-uniform grid has enabled an increase in the accuracy of the numerical solution by the CLS method by the factor of 3.05 in comparison with the case of a uniform grid with the same number of cells (in both cases, the grid of 40 × 40 cells was used).

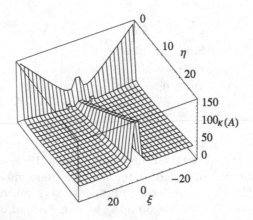

Fig. 1. Surface $\kappa = \kappa(A(\xi, \eta))$ obtained at Re = 1000, $c_{1L} = c_{1R} = 0.4$, $c_{2bot} = 1$, $c_{2top} = 0.35$

The computations of the problem under consideration were carried out on non-uniform grids also at Re = 1000. Because an increase in the Reynolds number leads to a growth of the number of iterations needed for numerical solution convergence, the importance of the application of optimal values ξ_{opt} and η_{opt} of the parameters ξ, η entering the preconditioner described in Sect. 3 also increases. In this connection, a search of the values ξ_{opt} and η_{opt} was implemented for Re = 1000. To this end, 400 iterations were done by the CLS method on the grid of 100 × 100 cells. After that, the matrix $A_{i,j}$ entering (17) was taken at $i = j = 50$ that is the cell was considered, which was located at the computational region center. The surface $\kappa = \kappa(\xi, \eta)$ was plotted, where κ is the condition number of the matrix $A_{i,j}$ according to (19). The values ξ_{opt} and η_{opt}, at which the function $\kappa(\xi, \eta)$ reached its minimum, were then found by the method of uniform search with variable step. Figure 1 shows the form of surface $\kappa = \kappa(\xi, \eta)$. It was found that $\xi_{opt} = \pm 7.20$, $\eta_{opt} = 1.60$. The points $(\xi_{opt}, \eta_{opt}, \kappa(\xi_{opt}, \eta_{opt}))$ lie in Fig. 1 at the centers of two small cubes. One can see that the surface $\kappa = \kappa(\xi, \eta)$ is symmetric with respect to the η axis. This result was predicted in Sect. 3 with the aid of the analytic investigation done by using a computer code written in the language of CAS *Mathematica*.

Table 3 presents some results of the influence of the grid stretching factors on the accuracy of the results obtained by the above-described CLS method. The first row of the table corresponds to the case of the uniform grid of

Table 3. Influence of the grid stretching factors c_{1L}, c_{1R}, c_{2bot}, c_{2top} on the accuracy of results, Re $= 1000$, $I = J = 160$

c_{1L}	c_{1R}	c_{2bot}	c_{2top}	$\| v_{1CLS} - v_{1BP} \|_{\infty}$
1	1	1	1	1.695E−02
1	1	1	0.35	9.740E−03
0.7	0.7	1	0.35	7.368E−03
0.4	0.4	1	0.4	6.398E−03

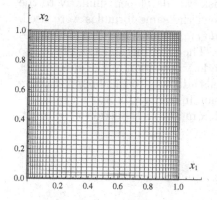

Fig. 2. Non-uniform computational grid of 40×40 cells obtained at $c_{1L} = c_{1R} = 0.4$, $c_{2bot} = 1$, and $c_{2top} = 0.4$

Fig. 3. Solution of the benchmark problem by the CLS method at Re $= 1000$: the profile of the velocity component v_1 along the centerline $x_1 = 0.5$ (solid line); ($\square\,\square\,\square$) Botella and Peyret [2]

160^2 square cells. The quantity $\| v_{1CLS} - v_{1BP} \|_{\infty}$ was computed along the cavity centerline $x_1 = 0.5$ with the use of tabular data from the work [2] as the benchmark data. In all computations summarized in Table 3, the same parameters of the preconditioner were used: $\xi = 0.1$, $\eta = 1.75$.

One can see from Table 3 that the error $\| v_{1CLS} - v_{1BP} \|_{\infty}$ is the least in the case of grid stretching factors presented in the last row. The above error obtained in the case of a uniform grid with the same number of cells exceeds by the factor of 2.65 the error shown in the last row of the table. We have also carried out a computation of the same fluid dynamics task by the CLS method on a uniform grid of 320^2 cells. In this case, $\| v_{1CLS} - v_{1BP} \|_{\infty} = 9.705E-03$, which exceeds by the factor of about 1.5 the error obtained in the case of a non-uniform grid of 160^2 cells with the grid stretching factors given in the last row of Table 3.

The result of the numerical solution by the CLS method of the square lid-driven cavity problem on the non-uniform rectangular grid shown in Fig. 2 is presented in Fig. 3 for the Reynolds number Re $= 1000$. We show in Fig. 2 only the 40×40 grid, which is part of the multigrid sequence of grids $5^2 \rightarrow 10^2 \rightarrow 20^2 \rightarrow 40^2 \rightarrow 80^2 \rightarrow 160^2$, to show more clearly the structure of the employed non-uniform grid. Solid line in Fig. 3 shows the numerical result obtained by the above-described CLS method.

The application of non-uniform grids at the numerical solution of the lid-driven cavity problem by the CLS method has enabled a lesser increase in the numerical solution accuracy (by a factor of about three) than in the case of the analytic test (24), and this is, of course, due to the presence of singularities in the given problem.

The problem of the flow over a backward-facing step at Re = 800 is frequently used for testing the computational algorithms of solving the Navier–Stokes equations [5]. The numerical algorithm for this task was developed similarly to the algorithm for solving the lid-driven cavity flow: the same formulas were used, which were generated with the aid of CAS *Mathematica* for computing the entries of matrix A in each cell of non-uniform grid. The CLS method has proved to be very efficient at the realization of the boundary condition at the open outlet boundary of a channel: no rows of fictitious cells adhering to this boundary were needed. This was achieved owing to the inclusion in the matrix A of a row corresponding to the constancy of the volumetric flux in any cross section $x_1 = $ const, including the outlet cross section.

7   Conclusion

An extension of the CLS method for the case of a non-uniform logically rectangular grid has been presented. As one can see from the work formulas of the method, which have been presented above, this has led to some complication of the CLS method as compared to the uniform grid case. On the smooth test solution (24), the application of a non-uniform grid has enabled a 47-fold reduction of the solution error in comparison with the uniform grid case. In the case of a problem with singularities, only a three-fold reduction of the solution error of the CLS method has proved possible owing to the use of a non-uniform grid.

A large amount of symbolic computations, which arose at the derivation of the basic formulas of the new version of the method, was done efficiently with *Mathematica*. It is very important that the application of CAS has facilitated greatly this work, reduced at all its stages the probability of errors usually introduced by the mathematician at the development of a new algorithm and also reduced the time needed for the development of new Fortran programs implementing the numerical stages of the CLS method.

It is also to be noted that the CLS method suits very well for parallelization. One can decompose the entire computational domain into any number of subdomains along cell boundaries, which is equal to the number of available processors. Each subdomain contains an approximately equal number of cells. One can then carry out the global problem computation in parallel in each subdomain, and the interaction of subdomains with one another is realized at each iteration from the requirement of the satisfaction of matching conditions on the boundaries between them.

References

1. Binzubair, H.: Efficient multigrid methods based on improved coarse grid correction techniques. Thesis. Delft Univ. of Technology, Delft, The Netherlands (2009)
2. Botella, O., Peyret, R.: Benchmark spectral results on the lid-driven cavity flow. Comput. Fluids **27**, 421–433 (1998)
3. Briggs, W.L., Henson, V.E., McCormick, S.F.: A Multigrid Tutorial, 2nd edn. SIAM, Philadelphia (2000)
4. Fedorenko, R.P.: The speed of convergence of one iterative process. USSR Comput. Math. Math. Phys. **4**(3), 227–235 (1964)
5. Gartling, D.K.: A test problem for outflow boundary conditions - flow over a backward-facing step. Int. J. Numer. Methods Fluids **11**, 953–967 (1990)
6. Isaev, V.I., Shapeev, V.P.: High-accuracy versions of the collocations and least squares method for the numerical solution of the Navier-Stokes equations. Comput. Math. Math. Phys. **50**, 1670–1681 (2010)
7. Knupp, P., Steinberg, S.: Fundamentals of Grid Generation. CRC Press, Boca Raton (1994)
8. Krylov, A.N.: On the numerical solution of the equation, which determines in technological questions the frequencies of small oscillations of material systems. Izv. AN SSSR, Otd. matem. i estestv. nauk **4**, 491–539 (1931). (in Russian)
9. Kudryavtseva, I.V., Rykov, S.A., Rykov, S.V., Skobov, E.D.: Optimization Methods in the Examples in the MathCAD 15 Package. Part I, NIU ITMO, St. Petersburg (2014). (in Russian)
10. Shapeev, A.V., Lin, P.: An asymptotic fitting finite element method with exponential mesh refinement for accurate computation of corner eddies in viscous flows. SIAM J. Sci. Comput. **31**, 1874–1900 (2009)
11. Shapeev, V.: Collocation and least residuals method and its applications. EPJ Web of Conferences 108, 01009. https://doi.org/10.1051/epjconf/201610801009
12. Shapeev, V.P., Vorozhtsov, E.V.: CAS application to the construction of the collocations and least residuals method for the solution of 3D Navier–Stokes equations. In: Gerdt, V.P., Koepf, W., Mayr, E.W., Vorozhtsov, E.V. (eds.) CASC 2013. LNCS, vol. 8136, pp. 381–392. Springer, Cham (2013). https://doi.org/10.1007/978-3-319-02297-0_31
13. Shapeev, V.P., Vorozhtsov, E.V.: Symbolic-numeric implementation of the method of collocations and least squares for 3D Navier–Stokes equations. In: Gerdt, V.P., Koepf, W., Mayr, E.W., Vorozhtsov, E.V. (eds.) CASC 2012. LNCS, vol. 7442, pp. 321–333. Springer, Heidelberg (2012). https://doi.org/10.1007/978-3-642-32973-9_27
14. Shapeev, V.P., Vorozhtsov, E.V.: Symbolic-numerical optimization and realization of the method of collocations and least residuals for solving the Navier–Stokes equations. In: Gerdt, V.P., Koepf, W., Seiler, W.M., Vorozhtsov, E.V. (eds.) CASC 2016. LNCS, vol. 9890, pp. 473–488. Springer, Cham (2016). https://doi.org/10.1007/978-3-319-45641-6_30
15. Shapeev, V.P., Vorozhtsov, E.V.: The method of collocations and least residuals combining the integral form of collocation equations and the matching differential relations at the solution of PDEs. In: Gerdt, V.P., Koepf, W., Seiler, W.M., Vorozhtsov, E.V. (eds.) CASC 2017. LNCS, vol. 10490, pp. 346–361. Springer, Cham (2017). https://doi.org/10.1007/978-3-319-66320-3_25

16. Sleptsov, A.G.: Collocation-grid solution of elliptic boundary-value problems. Modelirovanie v mekhanike 5(22), 101–126 (1991). (in Russian)
17. Thompson, J.F., Warsi, Z.U.A., Mastin, C.W.: Numerical Grid Generation - Foundations and Applications. Elsevier Science Publishing Co., New York (1985)
18. Wolfram, S.: The Mathematica Book, 5th edn. Wolfram Media, Inc., Champaign (2003)

Counting Roots of a Polynomial in a Convex Compact Region by Means of Winding Number Calculation via Sampling

Vitaly Zaderman[1] and Liang Zhao[2(✉)]

[1] Ph.D. Program in Mathematics, The Graduate Center of the City University of New York, New York, NY 10016, USA
vza52@aol.com
[2] Department of Computer Science, The Graduate Center of the City University of New York, New York, NY 10016, USA
liang.zhao1@lehman.cuny.edu

Abstract. In this paper, we propose a novel efficient algorithm for calculating winding numbers, aiming at counting the number of roots of a given polynomial in a convex region on the complex plane. This algorithm can be used for counting and exclusion tests in a subdivision algorithms for polynomial root-finding, and would be especially useful in application scenarios where high-precision polynomial coefficients are hard to obtain but we succeed with counting already by using polynomial evaluation with lower precision. We provide the pseudo code of the algorithm as well as a proof of its correctness.

Keywords: Polynomial root-finding · Winding number

1 Introduction

Let

$$p(z) = \sum_{k=0}^{d} p_k z^k = p_d \prod_{j=1}^{n} (z - z_j), p_d \neq 0 \tag{1}$$

be a polynomial of degree d with real or complex coefficients. Counting its roots (with their multiplicity) in a fixed domain (such as an interior of a polygon or a disc) is a fundamental problem with an important application to devising efficient root-finders for $p(z)$ on the complex plane, particularly subdivision algorithms, proposed by Hermann Weyl in [10] and then extended and improved in [1–4, 7, 8] [1] and recently implemented in [6].

We propose a new algorithm for counting the roots in a fixed convex region on the complex plane by expressing their number as the winding number computed

---

[1] The authors of [3, 4, 7] called it Quadtree algorithm, and under that name it was extensively used in computational geometry.

© Springer Nature Switzerland AG 2019
M. England et al. (Eds.): CASC 2019, LNCS 11661, pp. 451–457, 2019.
https://doi.org/10.1007/978-3-030-26831-2_29

along the boundary of the region, provided that the boundary was sufficiently isolated from the roots of $p(z)$.

Winding number algorithms have been proposed for counting roots in a disc as parts of root-finding algorithms by Henrici and Gargantini in [4], then by Henrici in [3] and by Renegar in [8]. Pan in [7] used root-radii algorithm by Schönhage [9] for counting roots in a disc, and Becker et al. in [1,2] performed counting based on Pellet's theorem.

Our winding number computation shares some techniques with the algorithms of [3] and [8], but there the algorithms have only devised in the special case of a disc rather than an arbitrary convex compact region, and unlike these papers we ensure numerical stability of our computation of the winding number. Another method using insertion technique, but not requiring isolation of the input region has been proposed by Zapata and Martin in [11,12]. We evaluate an input polynomial $p(z)$ at some additional auxiliary points that we insert a priori on the boundary of the input region. In this way we made our parametrization is smooth on the associated sub-segments of the boundary curve.

Our proposed root-counting algorithm has the following computational advantages:

- It does not involve polynomial coefficients: only polynomial evaluations are required. This is especially useful when polynomial evaluations can be provided as a fast "black box".
- Computational precision can be kept low: the algorithm outputs the winding number correctly as long as polynomial evaluations are precise enough to indicate correctly the quadrant of the complex plane in which the values of the polynomial lies.
- Besides evaluating polynomials, only integer calculations are involved.

We present our algorithm in the next section and then continue the paper in Sect. 3 by proving its correctness.

2   Winding Number Calculation via Sampling

Suppose that $p(z)$ is a polynomial of Eq. (1), $\gamma : [0, 1] \to \mathbb{C}$ is a simple convex closed piecewise-smooth curve, and Γ is the region enclosed by γ. The winding number $\omega_{p \circ \gamma}$ of a curve $p \circ \gamma$ is the number of counterclockwise turns that $p(\gamma(t))$ makes around the origin as t increases from 0 to 1. Namely,

$$\omega_{p \circ \gamma} = \frac{1}{2\pi} \oint_\gamma \frac{p'(z)}{p(z)} dz = \frac{1}{2\pi} \int_0^1 \frac{p'(\gamma(t))\gamma'(t)}{p(\gamma(t))} dt. \tag{2}$$

Hereafter we write $\omega := \omega_{p \circ \gamma}$ omitting the subscript $p \circ \gamma$. It is well-known by principle of argument, that if $(p \circ \gamma)(t) \neq 0$ for all $t \in [0, 1]$, then the winding number ω is a well-defined integer equals to the number of the roots of $p(z)$ inside the region bounded by Γ.

In this paper we aim at developing algorithm that calculates winding numbers of $p \circ \gamma$, where p is a univariate polynomial whose roots lie reasonably far from

γ. In particular, such algorithm can be applied to circles, squares or polygons. Before diving into the details of the algorithm, we should clarify the assumptions about the curve $\gamma : [0,1] \to \mathbb{C}$.

Assumptions on γ.

1. γ is the boundary of a connected region Γ on the complex plane. It is a convex closed curve, i.e., $\gamma(0) = \gamma(1)$.
2. There exists the continuous derivative $\gamma'(t)$ except for t lying in a finite subset $T \subset [0,1]$ (which is relatively small).
3. Furthermore the derivative $\gamma'(t)$ is bounded by L from above, that is,

$$L = \max_{t \in [0,1] \setminus T} |\gamma'(t)| \tag{3}$$

4. γ is $\frac{1}{3}r$-isolating the roots of polynomial $p(z)$, meaning that the minimum distance between a point on the curve γ and a root of $p(z)$ is at least $\frac{1}{3}r$ where r denotes the minimal distance between the origin and the curve $p \circ \gamma$, that is,

$$r = \min_{t \in [0,1]} |(p \circ \gamma)(t)|. \tag{4}$$

In particular $(p \circ \gamma)(t) \neq 0$ for all $t \in [0,1]$.

Remark 1. For a convex domain with a center (which covers a disc and an interior of a rectangle as particular cases) we can define its dilation with a coefficient $\theta > 1$. If the number of roots of $p(z)$ in the domain is invariant in its dilation with coefficients θ and $1/\theta$, then we call the domain θ-isolated. We can square isolation coefficient θ by performing Dandelin's root-squaring iteration $p(z) \to (-1)^d\, p(\sqrt{z}\,)p(-\sqrt{z}\,)$ (cf. [5]). s iterations

$$p_0(z) = p(x), \quad p_{j+1}(z) = (-1)^d\, p_j(\sqrt{x}\,)p_j(-\sqrt{z}\,), \quad j = 0,1,\dots,s$$

change that coefficient into θ^{2^s}.

The core idea of our winding number algorithm is to compute the number of turns of Γ around 0. We do it by computing polynomial in finite number of points $t_0, \dots, t_N \in [0,1]$. More precisely we correctly compute the number of roots in a given region if for every i the actual value $p(\gamma(t_i))$ and the computed value of $p(x)$ at the point $\gamma(t_i)$ lie in the same quadrant on the complex plane, labeled by the following integers $m(p(\gamma(t_i)))$.

Definition 1. Given polynomial $p(z)$, closed curve $\gamma : [0,1] \to \mathbb{C}$, and $t \in [0,1]$, the quadrant label $m((p \circ \gamma)(t))$ is defined as

$$m((p \circ \gamma)(t)) = \begin{cases} 0 & \text{if } Re((p \circ \gamma)(t)) \geq 0, Im((p \circ \gamma)(t)) \geq 0 \\ 1 & \text{if } Re((p \circ \gamma)(t)) < 0, Im((p \circ \gamma)(t)) \geq 0 \\ 2 & \text{if } Re((p \circ \gamma)(t)) < 0, Im((p \circ \gamma)(t)) < 0 \\ 3 & \text{if } Re((p \circ \gamma)(t)) \geq 0, Im((p \circ \gamma)(t)) < 0 \end{cases} \tag{5}$$

We simplify the notation letting the integers $m_0, ..., m_N$ denote the quadrant labels for a sequence $0 = t_0 < t_1 < \cdots < t_N \leq 1$.

We are going to prove that the winding number increases by 1 (respectively, decreases by 1) whenever a sub-sequence $(m_0, \ldots, m_l)$ goes through all four quadrants counterclockwise (respectively, clockwise).

$1 \to 2 \to 3 \to 3 \to 0 \to 1$ is an example of a full counterclockwise cycle, and $3 \to 2 \to 1 \to 2 \to 1 \to 0 \to 3$ is an example of a full clockwise cycle.

Notice that in the latter example the labels go counterclockwise at some point (the $1 \to 2$ part), but do not complete a full counterclockwise cycle and thus make no impact on the value of winding number.

To calculate the number of cycles in quadrant labels, we take the difference of each quadrant label with its preceding label modulo 4. For example, the difference between label 2 and its preceding label 1 is $2 - 1 \equiv 1 (mod\ 4)$; the difference between label 0 and its preceding label 3 is also 1, since $0 - 3 = -3 \equiv 1 (mod\ 4)$.

Notice that for a counterclockwise cycle, the overall sum of these differences must equal 4 (as there must be 4 net increases in quadrant labels); for a clockwise cycle, the overall sum of the label differences must be -4 (as there must be 4 net decreases in quadrant labels). As a result, if we construct sequence $m(0), \ldots, m(N)$ where $m(0) = m_0$ and $m(k)$ for $k = 1, \ldots, N$ are chosen such that $m(k) - m(k - 1) \in \{0, 1, 2, 3\}$ and $m(k) - m(k - 1) \equiv m_k - m_{k-1} (mod\ 4)$, then $(m(N) - m(0))/4$ will be the number of counterclockwise cycles minus the number of clockwise cycles.

In order to establish the link between winding number and the cycles of quadrant labels, we need to eliminate two possibilities: (1) a full cycle of the curve that does not correspond to a full cycle of quadrant labels (this may happen if the sampled points are too far apart, for instance only three first-quadrant points from a cycle are sampled, showing labels $0 \to 0 \to 0$), and (2) we cannot determine whether a full cycle of quadrant labels is a clockwise or counterclockwise cycle (this may happen when two consecutive quadrant labels differ by more than 1, e.g., if $0 \to 2 \to 0$). Our winding number algorithm ensures that the points are sampled properly so that neither bad scenario will occur, and so the winding number can be calculated correctly as

$$\omega = \frac{m(N) - m(0)}{4}. \tag{6}$$

3   Correctness of the Winding Number Algorithm

In this section we prove that our algorithm indeed produces correct winding number.

Theorem 1. For a degree-d univariate polynomial $p(z)$, a parametrized curve satisfying Assumption 1-5, and a sequence $0 = t_0 < t_1 < \cdots < t_N \leq 1$ such that $|t_i - t_{i-1}| \leq \frac{\pi r}{12dL}$ for all $i = 1, \ldots, N + 1, t_{N+1} := t_0$, construct a sequence of

---

Algorithm 1. The Winding Number Algorithm

Require: A polynomial $p(z) = \sum_{k=0}^{d} p_k x^k$, a region Γ with boundary parametrized
   as a piece-wise smooth curve $\gamma : [0,1] \to \mathbb{C}$, $r > 0$, $L > 0$.
Ensure: A positive integer ω such that if γ, r, L satisfy Assumption 1-4, then ω equals
   to the winding number of $p \circ \gamma$.
1: Sample $N = \lceil \frac{12dL}{\pi r} \rceil + |T|$ points $0 \leq t_0 < t_1 < \cdots < t_N \leq 1$ such that $T \subset \{t_i :$
   $0 \leq i \leq N\}$ and $t_i - t_{i-1} \leq \frac{\pi r}{12dL}$ for all $i = 1, \ldots, N+1, t_{N+1} := t_0$.
2: $m_0 \leftarrow$ the quadrant label of $(p \circ \gamma)(t_0)$.
3: **for** i=1 to N **do**
4:    $m_i \leftarrow$ the quadrant label of $(p \circ \gamma)(t_i)$
5:    Choose $m(i)$ such that $\{0,1,2,3\} \ni m(i) - m(i-1) \equiv m_i - m_{i-1} (\mathrm{mod}\ 4)$.
6: **end for** i
7: **return** $\frac{m(N)-m(0)}{4}$.

---

integers $m(0), \ldots, m(N)$ such that $m(0) = m_0$, $m(i) - m(i-1) \in \{0,1,2,3\}$, and
$m(i) - m(i-1) \equiv m_i - m_{i-1} (mod\ 4)$ for $i = 1, \ldots, N$, where m_i is the quadrant
label of $(p \circ \gamma)(t_i)$. Then the winding number ω of $p(z)$ along curve γ is equal to

$$\omega = \frac{m(N) - m(0)}{4}. \tag{7}$$

Proof. On each segment $[t_i, t_{i-1}], \gamma(t)$ is smooth. If a sequence of consecutive
labels $m(i), m(i+1), \ldots, m(j)$ completes a counterclockwise cycle, then the sum
of differences must equal to 4, i.e.,

$$m(j) - m(i) = \sum_{k=i}^{j-1} (m(k+1) - m(k)) = 4. \tag{8}$$

Similarly, a sequence of labels representing a clockwise cycle must satisfy
$m(j) - m(i) = -4$. Thus the overall sum $\frac{m(N)-m(0)}{4}$ is equal to the number
of counterclockwise cycles minus the number of clockwise cycles. Given this
property, it suffices to show that for any $i = 1, ..., N$ it holds that

1. It is impossible that the curve $p \circ \gamma$ can complete a full turn in $[t_i, t_{i+1}]$, that
   is,

$$\frac{1}{2\pi} \int_{t_{i-1}}^{t_i} \frac{p'(\gamma(t))\gamma'(t)}{p(\gamma(t))} dt < 1 \tag{9}$$

2. The quadrant labels m_i differs from m_{i-1} by at most 1, that is,

$$|m_i - m_{i-1}| \leq 1. \tag{10}$$

Proof of claim 1. Recall that $p(z) = p_d \prod_{j=1}^{d} (z - z_j)$ and that

$$\frac{p'(z)}{p(z)} = \sum_{j=1}^{d} \frac{1}{z - z_j}. \tag{11}$$

We will show that the integral in Eq. (9) is less than 2π. It follows that

$$\left|\int_{t_{i-1}}^{t_i}\frac{p'(\gamma(t))\gamma'(t)}{p(\gamma(t))}dt\right| \leq \int_{t_{i-1}}^{t_i}\left|\frac{p'(\gamma(t))}{p(\gamma(t))}\right|\,|\gamma'(t)|dt$$

$$\leq L\int_{t_{i-1}}^{t_i}\sum_{j=1}^{d}\left|\frac{1}{\gamma(t)-z_j}\right|dt$$

$$\leq L\int_{t_{i-1}}^{t_i}\frac{3d}{r}dt \tag{12}$$

$$=\frac{3dL}{r}(t_i-t_{i-1})$$

$$\leq\frac{3dL}{r}\cdot\frac{\pi r}{12dL}$$

$$=\frac{\pi}{4}<2\pi.$$

This verifies Eq. (9).

Proof of claim 2. If m_i differs from m_{i-1} by more than one, then the path $(p\circ\gamma)(t)$ would cross both the real axis and the imaginary axis as t increases from t_{i-1} to t_i. As a consequence, the argument of $(p\circ\gamma)(t)$ would change at least by $\pi/4$. Since

$$arg((p\circ\gamma)(t))=\sum_{j=1}^{d}arg((\gamma(t)-z_j), \tag{13}$$

there exists at least one j such that $arg(\gamma(t_i)-z_j)$ differs from $arg(\gamma(t_{i-1})-z_j)$ by more than $\pi/(4d)$. Next we will show that this is impossible, because according to the choice of samples, $\gamma(t_i)$ is very close to $\gamma(t_{i-1})$. On one hand,

$$|\gamma(t_i)-\gamma(t_{i-1})|\leq L|t_i-t_{i-1}|\leq\frac{\pi r}{12d}. \tag{14}$$

On the other hand, both $|\gamma(t_i)-z_j|$ and $|\gamma(t_{i-1})-z_j|$ are at least $r/3$ and their arguments differ by at least $\pi/4d$. Let $\theta_1=arg(\gamma(t_i)-z_j)$ and $\theta_2=arg(\gamma(t_{i-1})-z_j)$, $\theta_1\neq\theta_2$ then

$$|\gamma(t_i)-\gamma(t_{i-1})|=|(\gamma(t_i)-z_j)-(\gamma(t_{i-1})-z_j)|$$

$$\geq\left|\frac{r}{3}e^{\theta_1 i}-\frac{r}{3}e^{\theta_2 i}\right|$$

$$=\frac{r}{3}|e^{(\theta_1-\theta_2)i}-1| \tag{15}$$

$$>\frac{r}{3}\cdot|\theta_1-\theta_2||$$

$$\geq\frac{\pi r}{12d}.$$

A contradiction proves the claim.

Computation Complexity. The complexity of the algorithm is dominated by the evaluations of the polynomial at N sampled points. Besides polynomial evaluation, the algorithm only requires arithmetic of small integers (mostly less than 8). The value of N is proportional to the Lipschitz bound L defined in Assumption 3. Thus the speed of the algorithm is determined by how fast it can obtain polynomial evaluations at sampled points. If the region is the unit disc $\{z : |z| \leq 1\}$, then we can evaluate $p(x)$ at 2^h equally-spaced points on the unit boundary circle $\{z : |z| = 1\}$ and by using FFT, would correctly compute the number of roots of $p(x)$ in the disc at a arithmetic cost in $\tilde{O}(dL)$, which means $O(dL)$ up to poly-logarithmic factors in dL.

Acknowledgements. Our research has been supported by the NSF Grant CCF–1563942, NSF Grant CCF-1733834, and the PSC CUNY Award 69813 00 48.

References

1. Becker, R., Sagraloff, M., Sharma, V., Xu, J., Yap, C.: Complexity analysis of root clustering for a complex polynomial. In: Proceedings of the ACM on International Symposium on Symbolic and Algebraic Computation, pp. 71–78. ACM (2016)
2. Becker, R., Sagraloff, M., Sharma, V., Yap, C.: A near-optimal subdivision algorithm for complex root isolation based on the pellet test and newton iteration. J. Symb. Comput. **86**, 51–96 (2018)
3. Henrici, P.: Applied and Computational Complex Analysis. vol. 1, Power Series, Integration, Conformal Mapping, Location of Zeros. John Wiley, New York (1974)
4. Henrici, P., Gargantini, I.: Uniformly convergent algorithms for the simultaneous approximation of all zeros of a polynomial. In: Constructive Aspects of the Fundamental Theorem of Algebra, pp. 77–113. Wiley-Interscience New York (1969)
5. Householder, A.S.: Dandelin, Lobačevskii, or Graeffe? Am. Math. Mon. **66**(6), 464–466 (1959)
6. Imbach, R., Pan, V.Y., Yap, C.: Implementation of a near-optimal complex root clustering algorithm. In: Davenport, J.H., Kauers, M., Labahn, G., Urban, J. (eds.) ICMS 2018. LNCS, vol. 10931, pp. 235–244. Springer, Cham (2018). https://doi.org/10.1007/978-3-319-96418-8_28
7. Pan, V.Y.: Approximating complex polynomial zeros: modified weyl's quadtree construction and improved newton's iteration. J. Complex. **16**(1), 213–264 (2000)
8. Renegar, J.: On the worst-case arithmetic complexity of approximating zeros of polynomials. J. Complex. **3**(2), 90–113 (1987)
9. Schönhage, A.: The fundamental theorem of algebra in terms of computational complexity. Manuscript. Univ. of Tübingen, Germany (1982)
10. Weyl, H.: Randbemerkungen zu hauptproblem der mathematik. Mathematische Zeitschrift **20**, 131–150 (1924)
11. Zapata, J.L.G., Martín, J.C.D.: A geometric algorithm for winding number computation with complexity analysis. J. Complex. **28**(3), 320–345 (2012)
12. Zapata, J.L.G., Martín, J.C.D.: Finding the number of roots of a polynomial in a plane region using the winding number. Comput. Math. Appl. **67**(3), 555–568 (2014)

Determining the Heilbronn Configuration of Seven Points in Triangles via Symbolic Computation

Zhenbing Zeng[1(✉)] and Liangyu Chen[2(✉)]

[1] Shanghai University, 99 Shangda Road, Shanghai 200444, China
zbzeng@shu.edu.cn, zbzeng@picb.ac.cn
[2] East China Normal University, 3633 North Zhongshan Road,
Shanghai 200062, China
lychen@sei.ecnu.edu.cn

Abstract. In this paper we first recall some rigorously proved results related to the Heilbronn numbers and the corresponding optimal configurations of $n = 5, 6, 7$ points in squares, disks, and general convex bodies K in the plane, $n = 5, 6$ points in triangles and a bundle of approximate results obtained by numeric computation in the Introduction section. And then in the second section we will present a proof to a conjecture on the Heilbronn number for seven points in the triangle through solving a group of non-linear optimization problems via symbolic computation. In the third section we list three unsolved well-formed such non-linear programming problems corresponding to Heilbronn configurations for $n = 8, 9$ points in squares and 8 points in triangle, we expect they can be solved by similar method we used in the Section two. In the final section we mention two generalizations of the classic Heilbronn triangle problem. The paper aims to provide a concise guide to further studies on Heilbronn-type problems for small number of points in specific convex bodies.

Keywords: Heilbronn number ·
Combinatorial geometry optimization · Symbolic computation

1 Introduction

The Heilbronn triangle problem is to ask how to distribute n points $P_1, P_2, \ldots, P_n$ into a given convex region K in the plane so that the smallest area of the $\binom{n}{3}$ triangles formed the n points is maximal $H(K, n)$. It is easy to see that the asymptotics of $H(K, n)$ is not dependent on the exact shape of K. Hans Heilbronn initially conjectured that

$$H(n) \sim \text{Constant} \cdot n^{-2}.$$

Supported by the grant from the National Natural Science Foundation of China (No. 11471209).

M. England et al. (Eds.): CASC 2019, LNCS 11661, pp. 458–477, 2019.
https://doi.org/10.1007/978-3-030-26831-2_30

The best known result related to this conjecture is that

$$An^{-2} \cdot \log n \leq H(n) \leq Bn^{-8/7}e^{C\sqrt{\log n}},$$

where A, B, C are constants, proved and improved by Roth [20–24], Schmidt [25], and Kómlos, Pintz and Szemerédi [17,18] from 1951 to 1980's.

For a specified convex set K and integer n, the Heilbronn triangle problem can be considered as non-linear global optimization searching for Heilbronn configurations of finite points. Even for very simple regular convex sets like square, triangle and disk, and for very small number $n > 4$, the problem for finding the Heilbronn configurations and values are often very hard. We give a brief recall to the known results proved in literatures [9,10,29–35] as follows.

1. When K is a square $\square$, (the non-trivial) $H(\square, n)$ is known for $n = 5, 6, 7$, that is,

$$H(\square, 5) = \frac{1}{3\sqrt{3}}, \quad H(\square, 6) = \frac{1}{8}, \quad H(\square, 7) = \frac{1}{11.9247}.$$

All floating point values are given to 4 decimal places. The corresponding optimal configurations are shown in Fig. 1. (See [30,31,33,35]).

2. When K is a triangle $\triangle$, $H(\triangle, n)$ is known for $n = 5, 6$

$$H(\triangle, 5) = \frac{1}{3 + 2\sqrt{2}}, \quad H(\triangle, 6) = \frac{1}{8}.$$

The corresponding optimal configurations are shown in Fig. 2. (See [30,32, 34]).

3. When K is a disc D, $H(D, n)$ is known for $n = 5, 6, 7$. The corresponding optimal configurations are maximal regular convex n-gons inscribed in the disk. These results were proved mainly by small perturbation analysis to the optimal configurations without complicated symbolic computation. And published in Chinese journals ([33,34]) in the 1990's. See [28] for numeric results (corresponding to lower bounds of $H(D, n)$) for $n \leq 16$. We encourage readers to give a new proof to $n \leq 7$. To our experience we feel optimistic to see a symbolic-computation-based proof to $n \leq 9$.

4. It is also known that for all convex set K,

$$H(K, 6) \leq \frac{1}{6}, \quad H(K, 9) \leq \frac{1}{9}$$

holds in general. The optimal configuration for $H(K, 7) = 1/9$ is shown in Fig. 3. (For proofs see [9,10,29]).

Many works related to the lower bounds of the $H(\square, n)$ for n up to 22 have been done by Goldberg [14], Comellas and Yebra [7], Cantrell [4], Beyleveld (see [11]), Karpov [16], Pegg Jr [19] and Tal [27]. Figure 4 shows their conjectures on the Heilbronn configurations for $8 \leq n \leq 12$. Similar works also gave been done

460   Z. Zeng and L. Chen

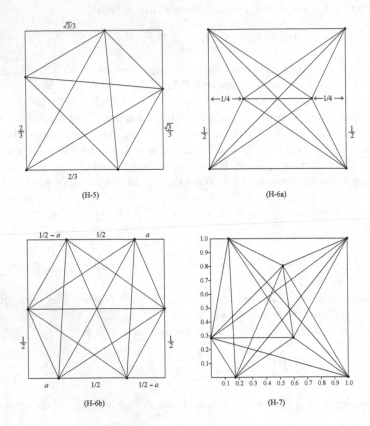

(H-5)   (H-6a)

(H-6b)   (H-7)

Fig. 1. The Heilbronn configurations of n points in the square, $n = 5, 6, 7$. Note that (H-5) is not an affine regular pentagon, (H-6b) is the greatest affine regular hexagon contained in the square, but the positions of the vertices are not fixed in the edges. $H(\Box, 7)$ is the smallest positive real root of the cubic equation $1 - 14z + 12z^2 + 152z^3 = 0$.

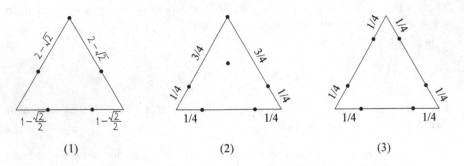

(1)   (2)   (3)

Fig. 2. The Heilbronn configurations of n points in the triangle, $n = 5, 6$.

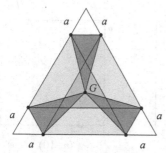

$a = \frac{1}{2} - \frac{\sqrt{3}}{6}$

G is the bary center of the triangle, and the optimial convex body K is a hexagon (in gray).

Fig. 3. The optimal convex planar body that maximizes the Heilbronn value $H(K, 7)$.

for K being triangles and disks by Friedman [12] and Cantrell [4]. The first few lower bounds for $H(\triangle, n)$ are as follows.

$$H(\triangle, 7) \geq \frac{7}{72}, \quad H(\triangle, 8) \geq 0.0677, \quad H(\triangle, 9) \geq \frac{43}{784},$$

$$H(\triangle, 10) \geq 0.0433, \quad H(\triangle, 11) \geq 0.0360,$$

$$H(\triangle, 12) > 0.0310, \quad H(\triangle, 12) \geq 0.0245.$$

More information can be found from web page [28].

For eight points in squares, the numeric result found by Comellas and Yebra in [7] can be "rounded" to the exact form as follows: for any 8 points $P_1, P_2, \ldots, P_8$ in the square, the smallest area of the $\binom{8}{3} = 56$ triangles formed by the 8 points is less than or equal to $1/(3 + 3\sqrt{13})$, and the equality holds if and only if the configuration is congruent to the following 8 points

$$(0,0), \quad (\tfrac{1+\sqrt{13}}{6}, 0), \quad (1, \tfrac{7-\sqrt{13}}{18}), \quad (1,1),$$

$$(\tfrac{5-\sqrt{13}}{6}, 1), \ (0, \tfrac{11+\sqrt{13}}{18}), \ (\tfrac{5-\sqrt{13}}{6}, \tfrac{7-\sqrt{13}}{9}), \ (\tfrac{1+\sqrt{13}}{6}, \tfrac{2+\sqrt{13}}{9}),$$

as shown in the Fig. 5.

2   A Proof to Heilbronn Number for Seven Points in Triangles

In 2009, Kahle [15] suggested a numeric searching method for finding the upper bounds of $H(\triangle, n)$ (see Fig. 6). Using this method De Comité and Delahaya [8] proved the following upper bound of Heilbronn's value for 7 points in a triangle,

$$H(\triangle, 7) \leq \frac{23}{200},$$

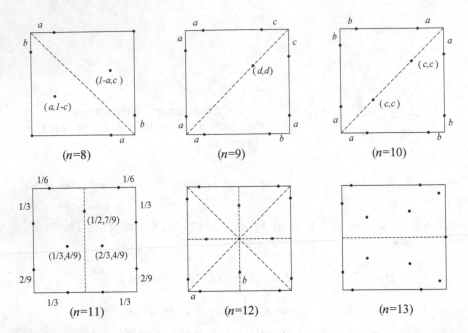

Fig. 4. The best found Heilbronn configurations for n points in the square, found by F. Comellas and J. Yebra ($n = 8, 9, 10, 12$), M. Goldberg ($n = 11$) and M. Beyleveld ($n = 13$). Note that for $n = 8$, the configuration is invariant under a composition of rotation and mirror reflection, where $a = (1 + \sqrt{13})/6$, $b = (7 - \sqrt{13})/18$, $c = (2 + \sqrt{13})/9$. For $n = 9, 10$, the configurations are symmetric with the $y = x$. The configuration for $n = 9$ is determined by $a = 3/8 - \sqrt{65}/40, b = \sqrt{65}/10, c = 7/16 + 3\sqrt{65}/80, d = 1/4 + \sqrt{65}/20$. For $n = 11, 13$, the configurations are symmetric with $x = 1/2$ and $y = 1/2$, with respectively. The configuration for $n = 12$ is full symmetry. In configurations for $n = 10, 12$, the parameters are real roots of cubic equations. The corresponding conjectures on $H(\Box, n)$ are: $H(\Box, 8) = 1/(3+3\sqrt{13}) = 0.0723, H(\Box, 9) = 7/(9\sqrt{65} + 55) = 0.0548, H(\Box, 10) = 0.0465, H(\Box, 11) = 1/27, H(\Box, 12) = 0.0325$ and $H(\Box, 13) = 0.0266$.

and that the Heilbronn configuration formed by 7 points in a triangle is contained in the 7-tuple of the small triangular regions shown in of the Fig. 7(a).

Though this upper bound is not very tight in the sense that $H(K, 7) \leq 1/9$ holds for all convex set K, the information on the position of the seven small triangular regions is helpful to transform the computing of $H(\triangle, 7)$ to a non-linear optimization problems with quadratic constraints. We explain this in the below. First we observed that a direct corollary of this result is that the convex hull of the Heilbronn configuration of 7 points in the triangle is a hexagon, and if $P_1, P_2, \ldots, P_6, P_7$ form a Heilbronn configuration such that $P_1, P_2, \ldots, P_6$ are the vertices of the convex hull, then $P_1, P_2, \ldots, P_6$ are contained in the edges of the triangle. This is based on the fact that the seven small triangular regions

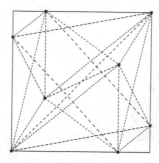

Fig. 5. The Heilbronn configuration of eight points in the square, found by F. Comellas and J. Yebra. The eight points are symmetry about one diagonal of the square. Two points are connected by dash line if they form an edge of one smallest triangle. The minimal triangle area is $(\sqrt{13} - 1)/36 = 0.0723 \in [\frac{1}{14}, \frac{1}{13}]$, and among the 56 triangles, there are 12 minimal ones.

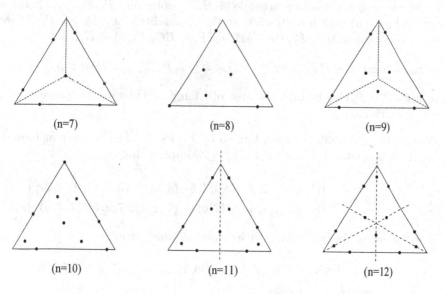

Fig. 6. The best found Heilbronn configurations for n points in the triangle found by D. Cantrell, $n = 7, 8, \ldots, 12$. Note that configurations are $2\pi/3$ rotationally symmetric for $n = 7, 8$, horizontally symmetric for $n = 11$, and completely symmetric for $n = 12$.

contain neither vertex nor midpoint of the triangle, and that if any edge AB satisfies the assumption

$$\#(AB \cap \{P_1, P_2, \ldots, P_6\}) \leq 1,$$

and

$$\{A, B, \texttt{midpoint}(AB)\} \cap \{P_1, P_2, \ldots, P_6\} = \emptyset,$$

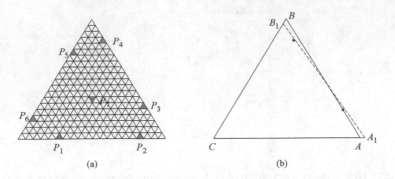

Fig. 7. Each of the seven small triangular regions contains one points of the Heilbronn configuration in the triangle.

then we can construct a new triangle A_1B_1C containing $P_1, P_2, \ldots, P_6, P_7$, as shown in Fig. 7(b), which contradicts to the optimality of $P_1, P_2, \ldots, P_7$. Therefore, we can assume that $P_1, P_2 \in AB$, $P_3, P_4 \in BC$, $P_5, P_6 \in CA$. Let

$$x_0 := \min\{\text{area}(P_iP_jP_k) | 1 \le i < j < k \le 7, P_1, P_2, \ldots, P_7 \in \triangle ABC\}.$$

Here $\text{area}(P_iP_jP_k)$ is the oriented area of triangles. Thus we can prove the following assertions.

1. Among the $\binom{7}{3} = 35$ triangles formed by $P_1, P_2, \ldots, P_7$ there are at most 12 smallest triangles, i.e., $\text{area}(P_iP_jP_k) = x_0$ implies that

$$\implies (i, j, k) \in \begin{cases} (1,2,3), (2,3,4), (3,4,5), (4,5,6), (5,6,1), (6,1,2); \\ (7,2,3), (7,4,5), (7,6,1), (7,1,4), (7,2,5), (7,3,6) \end{cases}.$$

2. The following six triangles must be those of smallest area.

$$(i, j, k) \in \{(7,2,3), (7,4,5), (7,6,1), (7,1,4), (7,2,5), (7,3,6)\}$$
$$\implies \text{area}(P_iP_jP_k) = x_0.$$

3. There are at most three smallest triangles among $\binom{6}{3} = 20$ triangles formed by $P_1, P_2, \ldots, P_6$.

$$\min\{\text{area}(P_2P_3P_4), \text{area}(P_4P_5P_6), \text{area}(P_6P_1P_2)\} \ge 1/9 > x_0.$$

The above inequality can been proved by estimating the interval computation result for $\text{area}(P_2P_3P_4)$ and so on under assuption that $P_1, P_2, \ldots, P_6$ are contained in the small triangular regions shown in the Fig. 7(a).

In the third step, we associate $P_1, P_2, \ldots, P_7$ to the following coordinates:

$$P_1 = (x_1, 0),\ P_2 = (x_2, 0),\ P_3 = (x_3, 1 - x_3),\ P_4 = (x_4, 1 - x_4),$$
$$P_5 = (0, x_5),\ P_6 = (0, x_6),\ P_7 = (x_7, x_8).$$

Then we can construct 6 equality constraints corresponding to the set of the smallest triangles,

$$f_1 := -x_7 + x_7 x_3 - x_2 x_8 + x_2 - x_2 x_3 + x_3 x_8 - x_0,$$
$$f_2 := x_7 - x_7 x_4 - x_7 x_5 - x_4 x_8 + x_4 x_5 - x_0,$$
$$f_3 := x_7 x_6 + x_1 x_8 - x_1 x_6 - x_0,$$
$$f_4 := x_7 - x_7 x_4 + x_1 x_8 - x_1 + x_1 x_4 - x_4 x_8 - x_0,$$
$$f_5 := -x_7 x_5 - x_2 x_8 + x_2 x_5 - x_0,$$
$$f_6 := -x_7 + x_7 x_3 + x_7 x_6 + x_3 x_8 - x_3 x_6 - x_0,$$

and construct 3 inequality constraints corresponding to the other three possible smallest triangles $P_1 P_2 P_3, P_3 P_4 P_5, P_5 P_6 P_1$,

$$f_7 := -x_1 + x_1 x_3 + x_2 - x_2 x_3 - x_0 \geq 0,$$
$$f_8 := x_3 - x_3 x_5 - x_4 + x_4 x_5 - x_0 \geq 0,$$
$$f_9 := x_1 (x_5 - x_6) - x_0 \geq 0.$$

Therefore, we have transformed the computing of $H(\triangle, 7)$ to the following non-linear optimization problem with quadratic constraints.

$$\max x_0$$

$$\text{s.t. } f_1 = 0, f_2 = 0, \ldots, f_6 = 0,$$

$$f_7 \geq 0, f_8 \geq 0, f_9 \geq 0,$$

$$\tfrac{1}{8} < x_1 < \tfrac{5}{24}, \tfrac{17}{24} < x_4, x_6 < \tfrac{19}{24},$$

$$\tfrac{19}{24} < x_3, x_5 < \tfrac{7}{8}, \tfrac{5}{24} < x_2 < \tfrac{7}{24},$$

$$\tfrac{7}{24} < x_7, x_8 < \tfrac{3}{8},$$

$$\tfrac{7}{72} \leq x_0 < \tfrac{1}{9}.$$

Below we briefly describe the main process for solving the non-linear programming related to the Heilbronn configuration of 7 points in triangles.

The first step is to decompose the problem into several subproblems which contain no constraints of the form $f_i \geq 0$.

$$(1): \quad \begin{array}{l} \max x_0 \\ \text{s.t. } f_1 = 0, \ldots, f_6 = 0, \\ \quad f_7 > 0, f_8 > 0, f_9 > 0, \\ H. \end{array}$$

$$(2): \quad \begin{array}{l} \max x_0 \\ \text{s.t. } f_1 = 0, \ldots, f_6 = 0, \\ \quad f_7 = 0, f_8 > 0, f_9 > 0, \\ H. \end{array}$$

$$\begin{array}{l} \max x_0 \\ \text{s.t. } f_1 = 0, \ldots, f_6 = 0, \\ \quad f_7, f_8, f_9 \geq 0, \\ H := \begin{cases} \frac{1}{8} < x_1 < \frac{5}{24}, \\ \frac{19}{24} < x_3, x_5 < \frac{7}{8}, \\ \frac{5}{24} < x_2 < \frac{7}{24}, \\ \frac{7}{24} < x_7, x_8 < \frac{3}{8}, \\ \frac{7}{72} \leq x_0 < \frac{1}{9}. \end{cases} \end{array} \quad \dashrightarrow \quad \vdots$$

$$(7): \quad \begin{array}{l} \max x_0 \\ \text{s.t. } f_1 = 0, \ldots, f_6 = 0, \\ \quad f_7 > 0, f_8 = 0, f_9 = 0, \\ H. \end{array}$$

$$(8): \quad \begin{array}{l} \max x_0 \\ \text{s.t. } f_1 = 0, \ldots, f_6 = 0, \\ \quad f_7 = 0, f_8 = 0, f_9 = 0, \\ H. \end{array}$$

The second step is to solve each subproblem by Lagrangian multiplier with symbolic computation. We refer the readers to [13] for a general introduction to the Lagrangian multiplier method, and [26] for a detail description on solving systems of polynomial equations in several unknowns. Here we show this for (1), (2), (3), (7) and (8). Note that

$$f_1 = 0, \quad f_2 = 0, \quad \ldots, \quad f_6 = 0$$

in our problems form a linear equation system with respect to $x_1, x_4, x_5, x_6, x_7,$ x_8, thus we can obtain

$$x_1 = \frac{4 x_2{}^2 x_3{}^2 - 4 x_2{}^2 x_3 + x_2{}^2 - 2 x_2 x_3{}^2 + x_2 x_3 + 2 x_3 x_0}{4 x_2 x_3{}^2 - 4 x_2 x_3 + x_2 - 2 x_3{}^2 + x_3 + 2 x_0},$$

$$x_4 = \frac{-2 x_2{}^2 x_3 + 4 x_2{}^2 x_3{}^2 - 4 x_2 x_3{}^2 + 2 x_0 x_2 + x_2 x_3 + x_3{}^2}{-2 x_2{}^2 + 4 x_2{}^2 x_3 + x_2 - 4 x_2 x_3 + x_3 + 2 x_0},$$

$$x_5 = -\frac{-2 x_2{}^2 + 2 x_2{}^2 x_3 + x_2 + x_0 - x_2 x_3}{2 x_2{}^2 - 2 x_2 x_3 - x_2 + x_3},$$

$$x_6 = -\frac{2 x_2 x_3{}^2 - 3 x_2 x_3 + x_2 + x_0}{(-1 + 2 x_3)(-x_3 + x_2)},$$

$$x_7 = x_2 - 2 x_2 x_3 + x_3,$$

$$x_8 = -\frac{-2 x_2 x_3 + 2 x_2 x_3{}^2 + x_3 - x_3{}^2 + x_0}{-x_3 + x_2}$$

and represent f_7, f_8, f_9 as

$$f_7 = -\frac{x_0 \left(-2 x_2 x_3 + 2 x_0 - 4 x_3{}^2 - x_2 + 3 x_3 + 4 x_2 x_3{}^2\right)}{4 x_2 x_3{}^2 - 4 x_2 x_3 + x_2 - 2 x_3{}^2 + x_3 + 2 x_0},$$

$$f_8 = -\frac{\left(8 x_2{}^3 x_3 - 4 x_2{}^3 - 8 x_2{}^2 x_3 + 4 x_2{}^2 - x_2 + 4 x_0 x_2 + x_3\right) x_0}{(2 x_2 - 1)\left(-2 x_2{}^2 + 4 x_2{}^2 x_3 + x_2 - 4 x_2 x_3 + x_3 + 2 x_0\right)},$$

$$\begin{aligned} f_9 = -x_0 \,\big(& 16 x_2{}^2 x_3{}^3 - 32 x_2{}^2 x_3{}^2 + 20 x_2{}^2 x_3 - 4 x_2{}^2 - 10 x_2 x_3 \\ & - 4 x_0 x_2 + 24 x_2 x_3{}^2 + x_2 + 8 x_0 x_2 x_3 - 16 x_2 x_3{}^3 + 4 x_3{}^3 \\ & - 4 x_3{}^2 + 2 x_0 - 8 x_3 x_0 + x_3 \big) / (-1 + 2 x_3)/(2 x_2 - 1) \\ & / \left(4 x_2 x_3{}^2 - 4 x_2 x_3 + x_2 - 2 x_3{}^2 + x_3 + 2 x_0\right). \end{aligned}$$

Thus the subproblem (1) can be transformed into the following one:

max x_0

$$f_7(x_0, x_2, x_3) > 0,\ f_8(x_0, x_2, x_3) > 0,\ f_9(x_0, x_2, x_3) > 0,$$
$$H' = \left\{ \tfrac{7}{72} \le x_0 < \tfrac{1}{9}, \tfrac{5}{24} < x_2 < \tfrac{7}{24}, \tfrac{19}{24} < x_3 < \tfrac{7}{8}, \right\}.$$

The feasible set of this problem is a polyhedron formed by algebraic surfaces

$$f_7 > 0,\ f_8 > 0,\ f_9 > 0$$

and the cube

$$\frac{7}{72} \le x_0 < \frac{1}{9},\ \frac{5}{24} < x_2 < \frac{7}{24},\ \frac{19}{24} < x_3 < \frac{7}{8},$$

in $\mathbb{R}^3$. It is clear that if this problem either has the optimal solution $x_0 = 7/72$ or has no local maximum in the interior of its feasible set.

For subproblem (2), it is known that any local maximal point (x_0, x_2, x_3) of this problem satisfies

$$\left\{ \text{numer}(f_7) = 0, \ \frac{\partial L_2}{\partial x_0} = 0, \ \frac{\partial L_2}{\partial x_2} = 0, \ \frac{\partial L_2}{\partial x_3} = 0 \right\},$$

where

$$\text{numer}(f_7) = x_0 \cdot (-2 x_2 x_3 + 2 x_0 - 4 x_3{}^2 - x_2 + 3 x_3 + 4 x_2 x_3{}^2),$$

and

$$L_2 = x_0 + a \cdot \text{numer}(f_7).$$

It is easy to see that this equation system can be transformed to the following ascending form

$$1 - 2 a = 0,$$
$$1 + 2x_3 - 4x_3{}^2 = 0,$$
$$(2 - 8 x_3) x_2 - 3 + 8 x_3 = 0,$$
$$-2 x_0 + x_2 - 3 x_3 + 2 x_2 x_3 + 4 x_3{}^2 - 4 x_2 x_3{}^2 = 0$$

and verify that it has no zero in the cube

$$H' = \left\{ \frac{7}{72} \le x_0 < \frac{1}{9}, \frac{5}{24} < x_2 < \frac{7}{24}, \frac{19}{24} < x_3 < \frac{7}{8}, \right\}.$$

This also implies that under the constraints $f_8 > 0, f_9 > 0$ and H', the problem (2) also either has optimal solution $x_0 = 7/72$ or has no stable point in the interior of the feasible set.

For subproblem (7), any local maximal point (x_0, x_2, x_3) in the interior of the feasible set satisfies

$$\left\{ \text{numer}(f_8) = 0, \ \text{numer}(f_9) = 0, \ \frac{\partial L_7}{\partial x_0} = 0, \ \frac{\partial L_7}{\partial x_2} = 0, \ \frac{\partial L_7}{\partial x_3} = 0 \right\},$$

where

$$\text{numer}(f_8) := x_0 \cdot (-8 x_2{}^3 x_3 + 4 x_2{}^3 + 8 x_2{}^2 x_3 - 4 x_2{}^2 + x_2 - 4 x_0 x_2 - x_3)$$
$$\text{numer}(f_9) := x_0 \cdot \big(-16 x_2{}^2 x_3{}^3 + 32 x_2{}^2 x_3{}^2 - 20 x_2{}^2 x_3 + 4 x_2{}^2 + 10 x_2 x_3$$
$$+ 4 x_0 x_2 - 24 x_2 x_3{}^2 - x_2 - 8 x_0 x_2 x_3 + 16 x_2 x_3{}^3 - 4 x_3{}^3$$
$$+ 4 x_3{}^2 - 2 x_0 + 8 x_3 x_0 - x_3 \big),$$

and

$$L_7 = x_0 + a \cdot \text{numer}(f_8) + b \cdot \text{numer}(f_9).$$

Therefore, any maximal point (x_0, x_2, x_3) in the interior satisfies the following univariate equation

$$q2 = \texttt{resultant}(\texttt{numer}(\texttt{resultant}(q1, h2, x0)), \texttt{resultant}(h3, h2, x0), x2)$$

$$= C_1 \cdot x_3{}^2 \left(6\,x_3{}^2 - 6\,x_3 + 1\right) \cdot \left(32\,x_3{}^5 - 64\,x_3{}^4 + 40\,x_3{}^3 - 16\,x_3{}^2 + 8\,x_3 - 1\right)^2$$

$$\cdot \left(512\,x_3{}^5 - 1344\,x_3{}^4 + 1312\,x_3{}^3 - 664\,x_3{}^2 + 188\,x_3 - 23\right) \cdot \left(-1 + 2\,x_3\right)^6 ;$$

according to the property of resultants, where $C_1 = 34359738368$ and

$$h_2 = \frac{\partial L_7}{\partial x_2}, \quad h_3 = \frac{\partial L_7}{\partial x_3},$$

$$q1 = \texttt{numer}\left(\texttt{subs}\left(\texttt{solve}\left(\{\frac{\partial L_7}{\partial x_0}, \frac{\partial L_7}{\partial x_2}\}, \{\texttt{a}, \texttt{b}\}\right), \frac{\partial L_7}{\partial x_3}\right)\right).$$

It is easy to verify that $q2$ has no real root in the interval $(19/24, 7/8)$ by computing $\texttt{sturm}(q2, x3, 19/24, 21/24)$ with Maple. This proves that the subproblem has also no local maximal point in its feasible set.

For subproblem (8), we can transform the equation system

$$\{\texttt{numer}(f_7) = 0, \quad \texttt{numer}(f_8) = 0, \quad \texttt{numer}(f_9) = 0\}$$

to the following two ascending chains:

$$(8a): \begin{cases} 2\,x_2\,x_3 - 2\,x_0 + 4\,x_3{}^2 + x_2 - 3\,x_3 - 4\,x_2\,x_3{}^2 = 0, \\ \left(8\,x_3{}^2 - 10\,x_3 + 3\right)x_2 + 6\,x_3 - 1 - 6\,x_3{}^2 = 0, \\ -\left(6\,x_3 - 5\right)\left(-1 + 2\,x_3\right)\left(8\,x_3{}^3 - 20\,x_3{}^2 + 12\,x_3 - 1\right) = 0, \end{cases}$$

$$(8b): \begin{cases} 2\,x_2{}^3 - 3\,x_2{}^2 + x_2 + x_0 = 0, \\ x_2 - x_3 = 0. \end{cases}$$

It is easy to see that (8a) has the unique zero point

$$x_0 = \frac{7}{72}, \quad x_2 = \frac{3}{4}, \quad x_3 = \frac{5}{6}$$

in the cube H' and (8b) has no real zero in H'. This proves that the feasible set of problem (8) contains only one point $(7/72, 3/4, 5/6)$ and therefore $x_0 = 7/72$ is its optimal solution.

We applied the similar computation to the following remaining subproblems

$$(3): \begin{array}{l} \max x_0 \\ \text{s.t. } f_1 = 0, \ldots, f_6 = 0, \\ \quad f_7 > 0, f_8 = 0, f_9 > 0, \\ H. \end{array} \qquad (4): \begin{array}{l} \max x_0 \\ \text{s.t. } f_1 = 0, \ldots, f_6 = 0, \\ \quad f_7 = 0, f_8 = 0, f_9 > 0, \\ H. \end{array}$$

$$(5): \begin{array}{l} \max x_0 \\ \text{s.t. } f_1 = 0, \ldots, f_6 = 0, \\ \quad f_7 > 0, f_8 > 0, f_9 = 0, \\ H. \end{array} \qquad (6): \begin{array}{l} \max x_0 \\ \text{s.t. } f_1 = 0, \ldots, f_6 = 0, \\ \quad f_7 = 0, f_8 > 0, f_9 = 0, \\ H. \end{array}$$

We can see that all above 4 optimization problems have no stable points in the interiors of the related feasible sets. Thus, $x_0 = 7/72$ is the global maximal of the original optimization problem

$$\max x_0$$
$$\text{s.t. } f_1 = 0, \ldots, f_6 = 0,$$
$$f_7 \geq 0, f_8 \geq 0, f_9 \geq 0,$$
$$H.$$

This proves the following theorem.

Theorem 1. *For any 7 points $P_1, P_2, \ldots, P_7$ in a triangle, the smallest area of the $\binom{7}{3} = 35$ triangles formed by the 7 points is less than or equal to 7/72 of the area of the triangle, and the equality holds if and only if the configuration is congruent to the following 7 points*

$$(1/6, 0), (3/4, 0), (5/6, 1/6), (1/4, 3/4),$$
$$(0, 5/6), (0, 1/4), (1/3, 1/3)$$

as shown in the Fig. 8 .

3   Three Unsolved Non-linear Programming Problems Related to Heilbronn Configurations

In this section we list some advances on Heilbronn Configuration on $H(\triangle, 8)$ and $H(\square, n)$ for $n = 8, 9$.

In [5], Chen et al. proved that the

$$0.067\,789 < H(\triangle, 8) < 0.067\,816,$$

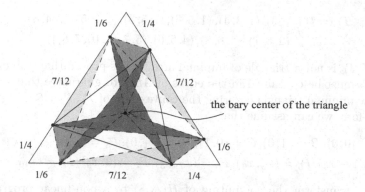

Fig. 8. The Heilbronn configuration of seven points in a triangle, found by David Cantrell. Two points are connected by dash line if they form an edge of one smallest triangle (in gray). There are 9 smallest triangles among $\binom{7}{3} = 35$ triangles formed by the seven points.

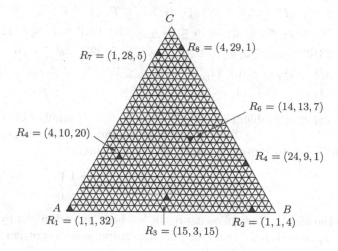

Fig. 9. The configuration formed by 8 small triangles $R_1, R_2, \ldots, R_8$ in a given triangle ABC. The triangle ABC is divided into 32×32 small triangles of equal size by 31 lines that parallel to AB, 31 lines parallel to BC, and 31 lines parallel to CA. Each small triangle can be represented by (i, j, k) where $1 \leq i, j, k \leq 32$ where i, j, k indicates the small triangle is located on the i-th strip that parallels to AB counting from bottom to top, the j-th strip that parallels to CA counting from left to right, ang the k-th strip that parallels to AB counting from right to left.

and that the Heilbronn configuration contained in the eight small triangles $R_1, R_2, \ldots, R_8$ is depicted in the Fig. 9. It is also proved that, assuming that $P_1, P_2, \ldots, P_8$ is any Heilbronn configuration of 8 points in the triangle ABC and $P_i \in R_i (i = 1, 2, \ldots, 8)$, then there are at most 11 triangles among all $P_i P_j P_k$ $(1 \leq i < j < k \leq 8)$ satisfies Area$(P_i P_j P_k) =$ minimal, here Area$(\cdot)$ represents the un-oriented area of triangles. Namely, if (i, j, k) is not a member of the following set

$$T_1 := \{(1,2,3),(1,3,4),(1,5,6),(1,5,8),(2,3,5),(2,4,8),$$
$$(2,6,7),(3,6,8),(4,5,6),(5,7,8),(6,7,8)\}.$$

Then $P_iP_jP_k$ is not a triangle of minimal area. It has proven that points P_1, P_2, P_4, P_7, P_8 must be contained in the edges of ABC to guarantee that the ABC is also the minimal triangle covering the convex hull of $P_1, P_2, \ldots, P_8$.

Therefore, we can assume that

$$P_1 = A = (0,0), B = (1,0), C = (0,1), P_2 = (x_1,0), P_3 = (x_2,x_3),$$
$$P_4 = (x_4, 1-x_4), P_5 = (x_5,x_6), P_6 = (x_7,x_8), P_7 = (0,x_9), P_8 = (x_{10}, 1-x_{10}),$$

and hence transform the computing of $H(\triangle, 8)$ to a non-linear programming problem as follows

$$\max x,$$
$$\text{s.t.} \quad 2\text{area}(P_iP_jP_k) \geq x, \quad (i,j,k) \in T_{11},$$
$$2\,\text{Area}(P_iP_jP_k)| > 1/14, \quad (i,j,k) \notin T_{11},$$
$$7/8 \leq x_1 \leq 29/32, \ 7/16 \leq x_2 \leq 15/32, \ 1/16 \leq x_3 \leq 3/32,$$
$$23/32 \leq x_4 \leq 3/4, \ 3/32 \leq x_5 \leq 1/8, \ 9/32 \leq x_6 \leq 5/16,$$
$$3/8 \leq x_7, x_8 \leq 13/32, \ 27/32 \leq x_9 \leq 7/8, \ 3/32 \leq x_{10} \leq 1/8,$$
$$1/15 < x < 1/14.$$

The second unsolved problem is related to the Heilbronn number of eight points in the square. Let $[0,1] \times [0,1]$ be the unit square and $k > 0$ an integer. We shall use notation $[i,j,k]$ to represent the small square region

$$\left\{ (x,y) \mid \frac{j}{k} \leq x \leq \frac{j+1}{k}, \frac{i}{k} \leq y \leq \frac{i+1}{k} \right\},$$

as shown in the Fig. 10. The following result has been proved in [36] by Zhenbing Zeng using `Maple` with pure symbolic numeric computation years ago in attempt to prove a stronger result

$$H(\square, 8) = (\sqrt{13} - 1)/36,$$

as indicated in the caption of Fig. 5. Since later Liangyu Chen showed in [6] that together with GPGPU computation, the numeric computation method can be extended to a more complicated case for nine points in the square, and that a final proof to the stronger theorem had not been completed yet, the manuscript is still kept unpublished.

Theorem 2. *If $P_1, P_2, \ldots, P_8 \in I = [0,1] \times [0,1]$ form a Heilbronn configuration in the unit square, then, up to a permutation,*

$$P_1, P_2 \in [0,1] \times \{0\}, \quad P_8 \in [0,1] \times \{1\},$$

and $\text{area}(P_i, P_j, P_k) \geq (\sqrt{13} - 1)/36$ *for all* i, j, k *with* $1 \leq i < j < k \leq 8$. *Further more, the following statements are true.*

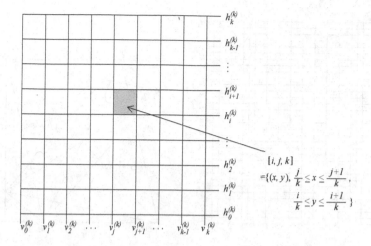

Fig. 10. The small square regions in the unit square formed by k horizontal lines and k vertical lines. Each of the obtained $k \times k$ small square is recorded by $[i, j, k]$.

1. The points $P_1, P_2, \ldots, P_8$ have the same combinatorial type with the following 8 points

$$Q_1 = (0,0), Q_2 = (3/4, 0), Q_3 = (1, 3/16), Q_4 = (1,1),$$
$$Q_5 = (1/4, 1), Q_6 = (0, 13/16), Q_7 = (1/4, 3/8), Q_8 = (3/4, 5/8),$$

shown in Fig. 11(b), in particular, the convex hull of $P_1, P_2, \cdots, P_8$ is a hexagon.

2. Up to a permutation, $P_1, P_2, \ldots, P_8$ are contained in the following eight small rectangle regions

$$U_1 = [0, 0, 32], U_2 = [0, 24, 32], U_3 = [5, 31, 32] \cup [6, 31, 32],$$
$$U_4 = [31, 31, 32], U_5 = [7, 31, 32], U_6 = [0, 25, 32] \cup [0, 26, 32],$$
$$U_7 = [7, 11, 32] \cup [7, 12, 32], U_8 = [19, 24, 32] \cup [20, 24, 32],$$

shown in Fig. 11(a).

3. Let

$$C_3 = \{(1,2,3), (1,4,7), (1,4,8), (1,5,7), (1,7,8), (2,3,8),$$
$$(2,4,8), (2,6,7), (3,5,8), (4,5,6), (4,7,8), (5,6,7)\}.$$

Then under the permutation σ that makes $P_{\sigma(i)} \in U_i (i = 1, 2, \ldots, 8)$, the 56 triangles formed by $P_1, P_2, \ldots, P_8$ satisfy
(a) If $(i, j, k) \notin C_3$ then $\mathrm{Area}(P_i, P_j, P_k) > 1/13$, and
(b) if $P_i P_j P_k$ is a triangle of the minimal area, then $(i, j, k) \in C_3$.

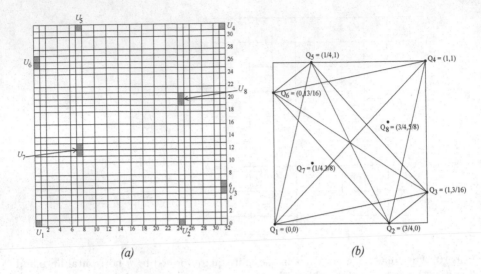

Fig. 11. Eight small square regions $U_1, U_2, \ldots, U_8$ and the eight points $Q_1, Q_2, \ldots, Q_8$, with $Q_1 = (0,0), Q_2 = (3/4,0), Q_3 = (1,3/16), Q_4 = (1,1),\ Q_5 = (1/4,1), Q_6 = (0,13/16), Q_7 = (1/4,3/8), Q_8 = (3/4,5/8)$ as indicated in the Theorem 2.

In view of the Theorem 2, the computing of $H(\square, 8)$ is also transformed to a non-linear programming problem. It is easy to derive that if $P_1, P_2, \ldots, P_8$ is an optimal configuration such that $P_i \in U_i$ for $1 \leq i \leq 8$, then $P_2, P_3, P_{5,6}$ must be contained in the edges of the square, otherwise the convex hull of $P_1, P_2, \ldots, P_8$ can be covered by a smaller parallelogram. Therefore, we may assume that

$$P_1 = (x_1, y_1), P_2 = (x_2, 0), P_3 = (1, y_3), P_4 = (x_4, y_4),$$
$$P_5 = (x_5, 1), P_6 = (0, y_6), P_7 = (x_7, y_7), P_8 = (x_8, y_8).$$

Notice that here we have 12 variables, each constrained in a small interval of length $1/32$, and the non-linear programming problem has 12 bi-linear polynomial inequality constrains. For saving space we will not list the complete form of the programming problem here.

The third unsolved problem is related to $H(\square, 9)$. Chen et al. proved in [6] that the Heilbronn configuration of 9 points in the unit square is determined by the following nine small squares

$$R_1 = \left[\frac{13}{80}, \frac{14}{80}\right] \times \left[0, \frac{1}{80}\right], R_2 = \left[\frac{59}{80}, \frac{60}{80}\right] \times \left[0, \frac{1}{80}\right], R_3 = \left[0, \frac{1}{80}\right] \times \left[\frac{15}{80}, \frac{16}{80}\right],$$
$$R_4 = \left[\frac{79}{80}, 1\right] \times \left[\frac{20}{80}, \frac{21}{80}\right], R_5 = \left[\frac{52}{80}, \frac{53}{80}\right] \times \left[\frac{27}{80}, \frac{28}{80}\right], R_6 = \left[0, \frac{1}{80}\right] \times \left[\frac{66}{80}, \frac{67}{80}\right],$$
$$R_7 = \left[\frac{79}{80}, 1\right] \times \left[\frac{66}{80}, \frac{67}{80}\right], R_8 = \left[\frac{13}{80}, \frac{14}{80}\right] \times \left[\frac{79}{80}, 1\right], R_9 = \left[\frac{64}{80}, \frac{65}{80}\right] \times \left[\frac{79}{80}, 1\right],$$

as each contains exactly one points of the optimal configuration. It is also proved that among the $\binom{9}{3} = 84$ triangles formed by the 9 points in the optimal configuration, there are at most 12 minimal ones. If $\{P_1, P_2, \ldots, P_9\}$ is a Heilbronn optimal configuration and $P_i \in R_i$ for $i = 1, 2, \ldots, 9$, then the possible minimal triangle $P_i P_j P_k$ satisfies that

$$(i, j, k) \in \{(1, 2, 3), (1, 6, 3), (1, 5, 7), (2, 4, 5), (2, 5, 8), (2, 5, 9),$$
$$(3, 4, 5), (3, 8, 6), (4, 6, 5), (4, 7, 9), (6, 9, 8), (7, 9, 8)\}.$$

Assume that $P_i = (x_i, y_i)$ for $i = 1, 2, \ldots, 9$, then we also obtain a non-linear programming problem for computing $H(\square, 9)$. See [6] for details of the constraints. Figure 12 shows the optimal configuration.

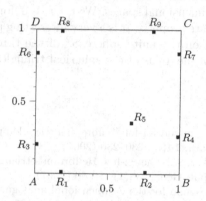

Fig. 12. The 9 small squares that contain the Heilbronn configuration of 9 points in the unit squares. The edge length of each square is 1/80. To get better print effect, the small squares has been enlarged to 161.8% of their real size.

As we have seen, the above three optimization problems are similar to the non-linear programming problem associated to $H(\triangle, 7)$, the non-linear constraints are indeed bi-linear polynomials, roughly half variables can be expressed as polynomial fractions of other variables. Same as in the last section, the simplified optimization can be divided to a finitely many (from hundreds to thousands) subproblems, so that every constraint of each subproblem is either a polynomial equation, or a strict polynomial inequality. Therefore, it could be possible to apply the methods we used in the last section with appropriate parallel implementation to solve them in larger computers.

4 Two Generalizations of the Heilbronn Triangle Problem

One Heilbronn-like problem is for given integer $n \geq 3$ and $k \geq 0$ to find a convex n-gon $P_1 P_2 \ldots P_n$ of unit area and l extra points $P_{n+1}, \ldots, P_{n+l}$ in the its interior to maximize the minimum

$$\min\{\, \text{Area}(P_i P_j P_k) \mid 1 \leq i < j < k \leq n + l\}.$$

An elegant result that is also easy to prove is that for $n = 4, l = 2$ the maximal minimum is $1/8$, which means, for any convex quadrilateral $ABCD$ and any two points $E, F \in ABCD$, there exists one triangle formed by $ABCD$ so that its area is less than or equal to $1/8$ of the area of $ABCD$. It is clear that computer algebra method could be very useful in the studies of this problem, say, for $n + l \leq 9$.

In particular, for $l = 0$, the above problem reduces to finding a convex n-gon $P_1 P_2 \ldots P_n$ of unit area so that $\min\{\text{Area}(P_i P_{i+1} P_{i+2}), i = 1, 2, \ldots, n\}$ gets the maximum. It is reasonable to conjecture that the affine regular polygons are the optimal configuration. Nevertheless, this is only solved for $n \leq 9$.

The second generalization of Heilbronn problem is to extend from the plane to the three or higher dimensional spaces. We refer [1–3] for further investigation on studies on approximate property. It is a very interesting problem to find points $P_1 P_2 \ldots P_n$ for small n on the unite sphere S^2 of the three dimensional space, so that the minimum of the areas of the spherical triangles is maximal.

References

1. Barequet, G.: A lower bound for for Heilbronn's triangle problem in d dimensions. SIAM J. Discrete Math. **14**(2), 230–236 (2001)
2. Barequet, G., Shaikhet, A.: The on-line Heilbronn's triangle problem in d dimensions. Discrete Comput. Geom. **38**(1), 51–60 (2007)
3. Brass, P.: An upper bound for the d-dimensional analogue of Heilbronn's triangle problem. SIAM J. Discrete Math. **19**(1), 192–195 (2006)
4. Cantrell, D.: The Heilbronn problem for triangles. http://www2.stetson.edu/efriedma/heiltri/. Accessed 5 Apr 2019
5. Chen, L., Zeng, Z., Zhou, W.: An upper bound of Heilbronn number for eight points in triangles. J. Comb. Optim. **28**(4), 854–874 (2014)
6. Chen, L., Xu, Y., Zeng, Z.: Searching approximate global optimal Heilbronn configurations of nine points in the unit square via GPGPU computing. J. Global Optim. **68**(1), 147–167 (2017)
7. Comellas, F., Yebra, J.L.A.: New lower bounds for Heilbronn numbers. Elec. J. Comb. 9, 10 (2002). http://www.combinatorics.org/Volume_9/PDF/v9i1r6.pdf. Accessed 5 Apr 2019
8. De Comité, F., Delahaye, J.-P.: Automated proofs in geometry : computing upper bounds for the Heilbronn problem for triangles. https://arxiv.org/pdf/0911.4375.pdf. Accessed 5 Apr 2019
9. Dress, A., Yang, L., Zeng, Z.: Heilbronn problem for six points in a planar convex body. Combinatorics and Graph Theory 1995. vol. 1 (Hefei), pp. 97–118, World Sci. Publishing, River Edge (1995)
10. Dress, A., Yang, L., Zeng, Z.: Heilbronn problem for six points in a planar convex body. In: Du, D.Z., Pardalos, P.M. (eds.) Minimax and Applications. Nonconvex Optimization and Its Applications, vol. 4, pp. 173–190. Springer, Boston (1995)
11. Friedman, E.: The Heilbronn problem for squares. https://www2.stetson.edu/friedma/heilbronn/. Accessed 5 Apr 2019
12. Friedman, E.: The Heilbronn problem for triangles. https://www2.stetson.edu/efriedma/heiltri/. Accessed 5 Apr 2019
13. Gavin, H., Scruggs, J.: Constrained optimization using Lagrange multipliers. http://people.duke.edu/hpgavin/cee201/LagrangeMultipliers.pdf

14. Goldberg, M.: Maximizing the smallest triangle made by points in a square. Math. Mag. **45**, 135–144 (1972)
15. Kahle, M.: Points in a triangle forcing small triangles. Geombinatorics **XVIII**, 114–128 (2008). arxiv.org/abs/0811.2449v1
16. Karpov, P.: Notable results. https://inversed.ru/Ascension.htm#results. Accessed 5 Apr 2019
17. Kómlos, J., Pintz, J., Szemerédi, E.: On Heilbronn's triangle problem. J. London Math. Soc. **24**, 385–396 (1981)
18. Kómlos, J., Pintz, J., Szemerédi, E.: A lower bound for Heilbronn's triangle problem. J. London Math. Soc. **25**, 13–24 (1982)
19. Pegg Jr., E.: Heilbronn triangles in the unit square. Wolfram Demonstrations Project. http://demonstrations.wolfram.com/HeilbronnTrianglesintheunitSquare/. Accessed 6 Apr 2019
20. Roth, K.F.: On a problem of Heilbronn. J. London Math. Soc. **26**, 198–204 (1951)
21. Roth, K.F.: On a problem of Heilbronn II. Proc. London Math Soc. **3**(25), 193–212 (1972)
22. Roth, K.F.: On a problem of Heilbronn III. Proc. London Math Soc. **3**(25), 543–549 (1972)
23. Roth, K.F.: Estimation of the area of the smallest triangle obtained by selecting three out of n points in a disc of unit area. Proc. Symp. Pure Mathematics 24, AMS, Providence, 251–262 (1973)
24. Roth, K.F.: Developments in Heilbronn's triangle problem. Adv. Math. **22**, 364–385 (1976)
25. Schmidt, W.: On a problem of Heilbronn. J. London Math. Soc. **4**, 545–550 (1971/1972)
26. Sturmfels, S: Solving systems of polynomial equations. https://math.berkeley.edu/bernd/cbms.pdf. Accessed 25 May 2019
27. Tal, A.: Algorithms for Heilbronn's triangle problem. Israel Institute of Technology. Master Thesis. May 2009
28. Weisstein E.W.: Heilbronn triangle problem. From MathWorld - A Wolfram Web Resource. http://mathworld.wolfram.com/HeilbronnTriangleProblem.html
29. Yang, L., Zeng, Z.: Heilbronn problem for seven points in a planar convex body. In: Du, D.Z., Pardalos, P.M. (eds.) Minimax and Applications. Nonconvex Optimization and Its Applications, vol. 4, pp. 191–218. Springer, Boston (1995)
30. Yang, L., Zhang, J., Zeng, Z.: Heilbronn problem for five points. Int'l Centre Theoret. Physics preprint IC/91/252 (1991)
31. Yang, L., Zhang, J., Zeng, Z.: On Goldberg's conjecture: Computing the first several Heilbronn numbers. Universität Bielefeld, Preprint 91–074 (1991)
32. Yang, L., Zhang, J., Zeng, Z.: On exact values of Heilbronn numbers for triangular regions. Universität Bielefeld, Preprint 91–098 (1991)
33. Yang, L., Zhang, J., Zeng, Z.: A conjecture on the first several Heilbronn numbers and a computation. Chinese Ann. Math. Ser. A **13**, 503–515 (1992)
34. Yang, L., Zhang, J., Zeng, Z.: On the Heilbronn numbers of triangular regions. Acta Math Sinica **37**, 678–689 (1994)
35. Zeng, Z., Chen, L.: On the Heilbronn optimal configuration of seven points in the square. In: Sturm, T., Zengler, C. (eds.) ADG 2008. LNCS (LNAI), vol. 6301, pp. 196–224. Springer, Heidelberg (2011). https://doi.org/10.1007/978-3-642-21046-4_11
36. Zeng, Z., Chen, L.: Heilbronn configuration of eight points in the square. Manuscript. 1–74

Author Index

Printed in the United States
by Baker & Taylor Publisher Services

Printed in the United States
By Bookmasters